STRENGTH OF MATERIALS

MJP PUBLISHERS

STRENGTH OF MATERIALS

J. PATTABIRAMAN

Dean and Professor of Engineering Design
Department of Mechanical Engineering
KCG College of Technology
Chennai

Chennai   Trichy   Tirunelveli   New Delhi

MJP
PUBLISHERS

ISBN 978-81-8094-156-6   **MJP Publishers**

    No. 44, Nallathambi Street,
Triplicane, Chennai 600 005

MJP   316   © Publishers, 2019

Publisher: C. Janarthanan

This book has been published in good faith that the work of the author is original. All efforts have been taken to make the material error-free. However, the author and publisher disclaim responsibility for any inadvertent errors.

PREFACE

Ever since his student days at IIT Madras and during his 43 years of his professional career that spanned Engineering Consulting and Industry, Research and Teaching, the author has been intensely involved with the subjects Strength of Materials, Design and Analysis of Mechanical components and Systems and Vibrations.

Inspired with a passion for these subjects, the author felt the need, to establish a close link between 'Strength of Materials' and 'Design Engineering' and that 'Strength of Materials' as a subject, needs to have a 'Mechanical Engineering' orientation, when the subject has been conventionally dealt under Civil Engineering. Through his book, the author was keen, to clearly and coherently impart the 'fundamentals' of Strength of Materials to the readers. He felt the requirement to establish a relationship between Engineering principles and Mathematics and relate Engineering theory with practical applications of the theories, for the readers to acquire a holistic knowledge on this subject. With these objectives in mind, the author chose to write this book *'Strength of Materials'*.

This book is intended to benefit different segments of target audience—right from under-graduate and post-graduate students and teachers of Mechanical Engineering, in Universities and Engineering Colleges across India, practicing professionals, Design Engineers and Engineering Consultants working in Industries and Consulting organizations.

All the above aspects have together made this book unique in several aspects. From a Mechanical Engineering Student's angle, this book covers the syllabus prescribed by Indian Universities extensively, with theory, practical applications of the theory, illustrated with several worked out examples and problems, along with 'chapter wise review questions' taken from standard university question papers. The engineering application of the theories along with the case study, solved by the author himself, present the inter-disciplinary nature of engineering problems and solutions, in the subject of 'Strength of Materials'. The book strives to relate well and establish a good connect among various fields of study like Materials, Design, Engineering Tables, Design Codes, Design Cycle, Role of Analysis, Theory of Elasticity, Finite Element Methods, Failure theory, Experimental techniques and Product Engineering. The author sincerely hopes that the book will be found immensely beneficial and will be well received by its intended target audience—the students and teachers of Mechanical Engineering, as well as practicing Design Engineers and Consultants.

The author sincerely acknowledges the authors of various text books, that he has referred to, during the writing of this book, which has been cited in the bibliography. The author wishes to thank and place on record, the kind support and facilities provided by the management of KCG College of Technology, Chennai. The author is immensely grateful to the noble realized saints—Pujya Papaji, Pujya Mathaji, Pujya Swami Satchitanandaji of Anandashram, Kanhangad, for their kind blessings and grace, in writing this book.

The author wishes to dedicate this book to the loving memory of his late father, Sri.S. Jagannadhan, a dedicated and inspiring teacher and his late mother Kamalambal. The author wishes to acknowledge the support, encouragement and love received from his family members, for writing this book. He would like to specially thank his wife Mrs.Saraswathi Pattabiraman (Kala) for her motivation, support and encouragement in writing this book. The author offers his sincere thanks to the editorial team and management of MJP Publishers for bringing out this book. Last, but not the least, he thanks all his teachers, who have contributed to his knowledge and expertise on this subject.

J. Pattabiraman

CONTENTS

4. GENERAL THEORY OF ELASTICITY — 115

1

REVIEW OF ENGINEERING DESIGN FUNDAMENTALS

AN INTRODUCTION TO ROLE OF MATHEMATICS IN ENGINEERING

The purpose of this introductory chapter for students of mechanical engineering is to make them feel the importance and requirement of knowledge in mathematics needed to understand the formulation of problems outlined in this text. It is also intended to make them feel easy and appreciate mathematical modelling which is a powerful tool for solving real-life engineering problems—be it in solid mechanics, fluid mechanics, vibrations, thermodynamics, heat transfer, machine dynamics or statistical quality control. When a logical link is established between mathematical applications perceiving mathematics as 'embedded' in engineering rather than as an isolated subject from engineering , there is every possibility that students will start developing a liking for mathematics when it is explained from an engineering professional's application point of view.

In what follows, a brief write-up is furnished to introduce mathematics as an aid to solve engineering problems. Therefore, we are placing the role of mathematics in the beginning of this text book and not as an appendix.

Mathematical Foundations—A Review

* Knowledge by observation (common sense), e.g., when an object is pushed it moves, if sufficient force is applied to overcome resisting force, holding the body in its position initially.

* Physical laws emerge from observed physical phenomena, e.g., Newton's laws of motion, Newton's law of gravitation, gas laws, laws of thermodynamics, laws of heat transfer.

* Expression of physical laws in words or statements, e.g., quantity of heat input (Q) is equal to mass (m) × specific heat (s) × rise in temperature (t).

* Expression of physical laws in terms of symbols, considered essential since lengthy descriptions need to be avoided. The above example becomes $Q = mst$.

* The last step gives rise to equations/ mathematical models. Equations represent one or more of the following laws, commonly used in physics and in engineering.

a. Balance of forces, otherwise known as equilibrium equations—Mechanical/ Structural applications

b. Conservation of energy known as energy balance—Heat transfer/ Mechanics

c. Conservation of mass/momentum—mass balance, momentum balance—Chemical engineering transport phenomena/fluid mechanics/gas dynamics.

Table 1.1   Static inputs and problems in engineering

Static inputs application	Mechanical/ structural problems
1. Loads (forces)	Normal/shear
(Moments)	Bending/twisting (torsion)
(Torques)	
2. Displacements	Linear displacements and rotational movements
(Linear/Rotational)	can take place along and about the three coordinate axes, totally six (6) in all for a rigid body.

Note   Although wind loads and earthquake (seismic) loads are time-varying, yet only a static analysis is conducted for initial design calculations and if found necessary only, a dynamic analysis is carried out for critical equipment and structures.

Mathematical Functions

As a most general case, in real life, physical parameters are varying with respect to space and time and variations in space may involve three dimensions. When rate processes are involved, the equation applications as in (a), (b), or (c) mentioned above, take the form of ordinary or partial differential equations.

1. Temperature values taken at different points in space at various times.

   $T(x, y, z, t)$ — An example of a scalar quantity. Other examples of scalars are mass, volume.

Table 1.2   Mathematical functions—types

	Point function	Discrete functions
Continuum/distributed system	Line functions	Variations in one dimension
	Surface functions	Variations in two dimensions
	Volume or space functions	Variations in three dimensions
	Dynamic states or real-time functions	Variations in three dimensions and in time

2. Position of an aircraft in space.

   $P = F(x, y, z, t)$ — An example of a vector called as position vectors. Other examples are displacement, velocity, acceleration, force, torque, etc.

   A functional representation allows a convenient method to express the relation. Coordinate systems help in establishing this. As a result, methematical functions are conceived and used for further calculation.

MATHEMATICAL APPLICATIONS

Basic knowledge of algebra, trigonometry, coordinate geometry, differentiation and integration are assumed on the part of the reader and form a prerequisite for further understanding.

Table 1.3 Mathematical applications in mechanical engineering

Areas of mathematics	Application area in mechanical engineering (not limited to)
Vectors	Engineering mechanics—statics
	Engineering mechanics—dynamics
	Fluid mechanics
Matrices	Matrix methods of structural analysis
	Vibration—multi degree of freedom systems Finite element methods
Ordinary differential equations	Elastic analysis of beams
	Vibration (discrete or lumped parameter system), steady-state heat conduction, fluid flow.
Partial differential equations	Vibrations of beams, plates and shells (continuous system) theory of elasticity, transient heat conduction, thermodynamics, acoustics.
Fourier series	Vibrations, control systems
Fourier transforms, Laplace transforms	
Complex variables	Vibrations, fluid mechanics
	Planar problems in theory of elasticity
	Control systems $(\sigma, j\omega$ — complex plane)
Probability theory and statistics	Treatment of random forces arising in wind engineering, earthquake engineering and ocean engineering and ride comfort.
	Quality control, $6\text{-}\sigma$ concepts in reliability analysis and quality assurance in manufacturing.
Optimization	Design optimization
	Operations research—linear programming Resource optimization
Variational calculus	Variational formulation of boundary value problem as a basic need for finite element techniques
	Lagrange's equations for dynamical systems Extremum value problems.

FIELD PROBLEMS IN ENGINEERING—AN OVERVIEW

We are presenting the application areas first and then create a need for understanding the subject of strength of materials from mechanical engineering point of view. Problems in engineering field arise in various stages such as

i. Conceptual design

ii. Design model

iii. Mathematical applications

iv. Experimental model for validating the analytical model

v. Manufacturing feasibility (Design for manufacture)

vi. Assembly feasibility (Design for assembly)/Finalization of CAD drawings

vii. Prototype testing

viii. Final production

The foundation of engineering is built on four strong columns, viz., Physics, chemistry, mathematics and drawing. While pure sciences explain the phenomena, the engineer applies them to practical situations to construct or fabricate an equipment or structure or a process/power plant. Many problems are due to interconnections and interactions and solution requires detailed analysis.

To quote some complex example problems in the area of design and analysis, following problems arise while performing engineering tasks related to construction of components or systems for power or process plants or automotive assemblies or aerospace structures.

1. Interconnected pressure vessels (Refineries, petrochemical plants, fertilizer plants, power plants)

2. Multilayered pressure vessels with openings and attachments (chemical plants)

3. Vessel to piping connections (chemical and power plants)

4. Automotive components/assemblies/new designs/competitive products (automotive engineering)

5. Nuclear plant structures (to ensure safety and integrity) (Nuclear power plant engineering)

6. Tall structures subjected to earthquake loads/wind loads (chemical plants, chimneys)

7. Offshore platform structures (soil structure and solid–liquid interactions)

8. Optimal design of aircraft and space structures (high reliability, low risk designs), wind tunnel studies.

9. Suspension bridges, transmission towers

10. High temperature components like steam and gas turbines, steam piping (power plants)

These problems require simultaneous consideration of several design issues such as safety of operation , minimum weight, adverse environment, fail safe design, cost effective option, ease of manufacture and assembly, selection of correct materials and manufacturing process.

Basic knowledge of behaviour of structures and machine components are important from the point of view of design. Design largely implies deciding on dimensions and selection of appropriate materials. Dimensions depend on the forces or torques that are expected to be safely carried by the members. In order to determine the forces and moments acting on the components, the flow of forces from the point (area /surface) of application through the component must be determined from the basic principles of statics. Determination and proper mathematical description of the forces themselves are difficult and form a significant part of solution to the problem for obtaining a correct solution. A correct description of inputs as well as correct modelling of the problem are equally important to arrive at correct design solutions. Many design procedures in the field of civil and mechanical engineering are covered in the design codes (both national and international).These design codes help the design engineer to carry out the design according to well-established procedures which will ensure the safety of design. In order to account for areas of doubts or uncertainty, regarding loads, homogeneity of materials, etc., factors of safety are used in design calculations. For complex problems of engineering as cited earlier, more advanced analysis techniques, like finite element method, are used for obtaining a design solution, as the design codes may provide a limited guidance to the solution.

PRELIMINARY DESIGN PROCESS—DESIGN CRITERIA

The following detailed steps are necessary to systematically evolve a design. They are

* Conceptual design (design sketch)
* Design intent
    * Power transmission
    * Motion transmission
    * Load transfer and load flow or power flow (Block diagrams)
* Design criteria clearly spelt out
* Strength basis
* Static (yield)
* Dynamic (fatigue) (cyclic and alternating loads on rotating and static components)
* Thermal creep (high temperature piping, steam generators, combustion chamber, steam and gas turbines)
* Deflection basis (machine tools, telescope-mounting structures)
* Stability basis (Buckling or Euler loads)—static and dynamic stability.
* Process requirements affecting design needs (corrosion or wear—marine structures, chemical plant equipment, food-processing equipment)
* Safety and reliability (Nuclear power plant components, aircraft components)
    * Minimal weight consideration (space crafts, aircraft)

- ❋ Minimal power consumption (rotating machines, machine tools and domestic appliances)
- ❋ Minimal wear (Journal bearings, brakes and clutches)
- ❋ Impacts and shocks (Crushing machinery, power presses, forging machines)
- ❋ Comfort (subjective) (automobiles, aircraft suspensions, landing gears)
- ❋ Silent operation (Machinery used in hospitals, studios)
- ❋ Ergonomical criterion (Seats, operation tables, equipment for differently abled persons) and many combinations of the above criteria.

MODELLING

Any physical problem needs to be modelled for carrying out an analysis prior to our taking up actual design. A model is an idealization of a real-world situation that aids in the analysis of a problem. Models have been employed in various subjects, and especially in mechanical engineering. For example, a free-body diagram, is a typical model to show how forces are represented in "parts" and in the "whole". With the help of a model, the engineer can easily visualize the force flow and the assembly of the integral structure.

A model may be either descriptive or predictive. A descriptive model enables us to understand the real-world system or phenomenon. An example is a cutaway model of an aircraft gas turbine. Such a model serves as a device for communicating ideas and information. However it does not help us to predict the behaviour of the system. Typical use is in teaching.

A predictive model helps us to understand as well as predict the behaviours of the real system. A model is always constructed with the scope of application.

We can classify models as follows

1. Static/dynamic
2. Deterministic or probabilistic
3. Iconic/ Analog/ Symbolic

In static models, properties and variables do not change with time. In a dynamic model, time-varying effects are considered.

In deterministic or probabilistic class of models, there is a differentiation in the prediction of outcome. A deterministic model describes the behaviour of a system in which the outcome of an event occurs with certainty and can be mathematically formulated with known functions.

However in many real world situation, the outcome of an event (for e.g., wave forces, wind forces, earthquake forces) is not known with certainty. These must be treated with probabilistic models which use statistical concepts to predict the outcome with a certain percentage of probalility and degree of confidence based on mean and standard deviation.

An iconic model is one that looks like the real object. Examples are a scale model of an aircraft or an automobile for wind tunnel test and an enlarged model of a polymer molecule. Iconic models are used primarily to describe the static characteristics of a system, and they are used to represent entities rather than phenomena. As geometric representation, they may be two-dimensional (maps, photographs, engineering drawings) or three-dimensional (wood or paper model of an airplane or plastic model of an automobile). 3-D models are specially important to communicate a complex design concept, gauge customer reaction to design styling, study of human engineering (ergonomic) aspects of the design and check for interference between parts or components of a large system (plastic models for power and chemical plant layout to check interference between piping and piping, structure and piping, structure and equipment, etc.).

Analog models are those, that behave like real systems. They are often used to compare something that is unfamiliar with something that is already familiar. Analog model need not look like the real system it represents. It must either obey the same physical principles as the physical system or simulate the behaviour of the system. Electrical analogies help a lot in understanding other systems, viz., mechanical, hydraulic or pneumatic.

Symbolic models are abstractions of the important quantifiable components of a physical system. A mathematical equation expressing the dependence of the system output parameter on the input parameters is a very common symbolic model. These provide the greatest generality in resolving a problem. These models can be theoretical, based on well-established laws of nature or empirical models which are the best approximate mathematical representation based on existing experimental data.

Modelling is the representation of a whole or part of the system that is suitable for demonstrating the system behaviour. Simulation involves subjecting models to various inputs or environmental condition to observe how they behave and thus explore the nature of results for the real-world system. Simulation is manipulation of model. It may involve system hardware (prototype models like primary heat transport system of a nuclear power plant) subjected to the actual physical conditions/environment (full load/part load/transients) or it may involve mathematical models subjected to mathematical disturbance functions that simulate the expected service condition (creation of malfunctions or postulated accident or upset conditions). These simulation models help us to arrive at optimal and safe design parameters on paper, prior to making use of them in actual construction, to avoid any possible risks on the real plant or component.

Mathematical Modelling

In mathematical modelling the components of a system are represented by idealized elements that have the essential characteristics of the real components and whose behaviour can be described by mathematical equations.

The key issue is the assumptions, which determine on the one hand the degree of realism of the model and on the other hand the practicality and simplicity of the model for achieving a numerical solution. Skill in modelling comes from the ability to devise simple yet meaningful models and to have sufficient breadth of knowledge and experience to know when the model may be misleading to yield unrealistic results, thus cautioning the engineer about model limitations.

The model can be at the system level, subsystem level or component level. The scope and extent of study required decide the level of model to be constructed. It is interesting to perceive a finite element model to clear with system level and models used in strength of materials help in solving component level problems.

Discrete and continuous models

We had indicated earlier that all systems, equipment, structures and components fall under continuous systems. All parameters are distributed continuously in the member. These are called continuous systems. But to treat them as continuous system, complexity increases and also for the results desired, such a treatment may not be warranted for all types and class of problems.

Hence, parameters are lumped at discrete points and the problem studied as a lumped parameter or discrete system. If a greater accuracy is needed, then more locations are identified where the parameters are lumped, so that the representation of the lumped parameter model is closer to the continuous system (reality).

Finite difference techinique is one such mathematical technique, wherein, the problem domain is divided into grids, where the solution is desired. The more the number of divisions, smaller is the interval and better the idealization. We had also used similar concepts in finding the area or moment of inertia of an irregular shaped object by dividing into smaller strips or sub areas and ultimately getting the value for the total system by summing up over smaller areas or integrating. Integration for an infinitesimal variable takes on addition in a discrete model and similarly differential coefficient in a continuous system becomes slope or gradient in a discrete model. The meanings are same but the process of obtaining results is different.

ORGANISATION CHART—APPLIED MECHANICS

The identification of the subject "Strength of Materials" or "Mechanics of materials" in the vast area of applied mechanics is presented in a flowchart. In the Figure 1.1 that follows, the process of constructing a simplified model on the basis of strength of materials is also presented along with the material behaviour model. Together, they combine to yield the basic mathematical model for design and analysis.

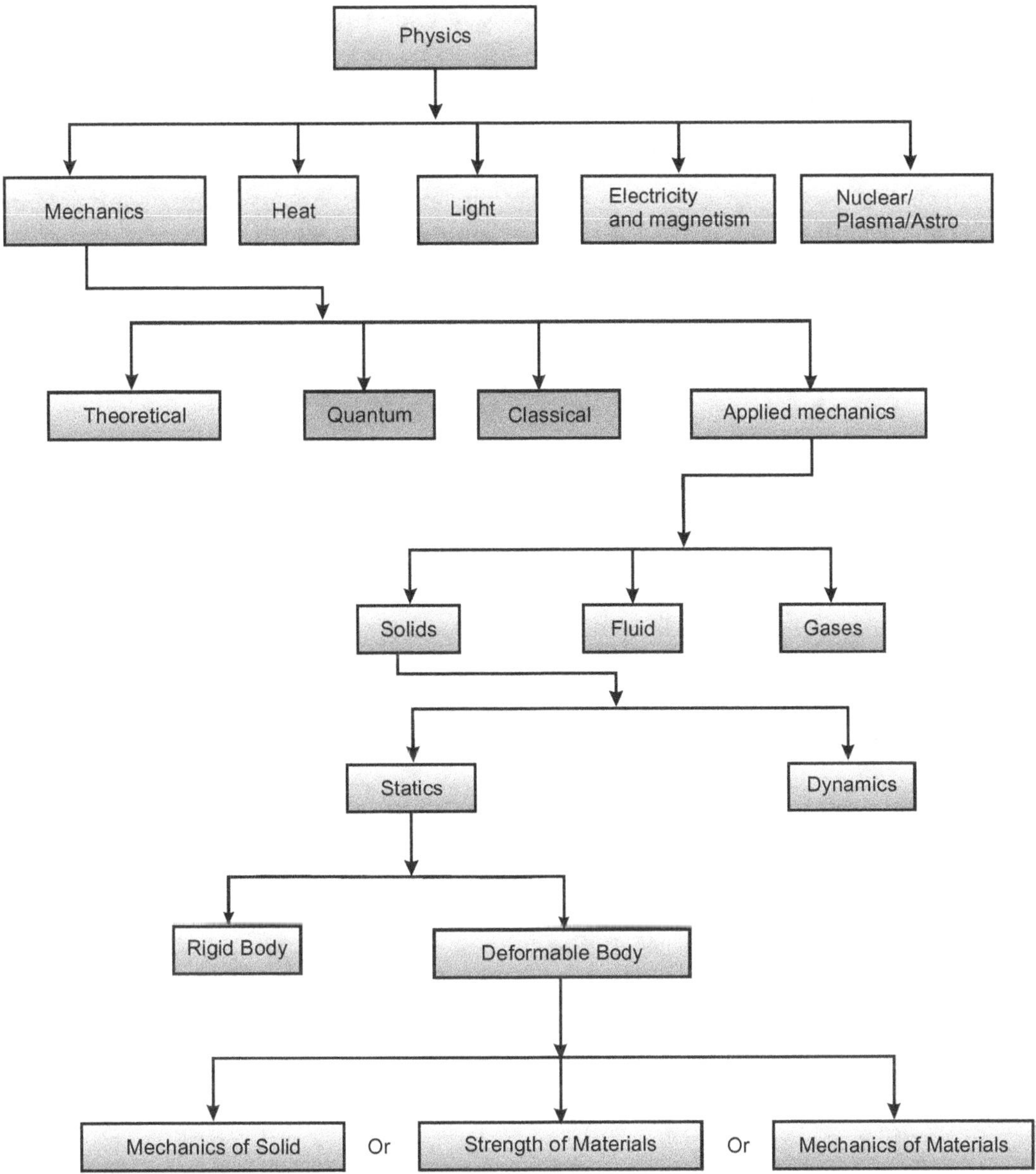

Figure 1.1 Organization chart – Applied mechanics

The subject on strength of materials as indicated in Figure 1.1 deals with simplified models to represent machine or structural members. This is illustrated in Table 1.4.

Table 1.4 Real System—Machine or structure

System model (simplified)	Material behaviour model system
Axially loaded bars—ID bars (Tensile + compressive)	Material behaviour law
Laterally loaded bars—Beams (Bending/Shear)	Experimentally determined and expressed mathematically
Twisting moments on shafts—torsion of shafts	
Bars with axial compression—columns Planes (biaxial loades—2D models)	Linear (small deformation) Non linear → Large deformation / Material behaviour
Solids (triaxial loads)—3D models Strings (Suspension Structures)—only tension	Elastic (brittle/ductile)
Membranes (2D)—only tension	Plastic
Plates Slabs—2D (Bending)	Visco-elastic
Shells, Pressure Vessels—3D (Stretching and bending)	Visco-plastic
Curved bars (Crane hooks)—Bending, Shear	Behaviour of materials with defects (Fracture mechanics)

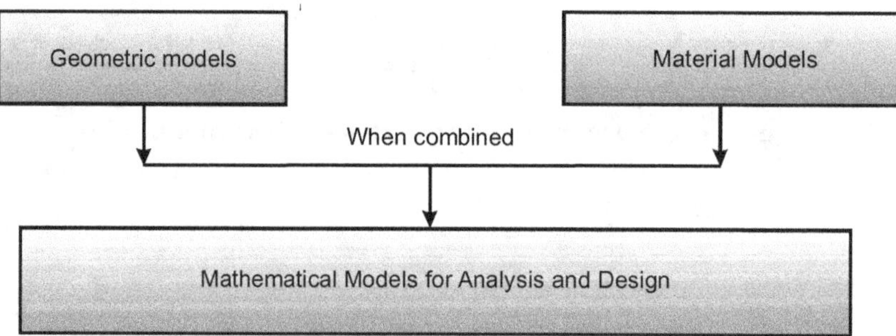

Figure 1.2 Complex structural models combining one or more of above models

Mathematics is Embedded in Engineering Design and Analysis (Figure 1.2)

It may also be seen that, any design activity is considered most fundamental and involves multidisciplinary inputs to design a products. Various stages of product development is given as follows:

1. Design calculations are made using principles of applied mechanics and various subjects such as heat transfer, fluid mechanics, thermodynamics, aerodynamics, electricity and magnetism, acoustics etc, and checked for acceptability

2. Drawings are then made considering production process, fitment/ assembly process, inspection process, fabrication/ construction process and testing needs

3. Keeping in view the safety/reliability/durability, customer needs, economics, innovation, creativity and market competition, the product is designed accordingly.

The reader may feel convinced now, how "Strength of materials" plays a key role in the mechanical or structural design activities and a "pass" of the design from strength point of view is absolutely essential for the design process to proceed to the next stage.

While dealing with strength of materials, the equations of equilibrium from statics are used extensively. This helps in deciding the flow of forces through the member. However, there is one major change in the free-body diagrams from what was used in rigid body mechanics. Namely, most free bodies are isolated by cutting through a member instead of simply removing a pin or some other connection. The internal force on the cut section is related to the stresses (force per unit area) generated by the cohesive forces holding the member together. The size and shape of the member must be adjusted to keep the stress below the limiting value for the type of material from which the member is constructed.

Once again, all the forces acting on the body are external forces. At every section along the member there also exists a system of equal magnitude, opposite direction, pairs of internal forces between the atoms on either side of the section. The study of mechanics of materials or mechanics of deformable bodies, depends on the calculation of these internal forces at various sections of a structure or machine element and how these forces are distributed over the sections. The determination of internal forces is discussed under various chapters that follow.

DESIGN AND ANALYSIS PROCESS

The block diagram prescuted in Figure 1.3 depicts the design process for static loads indicating forward path and feed back (checking) path. The final design after satisfying the codal and requirement of customers is converted to drawings and released for further procesing for production.

In our previous discussion of loads (forces), we saw that the loads might be concentrated forces, or distributed forces. Furthermore, we assumed that the forces did not vary with time. That is, they were static loads. In many cases, the loads may be dynamic, varying with time and this text confines only to static loads.

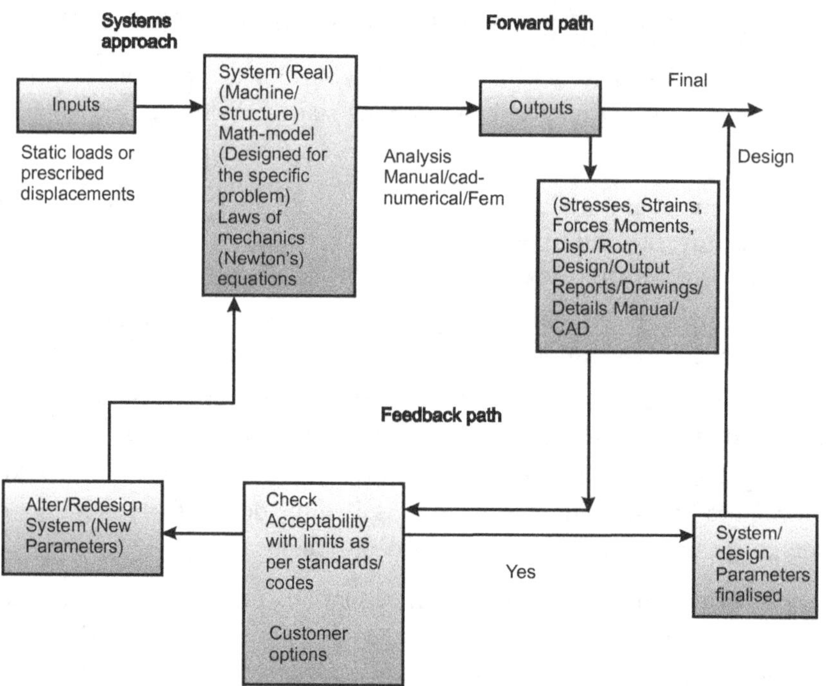

Figure 1.3   Block diagram for design process (Static loads)

In some instances, the specified maximum deformation and not the specified maximum stress will govern the maximum load that a member may carry. In other instances, it may be found that the equations of equilibrium (or motion) are not sufficient to determine all the unknown loads or reactions acting on a body. In such cases it is necessary to consider the geometry (the change in size or shape) of the body after the loads are applied. (The deformation per unit length in any direction or dimension, called strain, may be necessary to compute the loads).

It is necessary to always remember that the equilibrium equations are written only for forces. On a cut section of a solid, internal reactions are produced which are called stresses. Stresses have to be converted to forces for forming equations of equilibrium. Although, stresses may be equal and opposite on two end faces of an object, forces need not be equal as areas on which the forces act could be different, and may still produce the same numerical value of stress.

Knowledge of physical and mechanical properties of materials is required in order to create a design, to properly evaluate a given design, or even to write the correct relation between an applied load and the resulting deformation of a loaded member.

FMEA (FAILURE MODES AND EFFECTS ANALYSIS)

This is a very important phase in design when the product is conceived and modelled and analyzed. One must ensure at the very beginning of the design stage of a product itself, what are the likely nodes of failure, nature of severity of that failure and how to detect the same. This in short is FMEA.

* Carried out in product design and development organizations for design and process with a view to take care of possible impact of failures at early stages.
* Failures occuring after product launch are very "costly" to organizations. Early detections of potential failures at design stage itself is an intelligent and profitable approach.
* Failure implies " not meeting specified design/performance criteria".
* Fracture is one mode of failure.

FMEA has become almost " a must" for product organizations/ job outsourcing agents / supply materials /unfinished goods/ finished goods which " get into" the final product.

A sound design and analysis methodology practised in evolving the designs (product/process) minimize critical errors creeping in at early stages of design, ensuring a greater product/ process reliability.

A sound knowledge of strength of materials and its application to the first level or basic model, assist the design engineer to have a preliminary assessment of FMEA and improvize on the design, if the first level analysis itself show possibility of potential failures in the design. Thus it is emphasized adequately, how " strength of materials" is vital in the design process, be it preliminary or intermediate or final stages of design.

BASIC ISSUES TO BE ANSWERED FOR INPUTS TO DESIGN

In engineering design or analysis activity three different types of problems arise, in general

1. Given a certain function to perform, of what material should the machine or structure be constructed, and what should be the sizes and proportions of the various elements? This is the designer's task. There is no single solution to any given design problem.
2. Given the completed design, is it adequate? That is, does it meet the functional requirement economically and without excessive deformation? This is the task of analysis.
3. Given a completed structure or machine, what is its actual load-carrying capacity? The structure may have been designed for some purpose other than the one for which it is now to be used. Is it adequate for the revised use? This falls under category of task of certifying/approving/ rating authorities.

Factors affecting design with regard to mechanical stressing are listed below.

   i. Simple design for transmission of forces to avoid high localised stresses.

   ii. Place the material so that it follows the same direction as the lines of force.

   iii. Wherever possible use shapes which give low s.c.f (cylindrical, conical, spherical, etc.).

   iv. Determine internal and external forces present.

   v. Give consideration to additional inertia forces, elastic deformations and impact or shock loads if they arise.

   vi. Always design inside outwards. (from component level to gross level)

   vii. Calculate dimensions (approximately) using theory of elasticity/strength of materials.

   viii. Locate maximum stress and stress raisers.

   ix. Investigate ways of reducing maximum stress by choosing appropriate geometry.

   x. Use fatigue laws (Soderberg relations) and make final calculations and arrive at final dimensions (for rotating machinery).

DESIGN PROCEDURE

All the earlier discussions are now summarized in the following procedure adopted for design.

1. Load, type of component, fixity conditions (constraints)
2. Material selection—Environment and nature of load
3. Failure mode of selected material (tension/shear/ compression/fatigue)
4. Appropriate theory of failure

   For brittle materials—Rankine's theory.

   For ductile materials—Tresca, Haigh or Vonmises.
5. Factors for load and factor of safety for design (design data books/ handbooks)
6. Arrive at allowable (permissible) stress (σ allowable) (either derived or taken from hand books).
7. Develop a mathematical model for design analysis.
8. Determine stresses at critical locations using applicable formulae discussed (σ design for the model).
9. Use stress concentration factors (SCF) where necessary (notches, splines, grooves, holes, fillets, etc.) to raise σ design by factors greater than 1. Get revised σ design. (This book gives SCF for commonly used configurations).
10. Check whether σ design $\leq \sigma$ allowable
11. If design change is needed, then redesign, reanalyse, till acceptability criteria are met (refer design feedback referring to Figure 1.3).
12. Carry out FMEA. Assess risk factor using FMEA and reduce the same by redesign consideration and carry out steps (7)–(11) again for the revised design and obtain new risk factors. Ensure revised risk factors are lower.

All the above aspects which are essential for design based on strength is extensively covered in this book with adequate illustrative examples and explaining the underlying concepts.

The scope is restricted to a study of individual members and very simple structures or machines. The design courses that follow will consider the entire structures or machine, and will provide essential background for the complete analysis of the problems.

It is important for the reader to realize now, that applied mechanics or engineering mechanics deals with forces (cause) and motion or force (effect), not concerned about how these forces originate. However design activity encompasses study of how forces are created and the physics behind the cause of forces.

NEW PRODUCT DESIGN

All the discussions so far made are summarized below in the flow chart shown in Figure 1.4 integrating sepcification, analysis, design, drawing, manufacturing and testing.

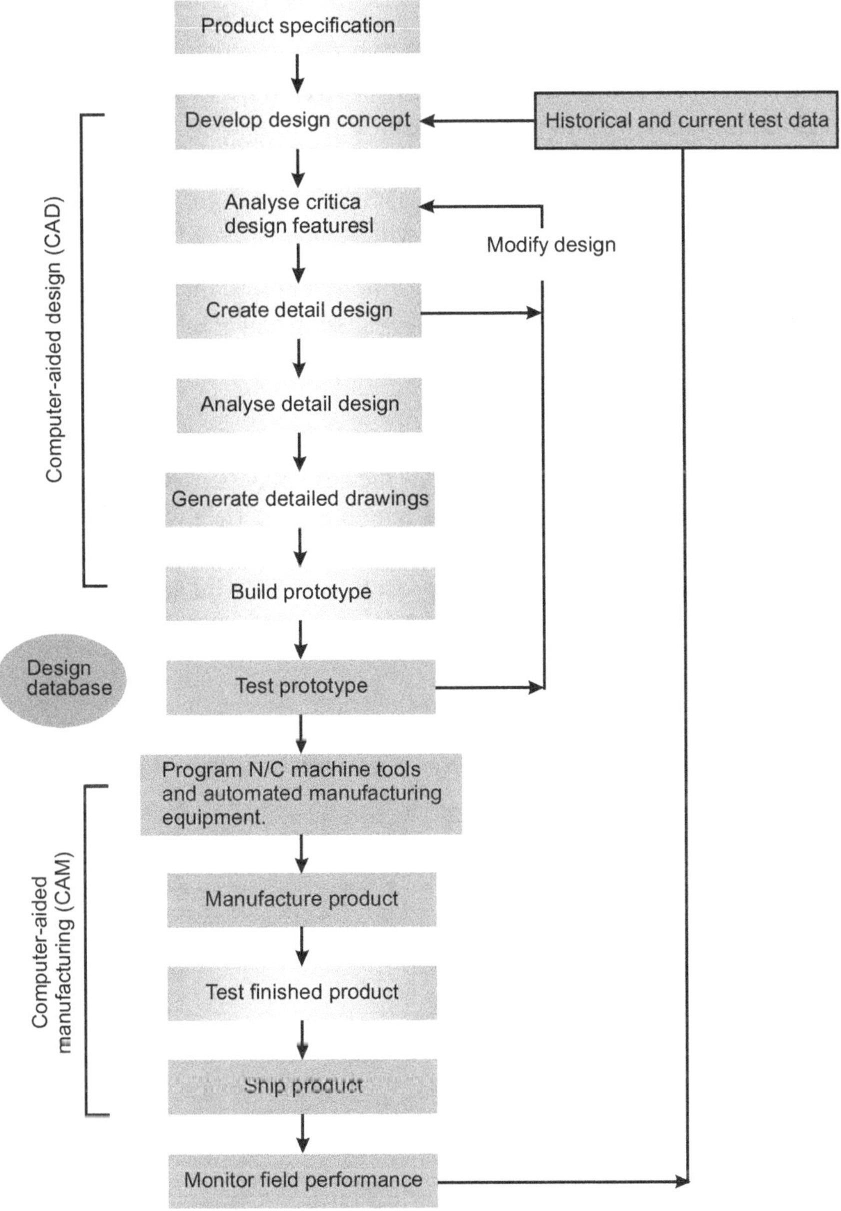

Figure 1.4   Flow chart for new product design

The above Figure 1.4 presents overall product design activity and shows how CAD/CAM/CAE are effectively employed. The box which indicates "Analyse detail design" includes all analytical processes including strength of materials. The feedback path "Modify design" is as a result of design check against design criteria set by codes or the company.

Frequently encountered difficulty in design is uncertain as to boundary conditions—whether a beam or flat plate should be calculated as freely supported or fixed, whether a load should be assumed uniformly or otherwise distributed, etc. In any such case it is a good plan to make assumptions, i.e., to calculate the desired quantity on the basis of each of two assumptions representing limits between which the actual conditions must lie. Thus for a beam with ends having an unknown degree of fixity, the bending moment at the centre cannot be more than if the ends were freely supported and the bending moments at the ends cannot be less than if the ends were truly fixed. If so designed as to be safe for either extreme condition, the beam will be safe for any intermediate degree of fixity.

In case of formulae used in design, especially empirical ones, they should be checked first for dimensional validity and also their limitations whether it is applicable for the case under consideration.

SUMMARY

This chapter comprehensively presents the entire activities forming part of design and establishes the need for check for stresses and displacement as the case may be. The determination of stresses, strains and displacements are in the purview of strength of materials which is the backbone for all strength-based treatment based on simple models yield the results directly or help in verification of results for complex problems in limited zones or location of the total structure.

2

FORCE, DEFORMATION AND EQUILBRIUM

CLASSIFICATION OF FORCES

Force is one of the most important of the basic concepts in the study of mechanics of materials (or the mechanics of deformable bodies). Force is the action of one body on another; forces always exist in equal magnitude, opposite direction pairs. Forces may result from direct physical contact between two bodies, or from two bodies that are not in direct contact. For example, consider a person standing on a sidewalk. The person exerts a force on the sidewalk through direct physical contact between the soles of his or her shoes and the sidewalk; the sidewalk in turn exerts an equal magnitude, opposite direction force on the soles of the person's shoes. If the person were to jump, the contact force would vanish but there would still be a gravitational attraction (force between two bodies not in direct contact) between the person and the earth. The gravitational attraction force exerted on the person by the earth is called the weight of the person; an equal magnitude, opposite direction, attraction force is exerted on the earth by the person. Another type of force that exists without direct physical contact is electromagnetic force.

Contact Forces

Contact forces are called surface forces, since they exist at surfaces of contact between two bodies. If the area of contact is small compared to the size of the body, the force is called a concentrated force; this type of force is assumed to act at a point. For example, the force applied by a car wheel to the pavement on a bridge is often modelled as a concentrated force. Also, a contact force may be distributed over a narrow region in a uniform or nonuniform manner. This situation would exist where floor decking contacts a floor joist. Here, the floor decking exerts a uniformly distributed load (force) on the joist. The intensity of the distributed load is W and has dimensions of force per unit length.

Other Common Types of Forces

Other common types of forces are external, internal, applied, and reaction. To illustrate, consider the beam loaded and supported. All forces acting on the free-body diagram are external forces; that is they represent the interaction between the beam and the external world (everything else that has been discarded). Force F is a concentrated force, whereas W is a uniformly distributed load with dimensions of force / length. The force F and W are applied forces or loads. They are the forces that the beam is designed to carry. Forces A_x, A_y and B are necessary to prevent movement of the beam. Such supporting forces are called reactions. Force distributions at supports are complicated, and reactions are usually modelled as concentrated forces.

EQUILIBRIUM OF A RIGID BODY

A rigid body (a body that does not deform or deforms insignificantly under the action of applied loads) is in equilibrium when the resultant of the system of forces acting on the body is zero. This condition is satisfied if

(Sum of all the forces) $\qquad\qquad\qquad\qquad \Sigma F = 0 \qquad\qquad\qquad\qquad$ (2.1)

(Sum of all the moments) $\qquad\qquad\qquad\qquad \Sigma M_0 = 0 \qquad\qquad\qquad\qquad$ (2.2)

The rigid body is a model adopted for solving problems where the deformation of the body is affecting the forces negligibly or in no way has an influence on the determination of forces.

Equation (2.1) states that the vector sum of all external forces acting on the body is zero. whereas equation (2.2) states that the vector sum of the moments of the external forces about any point 0 (on or off the body) is zero. Equations (2.1) and (2.2) are the necessary and sufficient conditions for equilibrium of a rigid body. The two vector equations of equilibrium may be written as six scalar equations. Selecting a right-handed, xyz-rectangular coordinate system, (Figure 2.1) the equations of equilibrium may be written in expanded form as,

$$\Sigma F_x = 0,\ \Sigma F_y = 0,\ \Sigma F_z = 0 \qquad\qquad (2.3)$$

$$\Sigma M_x = 0,\ \Sigma M_y = 0,\ \Sigma M_z = 0 \qquad\qquad (2.4)$$

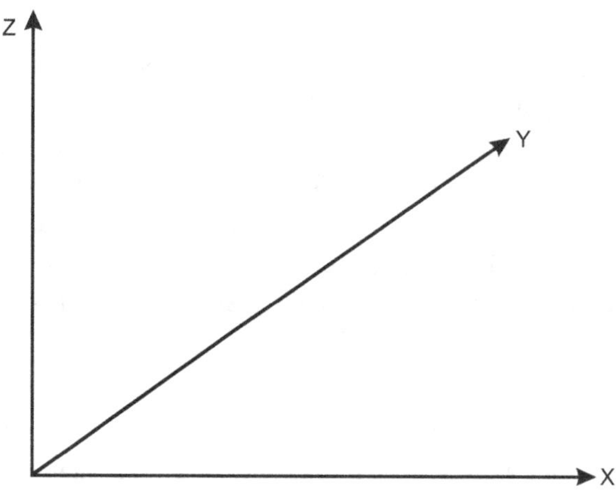

Figure 2.1   Cartesian coordinate system

In the right-handed system, if the fingers are kept pointing from X to Y direction, the positive Z is shown by the thumb pointed upwards. Equation (2.3) states that the sum of all external forces acting on the body in the X-,Y-, and Z-directions is zero. Equation (2.4) states that the sum of the moments of all of the external forces acting on a body about the X-, Y-, and Z-axes are zero. Many problems encountered in mechanics of materials can be treated as two-dimensional. Selecting the X- and Y- axes in the plane of the forces and the Z-axis perpendicular to the plane, the equations of equilibrium reduce to

$$\Sigma F_x = 0, \Sigma F_y = 0, \Sigma M_z = 0 \tag{2.5}$$

These equations of equilibrium would be applicable for the coplanar and concurrent force system. If the force system acting on the body is coplanar and concurrent, equation. (2.3) and (2.4) reduce to

$$\Sigma F_x = 0, \Sigma M_y = 0 \tag{2.6}$$

If the problem has more number of unknowns than the number of equations, then static equilibrium equations (6 in no.) alone cannot help. There is a definite requirement to consider the problem as statically indeterminate and take a longer route to determine deformation considering the body as a deformable body first, and then using the deformations, determine the forces by establishing relation between force and displacement.

Equations of equilibrium are applied to a system of forces. The system of forces may act on a single body, or on a system of connected bodies. A free-body diagram is a carefully prepared drawing that shows a "body of interest" separated from all other interacting bodies and that shows all external forces; both known and unknown, that are applied to the body. The word "free" in the name "free-body diagram" emphasizes the idea that all bodies exerting forces on the body of interest are removed and are replaced by the forces that they exert. At each position on the free-body diagram where other bodies have been removed, the equal magnitude, opposite direction, pairs of forces have been broken, and the forces which act on the free-body diagram must be shown. These forces may be either surface forces or body forces, or both. An important body force is the gravitational attraction of the earth, that is, the weight of the body. The term "force" is used in a generalized sense. Such a term includes "moments" and "torques" also. The free-body diagram clearly establishes which of the bodies are being studied. A correct free-body diagram clearly identifies all forces (known and unknown) that must be included in the equations of equilibrium.

The underlying basic principle is that the whole body or structure is in equilibrium. Therefore every part of it is also in equilibrium and hence, the removed part shall be acted upon by internal forces at the locations from where the part has been removed.

The readers would have already been made familiar with "free-body" diagrams in basic engineering mechanics course dealing with statics and dynamics and hence not dealt with any further to illustrate the concept.

INTERNAL FORCES

In the study of mechanics of materials, it is necessary that we examine the internal forces that exist throughout the interior of a body. We consider an arbitrary body in equilibrium (See Figure 2.2).

The forces F_1, F_2, F_3 and F_4 are applied loads and FS, the support reactions (found using the equations of static equilibrium). We pass an imaginary "cutting plane" *a-a* (henceforth called a section) through the body, and separate the body into two parts A and B. Considering a free-body diagram of part A we note that, in addition to the applied forces F_1 and F_2, the material of part B exerts forces on the material of part A. The forces on the section are distributed over the cut surface in an unknown fashion. However, we can replace the distributed force system by a resultant force R and a resultant couple C. In general, the couple C depends on where we place the force R. In mechanics of materials, we place R at the centroid G of the section. We use a double arrowhead to distinguish the couple C (vector) from the force R (vector). On the section, the distributed force system is statically equivalent to the force system R and C. The force system R and C will be referred to as an internal force system. We recognize that the internal force system depends on the orientation of the section.

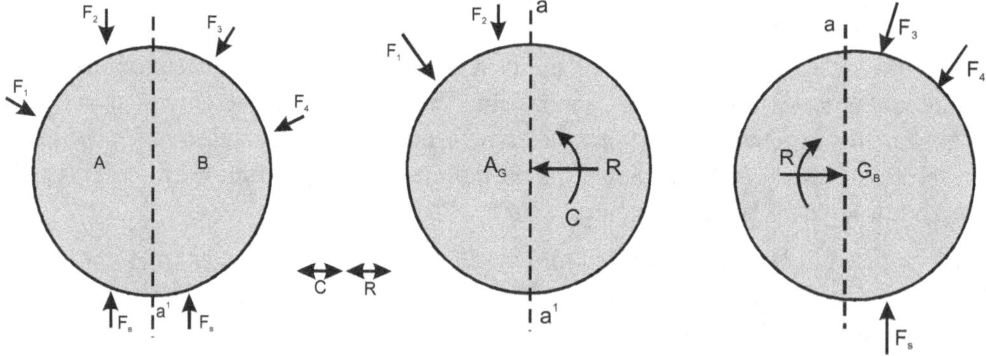

Fgiure 2.2   Forces on the body, A and B, are left and right portions

Instead of part A, we could have considered part B. By applying Newton's third law to every pair of particles on the section for parts A and B of the body, we have that the distributed force systems over the section of parts A and B are equal in magnitude but opposite in sense, and thus the resultant force R and the resultant couple C on the two parts of the body are equal in magnitude but opposite in sense. When the sections are merged, the internal forces cancel and what remains is the whole body with force F_1, F_2, F_3, F_4 and F_s acting upon it.

Because the body as a whole is in equilibrium, any portion of the body is also in equilibrium. Thus, using the equations of equilibrium and the force system we get

$$\Sigma F = O : F_1 + F_2 + F_s + R = 0$$
$$\Sigma Mc = O : M_1 + M_s + M_2 + C = 0$$

where M_1, M_2 and M_s are the moments of forces F_1, F_2 and F_s, respectively, about the centroid of the section. We would find the same result using the free-body diagram of the part B. Thus, we find the resultant of the internal forces system using the equations of equilibrium. However, we cannot find the exact distribution of the internal forces until we learn how to determine the deformation of the body.

Experience indicates that materials behave differently to forces trying to pull atoms apart than to forces trying to slide atoms past each other. Therefore, it is standard practice to resolve the resultants R and C into components along and perpendicular to the section. For convenience we select an XYZ-coordinate system in which X is perpendicular to the section and Y and Z lie in the section. The component of R which is perpendicular to the section, R_x, is called a normal force; this force tends either to pull the body apart or to compress the body. The symbol P is often used to denote the normal force. The components of R that lie in the section are called shear forces; these forces tend to slide part A of the body relative to part B. The symbol V is often used to denote shear forces; hence, the forces V_y and V_z.

The component T of couple C tends to twist the body and is called a twisting couple (or twisting moment or torque). The components M_y and M_z tend to bend the body and are called bending couples (or bending moments) (see Figure 2.3). Throughout this book we will examine the effects on a deformable body due to components of R and C. The components of internal force system can be found using the same equations of equilibrium equations (2.3) and (2.4).

$$\Sigma F_x = 0, \quad \Sigma F_y = 0, \quad \Sigma F_z = 0$$
$$\Sigma M_x = 0, \quad \Sigma M_y = 0, \quad \Sigma M_z = 0$$

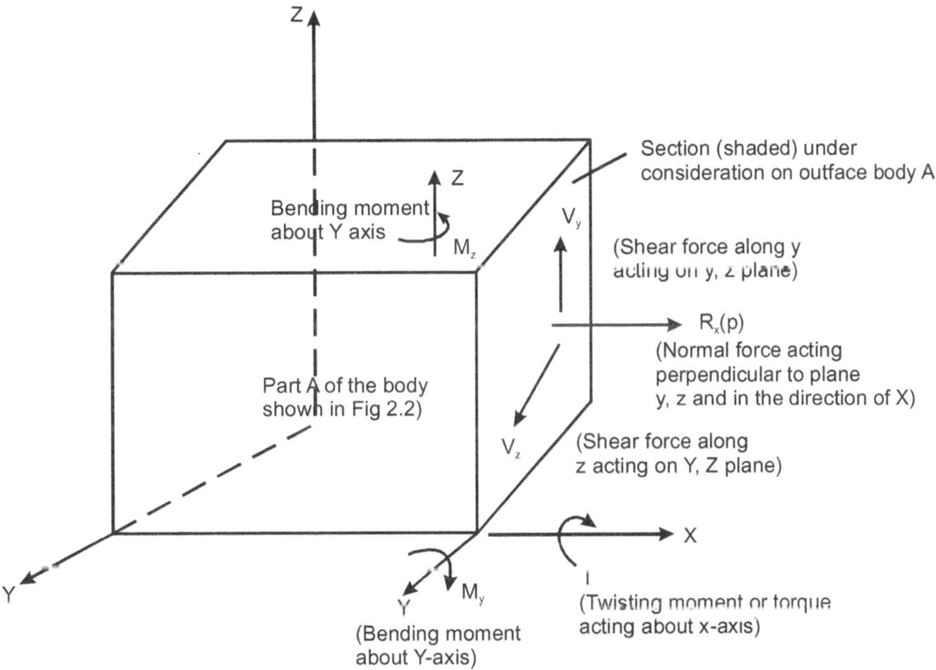

Figure 2.3   Components of internal reactions R and C acting on part A of the body shown in Figure 2.2

The equations of equilibrium should be applied to the body (or portion of the body) in its deformed state. However, as we saw earlier, the support reactions (or in this case internal forces) cannot be found until we know the relationship between the forces applied to the body and the deformation of the body. As we shall see in later chapters of this book, we can determine support reactions and internal forces using the undeformed configuration of a body. We accept this statement for now, and illustrate the determination of internal forces discussed in subsequent chapters.

EQUILIBRIUM OF A DEFORMABLE BODY

All of the equilibrium problems considered thus far have assumed that the bodies are rigid. That is, the shape of the body and its orientation relative to its surroundings were assumed to be independent of the loads applied to the body. However, no body is perfectly rigid. Wires subjected to tension forces will stretch. Beams carrying loads will bend. Shafts subjected to torques will twist. In fact, the main objective of the course in mechanics of materials is to develop relationships between the loads applied to a non-rigid body and the deformation of the body.

If the wire or beam or shaft is very stiff, the amount of deformation will be very small and the deformation will have a negligible effect on the solution of the equilibrium equations. If the wire or beam or shaft is not very stiff, however, the deformation can affect the geometry of the problem used to write the equilibrium equations which will in turn affect the solution of the equilibrium equations. The interaction between the loads acting on a body, the deformation of the body, and the geometry of the free-body diagram makes the solution of deformable body problems much more complex than the solution of rigid body problems. Frequently, the solution of deformable body problems requires either a trial and error solution or a numerical solution or an iterative solution method. Fortunately, most engineering structures and machines are designed "stiff" at the first level, that is, they do not deform very much. For such problems, the solution of the equilibrium equations often ignores the deformation and treats the structure as though it were rigid. The first level is very basic and needs to be made more realistic using deformation consideration which yields correct solutions for design. The discussion on mathematical models convey this idea regarding scope of rigid body model or deformable body model. Regardless of the type of the structure or machine component or the type of loading, solution process generally consists of three steps, namely setting up of

1.   Equations of equilibrium,
2.   Force–deformation relationship and establishing the
3.   Geometry of deformation

Since the equations of equilibrium must be applied to the forces acting on the deformed structure, the above three sets of equations are often interdependent. It is this interdependence that makes the solutions of deformable body problems more complex than the solution of rigid body problems. To reiterate, step (1) alone is sufficient for statically determinate type of problem and steps (1), (2) and (3) are all needed to solve problems which are of statically indeterminate type.

3

CONCEPT OF STRESS AND STRAIN

Strength of materials/mechanics of materials/mechanics of solids basic engineering science to understand material behaviour under action of load, to decide load-carrying capacities and to design a member or component to carry the specified loads. It is the basis of machine design and structural design from strength point of view.

The real world problems are modelled using concept of

* Rigid bodies and deformable bodies which are discussed under Chapter 2.

* Deformations—small, large and time-independent—which is dealt under strength of materials.

* Time-dependent deformation which is dealt with in plasticity, viscoelasticity, viscoplasticity, creep/ fatigue using flow rules and flow models.

* Linear–nonlinear (based on deformation and or based on material behaviour or constitutive law)—stress/strain relations—together define the mathematical model for analysis of bodies under the influence of forces.

* Generalized theory of elasticity—3dimensional problems are dealt with and can be reduced to 2-D wherever applicable and can be further reduced to 1-D for specific cases applicable.

CONCEPT OF STRESS

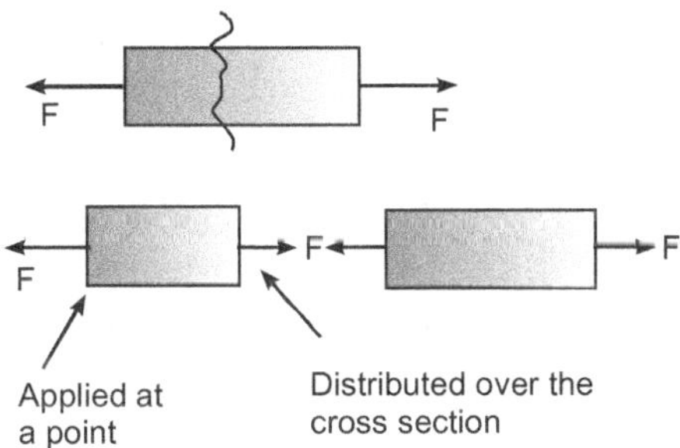

Figure 3.1 Free-body diagram

Different materials specimens have different load-carrying capacities. The same material with different cross-section can carry different loads before failure. But this can not be a design specification as we need to apply to members with different cross-section.

Definition of Failure: Failure implies inability of a material/component to perform its assigned function. Failure could be system failure or component failure. Fracture is one form of failure. Material yielding (transiting from elastic behaviour to plastic behaviour) is taken as the failure criterion for most engineering materials. Elastic behaviour is a "reversible" process and plastic behaviour is an "irreversible process". Following three points are to be borne in mind.

i. Given the same cross section, different materials fail in tension at different loads.

ii. Given the same material, different cross section, can carry different loads (prior to fracture).

iii. Therefore the engineer should have a basis for design based on material property only and the design itself implies arriving at appropriate cross-sectional area for the members identified to transmit or carry the loads.

A free-body diagram of a component gives an idea that in whatever manner the force is applied at the ends, at locations away from the ends, more or less it gets uniformly distributed at such sections. This is a phenomenon that happen in every piece of equipment that we use in day-to-day life. The cross-sectional area offers internal resistance by virtue of the strength of the material to the externally applied load/force. This internal resistance is termed as **Stress**. This is measured as load per unit area and the units are Newton /mm^2 in SI units (psi, ksi, pascal, etc. are also stress units). At present, this definition would suffice but in reality we talk of stress at a point. In this example, the stress is averaged out over the entire cross section and the average stress is being referred to as stress. Later when we deal with 3-D theory of elasticity, we will again revert to the concept of stress at a point.

$$\text{Stress}, \sigma = \frac{F}{A} \text{(definition)}$$

where,

F—Force is a vector (represented by the line of action)

A—Area is a vector (represented by the normal to the plane of the area)

σ—Stress is a tensor.

It has components in all the 3 directions, along normal as well as on the tangential planes. In reality it is defined by a column vector of six rows.

However, the average stress over a cross section is described qualitatively as tensile, or compressive, or shear and is given the magnitude in N/mm^2 for design purposes.

DESIGN FOR STRENGTH

Strength of materials approach

We take up the case of axially loaded bars (Normal stresses) as an example to start with. Consider a bar under tension, as a test specimen in UTM (Universal tensile testing machine). Note the following.

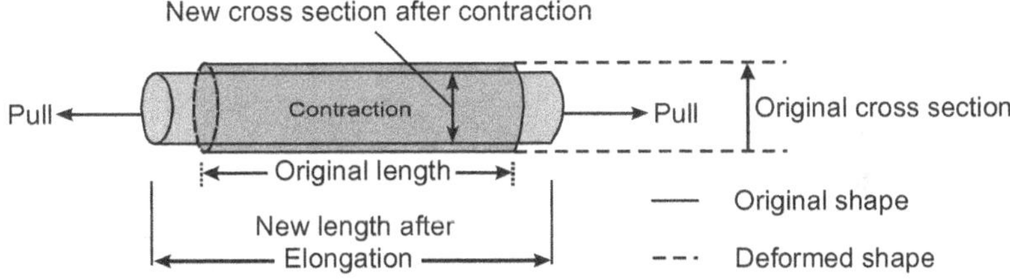

Figure 3.2

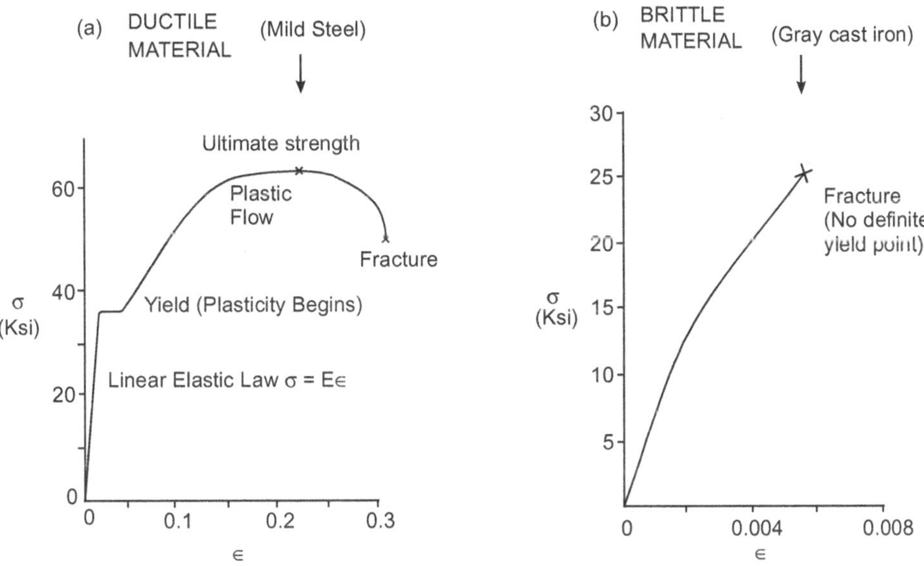

Figure 3.3   Stress–strain relations (a) Ductile, (b) Brittle materials

1. Uniaxial tension test (UTM)—Hydraulic actuation for loading

2. Deformation measurement—Extensometer (or strain gauge)

3. $$\sigma = \frac{\text{Load}}{\text{Original area}} = \text{stress}$$

$$\in = \frac{\text{Extension}}{\text{Original length}}, \sigma = E \in \text{ until yield or proportional limit}, \in = \text{strain}$$

E = Constant of proportionality called young's modulus, units N/mm².

4. Plots of σ $vs\in$ made and curve-fitted. This is called stress–strain curve which is shown in Figure 3.3 (a) and (b) for ductile and brittle behaviour.

5. Poisson's ratio $= \upsilon = \dfrac{\text{Lateral strain}}{\text{Longitudinal strain}}$ (usually $\upsilon = 0.25$ to 0.33, taken as 0.3 for most materials). For rubber $\upsilon = 0.5$ and for concrete it is 0.15.

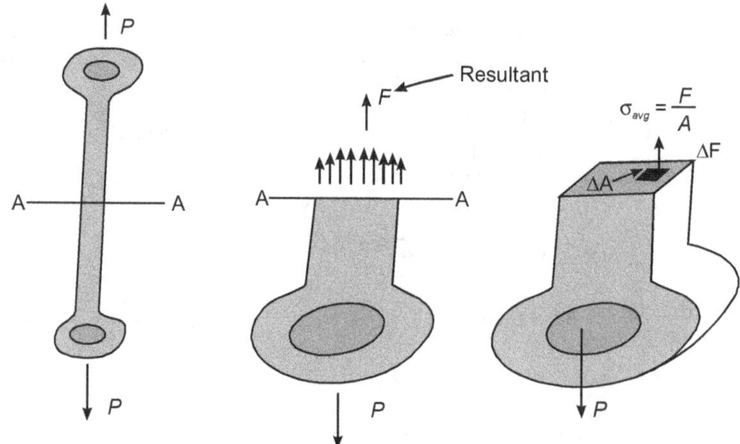

Figure 3.4 Normal stress distribution on section A-A

6. Let us consider normal stress σ uniformly distributed over an area of cross section A except in the vicinity of point of application of the loads. The normal stress distribution at cross section in section A is shown in Figure 3.4.

7. Normal strain, $\in_{avg} = \dfrac{\delta}{L}$ (elongation or contraction/original length)

8. Shearing stresses: Consider the example in Figure 3.5 of a riveted joint, joining three plates, acted upon by forces P on either side. The rivet is subjected to shear on both sides. The shear stress is $\tau_{avg} = \dfrac{V}{A}$. (V—shearforce, A—area of cross section over which V acts)

Note $V = \dfrac{P}{2}$

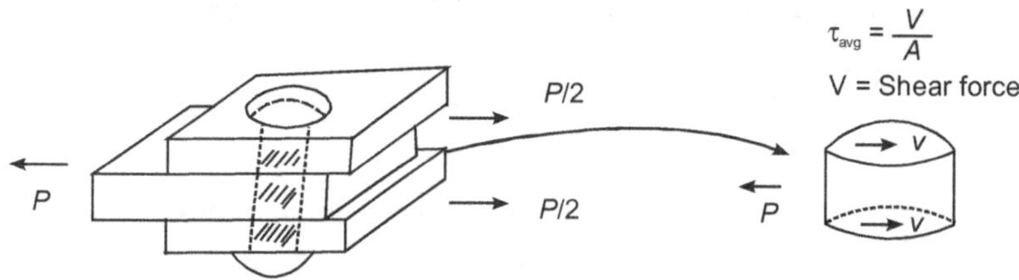

Figure 3.5 Shearing Stresses in Planes

Many problems in strength of materials can be modelled as plane problems as shown in Figure 3.6. Notice the action of normal stresses and σ_x and σ_y shear stresses τ_{xy} and τ_{yx}.

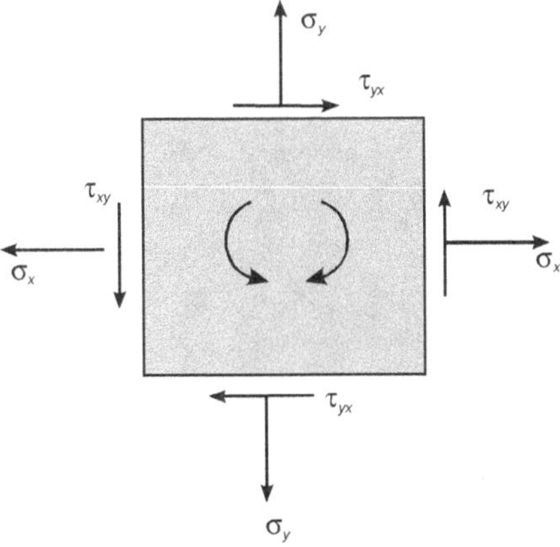

Figure 3.6   Combined normal and shearing stresses in planes

Observe in the diagram (Figure 3.4) that in the immediate vicinity of the load, the distribution of stress is nonuniform.

Uniform distribution can exist only in centrally loaded members, the load acting at the centre (centroid) of the section. Figure 3.7 shows a case of an accentrically loaded member where we cannot expect a uniform stress distribution in section *A–A*.

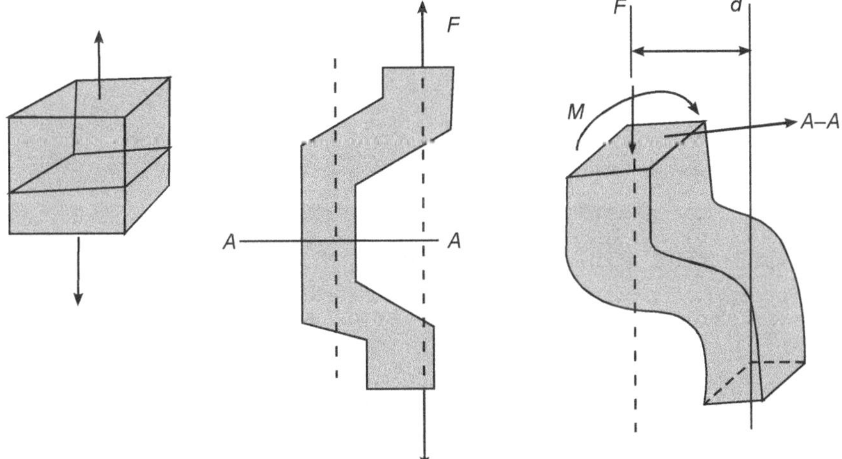

Figure 3.7   Non-uniform stresses in an eccentrically loaded member

SHEARING STRESSES

We discussed the internal reaction normal to the section. We also considered earlier the definition of shear stress. To explain, consider the following case of a riveted joint. We conclude that the internal force, must exist in the plane of the section and their resultant must be equal to the applied load. (Earlier the force considered was normal to the section) (Figure 3.8).

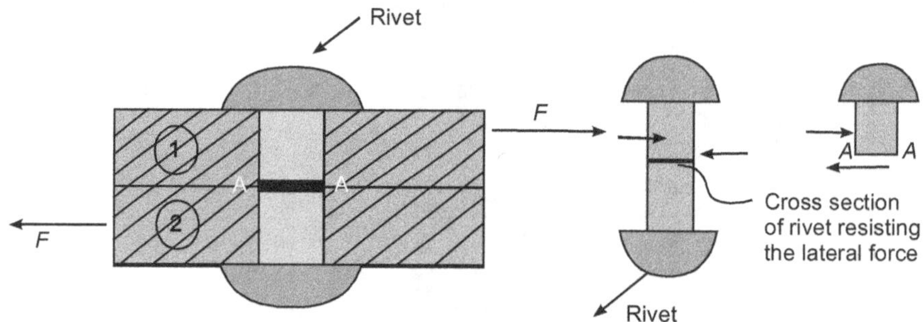

Figure 3.8 Rivet in single shear

Shear stress $= \dfrac{F}{A}$. The rivet considered is in single shear.

Consider the following case of double shear (Figure 3.9).

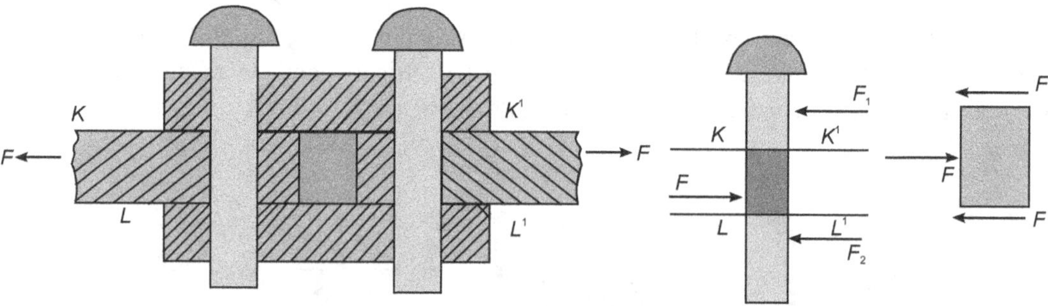

Figure 3.9 Rivet in double shear

The cross-section planes resisting the external load F are two—'KK' and 'LL' plane. Hence shear stress $= \dfrac{F}{2A}$, Since both sides with area A resist the external load F.

This is the case of a riveted joint subjected to double shear. Examples are steel plates used in construction of bridge and ship hull structures, boiler plates and base plates.

BEARING STRESSES

Components like bearings, pins, rivets create stresses in the members they connect along the bearing surface or surface of contact (Figure 3.10).

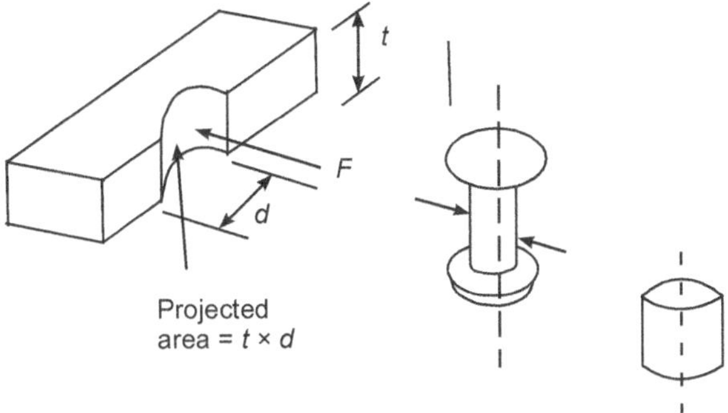

Figure 3.10 Bearing stress in components

$$\sigma_b = \frac{F}{td}$$

where,

t = Thickness,

d = Diameter of pin/rivet,

F = Force and

σ_b = Bearing stress

Stress distribution means how the stress is distributed at all points in the cross-section. The basic stress distribution is more complicated. However, we deal with average stress at this point in time, the above is the average value of σ_b.

Example

In the hanger shown, the upper portion of link ABC is 9 mm thick and the lower portions are each 6 mm thick. Epoxy resin is used to bond the upper and lower portion at B. The pin at A is 9 mm ϕ, while a 6 mm ϕ is used at C. Determine (a) the shearing stress in pin A, (b) the stress in pin C, (c) the largest normal stress in link ABC, (d) the average shearing stress on the bonded surface at B, (e) the bearing stress in the link at C.

Link ABC is a two force member. Reaction at A is vertical

Reaction at D is having two components D_x and D_y

Taking moments due to forces 2400 N and F_{AC} about D and using moment equilibrium equation $\Sigma MD = 0$, yields (2400 N) (120 + 240) – F_{AC}(240) – 0

F_{AC} = 3600 N (Tension)

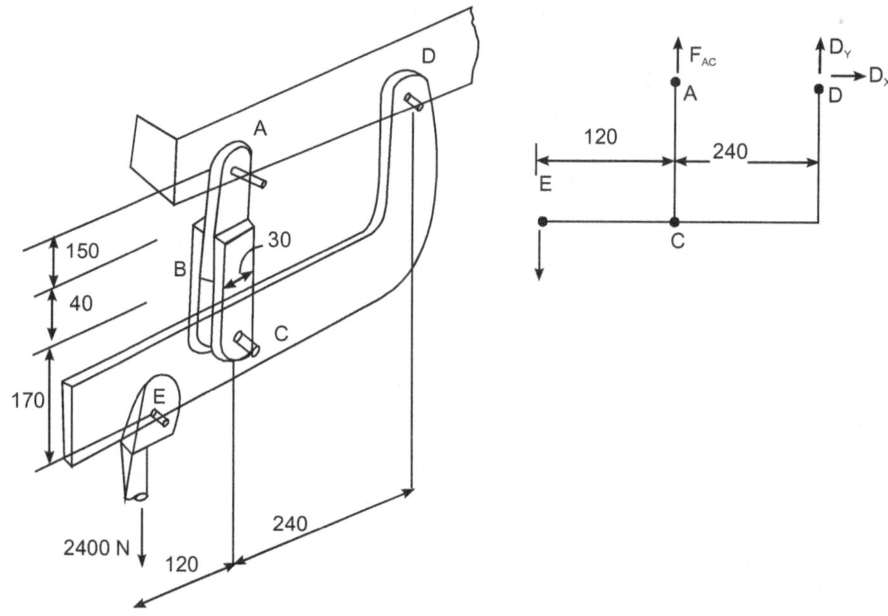

(a) *Shearing stress in Pin A*   Since the 9 mm ϕ pin is in single shear,

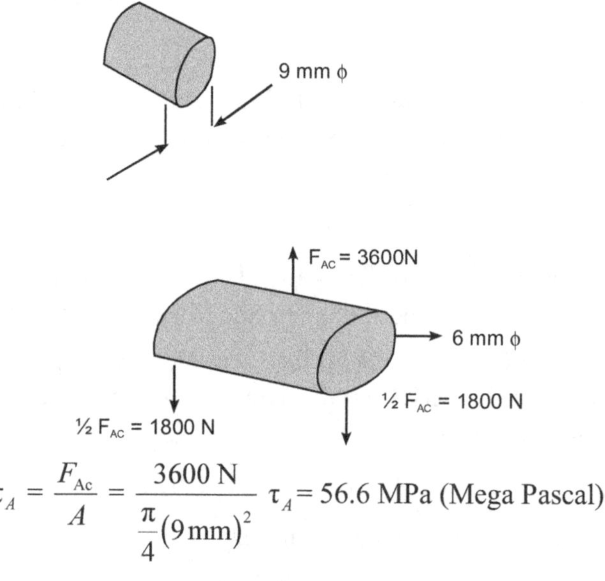

$$\tau_A = \frac{F_{Ac}}{A} = \frac{3600\ \text{N}}{\frac{\pi}{4}(9\,\text{mm})^2} \qquad \tau_A = 56.6\ \text{MPa (Mega Pascal)}$$

(b) *Shearing stress in Pin C*   Since the 6 mm ϕ pin is in double shear,
we take the load as half of F_{AC}.

$$\tau_C = \frac{1/2 F_{AC}}{A} = \frac{1800\ \text{N}}{\frac{1}{4}\pi(6\,\text{mm})^2} = 63.7\ \text{MPa}$$

(c) *Largest normal stress in link ABC*

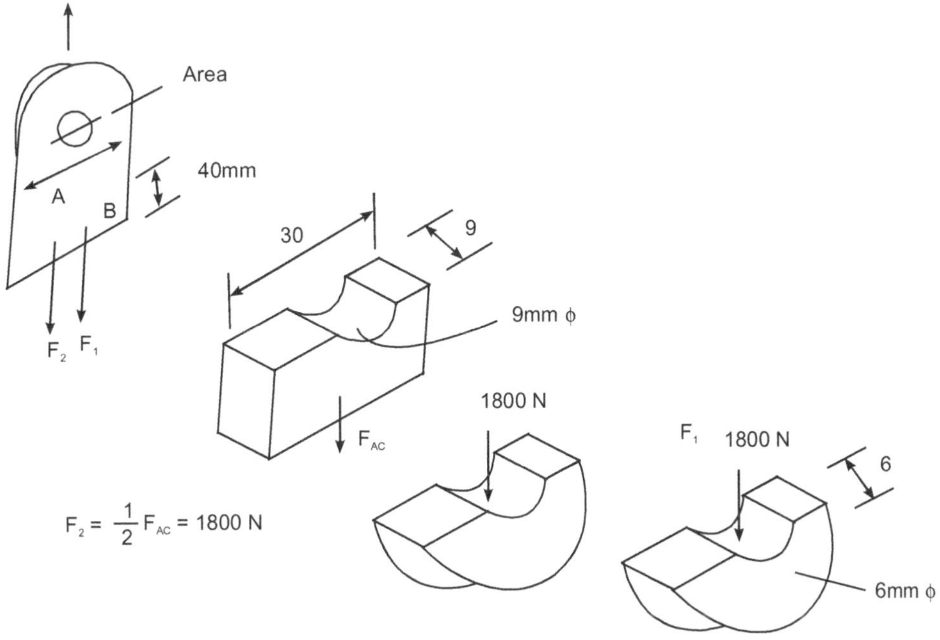

$$F_2 = \frac{1}{2}F_{AC} = 1800 \text{ N}$$

Largest stress occurs where area is smallest. This occurs at the cross section at A where the 9 mm hole is located.

$$\sigma_A = \frac{F_{AC}}{A\,net} = \frac{3600\,\text{N}}{(9\,\text{mm})(30-9)} = \frac{3600}{189} = 19.05 \text{ MPa}$$

(d) *Average shearing stress at B*

$$\tau_B = \frac{F_1}{A} = \frac{1800}{30\times40} = 1.5 \text{ MPa}$$

(e) *Bearing stress in link at C*

$$F_1 = 1800 \text{ N}$$

Bearing area = $6 \times 6 = 36$ mm²

$$\sigma_b = \frac{1800}{36} = 50\,\text{MPa}$$

STRESSES IN AN OBLIQUE PLANE

We considered internal reaction offered by the material in two cases, namely stresses normal to the surface and tangential or shear along the plane depending upon the direction of loading. The area of cross sections are so chosen, that the reaction is either normal or along the surface. This implies that if we choose the plane other than normal or tangential, say, an oblique plane, there will certainly be reaction, the sum total of reaction (resultants) being equal to the applied external load. Consider the case shown in Figure 3.11.

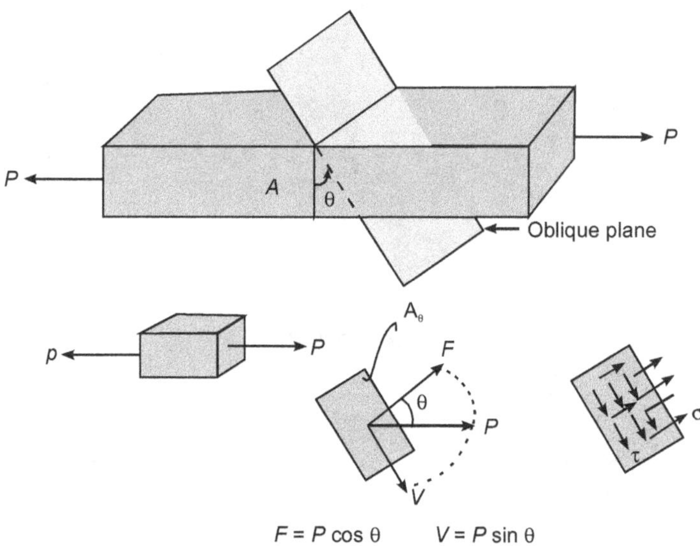

Figure 3.11   Stresses in an oblique plane

$$F = P \cos \theta, \ V = P \sin \theta$$

$$\sigma = \frac{F}{A_\theta}, \ \tau = \frac{V}{A_\theta}, \ A_\theta = \frac{A}{\cos\theta} \tag{3.1}$$

Substituting for A_θ in (3.1)

$$\sigma = \frac{P}{A} \cos^2\theta, \ \tau = \frac{P}{A} \sin\theta \ \cos\theta \tag{3.2}$$

When $\theta = 0$, $\sigma = \dfrac{P}{A}$, $\tau = 0$

When $\theta = 90°$, $\sigma = 0$, $\tau = 0$

When $\theta = 45°$, $\sigma = \dfrac{P}{2A}$, $\tau = \dfrac{P}{2A}$

Depending upon the orientation of the plane, the same load produces both direct and shear stresses on the plane.

COMPONENTS OF STRESS

Previous examples are limited to axial loading and combined with transverse loading. Most machine members and structural members are under more complex loadings.

We consider a 3D elastic solid (Figure 3.12). Take a small element of sides *dx, dy* and *dz*. The 3D solid is subjected to several loads P_1, P_2, P_3, etc.

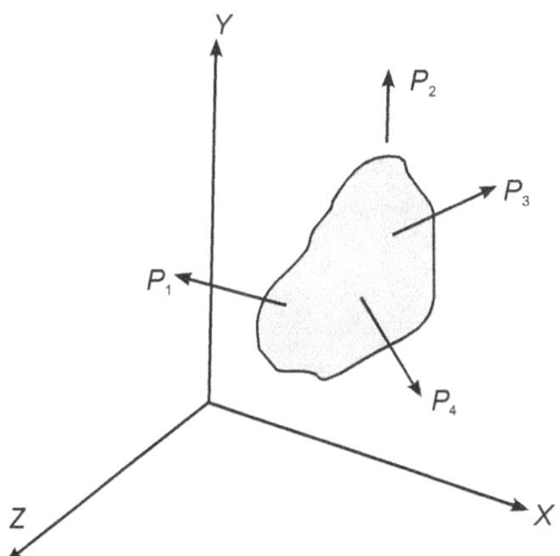

Figure 3.12 3-D solid subjected to forces

The elemental solid is in equilibrium under the action of internal forces. Let the direct stress be denoted by $\sigma_x, \sigma_y, \sigma_z$ (along the 3 axes) and the shear stress as $\tau_{xy}, \tau_{yz}, \tau_{zx}$ and $\tau_{yx}, \tau_{zy}, \tau_{xz}$

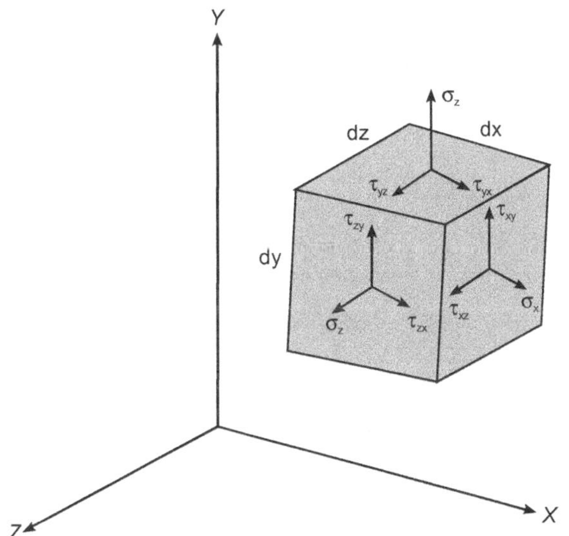

Figure 3.13

Sign convention: τ_{ij} represents a stress on a plane whose normal is along i and directed towards $+j$.

The cube is in equilibrium: Force equation $\Sigma F_x = 0; \ \Sigma F_y = 0; \ \Sigma F_z = 0$ satisfied. (force on sides shown and those on diametrically opposite faces not shown, are equal).

Moment equations: $\Sigma M_x = 0$; $\Sigma M_y = 0$; $\Sigma M_z = 0$ (about z-axis)

$$\tau_{yx}\, dA = \tau_{yx} dzdx$$

$$\tau_{xy} dA = \tau_{xy} dzdy$$

These form a couple.

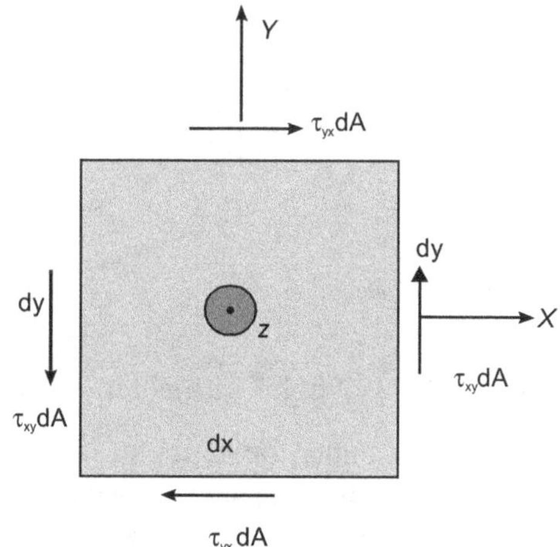

Figure 3.14   Shear stresses on x-y plane

$$\tau_{xy} dA \cdot dx - \tau_{yx} dA dy = 0 \quad (\tau_{xy} dzdydx = \tau_{yx} dzdxdy)$$

$$\therefore\ \tau_{xy} = \tau_{yx}$$

Similarly

$$\tau_{xz} = \tau_{zx}$$

$$\tau_{zy} = \tau_{yz}$$

These are called complementary shear stresses, i.e. τ_{xy} and τ_{yx} are complementary. Hence, although there are six components of shear stresses, in reality there are only three.

Note here that although the cut faces carry stresses, these stresses are converted to forces by multiplying them by the corresponding areas and then equating the forces for equilibrium conditions. **Equilibrium equations are always written for forces and moments only.**

To summarize, six components of stress are needed to specify a stress at a point and shear stresses cannot be present in one plane only as an equal shear stress must be present on another plane perpendicular to the first, called complementary shear stresses.

Note   Stress is a tensor having six components. Tensor is a higher order vector and obeys transformation law from one coordinate system to another.

ULTIMATE AND ALLOWABLE STRESSES

We had earlier, mentioned during the introduction, about the need of the design engineer to understand this concept. The knowledge of stresses are used by engineers in:

1. The **analysis** of existing or proposed machine or structure with the given loadings.

2. **Synthesis** (or design) of new machines and structures which will safely remove function under specified loads. Remember that loads include forces, moments, torques and prescribed displacements.

Let a test specimen be prepared and loaded in the UTM. The largest load which causes breaking (or causes the specimen to carry much less load) is called the ultimate load. The ultimate load divided by the original cross-sectional area is called ultimate normal stress of the material. This stress is also known as ultimate strength in tension or tensile strength of the material.

$$\sigma_u = \frac{P_u}{A}, \text{ where } P_u = P_{\text{ultimate}}$$

Similar ultimate values are also obtained for shear (ultimate shear strength of the material). The design should be done so that the ultimate load is considerably larger than the load it is carrying or allowed to carry under norrmal working conditions. This is called design load or allowable load.

$$\text{Factor of safety} = \text{F.S.} = \frac{\text{Ultimate load}}{\text{Design or allowable load}}$$

A linear relationship exists between the load and the stress caused by the load.

$$\text{F.S} = \frac{\text{Ultimate stress}}{\text{Allowable stress}}$$

Choosing an appropriate F.S is an important task in every design. Small F.S $\rightarrow$ risky and increased probability of failure, large F.S $\rightarrow$ uneconomical and nonfunctional designs.

Choice of F.S depends upon many factors. Engineering judgement is required in approximately choosing F.S for a specified design application. Following are some of the important factors to be kept in mind while choosing the factor of safety (F.S).

(a) Variations in material properties and residual stresses induced in fabrication or manufacturing process.

(b) No. of loadings, frequency of loading and amplitude of stress responsible for causing fatigue.

(c) Load planned at the design stage, actual load applied during the operation, i.e., operating loads, and test loads.

(d) Type of failure which may occur, e.g., brittle, ductile, shear, bending, tensile or compression.

Yielding – low F.S., warning provided for impending failure in ductile materials $F.S = \dfrac{\sigma_{yP}}{\sigma_{actual}}$

Brittle – large F.S. No warning provided and causes abrupt breakage in brittle materials $F.S = \dfrac{\sigma_u}{\sigma_{actual}}$

(e) Uncertainty in methods of analysis (inability of the model to reflect reality in totality).

    ❀ Assumption (properties, characteristics, constraints)

    ❀ Actual stress approximated over an area

(f) Deterioration during service and ageing of materials.

    ❀ time-dependent damage

(g) Integrity of the whole structure, important especially in applications like aircraft, nuclear power plants, absence of which causes disaster to humans and property. Nonuniform—bracings or secondary members with lower F.S. than for primary members with a larger F.S, where local member failure does not affect the structural integrity.

ADDITIONAL FACTORS

❀ Risk and safety of people—excessive F.S used for nuclear plants.

❀ Non-excessive F.S used in aircrafts (minimum weight design)—lower F.S is in view of accurate data on materials, loading, constraints and geometry and using advanced tools of design and testing methodology.

❀ F.S. specified by design specification or codes. Examples of International standards.

Steel: A I S C (American Inst. for Steel Construction)

Concrete: A C I (American Conrete Institute)

Pressure vessels: A S M E (American Society of Mechanical Engineers)

Piping: A N S I (American National Standards Institute)

Boilers: IBR/ ANSI/ASME (IBR—Indian Boiler Regulations)

Refineries (chemical) API, DIN, GOST, ASTM, IS—are other codes of importance.

CONCEPT OF STRAIN

❀ Normal strain under axial loading

❀ Stress–strain diagram

❀ Hooke's law

When we discussed stresses, we did not go into details of deformation. Another important aspect of design of machines and structures is deformation.

While static equation alone may be sufficient to determine force and hence stresses in some cases (statically determinate), most structures come under statically indeterminate problems. In this case deformation considerations (equations) are necessary to arrive at the stresses.

Deformation can also be a design requirement for the following cases.

(a) Machine tool structures—excessive deformation can cause poor machining. Good rigidity is required for machine tools. With good rigidity, stresses are automatically low and hence strength is assured.

(b) Turbine rotors—excessive (deformations) elongation of disc and blade (due to centrifugal stresses and thermal expansion) may cause rubbing of blade with the inner surface of casing of the turbine or compressor leading to failure.

(c) Transmission lines – excessive deformation (deflection) of the line due to wind loads can cause short-circuiting in transmission lines and lead to a total black out.

(d) Telescope-mounting structures—deflections and rotations more than permitted values, cause inaccuracy in the observed data relating to planets and stars.

Hence we need to know calculation of deformations and provide design changes to maintain them within limits prescribed.

DEFINITION

To recaptulate, we saw earlier that

normal strain ϵ = Deformation of the member per unit length $= \dfrac{\text{Change in length}}{\text{Original length}}$

Plotting stress vs. strain, we get the stress–strain diagram. From such diagrams, we can obtain important properties of materials like modulus of elasticity and also know whether the material is brittle or ductile. Load vs. deflection diagram is shown in Figure 3.15.

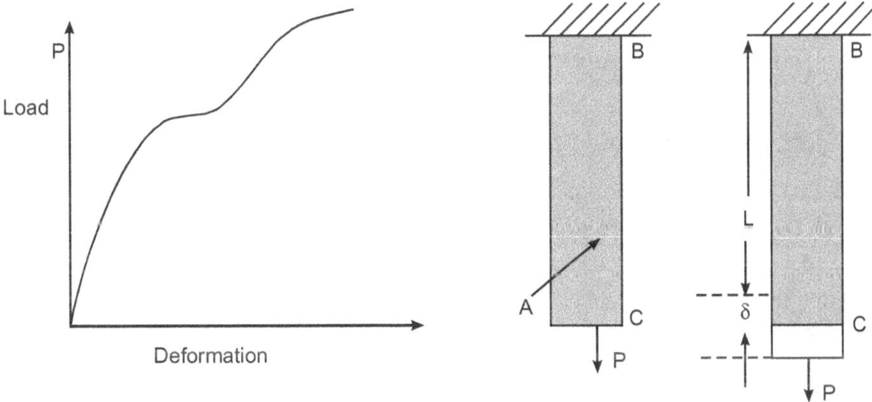

Figure 3.15 Load-deformation curve

Average strain, $\epsilon = \dfrac{\delta}{L}$ is same as Normal Strain

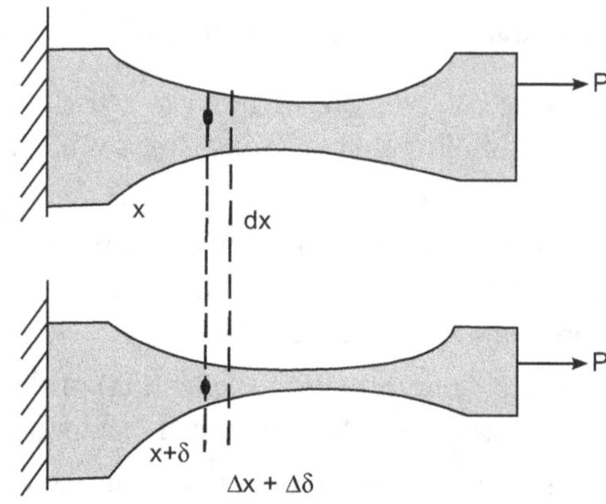

Figure 3.16 Incremental strain

However for strain at a point we need incremental strain. (see Figure 3.16)

$$\text{Incremental strain, } \epsilon = \lim_{\Delta x \to 0} \frac{\Delta \delta}{\Delta x} = \frac{d\delta}{dx}$$

ϵ is dimensionless (represented as %)

STRESS–STRAIN DIAGRAM

Universal testing machine is used for generating stress–strain diagram.

The specimen is as in Figure 3.17.

Sequence of Events in the Testing of Specimen in UTM

1. Linear stress–strain relation exhibited up to reaching yield point. This part of elongation is due to elasticity of the material and is a reversible phenomenon. It is time independent.

2. Then the deformation is increased for additional loading.

3. After σ_Y is reached, specimen undergoes large deformation with a relatively small increase in load. This deformation is caused by slippage of the material along oblique planes and is primarily due to shearing stresses. From elastic deformation, it proceeds to plastic behaviour. This is a time-dependant deformation and is irreversible. Elongation after the yield may be of the order of 200 times as large as the deformation before yield. Brittle materials rupture without yielding.

4. After a certain maximum value of load is reached, the diameter of the central portion of the specimen begins to decrease. This phenomenon is called necking.

5. After necking, somewhat lower loads are sufficient to keep the specimen elongating until it ruptures. Note that the rupture occurs along a cone-shaped surface which forms an angle of approximately 45° with the original surface of the specimen. This indicates that the shear is primarily responsible for the failure of ductile materials and confirm the fact that, under an axial load, shearing stresses are largest on surfaces forming an angle of 45° with the direction of load (we will prove this later).

σ_Y —Stress at which yielding initiated—yield strength (lower yield point determine this)

σ_U —Corresponds to maximum load applied ($\sigma_{Ultimate}$, ultimate strength)

σ_B—Stress corresponding to rupture—breaking strength. (For brittle materials $\sigma_U = \sigma_B$).

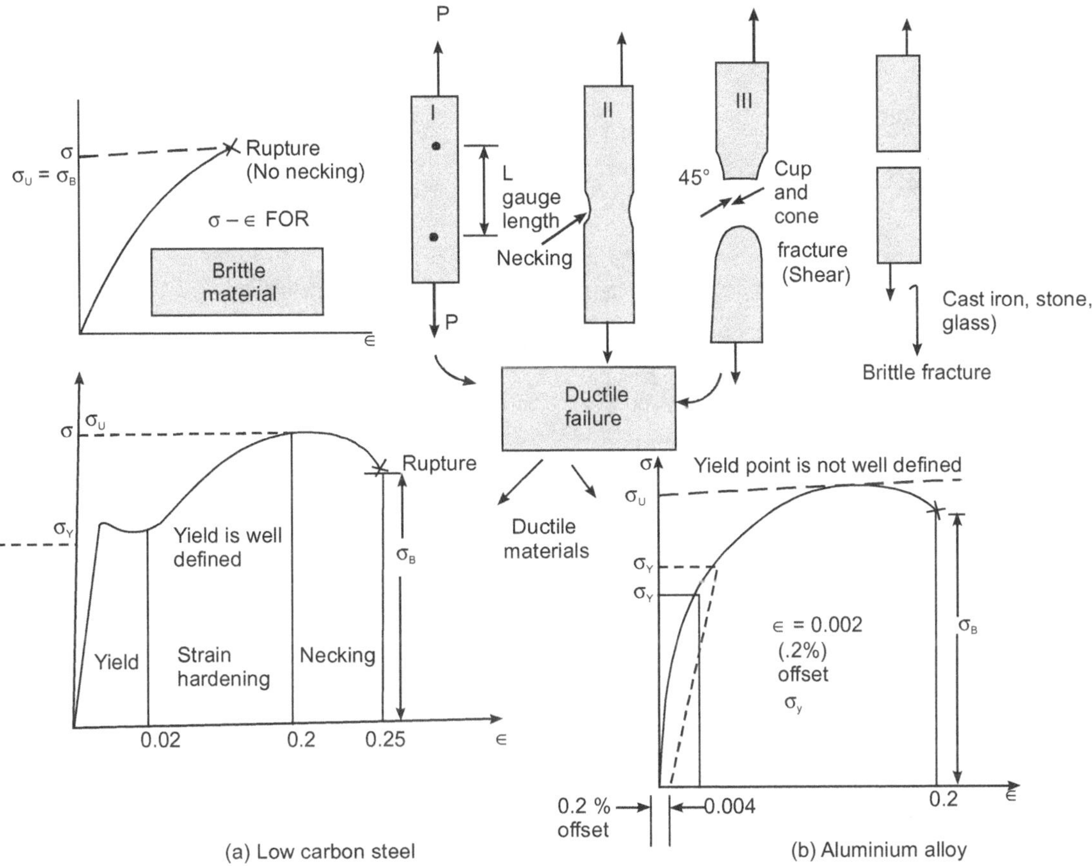

Figure 3.17 Stress–Strain diagram

Note

i. Time to failure for brittle materials is much smaller than for ductile materials.

ii. No necking also takes place in brittle materials where failure occurs along surface normal to the load. Hence normal stresses are responsible for their failure. (In fracture mechanics we will use normal stresses as these are responsible for opening up and propagating cracks).

iii. Ductility is measured in terms of % elongation or % reduction in area.

$$\% \text{ Elongation } = 100 \ \frac{L_B - L_0}{L_0} \ (20\% \text{ required for most structural steels}).$$

$$\% \text{ Reduction in area } = 100 \ \frac{A_0 - A_B}{A_0} \ (\text{average strain} = 0.2 \text{ at rupture, for structural steel} \approx$$

60–70%).

TRUE STRESS AND TRUE STRAIN

$$\text{Stress } = \frac{P}{A_0} = \text{Original cross-sectional area. A}_0 \text{ decreases as load } P \text{ increases.}$$

$$\text{Engineering stress} = \frac{P}{A_0}$$

$$\text{Thus stress} = \frac{P}{A_B}$$

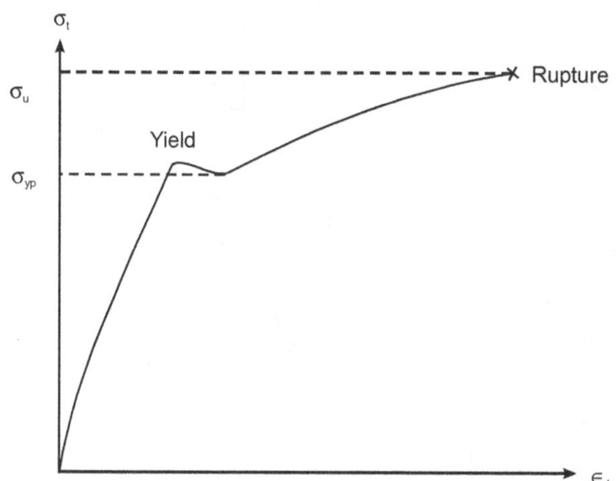

Figure 3.18   Yield and Rupture Stress for ductile material

Engineering design is based on original cross-sectional area to determine acceptable stress and deformation (Figure 3.18).

HOOKE'S LAW—MODULUS OF ELASTICITY

Most engineering structures are permitted to undergo relatively small deformation and are designed so, involving only the linear portion of the stress–strain diagram. For this linear portion, we can write $\sigma = E \in$

This relation is known as Hooke's law and E is known as Young's modulus or modulus of elasticity. Since $\in$ is dimensionless, σ and E have the same, units, viz., N/mm².

Common mistake : σ_y and E are taken as same. They are different.

Reason: σ_y is stress, so, σ_y divided by corresponding strain is E. E is a material property.

NOTE

1. *Generalized Hooke's law for 3D state of stress in generalized materials.* Six stress components are related to 6 strain components through a $[6 \times 6]$ matrix of elastic constants as shown in the following equation. The 36 elastic constants for linear, isotropic, homogenous conservative materials reduce to two in number namely E and ν, which are Young's modulus and Poisson's ratio respectively.

$$\begin{bmatrix} \sigma_{xx} \\ \sigma_{yy} \\ \sigma_{zz} \\ \sigma_{xy} \\ \sigma_{yz} \\ \sigma_{xz} \end{bmatrix} = \begin{bmatrix} E_{11} E_{12} E_{13} E_{14} E_{15} E_{16} \\ E_{21} \\ \\ \\ \\ \cdots\cdots\cdots\cdots\cdots E_{66} \end{bmatrix} \begin{bmatrix} \in_{xx} \\ \in_{yy} \\ \in_{zz} \\ r_{xy} \\ r_{yz} \\ r_{xz} \end{bmatrix} \tag{3.3}$$

2. *Plastic behaviour of materials—Strain hardening.* After loading up to yield and unloading and reloading proportional or elastic limit is increased. Point R remains unchanged. This phenomenon is called strain hardening (Figure 3.19).

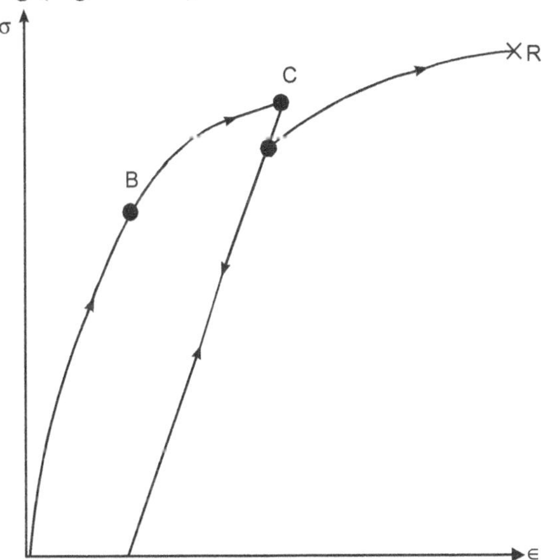

Figure 3.19 Strain-Hardening

3. *Reversal of load over a complete cycle—Plastic cycling and Bauschinger effect.*

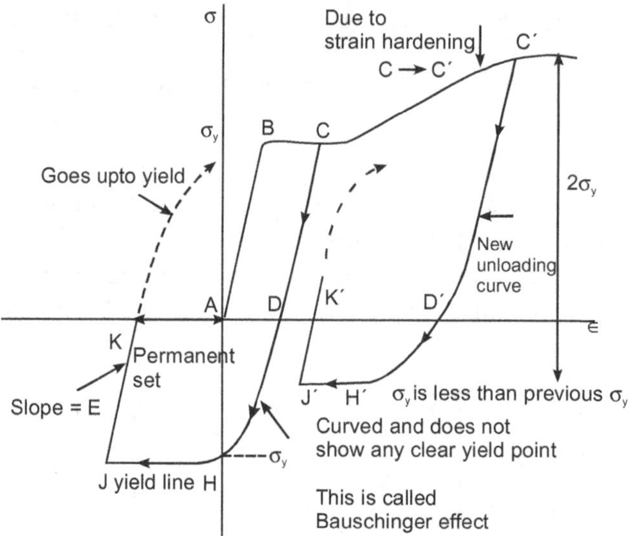

Figure 3.20

Plastic deformation cause internal changes in reverse loadings under plastic conditions are seldom allowed in design. Care is needed. (straightening and damaged material, etc.)

4. *Repeated loadings—Fatigue.* Vibrations can cause fatigue in members. Endurance limit for complete reversal of loading is shown in Figure 3.21. Corrosion reduces (salt water or moist salt in atmospheric air cause rapid reduction) fatigue life. Examples are structures, vehicles operating in coastal areas and endurance limit reduces by almost 50%).

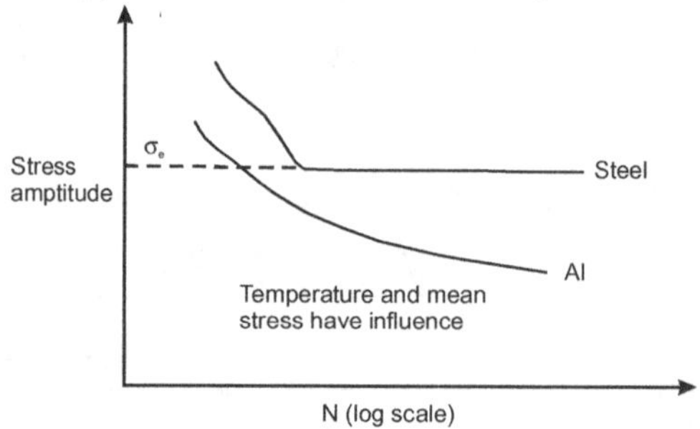

Figure 3.21   Fatigue curve (S–N curve)

σ corresponding to failure or rupture at 500 million cycle is called fatigue limit.

POISSON'S RATIO, ELASTIC CONSTANTS AND THEIR RELATIONS

Let us take the case of uniaxial loading of a specimen.

$$\in_x = \frac{\sigma_x}{E}$$

Normal stresses on other faces namely $\sigma_y = \sigma_z = 0$ (since loading is uniaxial).

We might be tempted to conclude that $\epsilon_y = \epsilon_z = 0$. This is not the case. Even the volume change might appear to be zero, but not correct.

In most engineering materials, elongation produced in the direction of loading is accompanied by contraction in the lateral direction or transverse direction. If we further assume that mechanical properties are uniform in all directions, i.e., the material is isotropic, then we can say $\epsilon_y = \epsilon_z$ and this is called lateral strain. The absolute value of lateral strain over longitudinal strain is called Poisson's ratio and denoted by Greek letter ν(Nu).

$$\nu = \left| \frac{\text{Lateral strain}}{\text{Longitudinal strain}} \right| = -\frac{\epsilon_y}{\epsilon_x} = -\frac{\epsilon_z}{\epsilon_x}$$

$$\epsilon_x = \frac{\sigma_x}{E}$$

$$\therefore \epsilon_y = \epsilon_z = -\frac{\nu \sigma_x}{E}$$

Generalized Hooke's law for isotropic materials (where elastic properties are uniform in all directions).

$$\epsilon_x = \frac{\sigma_x}{E} - \frac{\nu}{E}(\sigma_y + \sigma_z)$$

$$\epsilon_y = \frac{\sigma_y}{E} - \frac{\nu}{E}(\sigma_x + \sigma_z)$$

$$\epsilon_z = \frac{\sigma_z}{E} - \frac{\nu}{E}(\sigma_x + \sigma_y)$$

(3.4)

These laws are valid for multiaxial or 3D loading so long as loading does not cause yielding (within elastic limit).

DILATATION—BULK MODULUS

We shall now examine the effect of normal stresses on the volume change of the material (Refer earlier comments on volume change). Take a cube of unit volume with unit length along edges. Under the action of normal stresses $\sigma_x, \sigma_y, \sigma_z$ it deforms into a rectangular parallelopiped of volume V,

$$V = (1 + \epsilon_x)(1 + \epsilon_y)(1 + \epsilon_z)$$

If we omit products of strains in relation to unity as their values are very small,

$$V = 1 + \epsilon_x + \epsilon_y + \epsilon_z$$

Denoting by e the change in volume of the material,

$$e = V - 1 = + \in_x + \in_y + \in_z$$

Since original volume is unity, e also represents change in volume per unit volume. This is referred to as dilatation of the material. Substituting for $\in_y$ and $\in_z$ and $\in_x$

$$e = \frac{\sigma_x + \sigma_y + \sigma_z}{E} - \frac{2v}{E}(\sigma_x + \sigma_y + \sigma_z)$$

$$e = \frac{1-2v}{E}(\sigma_x + \sigma_y + \sigma_z) \tag{3.5}$$

A case of special interest is that when a body is subjected to hydrostatic pressure P, in which case each of the stress component is equal to $-p$ (Pascal's law)

$$\sigma_x = \sigma_y = \sigma_z = -p \tag{3.6}$$

$$\therefore \ e = \frac{(1-2v)3(-p)}{E}$$

$$e = -\frac{p}{K} \tag{3.7}$$

where, $K = \dfrac{E}{3(1-2v)}$,

where K = Bulk modulus or modulus of compression. The units of K are as that of E, i.e., (N/mm²).

$$1 - 2v > 0 \ \text{or} \ v < \frac{1}{2}$$

or

$$0 < v < \frac{1}{2}$$

Note   Under elasticity conditions $v \simeq 0.5$

SHEARING STRAIN–SHEAR MODULUS

In a general case, we sa w that both normal stresses and shearing stresses acting on a body. The shear stresses do not have any effect on normal strains as long as deformations are small. The shearing stresses will however tend to distort a cube into an oblique parallelepiped (changes shape) (Figure 3.22).

$$K = \frac{E}{3(1-2v)}$$

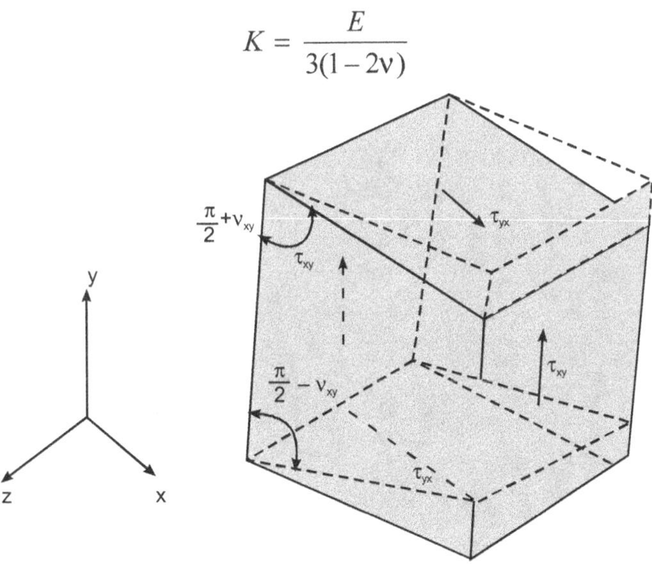

Figure 3.22

Consider the cube subjected to only τ_{yx} and τ_{xy} and no other stresses ($\tau_{xy} = \tau_{yx}$). The element is observed to deform into a rhomboid of sides equal to 1. Two of the angles formed by the four faces under stress are reduced from $\frac{\pi}{2}$ to $\frac{\pi}{2} - \gamma_{xy}$ while the other two are increased to $\frac{\pi}{2} + \gamma_{xy}$ from $\frac{\pi}{2}$. The small angle γ_{xy} (expressed in radians) defines the shearing strain corresponding to x and y.

When deformation involves reduction, γ_{xy} is +ve otherwise, γ_{xy} is − ve.

Superimposed rigid body rotations are not concerned with. Only elastic deformation are dealt with. We can use torsion test result to plot τ_{xy} vs γ_{xy}. The diagram is similar to what was obtained in normal stress–strain case. However, the values corresponding to yield and ultimate strength are about half of what was obtained in normal stress case. As was the case earlier, the initial portion of shearing stress–shearing strain diagram is linear. Hence we can write

$$\tau_{xy} = G\gamma_{xy}$$

This is Hooke's law for shearing stress and strain. (G—Modulus of rigidity or shear modulus)

G has same dimension as τ and γ is in radians (dimensionless). Similarly, for other faces of the cube also, we can write

$$\tau_{yz} = G\gamma_{yz}$$
$$\tau_{zx} = G\gamma_{zx}$$

where G is a material property.

The most generalized form of Hooke's law now can be written as

$$[\sigma] - [\in][\text{Elastic constants}]$$

This might indicate that there are three constants G, E and v. Actually there are only two independent constants for elastic, istropic materials, namely E and v.

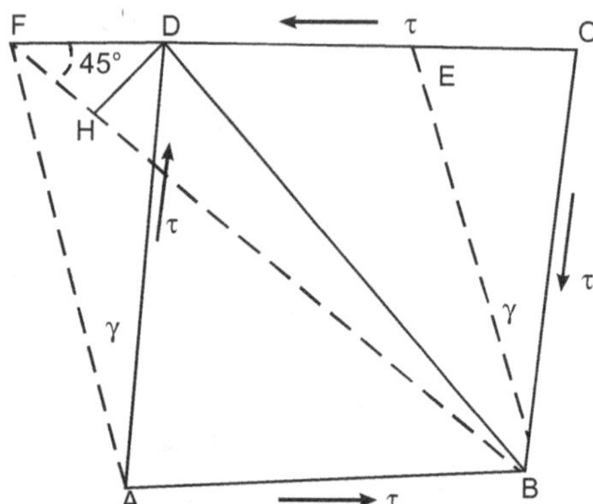

Figure 3.23   Distorted face of a cube due to shear force

ABCD – Square face, τ shear stress

ABEF – Distorted square face, γ, shear strain (Figure 3.23).

τ causes extension of the diagonal BD to BF (In perpendicular directions compressive stresses form). τ also causes shear strain γ, to distort BD to BF

$$\text{Strain in diagonal BD} = \frac{BF - BD}{BD} = \frac{BF - BH}{BD} \quad \text{(Distortion is small)}$$

$$= \frac{FH}{BD}$$

As the distortion is small we can assume $\angle BFD$ is 45°.

$$\text{Strain in diagonal BD} = \frac{FD\cos 45°}{\sqrt{2}\,AD} = \frac{1}{\sqrt{2\sqrt{2}}}\frac{FD}{AD} = \frac{1}{2}\frac{FD}{AD}$$

Since γ is small, $\tan\gamma = \gamma = \dfrac{FD}{AD} = $ Shear strain

$$\text{Strain in diagonal } BD = \frac{\gamma}{2} = \frac{1}{2}\left(\frac{\tau}{G}\right)$$

Due to tensile stress strain in diagonal BD,

$$\frac{\tau}{E} - v\left(\frac{-\tau}{E}\right)$$

$$= \frac{\tau}{E}(1+v) \quad \text{(Shear stress = normal stress in 45° plane).}$$

Refer to previous derivation equation 3.2.

Equating both the strains,

$$\frac{\tau}{E}(1+v) = \frac{1}{2}\frac{\tau}{G}$$

$$E = 2G(1+v) \tag{3.8}$$

MECHANICAL AND THERMAL STRESSES IN SIMPLE AND COMPOSITE MEMBERS

Thermal Stresses and Strains

If the temperature of a body is lowered or raised, its dimension will decrease or increase correspondingly. If these changes, however, are controlled or constrained, the stresses thus developed in the body are called temperature stresses and the corresponding strains are called temperature strains. Extension in the bar due to raise in temperature will be $= \alpha(t_2 - t_1)l$, where,

where,

l = Length of bar of uniform cross section

t_1 = Initial temperature of the bar

t_2 = Final temperature of the bar

α = Coefficient of linear expansion.

If this elongation in the bar is prevented by some external force or by fixing or constraining the ends of the bar, the temperature strain thus produced is given by,

$$\epsilon_t = \frac{\alpha\,(t_2 - t_1)l}{l} = \alpha\,(t_2 - t_1) \text{ which is compressive} \tag{3.9}$$

$$\sigma = E\epsilon$$

$$\therefore \sigma_t = \alpha(t_2 - t_1)E \text{ (compressive)}$$

Example: Plant piping in thermal power or chemical plants are subjected to thermal stresses. In practical plant design, sufficient flexibility in the piping system is provided by incorporating bends, loops or expansion joints to keep the pipe stresses with in allowable limits. If however the temperature of the bar is lowered, the temperature stress and strain will be tensile.

PROBLEM 3.1

A steel rod 15 m long is at a temperature of 15°C. Find the free expansion of the length when the temperature is raised to 65°C. Find the temperature stress produced (i) when the expansion of the rod is prevented (ii) when the rod is permitted to expand by 6 mm.

Take $\alpha = 12 \times 10^{-6}\,°C$ and $E = 200\,\text{GN/m}^2$

Solution

Free expansion of the rod $= l\alpha(t_2 - t_1)$

$$= 15 \times 12 \times 10^{-6} \times (65 - 15)$$

$$= 0.009 \text{ m} = 9 \text{ mm}$$

i.   When expansion is fully prevented

$$\text{Temperature stress} = \alpha\,(t_2 - t_1)\,E$$

$$= 12 \times 10^{-6} \times (65 - 15) \times 200 \times 10^9$$

$$= 120 \times 10^6\,\text{N/m}^2$$

$$= 120 \text{ MN/m}^2$$

ii.   When the rod is permitted to expand by 6 mm

Amount of expansion pevented $= 9 - 6 = 3$ mm.

$$\epsilon = \text{Strain} = \frac{\text{Expansion prevented}}{\text{Original length}} = \frac{3}{15 \times 1000} = 0.0002$$

$$\text{Temperature stress} = \epsilon \times E = 0.0002 \times 200 \times 10^9 = 40 \times 10^6 \text{ N/m}^2$$

$$= 40\,\text{MN/m}^2$$

Check: 9 mm expansion prevented: $\sigma_t = 120\,\text{MN/m}^2$

3 mm expansion prevented: $\frac{1}{3}\sigma_t = 40\,\text{MN/m}^2$

(Linear law assumed within proportionality limit)

COMPOSITE MEMBERS

The principle of stresses and strain due to temperature raise in composite bars is only an extension of what was discussed in the case of simple bars. The difference is owing to the differences in values of thermal expansion coefficients and Young's modulus of elasticity. The following problem will best illustrate the principle.

PROBLEM 3.2

A copper flat measuring 60 mm × 30 mm is brazed to another flat measuring 60 mm × 60 mm mild steel flat as shown in Figure. If the combination is heated through 120° C, determine the following if length of each flat = 400 mm.

i. The stress produced in each of the bars

ii. Shear force which tends to rupture the brazing; and

iii. Shear stress

Take, $\quad \alpha_c = 18.5 \times 10^{-6} \, / \, {}^\circ C$

$\qquad \alpha_s = 12 \times 10^{-6} \, / \, {}^\circ C$

$\qquad E_c = 110 \text{ GN} / \text{m}^2$

$\qquad E_s = 220 \text{ GN} / \text{m}^2$

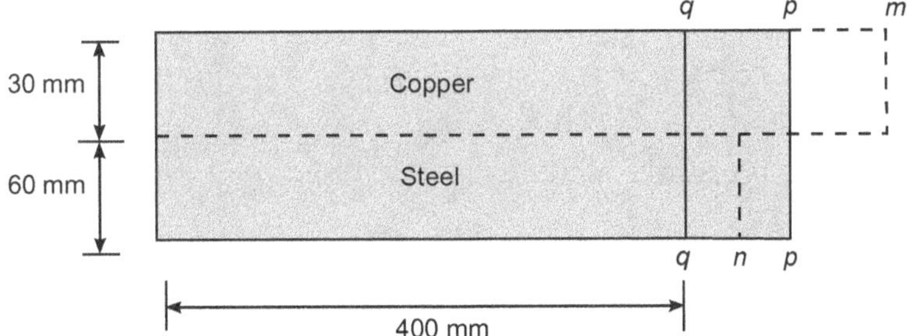

SOLUTION

i. Stress in each bar.

As is α_c greater than α_s, copper flat will tend to extend more than the steel flat, but since they are brazed together, the copper flat will try to pull the steel flat and the steel flat will try to push the copper flat. Finally, however, they will combinedly expand to a certain position "pp", a common position,

Increase in length of copper flat, $q_m = \alpha_c (t_2 - t_1) l$

Increase in length of steel flat, $q_n = \alpha_s (t_2 - t_1) l$

Compressive strain in copper flat $= e_c$

$$e_c = \frac{pm}{l} = \frac{qm - qp}{l}$$

or

$$e_c = \frac{qm}{l} - \frac{qp}{l}$$

$$= \alpha_c (t_2 - t_1) - e \tag{i}$$

where is $\dfrac{qp}{l}$ common strain $= e$

Tensile strain in steel

$$e_s = \frac{np}{l} = \frac{qp - qn}{l}$$

$$- e - \alpha_s (t_2 - t_1) \tag{ii}$$

Adding (i) and (ii) we get

$$e_c + e_s = (\alpha_c - \alpha_s)(t_2 - t_1)$$

But

$$e_c = \frac{\sigma_c}{E_c} = (\alpha_c - \alpha_s)(t_2 - t_1)$$

$$\frac{\sigma_c}{E_c} + \frac{\sigma_s}{E_s} = (\alpha_c - \alpha_s)(t_2 - t_1)$$

$$\frac{\sigma_c}{110 \times 10^9} + \frac{\sigma_s}{220 \times 9} = (18.5 - 12) \times 10^{-6} \times 120$$

$$2\sigma_c + \sigma_s = 220 \times 10^9 \times 120 \times 6.5 \times 10^{-6}$$

At final position PP, $= 171.6 \times 10^6$

Pull on steel = push in copper, indicates force equilibrium.

$$\sigma_s . A_s = \sigma_c . A_c$$

$$\sigma_s \times \frac{60}{1000} \times \frac{60}{1000} = \sigma_c \times \frac{60}{1000} \times \frac{30}{1000}$$

$$\sigma_s = \frac{\sigma_c}{2} = 0.5\sigma_c$$

Substituting this in eqn (iii),

$$2\sigma_c + 0.5\sigma_c = 171.6 \times 10^6$$

$$\sigma_c = 68.64 \times 10^6 \, N/m^2 = 68.64 \, MN/m^2$$

$$\sigma_s = 0.5\sigma_c = 34.32 \, MN/m^2$$

ii. Shear force:

$$\text{Shear force} = \sigma_s A_s = \sigma_c A_c = 34.32 \times 10^6 \times \frac{60}{1000} \times \frac{60}{1000}$$

$$= 123.552 \, KN$$

iii. Shear stress:

$$\text{Shear stress: } \frac{\text{Shear force}}{\text{Shear area}} = \frac{123.552 \times 10^3}{\left(\dfrac{400}{1000}\right) \times \left(\dfrac{60}{1000}\right)} = 5.148 \times 10^6 \, N/M^2$$

$$= 5.148 \, MN/m^2$$

PROBLEM 3.3

A composite bar made up of aluminium and steel is held between two supports as shown in Figure. The bars are stress free at a temperature of 40°C. What will be the stresses in the two bars when the temperature is 20°C, if

i. the supports are nonyielding

ii. the supports come nearer to each other by 0.1 mm. It can be assumed that the change of temperature is uniform all along the bar length.

Take $E_s = 210$ GN/m²

$$E_{al} = 74 \text{ GN/m}^2$$

$$\alpha_s = 11.7 \times 10^{-6} \, /°C$$

$$\alpha_{al} = 23.4 \times 10^{-6} \, /°C$$

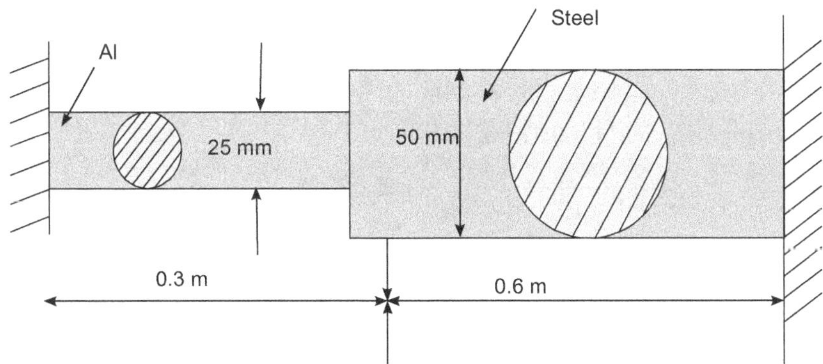

Solution

Let us assume that the support at L is removed so that the contraction of the bar is allowed freely.

Fall in temperature = 40 – 20°C = 20°C

Contraction of steel bar $= \alpha_s (t_2 - t_1) l_s$

$$= 11.7 \times 10^{-6} \times 20 \times 0.6 = 0.0001404 \, \text{m}$$

Contraction of aluminium bar $= \alpha_{al} (t_2 - t_1) l_{al}$

$$= 23.4 \times 10^{-6} \times 20 \times 0.3 = 0.0001404 \, \text{m}$$

∴   Total contraction = 0.0001404 + 0.0001404

$$= 0.0002808 \, \text{m}$$

Now let a force P be applied to end L till this end is brought in contact with the left hand support. Let this force cause a stress in the steel bar. The stress in the aluminium bar will be 4σ, (since the area of the aluminium bar is ¼ of area of steel bar).

Extension of steel bar due to $P = \dfrac{\sigma}{210 \times 10^9} \times 0.6\,\text{m}$

Extension of aluminium bar due to $P = \dfrac{4\sigma}{74 \times 10^9} \times 0.3\,\text{m}$

$$\text{Total extension} = \dfrac{0.6\sigma}{210 \times 10^9} + \dfrac{1.2\sigma}{74 \times 10^9}\,\text{m}$$

i. When the supports do not yield:

$$\dfrac{0.6\sigma}{210 \times 10^9} + \dfrac{1.2\sigma}{74 \times 10^9} = 0.0002808$$

$$0.0002857\sigma + 0.0162\sigma = 0.0002808 \times 10^9 = 280800$$

$$\sigma = 14.73 \times 10^6 \,\text{N/m}^2 = 14.73\,\text{MN/m}^2$$

Stress in steel bar = 14.73 MN / m²

Stress in Al bar = 4 × 14.73 = 58.92 MN / m²

ii. When the supports yield by 0.1mm

$$\dfrac{0.6\sigma}{210 \times 10^9} + \dfrac{1.2\sigma}{74 \times 10^9} = 0.0002808 - \dfrac{0.1}{1000}$$

$$0.002857\sigma + 0.0162\sigma$$

$$= 10^9(0.0002808 - .0001)$$

$$0.0190576\sigma = 180800$$

$$\sigma = 9.49 \times 10\,6\,\text{N/m}^2 \text{ or } 9.49\,\text{MN/m}^2$$

Stress in the steel bar = 9.49 MN / m²

∴ Stress in Al bar = 4 × 9.49

$$= 37.96 \text{ MN / m}^2$$

PROBLEM 3.4

A circle of diameter d = 200 mm is scribed on an unstressed aluminium plate of thickness, t = 18 mm. Forces acting in the plane of the plate later cause normal stresses σ_x = 85 MPa and σ_z = 150 MPa. For E = 70 GPa and $v = \dfrac{1}{3}$, determine the change in

(a) the length of diameter *AB*

(b) the length of diameter *CD*

(c) the thickness of the plate

(d) the volume of the plate

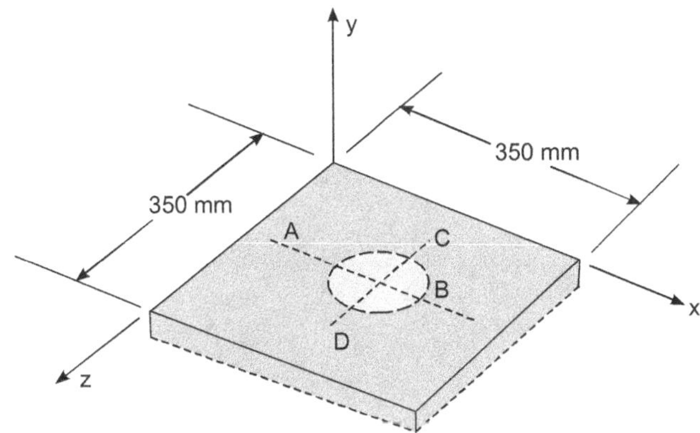

Solution

We note that, $\sigma_y = 0$

$$\in_x = \frac{\sigma_x}{E} - \frac{\nu\sigma_y}{E} - \frac{\nu\sigma_z}{E}$$

$$= \frac{1}{70\,\text{GPa}}\left[85\,\text{MPa} - 0 - \left(\frac{1}{3}\times 150\,\text{MPa}\right)\right] = 0.5 \times 10^{-3}$$

$$\in_y = -\frac{\nu\sigma_x}{E} + \frac{\sigma_y}{E} - \frac{\nu\sigma_{hg}}{E}$$

$$= \frac{1}{70\,\text{GPa}}\left[-\left(\frac{1}{3}85\,\text{MPa}\right) + 0 - \left(\frac{1}{3}150\,\text{MPa}\right)\right] = -1.119\times 10^{-3}$$

$$\in_z = -\frac{\nu\sigma_x}{E} - \frac{\nu\sigma_y}{E} + \frac{\sigma_z}{E} = +1.738\times 10^{-3}$$

(a) Diameter *AB*

The change in length is $\delta_{BA} = \in_x . d$

$= + 0.5 \times 10^{-3} \times 200$ mm

$= 100$ μm

(b) Diameter *CD*

$$\delta_{CD} = \in_z . d = +1.738\times 10^{-3} \times 200$$

$= 348$ μm

(c) Thickness

$$\delta t = \in_y t = -1.119\times 10^{-3} \times 18$$

$-- 20.1$ μm

(d) Volume of the plate

$$e = \epsilon_x + \epsilon_y + \epsilon_z$$

$$= (+0.5 - 1.119 + 1.738)10^{-3} = 1.119 \times 10^3$$

$$\Delta v = ev = +1.119 \times 10^{-3} \left[0.35 \times 0.35 \times 0.018 \right]$$

$$= 2.47 \times 10^{-6} \, m^3$$

$$= 2470 \, mm^3$$

PROBLEM 3.5

A circular section tapered bar is rigidly fixed at both the ends as shown in Figure. If the temperature is raised by 40°C, calculate the stress in the bar. Take $E = 200$ GN/m² and $\alpha = 12 \times 10^{-6}$/°C.

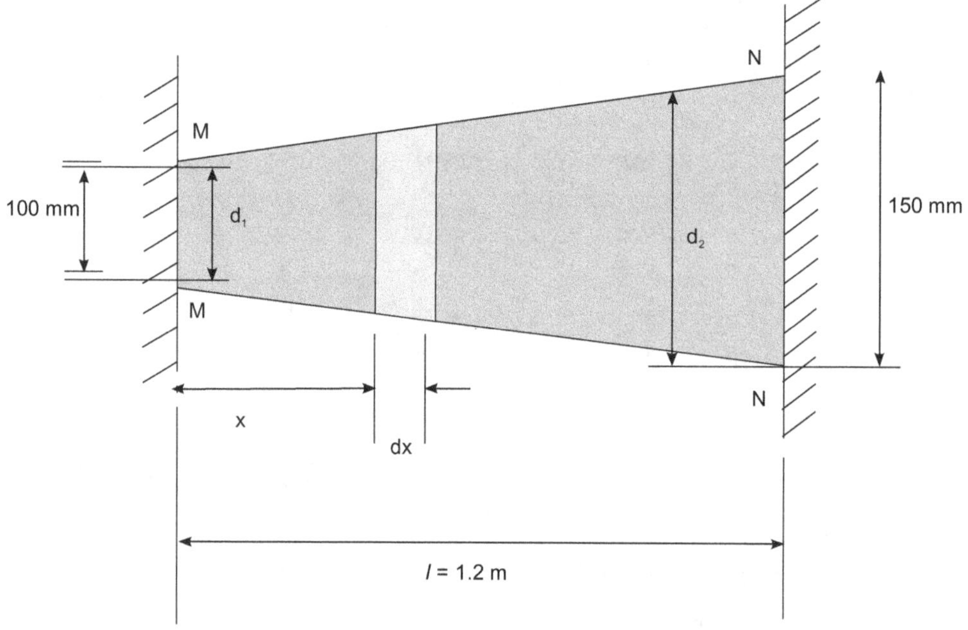

Solution

When the temperature is raised by t°C, a compressive force P is induced. Since this force is same for all cross sections, maximum stress will be at section MM (min. cross-section area). If the bar were free to expand, its expansion would be, $\delta l = \alpha l (t_2 - t_1) = 12 \times 10^{-6} \times 1.2 \times 40 = 0.000576 \, m$

The force induced will be a force P which is required to prevent this free expansion of 0.000576 m. For an element dx at a distance x from end MM, the expansion is

$$\delta_x = \frac{Pdx}{AxE}$$

Total extension is $\delta l = \int\limits_{0}^{l} \dfrac{Pdx}{AxE}$

$$Ax = (\pi/4)\left(d_1 + \dfrac{d_2 - d_1}{l}.x\right)^2 = \left(\dfrac{\pi}{4}\right)(d_1 + kx)^2$$

where, $k = (d_2 - d_1)/l$

$$\delta l = \dfrac{4P}{\pi E}\int\limits_{0}^{l}\dfrac{dx}{(d_1 + kx)^2} = \dfrac{-4P}{\pi kE}\left[\dfrac{1}{d_1 + kx}\right]_0^l$$

$$= -\dfrac{4Pl}{E(d_2 - d_1)}\left[\dfrac{1}{d_1 + d_2 - d_1} - \dfrac{1}{d_1}\right]$$

$$= \dfrac{4Pl}{\pi Ed_1 d_2}$$

Equating $\alpha l(t_2 - t_1) = \dfrac{4Pl}{\pi Ed_1 d_2}$

We get $P = \dfrac{\pi E\alpha(t_2 - t_1)d_1 d_2}{4}$

Maximum stress induced is $\sigma = \dfrac{P}{(\pi/4d_1^2)} = E\alpha(t_2 - t_1).\dfrac{d_2}{d_1}$

Substituting the numerical values, $\sigma = 200 \times 10^9 \times 12 \times 10^{-6} \times 40 \times \dfrac{0.150}{0.100}$

$$- 144 \times 10^6 \text{ N/m}^2$$

$$= 144 \text{ MN/m}^2$$

STRAIN ENERGY, RESILIENCE AND IMPACT LOAD

Strain Energy or Resilience

When an elastic body is loaded, it undergoes deformation, i.e., its dimension change, and when it is relieved of the load it regains its original shape (like reversible processes in themodynamics). For the time loaded energy is stored in it, the same is given up or released when the load is removed. This energy is called strain energy. The strain energy stored in the body within elastic limit, when loaded externally is called resilience and the maximum energy which a body stores up to elastic limit is called proof resilience.

Consider this familiar examples in sports. A pole vaulter runs fast and gains kinetic energy. He stops and quickly applies a compressive load on the flexible bamboo pole and causes a bending of the pole. The kinetic energy is converted to strain energy in bending. He uses this energy to raise himself to the height of the hurdle having converted strain energy to potential energy. Once he reaches the height without much effort, he crosses the hurdle bar and jumps to the otherside.

Proof resilience is the mechanical property of materials and it indicates their capacity to bear shocks. Proof resilience per unit volume is called modulus of resilience.

STRAIN ENERGY IN SIMPLE TENSION AND COMPRESSION

Let us take the case of a bar of cross-sectional area A and length L and subjected to an axial load W. Suppose this load extends the bar by an amount δl and produces a maximum stress σ.

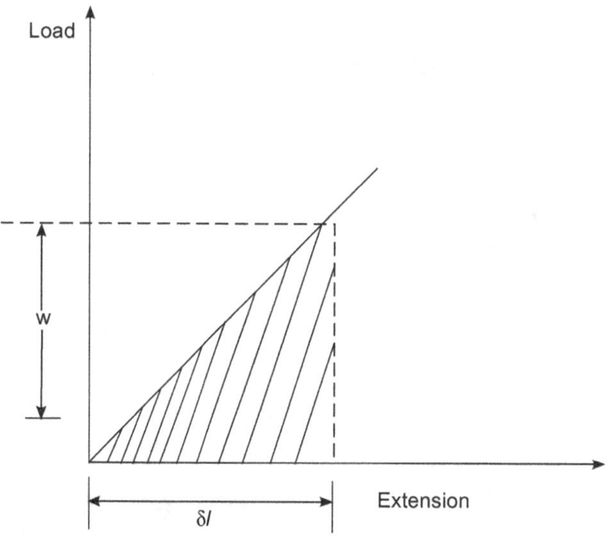

Figure 3.24 Load-extension diagram

The work done by force W and hence the strain energy U, stored in the material is equal to the area (shaded) under the force–extension curve.

Strain energy stored in the bar = Work done by the load

$$U = \frac{1}{2} . W. \delta l \left(\delta l = \frac{\sigma l}{E} \right)$$

$$\therefore \quad U = \left(\frac{W}{2} \right) . \left(\frac{\sigma l}{E} \right) (W = \sigma A)$$

$$= \frac{\sigma A}{2} . \frac{\sigma l}{E} = \frac{\sigma^2 Al}{2E} = \frac{\sigma^2 V}{2E} \quad (V = Al = Volume) \qquad (3.10)$$

$$\text{(Simlarly strain energy in shear} = \frac{\tau^2 V}{2G}; V = Volume)$$

If σ_p be the proof stress or maximum stress to which the bar is stressed upto the elastic limit, then,

$$\text{Proof resilience } U_p = \frac{\sigma_p^2}{2E} \times V$$

And modulus of resilience $= \dfrac{\sigma_p^2}{2E}$ (U_p per unit volume is also known as strain energy density).

Stresses due to different types of loads: A body may be subjected to the following types of loads.

(a) Gradually applied load

(b) Suddenly applied load

(c) Falling or impact load

(a) Gradually applied load

The load is increased steadily or in steps from zero to its final value. Let W be the final value of load and let δl and σ be the corresponding change in length and stress induced in the body.

$$\text{Energy due to external load} = \frac{1}{2}.\sigma.A.\delta l$$

$$\text{Workdone on the body} = \frac{1}{2}W.\delta l$$

Strain energy stored = Workdone

$$= \frac{1}{2}\sigma A\delta l = \frac{1}{2}W\delta l$$

$$\sigma = \frac{W}{A} \tag{3.11a}$$

Unless specified, the load is always applied gradually. (Assumption usually made unless stated otherwise).

(b) Suddenly applied loads

When the load is applied all of a sudden and not stepwise, it is called suddenly applied load. Let W be the load applied all of a sudden and let the maximum stress thus produced be σ_{su}. The extension being the same as δl.

Energy stored = External work done

$$w \times \delta l = \frac{1}{2}\sigma_{su} A \times \delta l$$

$$\sigma_{su} = \frac{2W}{A} \tag{3.11b}$$

This shows that the instantaneous stress produced due to a suddenly applied load is twice that when gradually applied.

(c) Freely dropped load

Let a load W fall freely through a distance h before it strikes the collar attached at the bottom of the bar as shown in figure.

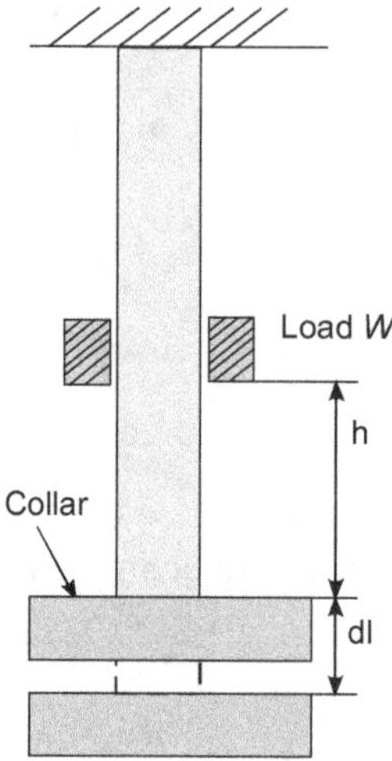

Figure 3.25

Workdone by the load $= W\,(h + dl)$, where dl is the instantaneous extension.

Equating this to the strain energy,

$$
\left(
\begin{aligned}
&\frac{dl}{l} = \in \;;\; dl = \in l \\[4pt]
&\frac{\sigma}{\in} = E; \; dl = \frac{\sigma}{\in}\, l \\[4pt]
&V = Al
\end{aligned}
\right)
$$

$$
\frac{1}{2}\sigma \cdot \in \cdot V = \frac{1}{2}\sigma \cdot \frac{dl}{l} \cdot Al = \frac{1}{2}\sigma\, Adl
$$

$$
= \frac{1}{2}\sigma A \frac{\sigma}{E} l
$$

$$
\frac{1}{2}\sigma A \frac{\sigma l}{E} = w\!\left(h + \frac{\sigma l}{E}\right)
$$

$$
= \text{workdone by } W
$$

$$\therefore \quad \frac{\sigma^2}{2E}.Al - wh - \frac{w\sigma l}{E} = 0$$

Multiplying throughout by $\left(\dfrac{2E}{Al}\right)$

$$i.e., \sigma_2 - \sigma \frac{2w}{A} - \frac{2whE}{AL} = 0$$

$$\sigma = \frac{w}{A}\left[1 + \sqrt{1 + \frac{2AEh}{wl}}\right] \qquad\qquad (3.12)$$

Thus σ freely dropped load $> \sigma$ suddenly applied load

If $h = 0$,

$$\sigma_i = \frac{2W}{A}$$

Static deflection due to load W

$$\delta l = \frac{Wl}{AE}$$

$$\sigma_i = \frac{W}{A}\left(1 + \sqrt{1 + \frac{2h}{\delta l}}\right)$$

Since $\sigma = \dfrac{W}{A}$

$$\sigma_i = \sigma\left(1 + \sqrt{1 + 2h/\delta l}\right)$$

$$\frac{\sigma_i}{\sigma} = 1 + \sqrt{1 + \frac{2h}{\delta l}}$$

This dimensionless ratio $\dfrac{\sigma_i}{\sigma}$ is usually called load factor n.

$$n = 1 + \sqrt{1 + \frac{2h}{\delta l}}$$

Let W_e be the equivalent static load that produces the same stress σ_i as produced by the load W with impact.

$$\sigma i = \frac{We}{A}, \sigma = \frac{W}{A}$$

$$\frac{\sigma i}{\sigma} = \frac{We}{W}\left(n = \frac{We}{W} \text{ or } \frac{\sigma i}{\sigma}\right)$$

$$W_e = nW$$

The load factor is important because it gives the factor by which the load producing impact should be multiplied to give an equivalent static load for design. Once *We* is known, the solution of a design problem can be carried out as usual by taking *We* instead of the load *W*.

In machine design, we use such factors for design of components subjected to impact loads, like in power presses, forging machine or in package design.

Example 1

A rod consists of two portions *BC* and *CD* of the same material and same length, but of different cross-sections. Determine the strain energy of the rod when it is subjected to a centric axial load P, expressing the result in terms of *P, L, E*, the cross-sectional area *A* of portion *CD*, and the ratio *n* of the two diameters.

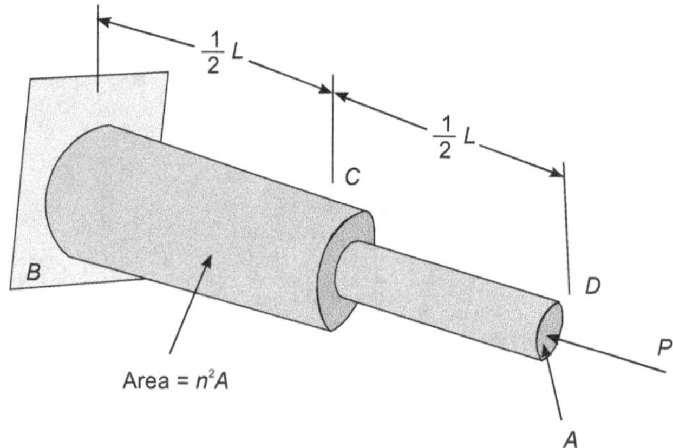

We use equation for *U* to compute the strain energy of each of the two portions, and the expressions obtained.

$$U_n = \frac{P^2(\frac{1}{2}L)}{2AE} + \frac{P^2(\frac{1}{2}L)}{2(n^2 A)E} = \frac{P^2(L)}{4AE}(1 + \frac{1}{n^2}) \qquad (3.13)$$

or

$$U_n = \frac{1+n^2}{2n^2} \; \frac{P^2 L}{2AE}$$

We check that, for *n* = 1, we have

$$U_1 = \frac{P^2 L}{2AE} \qquad (3.14)$$

which is the expression derived earlier for strain–energy (equation 3.10) for a rod of length *L* and uniform cross section of area *A*. We also note that, for $n > 1$, we have $U_n < U_1$; for example, when n = 2, we have $U_2 = \left(\dfrac{5}{8}\right)U_1$. Since the maximum stress occurs in port on *CD* of the rod and is equal to $\sigma_{max} = P/A$, it follows that, for a given allowable stress, increasing the diameter of portion *BC* of the rod results in a decrease of the overall energy-absorbing capacity of the rod. Unnecessary changes in cross-sectional area should therefore be avoided in the design of members which may be subjected to loadings, such as impact loadings, where the energy-absorbing capacity of the member is critical.

Example 2

A load P is supported at *B* by two rods of the same material and of the same uniform cross section of area *A*. Determine the strain energy of the system.

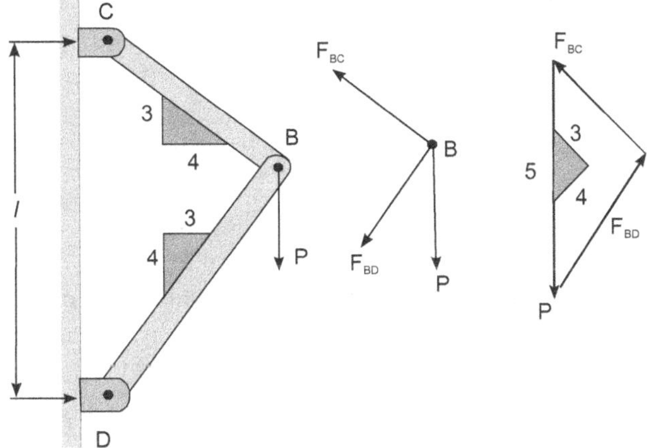

Denoting by F_{BC} and F_{BD} respectively, the forces in members *BC* and *BD*, and recalling (Equation 3.10) we express the strain energy of the system as

$$U = \frac{F_{BC}^{2}\,(BC)}{2AE} + \frac{F_{BD}^{2}\,(BD)}{2AE}$$

But we note from figure that BCD is a right angled Δ and

hence,
$$BC = 0.6l \quad BD = 0.8l$$

and from the free-body diagram of pin *B* and the corresponding force triangle shown in figure that

$F_{BC} = +0.6P,\ F_{BD} = -0.8P.$ Substituting in above equation for U.

$$\text{we have } U = P^2 l\,\frac{[(0.6)^3 + (0.8)^3]}{2AE}$$

$$= 0.364\frac{P^2 l}{AE}$$

STRAIN ENERGY IN BENDING

Consider a beam *AB* subjected to a given loading and let *M* be the bending moment at a distance *x* from end *A*. Neglecting for the time being the effect of shear, and taking into account only the normal stresses $\sigma_2 = My/I$, we substitute this expression into equation 3.10 and after integration over the volume total strain energy is given by

$$U = \int \frac{\sigma_x^2}{2E} dv = \int \frac{M^2 y^2}{2EI^2} dv \qquad (3.15)$$

ST. VENANT'S PRINCIPLE AND STRESS CONCENTRATION

Please refer to our earlier discussions on stresses in axially loaded members. We have been emphasising that the concept of average stress in a member cross-section is valid only from sections away from the point of application of load. If we really want to determine the stresses close to the point of application of load, we need to resort to the concept of 3D elasticity and stress at a point, and apply equations compatible with indeterminate structures. Deformation considerations have to necessarily play a role in such problems. Consider the case of a compression as shown in Figure 3.26.

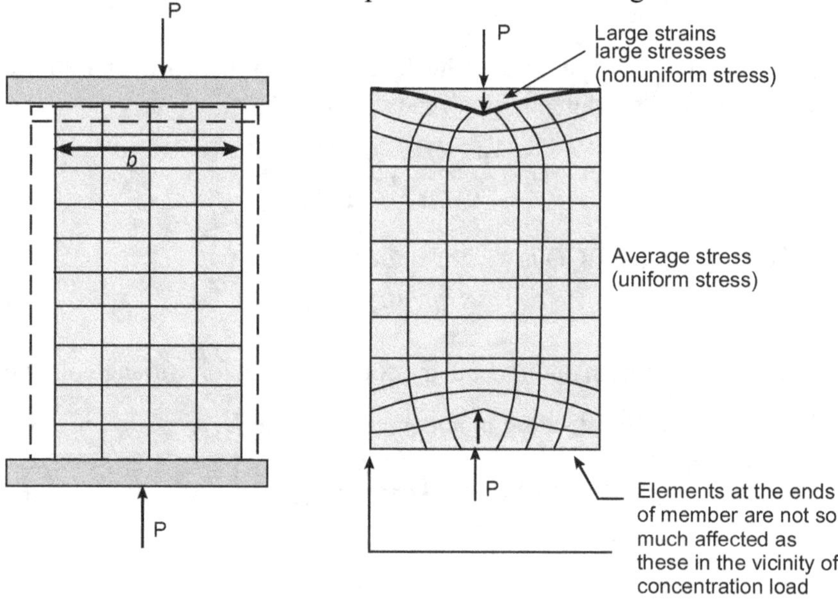

Figure 3.26 Actual stress distribution near vicinity of load

We note that at a distance b from either end (b is the width of the plate), the stress distribution is nearly uniform. Thus at distances $\geq b$, = the stress across the section is nearly same.

ST.VENANT'S PRINCIPLE

Except in the immediate vicinity of the points of application of the loads, the stress distribution may be assumed independent of the actual mode of application of the loads, so long as the loads are statically equivalent.

This statement which applies not only to axial loadings, but to any type of load practically possible is known as St. Venant's principle. (Adhemir Barre de Saint Venant). While St. Venant's principle makes it possible to replace a given loading by a simpler one for the purpose of computing the stresses in a structural member, we should keep in mind two important points when applying this principle.

1. The actual loading and the loading used to compute the stresses must be statically equivalent.

2. Stresses cannot be computed in this member in the immediate vicinity of the points of application of the loads. Advanced theoretical or experimental methods must be used to determine the distribution of stresses in these zones.

STRESS CONCENTRATIONS

So long as sections of a component change gradually, the state of stress will be uniform. If there is an abrupt or sudden change in cross section or sudden change in material (like in welded joints), some extra amount of stress is accumulated in a place or zone like a shoulder, V-notch, groove, hole or key-way. This phenomenon of local accumulation of stress is called stress concentration. The maximum stress at such locations divided by the nominal stresses (that would have been present had there been no abrupt change in cross section) is called stress concentration factor, usually denoted by 'K'.

The stress concentration factor (SCF) under dynamic loading is called fatigue stress concentration factor (denoted by K_f).

Design engineers can refer to the table of S.C.F's for the appropriate cases which pertain to or close to their designs for geometrics and loadings given in the book. Large compilation of SCFs by RE Peterson.

$$K = \frac{\sigma_{max}}{\sigma_{av}}$$

Concentration of stress will cause failures more rapidly. Hence the designer must attempt to improve the design so that the stress concentration is reduced (cannot be avoided fully).

Rounding off of sharp corners, providing liberal fillet radii, providing a smooth transition at zones where two different cross section meet, are some of the measures which a designer incorporates in design to reduce the stress-concentration effects. See Figure 3.27 for illustration.

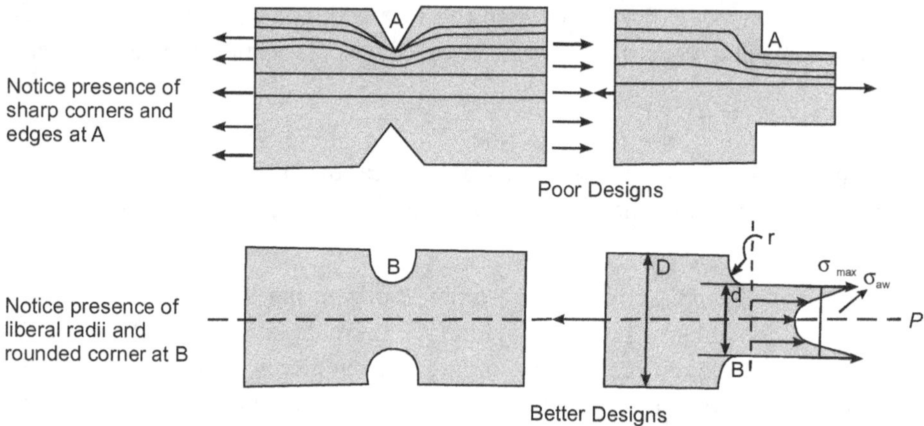

Figure 3.27   Example of poor and better designs

Here we see that 'stress at a point' concept is necessary and 'average stress' concept fails to explain the stress-concentration phenomenon. Defects produced in casting, welding, etc. cause stress concentration. By X-ray radiography or ultrasonic inspection, such defects can be detected.

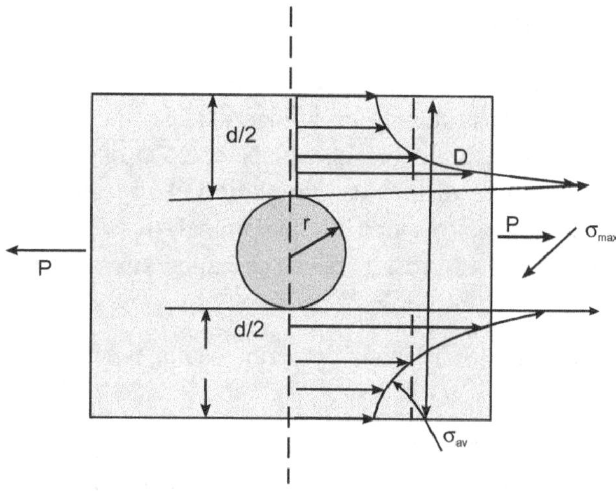

Figure 3.28

The results for SCF are obtained by photoelastic technique (Refer to chapter on experimental methods). The results were independent of the material used and is a function of geometry only, like *r/d* and *d/D* in the case of fillets. Further the designer is more interested in the maximum value of σ than the actual stress distribution, whether $\sigma_{max} \leq \sigma_{all}$ is the concern of the designer and not where this occurs.

The stress concentration factor gives a feel to the designer of how much the nominal stress is increased. He needs to know only the nominal stress and multiplies by appropriate SCF to get the actual stress.

Some typical cases along with sketches are furnished in Table 3.1. Arrive at SCF and multiply the nominal stress by SCF to obtain maximum stress for design purposes.

Assumption is that $\sigma_{max} \leq \sigma_{yp}$. (Providing reinforcement around holes or cut outs, locally increase the area for resistance and hence reduce the stresses. This is a designer's job).

PROBLEM 3.6

Determine the largest axial load P which may be safely supported by a flat steel bar consisting of two portions, both 10 mm thick and respectively 40 mm and 60 mm wide, connected by fillets of radius = 8 mm. Assume an allowable normal stress of 165 MPa.

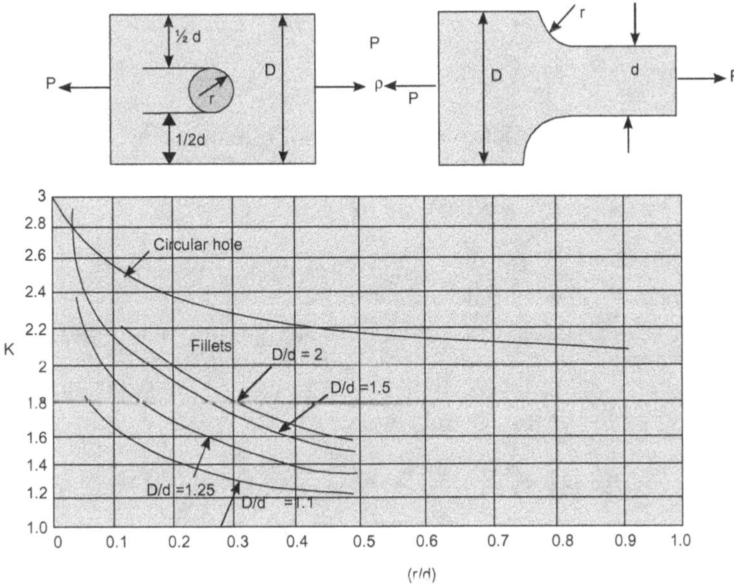

We first compute the ratio

$$\frac{D}{d} = \frac{60}{40} = 1.5$$

$$\frac{r}{d} = \frac{8}{40} = 0.2$$

Using the curve , corresponding to $\dfrac{D}{d} = 1.5$

We find SCF corresponding to $\dfrac{r}{d} = 0.2$

$$K = 1.72$$

$$\sigma_{av} = \frac{\sigma_{max}}{1.72}$$

$$\sigma_{av} = 165\,\text{MPa} = \sigma_{max}$$

$$\sigma_{av} = \frac{165}{1.72} = 96\,\text{MPa}$$

$$\sigma_{av} = \frac{P}{A} \quad (A \text{ should be taken for narrower portion.})$$

$$P = A\sigma_{av} = (40\,\text{mm})(10\,\text{mm})(96\,\text{Mpa})$$

$$P = 38.4 \times 10^3\,\text{N}$$

$$P = 38.4\,\text{KN}$$

The elastic stress concentration factor k is the ratio of the maximum stress in the stress raiser to the nominal stress computed by the ordinary strength-of-materials formulas, using the dimensions of the net cross section unless defined otherwise in specific cases.

For those data presented in the form of equations, the equations have been developed to fit as possible the many data points given in the lite rature referenced in each case. Over the majority of the ranges specified for the variables, the curves fit the data points with much less than a 5% error.

It is not possible to tabulate all the available values of stress concentration factors found in the literature, but the following list of topics and sources of data will be helpful.

INTERESTING POINTS TO PONDER

One important theorem under strain energy is Castigliano's theorem. This helps in determining deflections or deformations under different loading conditions. The two elements of this theorem are stated as under.

First theorem

Partial Derivative of strain energy, within yield limit, with respect to any displacement gives the forces or loads in the direction of displacement

$$\text{Force, } F_k = \frac{\partial V}{\partial \delta_k}$$

V-strain energy, δ_k -Displacement in the 'k' direction.

Second theorem

Partial Derivative of strain energy, within yield limits, with respect to a force producing it, gives the displacement in the direction of force

$$\delta_k = \frac{\partial V}{\partial F_k}$$

Table 3.1 Factors of stress concentration for elastic stress (k)

Type of form irregularity or stress raiser	Stress condition and manner of loading	Factor of stress concentration k for various dimensions
1. Two U-notches in a member of rectangular section	1a. Elastic stress, axial tension	$k = K_1 + K_2 \left(\dfrac{2h}{D}\right) + K_3 \left(\dfrac{2h}{D}\right)^2 + K_4 \left(\dfrac{2h}{D}\right)^3$ where $0.1 \le h/r \le 2.0 \qquad\qquad 2.0 \le h/r \le 50.0$ $K_1\ \ 0.850 + 2.628\sqrt{h/r} - 0.413h/r \qquad 0.833 + 2.069\sqrt{h/r} - 0.009h/r$ $K_2\ -1.119 - 4.826\sqrt{h/r} + 2.57h/r \qquad 2.732 - 4.157\sqrt{h/r} + 0.176h/r$ $K_3\ \ 3.563 - 0.514\sqrt{h/r} - 2.402h/r \qquad -8.859 + 5.327\sqrt{h/r} - 0.320h/r$ $K_4\ -2.294 + 2.713\sqrt{h/r} + 0.240h/r \qquad 6.294 - 3.239\sqrt{h/r} + 0.154h/r$ For the semicircular notch ($h/r = 1$) $k = 3.065 - 3.370\left(\dfrac{2h}{D}\right) + 0.647\left(\dfrac{2h}{D}\right)^2 + 0.658\left(\dfrac{2h}{D}\right)^3$
	1b. Elastic stress, in - plane bending	$k = K_1 + K_2 \left(\dfrac{2h}{D}\right) + K_3 \left(\dfrac{2h}{D}\right)^2 + K_4 \left(\dfrac{2h}{D}\right)^3$ where $0.25 \le h/r \le 2.0 \qquad\qquad 2.0 \le h/r \le 50.0$ $K_1\ \ 0.723 + 2.845\sqrt{h/r} - 0.504h/r \qquad 0.833 + 2.069\sqrt{h/r} - 0.009h/r$ $K_2\ -1.836 - 5.746\sqrt{h/r} + 1.314h/r \qquad 0.024 - 5.383\sqrt{h/r} + 0.126h/r$ $K_3\ \ 7.254 - 1.885\sqrt{h/r} - 1.646h/r \qquad -0.856 + 6.460\sqrt{h/r} - 0.199h/r$ $K_4\ -5.140 + 4.785\sqrt{h/r} - 2.456h/r \qquad 0.999 - 3.146\sqrt{h/r} + 0.082h/r$ For the semicircular notch ($h/r = 1$) $k = 3.065 - 6.269\left(\dfrac{2h}{D}\right) + 7.015\left(\dfrac{2h}{D}\right)^2 - 2.812\left(\dfrac{2h}{D}\right)^3$

Contd.

Table 3.1 Factors of stress concentration for elastic stress (k)

Type of form irregularity or stress raiser	Stress condition and manner of loading	Factor of stress concentration k for various dimensions
	1c. Elastic stress, out-of-plane bending	$k = K_1 + K_2\left(\dfrac{2h}{D}\right) + K_3\left(\dfrac{2h}{D}\right)^2 + K_4\left(\dfrac{2h}{D}\right)^3$ where for $0.25 \leq h/r \leq 4.0$ and h/t is large $K_1 = 1.031 + 0.831\sqrt{h/r} + 0.014h/r.$ $K_2 = -1.227 - 1.646\sqrt{h/r} + 0.117h/r$ $K_3 = 3.337 - 0.750\sqrt{h/r} + 0.469h/r$ $K_4 = -2.141 + 1.566\sqrt{h/r} - 0.600h/r$ For the semi-circular notch ($h/r = 1$) $k = 1.876 - 2.756\left(\dfrac{2h}{D}\right) + 3.056\left(\dfrac{2h}{D}\right)^2 - 1.175\left(\dfrac{2h}{D}\right)^3$
2. Two V-notches in a member of rectangular section	2a. Elastic stress, axial tension	The stress concentration factor for the V-notch, k_θ, is the smaller of the values $k_\theta = k_U$ or $k_\theta = 1.11k_U - \left[0.0275 + 0.000145\theta + 0.0164\left(\dfrac{\theta}{120}\right)^8\right]k_U^2$ for $\dfrac{2h}{D} = 0.40$ and $\theta \leq 120°$ or $k_\theta = 1.11k_U - \left[0.0275 + 0.00042\theta + 0.0075\left(\dfrac{\theta}{120}\right)^8\right]k_U^2$ for $\dfrac{2h}{D} = 0.667$ and $\theta \leq 120°$ where k_U is the stress concentration factor for a U-notch, case la, when the dimensions h, r, and D are the same as for the V-notch and θ is the notch angle in degrees.

Contd.

Table 3.1 *(Continues)*

Type of form irregularity or stress raiser	Stress condition and manner of loading	Factor of stress concentration k for various dimensions
3. One U-notch in a member of rectangular section	3a. Elastic stress, axial tension	$k = K_1 + K_2\left(\dfrac{h}{D}\right) + K_3\left(\dfrac{h}{D}\right)^2 + K_4\left(\dfrac{h}{D}\right)^3$ where for $0.5 \le h/r \le 4.0$ $K_1 = 0.721 + 2.394\sqrt{h/r} - 0.127h/r$ $K_2 = 1.978 - 11.489\sqrt{h/r} + 2.211h/r$ $K_3 = -4.413 + 18.751\sqrt{h/r} - 4.569h/r$ $K_4 = 2.714 + 9.655\sqrt{h/r} + 2.512h/r$ For the semicircular notch ($h/r = 1$) $k = 2.988 - 7.300\left(\dfrac{h}{D}\right) + 9.742\left(\dfrac{h}{D}\right)^2 - 4.429\left(\dfrac{h}{D}\right)^3$
	3b. Elastic stress, in-plane bending	$k = K_1 + K_2\left(\dfrac{h}{D}\right) + K_3\left(\dfrac{h}{D}\right)^2 + K_4\left(\dfrac{h}{D}\right)^3$ where for $0.5 \le h/r \le 4.0$ $K_1 = 0.721 + 2.394\sqrt{h/r} - 0.127h/r$ $K_2 = -0.426 - 8.827\sqrt{h/r} + 1.518h/r$ $K_3 = 2.161 + 10.968\sqrt{h/r} - 2.455h/r$ $K_4 = -1.456 - 4.535\sqrt{h/r} + 1.064h/r$ For the semicircular notch ($h/r = 1$) $k = 2.988 - 7.735\left(\dfrac{h}{D}\right) + 10.674\left(\dfrac{h}{D}\right)^2 - 4.927\left(\dfrac{h}{D}\right)^3$
4. One V-notch in a member of rectangular section	Elastic stress, in-plane bending	The stress concentration factor for the V-notch, k_θ, is the smaller of the values $k_\theta = k_U$

Table 3.1 Factors of stress concentration for elastic stress (k)

Type of form irregularity or stress raiser	Stress condition and manner of loading	Factor of stress concentration k for various dimensions
		or $k_\theta = 1.11 k_u - \left[0.0275 + 0.1125 \left(\dfrac{\theta}{150} \right)^4 \right] k_U^2$ for $= \theta \leq 150°$

where $_k U$ is the stress concentration factor for a U-notch, case 3b, when the dimensions h, r, and D are the same as for the V-notch and θ is the notch angle in degrees.

5. Square shoulder with fillet in a member of rectangular section

5a. Elastic stress, axial tension

$$k = K_1 + K_2 \left(\frac{2h}{D} \right) + K_3 \left(\frac{2h}{D} \right)^2 + K_4 \left(\frac{2h}{D} \right)^3$$

where $\dfrac{L}{D} > \dfrac{3}{[r/(D-2h)]^{1/4}}$

	$0.1 \leq h/r \leq 2.0$	$2.0 \leq h/r \leq 20.0$
K_1	$1.007 + 1.000\sqrt{h/r} - 0.031 h/r$	$1.042 + 0.982\sqrt{h/r} - 0.036 h/r$
K_2	$-0.114 - 0.585\sqrt{h/r} + 0.314 h/r$	$-0.074 - 0.156\sqrt{h/r} - 0.010 h/r$
K_3	$0.241 - 0.992\sqrt{h/r} - 0.271 h/r$	$-3.418 + 1.220\sqrt{h/r} - 0.005 h/r$
K_4	$-0.134 + 0.577\sqrt{h/r} - 0.012 h/r$	$3.450 - 2.046\sqrt{h/r} + 0.051 h/r$

For cases where $\dfrac{L}{D} < \dfrac{3}{[r/(D-2h)]^{1/4}}$

5b. Elastic stress, in-plane bending

$$k = K_1 + K_2 \left(\frac{2h}{D} \right) + K_3 \left(\frac{2h}{D} \right)^2 + K_4 \left(\frac{2h}{D} \right)^3$$

where for $\dfrac{L}{D} > \dfrac{3}{[r/(D-2h)]^{1/4}}$

Contd.

Table 3.1 *(Continues)*

Type of form irregularity or stress raiser	Stress condition and manner of loading	Factor of stress concentration k for various dimensions	

		$0.1 \le h/r \le 2.0$	$2.0 \le h/r \le 20.0$
		K_1 $1.007 + 1.000\sqrt{h/r} - 0.031h/r$	$1.042 + 0.982\sqrt{h/r} - 0.036h/r$
		K_2 $-0.270 - 2.404\sqrt{h/r} + 0.749h/r$	$-3.599 - 1.619\sqrt{h/r} - 0.431h/r$
		K_3 $0.677 + 1.133\sqrt{h/r} - 0.904h/r$	$6.084 - 5.607\sqrt{h/r} + 1.158h/r$
		K_4 $-0.414 + 0.271\sqrt{h/r} - 0.186h/r$	$-2.527 - 3.006\sqrt{h/r} + 0.691h/r$

For cases where

$$\frac{L}{D} < \frac{3}{[r/D - 2h]^{1/4}}$$

6. Circular hole in an infinite plate

6a. Elastic stress, in-plane normal stress

(a1) Uniaxial stress, $\sigma_2 = 0$

$$\sigma_A = 3\sigma_1 \quad \sigma_B = -\sigma_1$$

(a2) Biaxial stress, $\sigma_2 = \sigma_1$

$$\sigma_A = \sigma_B = 2\sigma_1$$

(a3) Biaxial stress, $\sigma_2 = -\sigma_1$ (pure shear)

$$\sigma_A = -\sigma_B = 4\sigma_1$$

6b. Elastic stress, out-of-plane bending

(b1) simple bending, $M_2 = 0$ $\quad \sigma_A = k\dfrac{6M_1}{t^2}$

where $\quad k = 1.79 + \dfrac{0.25}{0.39 + (2r/t)} + \dfrac{0.81}{1 + (2r/t)^2} - \dfrac{0.26}{1 + (2r/t)^3}$

(b2) Cylindrical bending, $M_2 = \nu M_1$

$$\sigma_A = k\frac{6M_1}{t^2}$$

where $k = 1.85 + \dfrac{0.509}{0.70 + (2r/t)} - \dfrac{0.214}{1 + (2r/t)^2} - \dfrac{0.335}{1 + (2r/t)^3}$ for $\upsilon = 0.3$

Note: M_1 and M_2 are unit moments.

Contd.

Table 3.1 Factors of stress concentration for elastic stress (k)

Type of form irregularity or stress raiser	Stress condition and manner of loading	Factor of stress concentration k for various dimensions
7. Central circular hole in a member of rectangular cross section	7a. Elastic stress, axial tension	(b3) Isotropic bending, $M_2 = M_1$ $$\sigma_A = k\frac{6M_1}{t^2}$$ where $k = 2$ (independent of r/t) $$\sigma_{max} = \sigma_A = k\sigma_{nom}$$ where $\sigma_{nom} = \dfrac{P}{t(D-2r)}$ $$k = 3.00 - 3.13\left(\frac{2r}{D}\right) + 3.66\left(\frac{2r}{D}\right)^2 - 1.53\left(\frac{2r}{D}\right)^3$$
	7b. Elastic stress, in-plane bending	The maximum stress at the edge of the hole is $$\sigma_A = k\sigma_{nom}$$ where $\sigma_{nom} = \dfrac{12Mr}{[tD^3 - (2r)^3]}$ (at the edge of the hole) $K = 2$ (independent of r/D) The maximum stress at the edge of the plate is not directly above the hole but is found a short distance away on either side, points X. $$\sigma_x = \sigma_{nom}$$ where $\sigma_{nom} = \dfrac{6MD}{t[D^3 - (2r)^3]}$ (at the edge of the plate)

Contd.

Table 3.1 *(Continues)*

Type of form irregularity or stress raiser	Stress condition and manner of loading	Factor of stress concentration k for various dimensions
	7c. Elastic stress, out-of-plane bending	(c1) Simple bending, $M_2 = 0$ $\sigma_{max} = \sigma_A = k\dfrac{6M_1}{t^2(D-2r)}$ where, $k = \left[1.79 + \dfrac{0.25}{0.39+(2r/t)} + \dfrac{0.81}{1+(2r/t)^2} - \dfrac{0.26}{1+(2r/t)^3}\right]$ $\left[1-1.04\left(\dfrac{2r}{D}\right)+1.22\left(\dfrac{2r}{D}\right)^2 + 1.22\left(\dfrac{2r}{D}\right)^2\right]$ for $\dfrac{2r}{D} < 0.3$ (c2) Cylindrical bending (plate action), $M_2 = \nu M_1$ $\sigma_{max} = \sigma_A = k\dfrac{6M_1}{t^2(D-2r)}$ where $k = \left[1.85 + \dfrac{0.509}{0.70+(2r/t)} - \dfrac{0.214}{1+(2r/t)^2} + \dfrac{0.335}{1+(2r/t)^3}\right]$ $\left[1-1.04\left(\dfrac{2r}{D}\right)+1.22\left(\dfrac{2r}{D}\right)^2\right]$ for $\dfrac{2r}{D} < 0.3$ and $\nu = 0.3$
8. Off-centre circular hole in a member of rectangular cross section	8a. Elastic stress, axial tension	$\sigma_{max} = \sigma_A = k\sigma_{nom}$ where $\sigma_{nom} = \dfrac{P\sqrt{1-(r/c)^2}}{Dt\ 1-(r/c)}\dfrac{1-(c/D)}{1-(c/D)\left[2-\sqrt{1-(r/c)^2}\right]}$ $k = 3.00 - 3.13\left(\dfrac{r}{c}\right) + 3.66\left(\dfrac{r}{c}\right)^2 - 1.53\left(\dfrac{r}{c}\right)^3$

Contd.

Table 3.1 Factors of stress concentration for elastic stress (k)

Type of form irregularity or stress raiser	Stress condition and manner of loading	Factor of stress concentration k for various dimensions
	8b. Elastic stress, in-plane bending 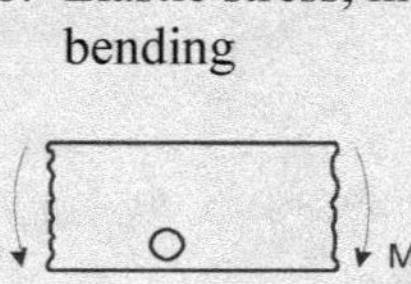	$\sigma_{max} = \sigma_A = k\sigma_{nom}$ where $\sigma_{nom} = \dfrac{12M}{tD^3}\left(\dfrac{D}{2} - c + r\right)$ and $k = 3.0$ if $r/c < 0.05$ or $k = K_1 + K_2\left(\dfrac{2c}{D}\right) + K_3\left(\dfrac{2c}{D}\right)^2 + K_4\left(\dfrac{2c}{D}\right)^3$ where for $0.05 \le r/c \le 0.5$ $K_1 = 3.022 - 0.422\, r/c + 3.556\,(r/c)^2$ $K_2 = -0.569 + 2.664\, r/c + 4.397\,(r/c)^2$ $K_3 = 3.138 - 18.367\, r/c + 28.093\,(r/c)^2$ $K_4 = -3.591 - 16.125\, r/c - 27.252\,(r/c)^2$
9. Elliptical hole in an infinite plate 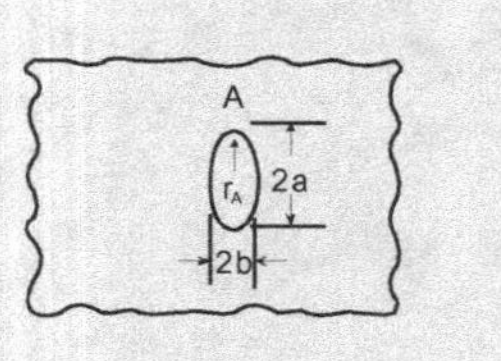 $r_A = \dfrac{b^2}{a}$	9a. Elastic stress, in-plane normal stress 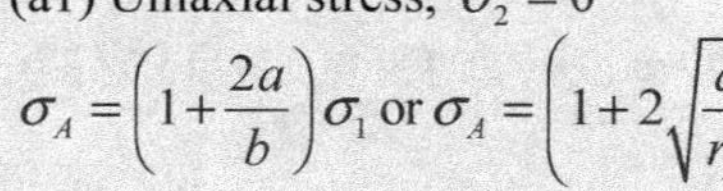	(a1) Uniaxial stress, $\sigma_2 = 0$ $\sigma_A = \left(1 + \dfrac{2a}{b}\right)\sigma_1$ or $\sigma_A = \left(1 + 2\sqrt{\dfrac{a}{r_A}}\right)\sigma_1$ $\sigma_B = -\sigma_1$ (a2) Biaxial stress, $\sigma_2 = \sigma_1$ $\sigma_A = 2\dfrac{a}{b}\sigma_1$ $\sigma_B = 2\dfrac{a}{b}\sigma_1$ (a3) Biaxial stress, $\sigma_2 = -\sigma_1$ $\sigma_A = 2\left(1 + \dfrac{a}{b}\right)\sigma_1$ $\sigma_B = -2\left(1 + \dfrac{a}{b}\right)\sigma_1$ This stress condition would also be created by pure shear inclined at 45° to the axes of the ellipse.

Contd.

Table 3.1 *(Continues)*

Type of form irregularity or stress raiser	Stress condition and manner of loading	Factor of stress concentration k for various dimensions
10. Central elliptical hole in a member of rectangular cross section $r_A = \dfrac{b^2}{c}$	10a. Elastic stress, axial tension	$\sigma_{max} = \sigma_A = k\sigma_{nom}$ where $\sigma_{nom} = \dfrac{P}{t(D-2a)}$ $k = K_1 + K_2\left(\dfrac{2a}{D}\right) + K_3\left(\dfrac{2a}{D}\right)^2 + K_4\left(\dfrac{2a}{D}\right)^3$ where $0.5 \le a/b \le 10.0$ $K_1 = 1.000 + 0.000\sqrt{a/b} + 2.000\,a/b$ $K_2 = -0.351 - 0.021\sqrt{a/b} - 2.483\,a/b$ $K_3 = 3.621 - 5.183\sqrt{a/b} + 4.494\,a/b$ $K_4 = -2.270 + 5.204\sqrt{a/b} - 4.011\,a/b$
	10b. Elastic stress, in-plane bending	The maximum stress at the edge of the hole is $\sigma_A = k\sigma_{nom}$ Where $\sigma_{nom} = \dfrac{12Ma}{t[D^3 - (2a)^3]}$ (at the edge of the hole) $k = K_1 + K_2\left(\dfrac{2a}{D}\right) + K_3\left(\dfrac{2a}{D}\right)^2$ where for $1.0 \le a/b \le 2.0$ and $0.4 \le 2a/D \le 1.0$ $K_1 = 3.465 - 3.739\sqrt{a/b} + 2.274\,a/b$ $K_2 = -3.841 + 5.582\sqrt{a/b} - 1.741\,a/b$ $K_3 = 2.376 - 1.843\sqrt{a/b} - 0.534\,a/b$

Contd.

Table 3.1 Factors of stress concentration for elastic stress (k) (*Continues*)

Type of form irregularity or stress raiser	Stress condition and manner of loading	Factor of stress concentration k for various dimensions
11. Off-centre elliptical hole in a member of rectangular cross section	11a. Elastic stress, axial tension	$$\sigma_{max} = \sigma_A = k\sigma_{nom}$$ The expression for σ_{nom} from case 8a can be used by substituting a/c for r/c. Use the expression for k from case 10a by substituting a/c for $2a/D$.
12. Rectangular hole with round corners in an infinite plate	12a. Elastic stress, axial tension	$$\sigma_{max} = k\sigma_1 \text{ and } k = K_1 + K_2\left(\frac{b}{a}\right) + K_3\left(\frac{b}{a}\right)^2 + K_4\left(\frac{b}{a}\right)^3$$ where for $0.2 \leq r/b \leq 1.0$ and $0.3 \leq b/a \leq 1.0$ $$K_1 = 14.815 - 15.744\sqrt{r/b} + 8.149\,r/b$$ $$K_2 = -11.201 - 9.750\sqrt{r/b} + 9.600\,r/b$$ $$K_3 = 0.202 + 38.622\sqrt{r/b} - 27.374\,r/b$$ $$K_4 = 3.232 - 23.002\sqrt{r/b} + 15.482\,r/b$$

Contd.

Table 3.1 *(Continues)*

Type of form irregularity or stress raiser	Stress condition and manner of loading	Factor of stress concentration k for various dimensions
13. Lateral slot with circular ends in a member of rectangular section The equivalent ellipse has a width $2b_{eq}$ where $b_{eq} = \sqrt{r_A a}$	13a. Elastic stress, axial tension 13b. Elastic stress in-plane bending	A very close approximation to the maximum stress σ_A can be obtained by using the maximum stress for the given loading with the actual slot replaced by an ellipse having the same overall dimension normal to the loading direction, $2a$, and the same end radius r_A. *See* cases 9a, 10a and 11a. As above, but see case 10b.
14. Reinforced circular hole in a wide plate	14a. Elastic stress, axial tension	$\sigma_{max} = k\sigma_1$ where for $r_f \geq 0.6t$ and $w \geq 3t$ $$k = 1.0 + \frac{1.66}{1+A} - \frac{2.182}{(1+A)^2} + \frac{2.521}{(1+A)^3}$$ A is the ratio of the transverse area of the added reinforcement to the transverse area of the hole: $$r = \frac{(r_b - r)(w - t) + 0.429 r_f^2}{rt}$$

Contd.

Table 3.1 Factors of stress concentration for elastic stress (k) (*Continues*)

Type of form irregularity or stress raiser	Stress condition and manner of loading	Factor of stress concentration k for various dimensions
15. U-notch in a circular shaft	15a. Elastic stress, axial tension	$k = K_1 + K_2\left(\dfrac{2h}{D}\right) + K_3\left(\dfrac{2h}{D}\right)^2 + K_4\left(\dfrac{2h}{D}\right)^3$

where

	$0.25 \le h/r \le 2.0$	$2.0 \le h/r \le 50.0$
K_1	$0.455 + 3.354\sqrt{h/r} - 0.769\,h/r$	$0.935 + 1.922\sqrt{h/r} + 0.004\,h/r$
K_2	$3.129 - 15.955\sqrt{h/r} + 7.404\,h/r$	$0.537 - 3.708\sqrt{h/r} + 0.040\,h/r$
K_3	$-6.909 + 29.286\sqrt{h/r} - 16.104\,h/r$	$-2.538 + 3.438\sqrt{h/r} - 0.012\,h/r$
K_4	$4.325 - 16.685\sqrt{h/r} + 9.469\,h/r$	$2.066 - 1.652\sqrt{h/r} - 0.031\,h/r$

For the semicircular notch ($h/r = 1$)

$$k = 3.04 - 5.42\left(\frac{2h}{D}\right) + 6.27\left(\frac{2h}{D}\right)^2 - 2.89\left(\frac{2h}{D}\right)^3$$

15b. Elastic stress, bending

$$k = K_1 + K_2\left(\frac{2h}{D}\right) + K_3\left(\frac{2h}{D}\right)^2 + K_4\left(\frac{2h}{D}\right)^3$$

	$0.25 \le h/r \le 2.0$	$2.0 \le h/r \le 50.0$
K_1	$0.455 + 3.354\sqrt{h/r} - 0.769\,h/r$	$0.935 + 1.922\sqrt{h/r} + 0.004\,h/r$
K_2	$0.891 - 12.721\sqrt{h/r} + 4.593\,h/r$	$-0.552 - 5.327\sqrt{h/r} + 0.086\,h/r$
K_3	$0.286 + 15.481\sqrt{h/r} - 6.392\,h/r$	$0.754 + 6.281\sqrt{h/r} - 0.121\,h/r$
K_4	$-0.632 - 6.115\sqrt{h/r} + 2.568\,h/r$	$-0.138 - 2.876\sqrt{h/r} + 0.031\,h/r$

For the semicircular notch ($h/r = 1$)

$$k = 3.04 - 7.236\left(\frac{2h}{D}\right) + 9.375\left(\frac{2h}{D}\right)^2 - 4.179\left(\frac{2h}{D}\right)^3$$

Contd.

Table 3.1 *(Continues)*

Type of form irregularity or stress raiser	Stress condition and manner of loading	Factor of stress concentration k for various dimensions
	15c. Elastic stress, torsion	$k = K_1 + K_2\left(\dfrac{2h}{D}\right) + K_3\left(\dfrac{2h}{D}\right)^2 + K_4\left(\dfrac{2h}{D}\right)^3$ where $0.25 \le h/r \le 2.0$ and $2.0 \le h/r \le 50.0$ $1.245 + 0.264\sqrt{h/r} + 0.491 h/r$ $1.651 + 0.614\sqrt{h/r} + 0.040 h/r$ $-3.030 + 3.269\sqrt{h/r} - 3.633 h/r$ $-4.794 - 0.314\sqrt{h/r} - 0.217 h/r$ $-7.199 - 11.286\sqrt{h/r} + 8.318 h/r$ $8.457 - 0.962\sqrt{h/r} + 0.389 h/r$ $-4.414 + 7.753\sqrt{h/r} - 5.176 h/r$ $-4.314 + 0.662\sqrt{h/r} - 0.212 h/r$ For the semicircular notch ($h/r = 1$) $k = 2.000 - 3.394\left(\dfrac{2h}{D}\right) + 4.231\left(\dfrac{2h}{D}\right)^2 - 1.837\left(\dfrac{2h}{D}\right)^3$
16. V-notch in a circular shaft	16c. Elastic stress, torsion	The stress concentration factor for the V-notch k_θ, is the smaller of the values $k_\theta = k_U$ or $k_\theta = 1.065 k_U - \left[0.022 + 0.137\left(\dfrac{\theta}{135}\right)^2\right](k_U - 1)k_U$ for $\dfrac{r}{D - 2h} \le 0.01$ and $\theta \le 135°$ where k_U is the stress concentration factor for a U-notch, case 15c, when the dimensions h, r, and D are the same as for the V-notch and θ is the notch angle in degrees.

Contd.

Table 3.1 Factors of stress concentration for elastic stress (k) *(Continues)*

Type of form irregularity or stress raiser	Stress condition and manner of loading	Factor of stress concentration k for various dimensions
17. Square shoulder with fillet in circular shaft	17a. Elastic stress, axial tension	$$k = K_1 + K_2\left(\frac{2h}{D}\right) + K_3\left(\frac{2h}{D}\right)^2 + K_4\left(\frac{2h}{D}\right)^3$$ where k_1 $0.25 \le h/r \le 2.0$ $\qquad\qquad$ $2.0 \le h/r \le 20.0$ k_2 $0.927 + 1.149\sqrt{h/r} - 0.086h/r$ $\quad$ $1.225 + 0.831\sqrt{h/r} - 0.010h/r$ k_3 $0.011 - 3.029\sqrt{h/r} + 0.948h/r$ $-1.831 + 0.318\sqrt{h/r} - 0.049h/r$ k_4 $-0.304 + 3.979\sqrt{h/r} - 1.737h/r$ $2.236 - 0.522\sqrt{h/r} + 0.176h/r$ k_5 $-0.366 + 2.098\sqrt{h/r} - 0.875h/r$ $-0.630 + 0.009\sqrt{h/r} - 0.117h/r$
	17b. Elastic stress, bending	$$k = K_1 + K_2\left(\frac{2h}{D}\right) + K_3\left(\frac{2h}{D}\right)^2 + K_4\left(\frac{2h}{D}\right)^3$$ where k_1 $0.25 \le h/r \le 2.0$ $2.0 \le h/r \le 20.0$ k_2 $0.927 + 1.149\sqrt{h/r} - 0.086h/r$ $\quad$ $1.225 + 0.831\sqrt{h/r} - 0.010h/r$ k_3 $0.015 - 3.281\sqrt{h/r} + 0.837h/r$ $-3.790 + 0.958\sqrt{h/r} - 0.257h/r$ k_4 $0.847 + 1.716\sqrt{h/r} - 0.506h/r$ $\quad$ $7.374 - 4.834\sqrt{h/r} + 0.862h/r$ k_5 $-0.790 + 0.417\sqrt{h/r} - 0.246h/r$ $-3.809 + 3.046\sqrt{h/r} - 0.595h/r$
	17c. Elastic stress, torsion	$$k = K_1 + K_2\left(\frac{2h}{D}\right) K_3\left(\frac{2h}{D}\right)^2 + K_4\left(\frac{2h}{D}\right)^3$$ Where for $0.25 \le h/r \le 4.0$ $K_1 = 0.953 + 0.680\sqrt{h/r} - 0.053h/r$ $K_2 = -0.493 - 1.820\sqrt{h/r} + 0.517h/r$ $K_3 = 1.621 - 0.908\sqrt{h/r} - 0.529h/r$ $K_4 = -1.081 + 0.232\sqrt{h/r} + 0.065h/r$

Contd.

Table 3.1 *(Continues)*

Type of form irregularity or stress raiser	Stress condition and manner of loading	Factor of stress concentration k for various dimensions
18. Radial hole in a hollow or solid circular shaft For a solid shaft $d = 0$	18a. Elastic stress, axial tension	$\sigma_{max} = k\,\dfrac{4P}{\pi(D^2 - d^2)}$ where $k = K_1 + K_2\left(\dfrac{2r}{D}\right) + K_3\left(\dfrac{2r}{D}\right)^2 + K_4\left(\dfrac{2r}{D}\right)^3$ and where for $d/D \le 0.9$ and $2r/D \le 0.45$ $K_1 = 3.000$ $K_2 = 2.773 + 1.529\,d/D - 4.379\,(d/D)^2$ $K_3 = -0.421 - 12.782\,d/D + 22.781\,(d/D)^2$ $K_4 = 16.841 + 16.678\,d/D - 40.007\,(d/D)^2$
	18b. Elastic stress, bending when hole is farthest from bending axis	$\sigma_{max} = k\,\dfrac{32MD}{\pi(D^4 - d^4)}$ where $k = K_1 + K_2\left(\dfrac{2r}{D}\right)^2 + K_3\left(\dfrac{2r}{D}\right) + K_4\left(\dfrac{2r}{D}\right)^3$ and where for $d/D \le 0.9$ and $2r/D \le 0.3$ $K_1 = 3.000$ $K_2 = -6.690 - 1.620\,d/D + 4.432\,(d/D)^2$ $K_3 = 44.739 + 10.724\,d/D - 19.927\,(d/D)^2$ $K_4 = -53.307 - 25.998\,d/D + 43.258\,(d/D)^2$

Contd.

Table 3.1 Factors of stress concentration for elastic stress (k) (*Continues*)

Type of form irregularity or stress raiser	Stress condition and manner of loading	Factor of stress concentration k for various dimensions
	18c. Elastic stress, torsional loading	$\sigma_{max} = k\dfrac{16TD}{\pi(D^4 - d^4)}$ where $k = K_1 + K_2\left(\dfrac{2r}{D}\right) + K_3\left(\dfrac{2r}{D}\right)^2 + K_4\left(\dfrac{2r}{D}\right)^3$ and where for $d/D \leq 0.9$ and $2r/D \leq 0.4$ $_K 1 = 4.000$ $_K 2 = -6.793 + 1.133\, d/D - 0.126\, (d/D)^2$ $K_3 = 38.382 - 7.242\, d/D + 6.495\, (d/D)^2$ $K_4 = -44.576 - 7.428\, d/D + 58.656\, (d/D)^2$
19. Multiple U-notches in a member of rectangular section	19.a. Elastic stress, axial tension, semicircular notches only, i.e., $h = r$	The stress concentration factor for the multiple semicircular U-notches, k_m is the smaller of the values $k_M = k_U$ or $k_M = \left\{ \begin{array}{l} 1.1 - \left[0.88 - 1.68\left(\dfrac{2r}{D}\right)\right]\dfrac{2r}{L} + \\[2mm] \left[1.3\left(0.5 - \dfrac{2r}{D}\right)^2\right]\left(\dfrac{2r}{L}\right)^3 \end{array} \right\} k_U$ for $\dfrac{2r}{L} < 1$ where k_U is the stress concentration factor for a single pair of semicircular U-notches, case 1a.

PROBLEM 3.7

The concrete pier shown in Figure is loaded at the top with a uniformly distributed load of 20 kN/m². Investigate the state of stress at a level 1 m above the base. Concrete weighs approximately 25 kN/m³.

Solution

Weight of the whole pier:

$$W = [(0.5 + 1.5)/2] \times 0.5 \times 2 \times 25 = 25 \text{ kN}$$

Total applied force:

$$P = 20 \times 0.5 \times 0.5 = 5 \text{ kN}$$

From $\Sigma F_y = 0$, reaction at the base:

$$R = W + P = 30 \text{ kN}$$

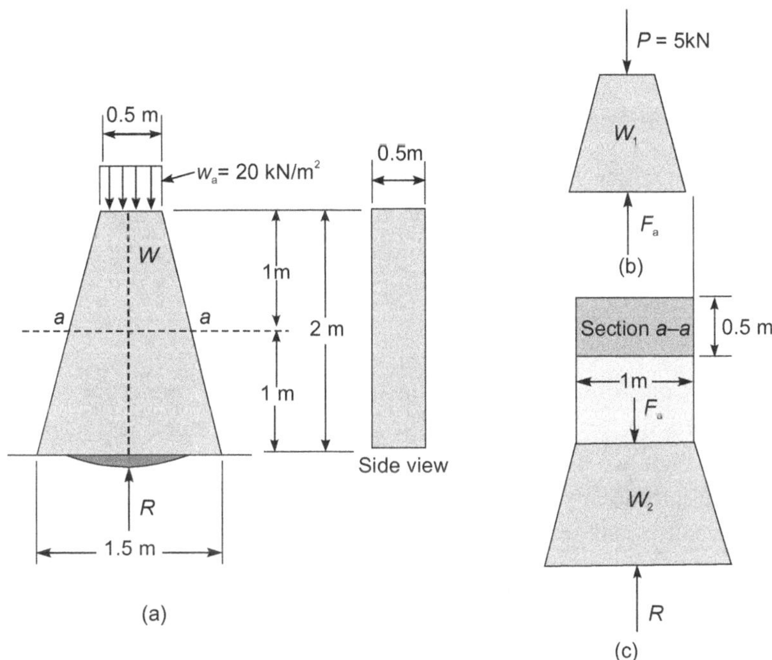

Those forces are shown schematically in the diagrams as concentrated forces acting through their respective centroids. Then, to determine the stress at the desired level, the body is cut into two separate parts. A free-body diagram for either part is sufficient to solve the problem. For comparison, the problem is solved both ways.

Using the upper part of the pier as a free body, the weight of the pier above the section:

$$W_1 = (0.5 + 1) \times 0.5 \times 1 \times 25/2 = 9.4 \text{ kN}$$

From $\Sigma F_y = 0$, the force at the section:

$$F_a = P + W_1 = 14.4 \text{ kN}$$

the normal stress at the level a–a is

$$\sigma_a = \frac{F_a}{A} = \frac{14.4}{0.5 \times 1} = 28.8 \text{kN/m}^2$$

This stress is compressive as F_a acts on the section.

Using the lower part of the pier as a free body, the weight of the pier below the section:

$$W_2 = (1 + 1.5) \times 0.5 \times 1 \times 25/2 = 15.6 \text{ kN}$$

From $\Sigma F_y = 0$, the force at the section:

$$F_a = R - W_2 = 14.4 \text{ kN}$$

The remainder of the problem is the same as before. Note that the pier considered here has a vertical axis of symmetry.

PROBLEM 3.8

A bracket of negligible weight shown in Figure is loaded with a vertical force P of 3 kips. For interconnection purposes, the bar ends are clevised (forked). Pertinent dimensions are shown in the figure. Find the axial stresses in members AB and BC and the bearing and shear stresses for pin C. All pins are 0.375 in diameter.

Solution

First an idealized free-body diagram consisting of the two bars pinned at the ends is prepared. As there are no intermediate forces acting on the bars and the applied force acts through the joint at B, the forces in the bars are directed along the lines AB and BC, and the bars AB and BC are loaded axially. The magnitude of the forces are unknown and are labelled F_A and F_C in the diagram. These forces can be determined graphically by completing a triangle of forces F_A, F_C, and P.

These forces may also be found analytically from two simultaneous equations $\Sigma F_y = 0$ and $\Sigma F_x = 0$, written in terms of the unknowns F_A and F_C, a known force P and two known angles a and b. Both these procedures are possible. Instead of treating forces F_A and F_C directly, their components are used instead of $\Sigma F = 0$ and $\Sigma M = 0$.

Any force can be resolved into components. For example, F_A can be resolved into F_{Ax} and F_{Ay} as in Figure. Conversely, if any one of the components of a directed force is known. the force itself can be determined. This follows from similarity of dimensions and force triangles. In Figure the triangles Akm and BAD are similar triangles (both are shaded in the diagram).Hence, if F_{Ax} is known

$$F_A = (AB/DB) F_{Ax}$$

Similarly, $F_{Ay} = (AD/DB)\, F_{AX}$. Note further that AB/DB or AD/DB are ratios; hence, relative dimensions of members can be used. Such relative dimensions are shown by a little triangle on member AB and again and BC. In the problem at hand.

$$F_A = \left(\sqrt{5}/2\right) F_{Ax} \quad \text{and} \quad F_{Ay} = F_{Ax}/2$$

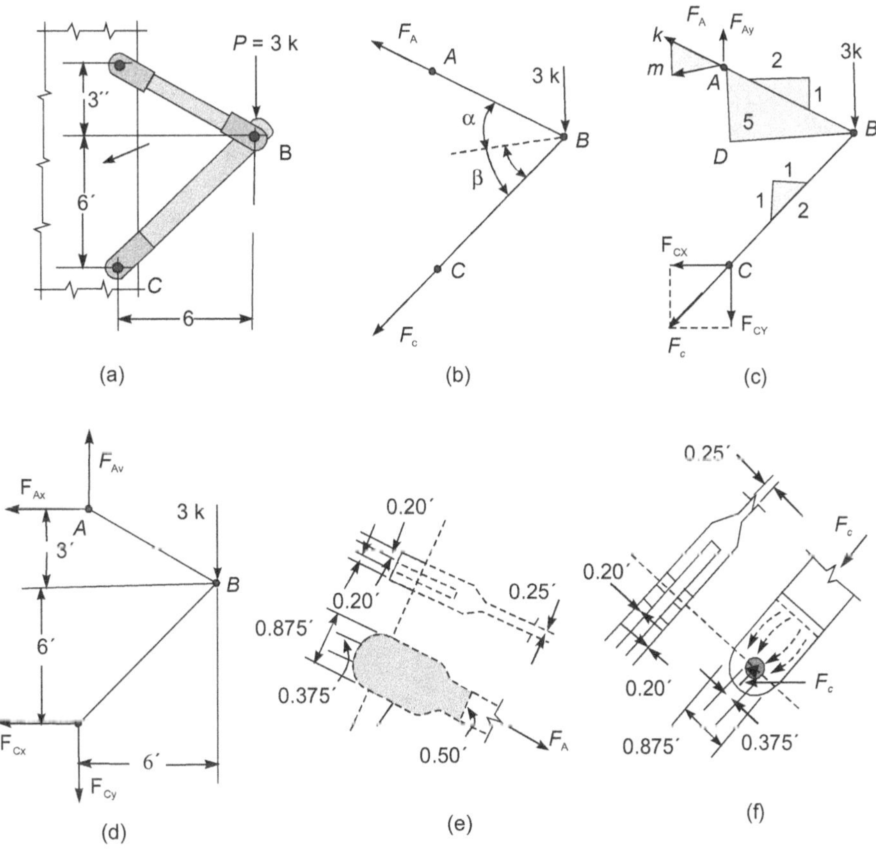

Adopting the procedure of resolving forces, a revised free-body diagram is prepared. Two components of force are necessary at the pin joints. After the forces are dertermined by statics, using the equilibrium equations namely $\Sigma F = 0$ and $\Sigma M = 0$, we obtain

$$\Sigma M_{_d} = 0 + F_{Ax}(3+6) - 3(6) = 0 \quad F_{Ay} = +2\,k$$

$$F_{Ay} = F_{Ax}/2 - 2/2 = +1\,k$$

$$F_A = 2(\sqrt{5}/2) = +2.23 k$$

$$\Sigma M_A = 0 + \quad + 3(6) + F_{Cx}(9) = 0 \quad F_{cx} = -2\,k$$

$$F_{Cy} = F_{Cx} = -2k$$

$$F_C = \sqrt{2}(-2) = -2.83\,k$$

$$\text{Check } \Sigma F_x = 0 \qquad F_{Ax} + F_{Cx} = 2 - 2 = 0$$
$$\Sigma F_y = 0 \qquad F_{Ay} - F_{Cy} - P = 1 - (-2) - 3 = 0$$

Tensile stress in main bar *AB*:

$$\sigma_{AB} = \frac{F_A}{A} = \frac{2.23}{0.25 \times 0.50} = 17.8 \text{ ksi}$$

Pensile stress in clevis of bar *AB*, Figure:

$$(\sigma_{AB})_{clevis} = \frac{F_A}{A_{net}} = \frac{2.23}{2 \times 0.20 \times (0.875 - 0.375)} = 11.2 \text{ ksi}$$

Compressive stress in main bar *BC*:

$$\sigma_{BC} = \frac{F_C}{A} = \frac{2.83}{0.875 \times 0.25} = 12.9 \text{ ksi}$$

In the compression member, the net section at the clevis need not be investigated; see Figure for the transfer of forces. The bearing stress at the pin is more critical. Bearing between pin *C* and the clevis:

$$\sigma_b = \frac{F_C}{A_{bearing}} = \frac{2.83}{0.375 \times 0.20 \times 2} = 18.8 \text{ ksi}$$

Bearing between the pin *C* and the main plate:

$$\sigma_b = \frac{F_C}{A} = \frac{2.83}{0.375 \times 0.25} = 30.2 \text{ ksi}$$

Double shear in pin *C:*

$$\tau = \frac{F_C}{A} = \frac{2.83}{2\pi(0.375/2)^2} = 12.9 \text{ ksi}$$

For a complete analysis of this bracket, other pins should be investigated. However, it can be seen by inspection that the other pins in this case are stressed either the same amount as computed or less.

The advantages of the method used in this example for finding forces in members should now be apparent. It can also be applied with success in a problem such as the one shown in Figure. The force F_A transmitted by the curved member *AB* acts through points *A* and *B*, since the forces applied at *A* and *B* must be collinear. By resolving this force at A′, the same procedure can be followed. Wavy lines through F_A and F_C indicate that these forces are replaced by the two components shown. Alternatively, the force F_A can be resolved at *A*, and since $F_{Ay} = (y/x)\, F_{Ax}$, the application of $\Sigma M_C = 0$ yields F_{Ax}.

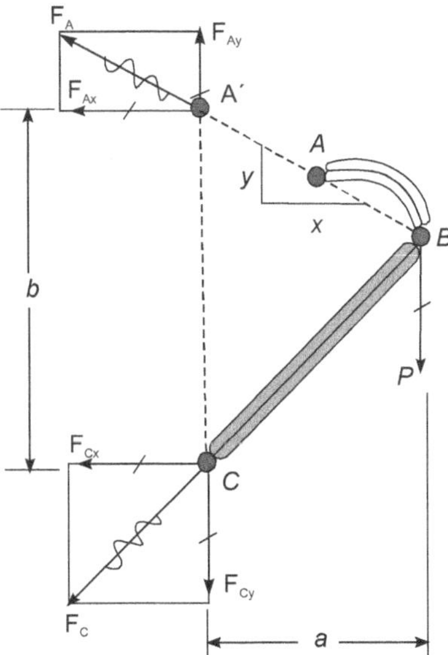

In frames, where the applied forces do not act through a joint, proceed as done in problem 3.8 as far as possible. Then isolate an individual member and using its free-body diagram, complete the determination of forces. If inclined forces are acting on the structure, resolve them into convenient components.

PROBLEM 3.9

Consider the idealized system shown in Fig., where a 5-kg mass is to be spun on a frictionless plane at 10 Hz. If a light rod CD is attached at C, and its stress can reach 200 MPa, what is the required size of the rod? Neglect the weight of the rod and assume that the rod is enlarged at the ends to compensate for the threads.

Solution

The rod angular velocity ω is 20π rad/s. The acceleration a of the mass toward the centre of rotation is $\omega^2 R$, where R is the distance CD. By multiplying the mass m by the acceleration, the force F acting on the rod is obtained. As shown in the figure, according to the d'Alembert principle, this force acts in the opposite direction to that of the acceleration. Therefore,

$$F = ma = m\omega^2 R = 5 \times (20\pi)^2 \times 0.0.500 = 9870 \text{ kg.m/s}^2 = 9870\,\text{N}$$

$$A_{net} = \frac{9870}{200} = 49.3 \text{ mm}^2$$

An 8-mm round rod having an area $A = 50.3$ mm² would be satisfactory.

The additional pull at C caused by the mass of the rod, which was not considered, is

$$F_1 = \int_0^R (m_1\, dr)\omega^2 r$$

where M_1 is the mass of the rod per unit length, and $(m_1\, dr)$ is its infinitesimal mass at a variable distance r from the vertical rod AB. The total pull at C caused by the rod and the mass of 5 kg at the ends is $F + F_1$.

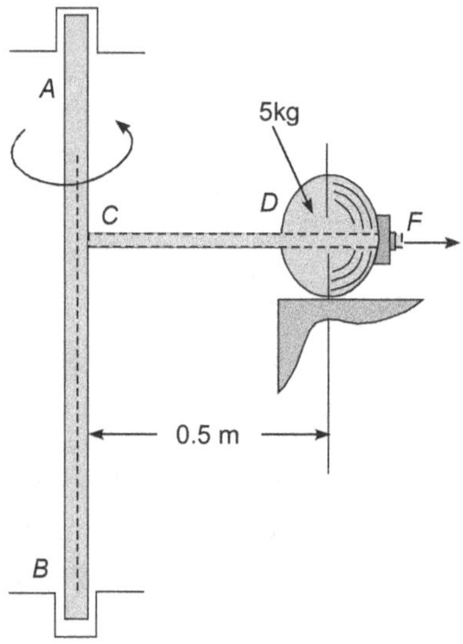

MEMBER STRENGTH AS A DESIGN CRITERION

We had have discussed this earlier under various criteria for design. We will review this again. The purpose for calculating stresses in members of a structural system is to compare them with the experimentally determined material strengths in order to assure desired performance. Physical testing of materials in a laboratory provides information regarding a material's resistance to stress. In a laboratory, specimens of known material, manufacturing process, and heat treatment are carefully prepared to desired dimensions. Then these specimens are subjected to successively increasing known forces. In the most widely used test, a round rod or a rectangular bar is subjected to tension and the specimen is loaded until it finally ruptures. The force necessary to cause rupture is called the ultimate load. By dividing this ultimate load by the original cross-sectional area of the specimen, the ultimate strength (stress) of a material is obtained. The tensile test is used most widely. However, compression, bending, torsion, and shearing tests are also employed. Tables 1 A and B of the Appendix give ultimate strengths and other physical properties for a few materials.

For applications where a force comes on and off the structure a number of times, the materials cannot withstand the ultimate stress of a static test. In such cases, the "ultimate strength" depends on the number of times the force is applied as the material works at a particular stress level. Such experimental points indicate the number of cycles required to break. The constant f represents the deflection resulting from the application of a unit force (i.e., $P = 1$)

For the particular case of an axially loaded ith bar of constant cross section,

$$f_i = \frac{L_i}{A_i E_i}$$

The concepts of structural stiffness and flexibility are widely used in structural analysis, including mechanical-vibration problems. For more complex structural systems, the expressions for k and f become more involved.

PROBLEM 3.10

Determine the relative displacement of point D from O for the elastic steel bar of variable cross section shown in Figure 3-3(a) caused by the application of concentrated forces $P_1 = 100$ kN and $P_3 = 200$ kN acting to the left, and $P_2 = 250$ kN and $P_4 = 50$ kN acting to the right. The respective areas for bar segments OB, BC and CD are 1000, 2000, and 1000 mm². Let $E = 200$ GPa.

Solution

By inspection, it can be seen that the bar is in equilibrium. *Such a check must always be made before starting a problem.* The variation in P_x along the length of the bar is determined by taking three sections, a–a, b–b, and c–c, in Figure and determining the necessary forces for equilibrium in the free-body diagrams in Figures. This leads to the conclusion that *within each bar segment, the forces are constant,* resulting in the axial force diagram shown. Therefore, the solution of the deformation problem consists of adding algebraically the individual deformations for the three segments $\Delta\dfrac{PL}{AE}$ can be written for each segment. Hence, the total axial deformation for the bar is

$$\Delta = \sum_i \frac{P_i L_i}{A_i E} = \frac{P_{OB} L_{OB}}{A_{OB} E} + \frac{P_{BC} L_{BC}}{A_{BC} E} + \frac{P_{CD} L_{CD}}{A_{CD} E}$$

where the subscripts identify the segments.

Using this relation, the relative displacement between O and D is

$$\Delta = +\frac{100 \times 10^3 \times 2000}{1000 \times 200 \times 10^3} - \frac{150 \times 10^3 \times 1000}{2000 \times 200 \times 10^3} + \frac{50 \times 10^3 \times 1500}{1000 \times 200 \times 10^3}$$

$$= +1.000 - 0.375 + 0.375 = +1.000 \text{ mm}$$

Note that in spite of large stresses in the bar, the elongation is very small.

A graphic interpretation of the solution is shown in the same Figure.

By dividing the axial forces in the bar segments by the corresponding AE, the axial strains along the bar are obtained. These strains are constant within each bar segment. The area of the strain diagram for each segment of the bar gives the change in length for that segment. These values correspond to those displayed numerically before.

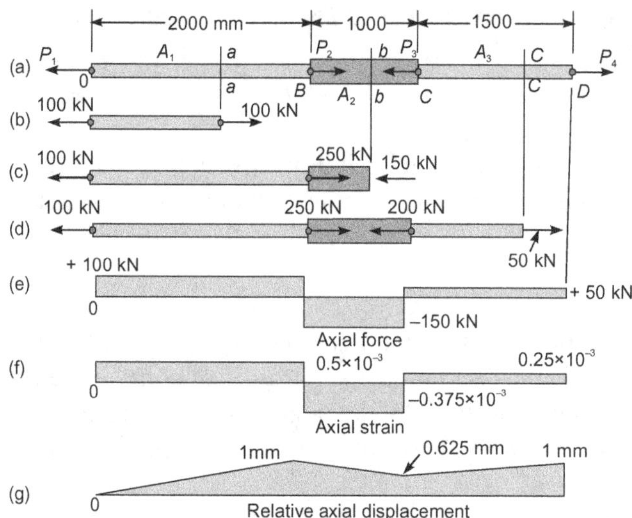

PROBLEM 3.11

Determine the deflection of free end B of elastic bar OB caused by its own weight w N/mm. The cross-sectional area is constant and is A. Young's modulus is E N/mm^2.

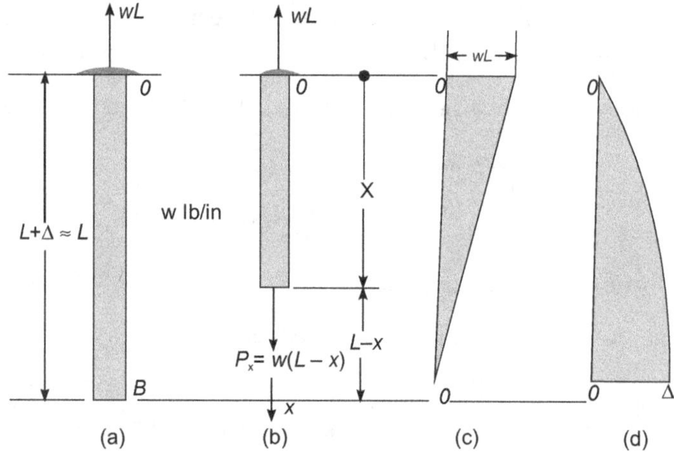

Solution

The free-body diagrams of the bar and its truncated segment are shown, respectively, in the Figures. These two steps are essential in the solution of such problems. The graph for the axial force $P_x = w(L-x)$ is in Figure. By applying equation for axial deformation, the change in bar length $\Delta(x)$ at a generic point x is given by,

$$\Delta(x) = \int_0^x \frac{P_x d_x}{A_x E} = \frac{1}{AE} \int w(L-x)dx = \frac{w}{AE}\left(Lx - \frac{x^2}{2}\right)$$

A plot of this function is shown in Figure, with its maximum as B.

The deflection of B is

$$\Delta = \Delta(L) = \frac{w}{AE}\left(L^2 - \frac{L^2}{2}\right) = \frac{wL^2}{2AE} = \frac{WL}{2AE}$$

where $W = wL$ is the total weight of the bar.

If a concentrated force P, in addition to the bar's own weight, were acting on bar OB at end B, the total deflection due to the *two causes* would be obtained by *superposition* as

$$\Delta = \frac{PL}{AE} + \frac{WL}{2AE} = \frac{[P+(W/2)]L}{AE}$$

In problems where the area of a rod is variable, a proper *function* for it must be substituted into equation for $\Delta(x)$ to determine deflections. In practice, it is sometimes sufficiently accurate to analyse such problems by approximating shape of a rod by a *finite number* of elements, as shown in Figure. The deflections for each one of these elements are added to obtain the total deflection. Because of the rapid variation in the cross section shown, the solution would be approximate.

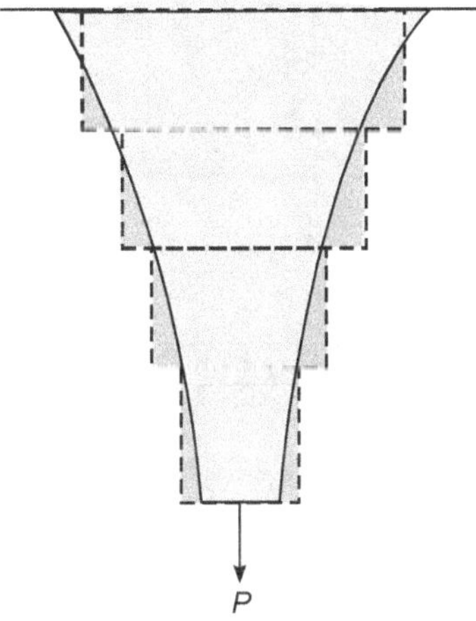

PROBLEM 3.12

An elastic bar fixed at both ends is loaded as shown in Figure. The known flexibility coefficients f and $2f$ for each of the three bar segments are shown in the figure. Determine the reactions and plot the axial force and the axial displacement diagrams for the bar.

Solution

Note that flexibility implies deflection per unit force. It is opposite to the term stiffness, which implies force per unit deflection. Remove the lower support to obtain the free-body diagram shown in Figure and calculate Δ_0. Since the applied forces act downward, because of the sign convention adopted in Figure, they carry negative signs. The deflection caused by R_2 on an unloaded system is calculated next. Then, the reaction R_2 is determined.

$$\Delta_0 = \sum_i f_i P_i = -2fP - f(2P+P) = -5fP$$

and

$$\Delta_1 = (2f + f + f)R_2 = 4f\,R_2$$

Since

$$\Delta_1 + \Delta_2 = 0 \qquad R_2 = 1.25P$$

Note that the applied forces are supported by a compressive reaction at the bottom and a tensile reaction at the top. In problems where the bar lengths and the cross-sectional areas together with the elastic moduli E for the materials, are given, the flexibilities are determined using.

$$f_i = L_i / A_i E_i \quad \text{for each section } i$$

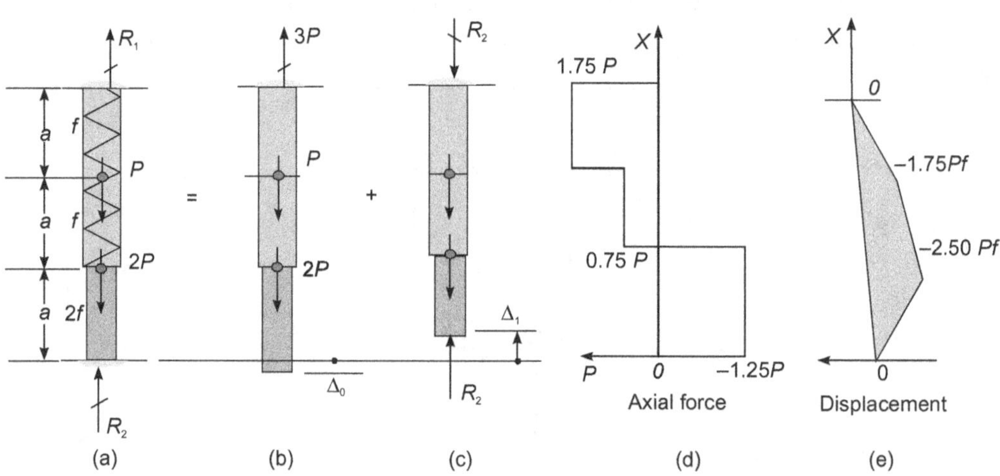

The axial force diagram is plotted in Figure. The compressive force in the bottom third of the bar causes a *downward* deflection of $1.25P \times 2f = 2.5Pf$. The tensile forces stretch the remainder of the bar $0.75Pf + 1.75Pf$ such that displacement at the top is zero. In this manner, the geometric boundary conditions are satisfied at both ends of the bar.

PROBLEM 3.13

An elastic bar is held at both ends, as shown in Figure. If the bar temperature increases by ΔT, what axial force develops in the bar? AE for the bar is constant.

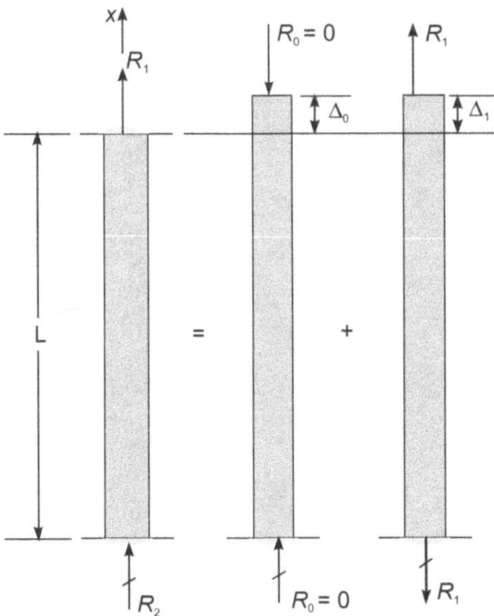

Solution

First the upper support is removed and Δ_0 determined. The raising of the temperature causes no axial force in the bar. Thus, Δ_1 is calculated. The axial force in the R_1, caused by the rise in temperature is found. The sequence is shown below.

$$\Delta_0 = \alpha\,\Delta T L_0$$

$$\Delta_1 = R_1 f = \frac{R_1 L}{AE}$$

$$\Delta_0 + \Delta_1 = 0, \quad R_1 = -\alpha \Delta T A E$$

PROBLEM 3.14

The planar system of the three elastic bars are shown in Figure. Determine the forces in the bars caused by applied force P. The cross-sectional area of each bar is the same, and their elastic modulus is E.

Solution

A free-body diagram of the assumed primary system with the support from the middle bar removed by cutting it at point B shown in Figure. Then, by using statics, the forces in the bars are determined, and the deflection of point D is calculated using the procedure illustrated below. Since BD carries no force, deflection Δ_0 at point B is the same as it is at point D. Recognizing symmetry,

$$F_{10} = 0 \qquad \text{and} \qquad 2F_{20}\cos\alpha = P$$

Therefore,

$$F_{20} = \frac{P}{2\cos\alpha}$$

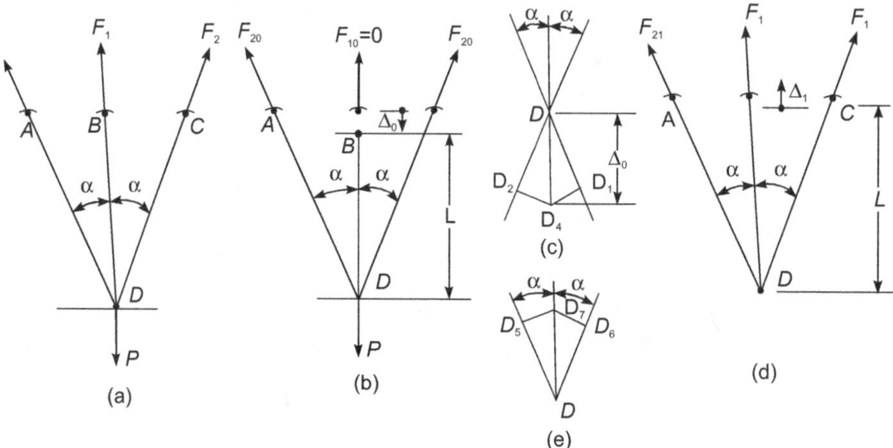

Since

$$L_{AD}\cos\alpha = L, \qquad L_{AD} = L/\cos\alpha$$

Hence, the stretch of bar *AD* in the primary system is

$$(\Delta_{AD})_0 = \frac{PL}{2AE\cos^2\alpha}$$

However, Since Δ_0 Equals *DD* in Figure.

$$\Delta_0\cos\alpha = (\Delta_{AD})_0 \qquad \text{and} \quad \Delta_0 = -\frac{PL}{2AE\cos^3\alpha}$$

where the negative sign signifies that the deflection is downward.

The same kind of relationship applies to the upward deflection of point *D* caused by the force F_1; see Figures. However, the deflection of point *B* is increased by the stretch of the bar *BD*. The latter quantity is calculated using $\Delta = \dfrac{FL}{AE}$. On this basis.

$$\Delta_1 = \frac{F_1 L}{AE} + \frac{F_1 L}{2AE\cos^3\alpha}$$

By applying Eq (i.e., $\Delta_0 + \Delta_1 = 0$, and noting from statics that $F_1 + 2F_2\cos\alpha = P$, on simplification,

$$F_1 = \frac{P}{2\cos^3\alpha + 1} \qquad \text{and} \qquad F_2 = \frac{P}{2\cos^3\alpha + 1}\cos^2\alpha$$

PROBLEM 3.15

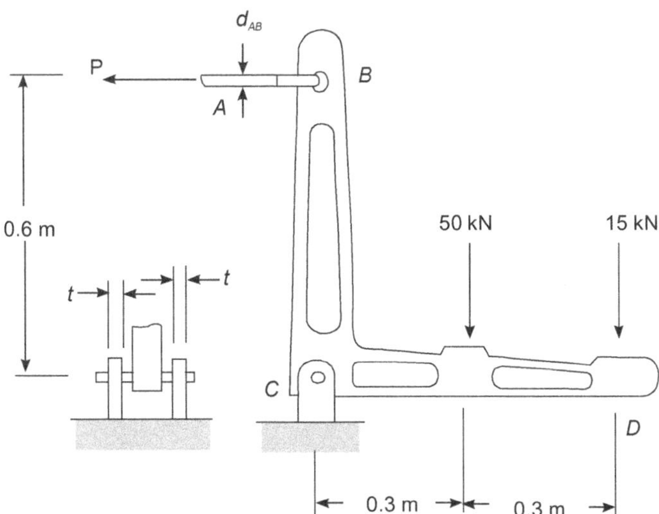

Two forces are applied to the bracket *BCD* as shown. (a) Knowing that the control rod *AB* is to

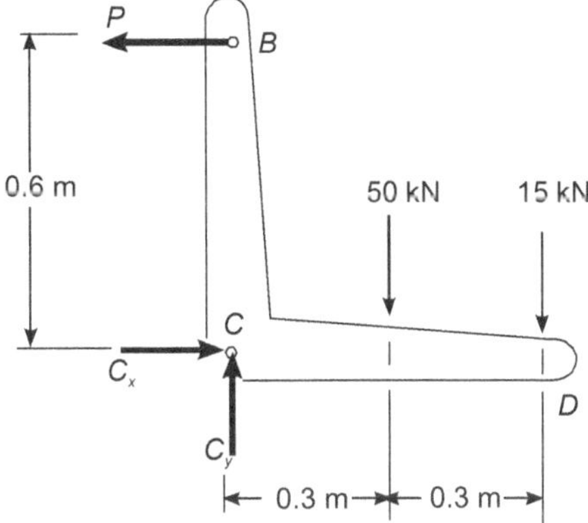

be made of a steel having an ultimate normal stress of 600 MPa determine the diameter of the rod for which the factor of safety with respect failure will be 3.3. (b) The pin at *C* is to be made of a steel having an ultimate shearing stress of 350 MPa. Determine the diameter of the pin *C* for which the factor of safety with respect to shear will also be 3.3. (c) Determine the required thickness of the bracket supports at *C* knowing that the allowable bearing stress for the steel used is 300 MPa.

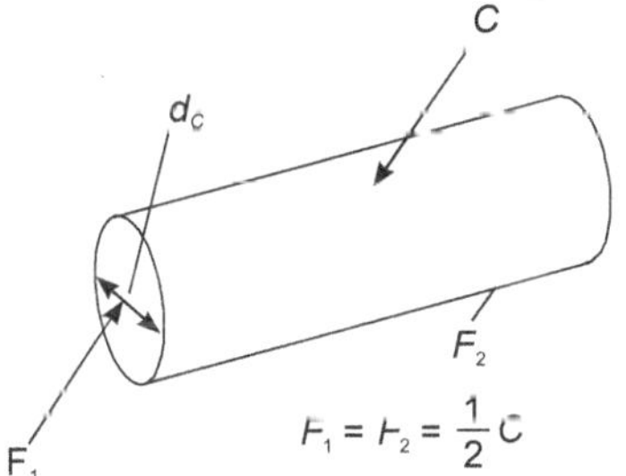

Free Body: Entire Bracket. The reaction at C is represented by its components C_x and C_y

$$+ \ \Sigma M_C = 0: \quad P(0.6\text{ m}) - (50\text{kN})(0.3\text{m}) - (15\text{kN})(0.6\text{m}) = 0 \qquad P = 40\text{kN}$$

$$\Sigma F_x = 0: \qquad C_x = 40\text{kN}$$

$$\Sigma F_y = 0: \qquad C_y = 65\text{ kN} \qquad C = \sqrt{C_x^2 + C_y^2} = 76.3\text{kN}$$

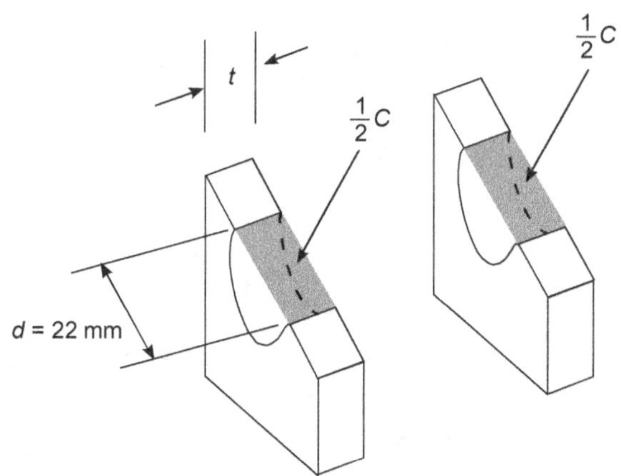

a. Control Rod AB: Since the factor of safety is to be 3.3, the allowable stress is

$$\sigma_{all} = \frac{\sigma U}{F.S} = \frac{600\,MPa}{3.3} = 181.8\,MPa$$

For $P = 40$ kN the cross-sectional area required is

$$A_{req} = \frac{P}{\sigma_{all}} = \frac{40\text{kN}}{181.8\,MPa} = 220 \times 10^{-6}\,m^2$$

$$A_{req} = \frac{\pi}{4} d_{AB}^2 = 220 \times 10^{-6}\,m^2 \qquad d_{AB} = 16.74\text{ mm}$$

b. Shear in Pin C: For a factor of safety of 3.3, we have

$$\tau_{all} = \frac{\tau_U}{F.S} = \frac{350\,MPa}{3.3} = 106.1\text{ MPa}$$

Since the pin is in double shear, we write

$$A_{req} = \frac{C/2}{\tau_{all}} = \frac{(76.3\text{kN})/2}{106.1\,MPa} = 360\,mm^2$$

$$A_{req} = \frac{\pi}{4} d_C^2 = 360\text{ mm}^2$$

$$d_C = 21.4\,mm$$

The next larger size pin available is of 22-mm diameter and should be used.

c. Bearing at C: Using $d = 22$ mm, the nominal bearing area of each bracket is 22 t. Since the force carried by each bracket is $C/2$ and the allowable bearing stress is 300 MPa, we write

$$A_{req} = \frac{C/2}{\sigma_{all}} = \frac{(76.3\,\text{kN})/2}{300\,\text{MPa}} = 127.2\,\text{mm}^2$$

Thus

$22t = 127.2$ $\qquad$ $t = 5.78\,\text{mm}$ $\qquad$ Use: $t = 6$ mm

PROBLEM 3.16

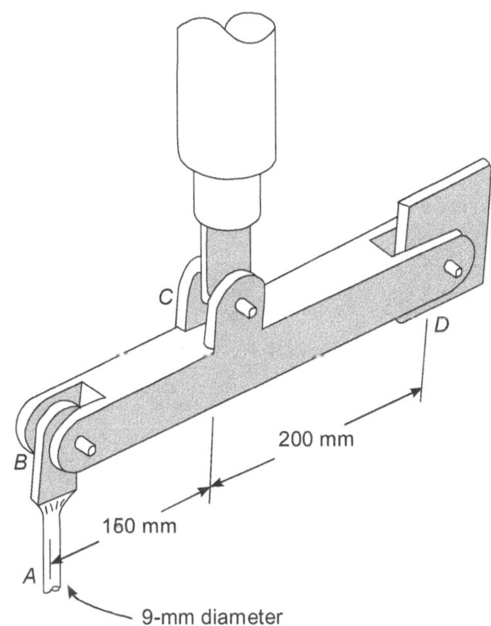

The rigid beam BCD is attached by bolts to a control rod at B, to a hydraulic cylinder at C, and to a fixed support at D. The diameters of the bolts used are: $d_B = d_D = 8$ mm, $d_C = 12$ mm. Each bolt acts in double shear and is made from a steel for which the ultimate shearing stress is $\tau_U = 300$ MPa. The 9-mm-diameter control rod AB is made of a steel for which the ultimate tensile stress is $\sigma_U = 450$ MPa. If the minimum factor of safety is to be 3.0 for the entire unit, determine the largest upward force which may be applied by the hydraulic cylinder at C.

Solution

The factor of safety with respect to failure must be 3.0 or more in each of the three bolts and in the control rod. These four independent criteria will be considered separately.

Free-body: Beam *BCD*   We first determine the force at C in terms of the force at B and in terms of the force at D.

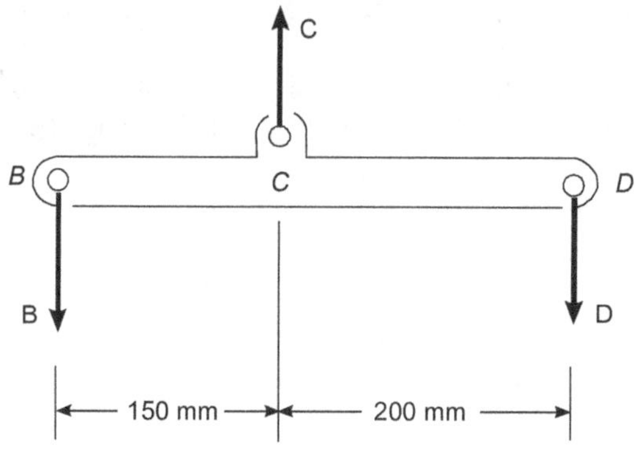

$$+ \ \Sigma M_D = 0: \qquad B(350\,\text{mm}) - C(200\,\text{mm}) = 0 \qquad C = 1.750\,B \qquad (1)$$

$$+ \ \Sigma M_B = 0: \qquad -D(350\,\text{mm}) + C(150\,\text{mm}) = 0 \qquad C = 2.33D \qquad (2)$$

Control rod For a factor of safety of 3.2 we have

$$\sigma_{all} = \frac{\sigma_U}{F.S} = \frac{450\,\text{MPa}}{3.0} = 150 \ \text{MPa}$$

The allowable force in the control rod is

$$B = \sigma_{all}(A) = (150\,\text{MPa})\tfrac{1}{4}\pi(9\,\text{mm})^2 = 9.54\text{kN}$$

Using Eq. (1) we find the largest permitted value of C:
$$C = 1.750 \ B = 1.750(9.54\,\text{kN}) \qquad\qquad C = 16.70\,\text{kN}$$

Bolt at B $\quad \tau_{all} = \tau_U / F.S = (300\,\text{MPa}) / 3 = 100\,\text{MPa}$

Since the bolt is in double shear, the allowable value of force B is

$$B = \tau_{all}(2A) = (100\,\text{MPa})\frac{2\pi}{4}(8\,\text{mm})^2 = 10.05\,\text{kN}$$

From Eq.(1): $\qquad\qquad C = 1.750 \ B = 1.750(10.05\,\text{kN}) \qquad C = 17.59\,\text{kN}$

Bolt at D Since this bolt is the same as bolt B, the allowable force is $D = B = 10.05$ kN. Using Eq. (2):
$$C = 2.33 \ D = 2.33(10.05\,\text{kN}) \qquad\qquad C = 23.4\,\text{kN}$$

Bolt at C

$$C = \tau_{all}(2A) = (100\,\text{MPa})\frac{2\pi}{4}(12\,\text{mm})^2 \qquad\qquad C = 22.6\,\text{kN}$$

Summary: We have found separately four maximum allowable values of the force C. In order to satisfy all these criteria we must choose the smallest value namely: $C = 16.70$ kN

PROBLEM 3.17

Determine the deformation of the steel rod shown in Figure under the given load ($E = 200$ GPa)

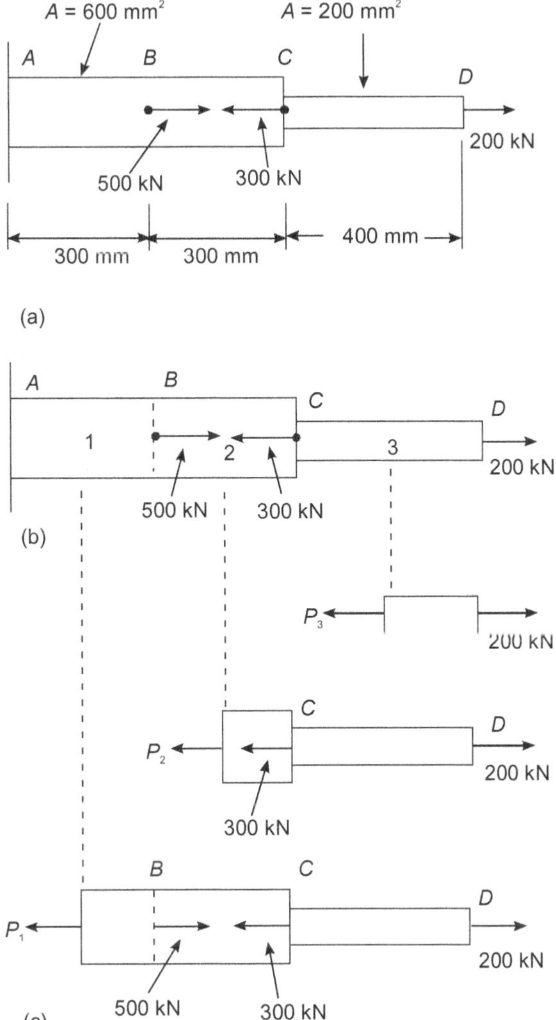

We divide the rod into the three component parts shown in Figure and write

$$L_1 = L_2 = 0.300\,\text{m} \qquad\qquad L_3 = 0.400\,\text{m}$$

$$A_1 = A_2 = 600 \times 10^{-6}\,\text{m}^2 \qquad\qquad A_3 = 200 \times 10^{-6}\,\text{m}^2$$

To find the internal forces P_1, P_2 and P_3, we must pass sections through each of the component parts, drawing each time the free-body diagram of the portion of rod located to the right of the section. Expressing that each of the free bodies is in equilibrium, we obtain successively

$$P_1 = 400\,\text{kN} = 400 \times 10^3\,\text{N}$$

$$P_2 = -100\,\text{kN} = -100 \times 10^3\,\text{N}$$

$$P_3 = 200\,\text{kN} = 200 \times 10^3\,\text{N}$$

Carrying the values obtained into Equation for deformation, we have:

$$\delta = \sum_i \frac{P_i L_i}{A_i E_i} = \frac{1}{E}\left(\frac{P_1 L_1}{A_1} + \frac{P_2 L_2}{A_2} + \frac{P_3 L_3}{A_3}\right)$$

$$= \frac{1}{200 \times 10^9}\left[\frac{(400 \times 10^3)(0.300)}{600 \times 10^{-6}} + \frac{(-100 \times 10^3)(0.300)}{600 \times 10^{-6}} + \frac{(200 \times 10^3)(0.400)}{200 \times 10^{-6}}\right]$$

$$\delta = 2.75 \times 10^{-3}\,\text{m} = 2.75\,\text{mm}$$

In this example, where one end is fixed, the deformation δ of the rod was equal to the displacement of its free end. When both ends of a rod move, however, the deformation of the rod is measured by the relative displacement of one end of the rod with respect to the other. Consider, for instance, the assembly shown in Figure below which consists of three elastic bars of length L connected by a rigid pin at A.

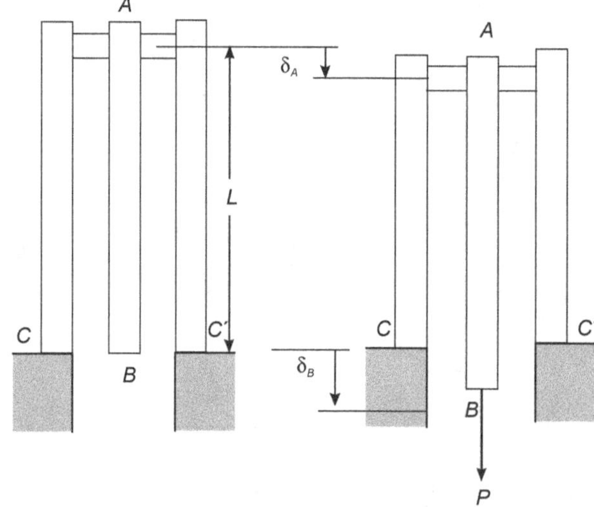

If a load P is applied at B, each of the three bars will deform. Since the bars AC and AC' are attached to fixed supports at C and C', their common deformation is measured by the displacement δ_A of point A. On the other hand, since both the ends of bar AB move, the deformation of AB is measured by the difference between the displacements δ_A and δ_B of points A and B, i.e., by the relative displacement of B with respect to A. Denoting this relative displacement by $\delta_{B/A}$, we write

$$\delta_{B/A} = \delta_B - \delta_A = \frac{PL}{AE}$$

where A is the cross-sectional area of AB and E its modulus of elasticity.

PROBLEM 3.18

The rigid castings A and B are connected by two 18 mm-diameter steel bolts CD and GH and are in contact with the ends of a 36 mm-diameter aluminum rod EF. Each bolt is single-threaded with a pitch of 2 mm, and after being snugly fitted, the nuts at D and H are both tightened one-quarter of a turn. Knowing that E is 200 GPa for steel and 70 GPa for aluminium, determine the normal stress in the rod.

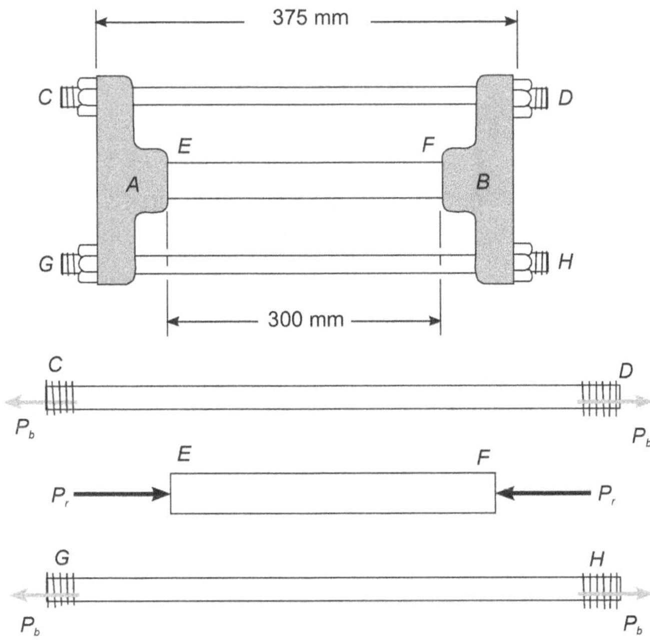

Deformations   Bolts CD and GH: Tightening the nuts causes tension in the bolts. Because of symmetry, both are subjected to the same internal force P_b and undergo the same deformation δ_b. We have

$$\delta_b = + \frac{P_b L_b}{A_b E_b} = + \frac{P_b(0.375\,\text{m})}{\frac{1}{4}\pi(0.018\,\text{m})^2(200\,\text{GPa})} = 7.368 \times 10^{-9} P_b \tag{1}$$

Rod EF: The rod is in compression. Denoting by P_r the magnitude of the force in the rod and by δ_r the deformation of the rod, we write

$$\delta_r = -\frac{P_r L_r}{A_r E_r} = -\frac{P_r(0.300\,\text{m})}{\frac{1}{4}\pi(0.036\,\text{m})^2(70\,\text{GPa})} = -4.210 \times 10^{-9} P_r \tag{2}$$

Displacement of D relative to B: Tightening the nuts one-quarter of a turn causes ends D and H of the bolts to undergo a displacement of (2 mm) relative to casting B.

Considering end D, we write

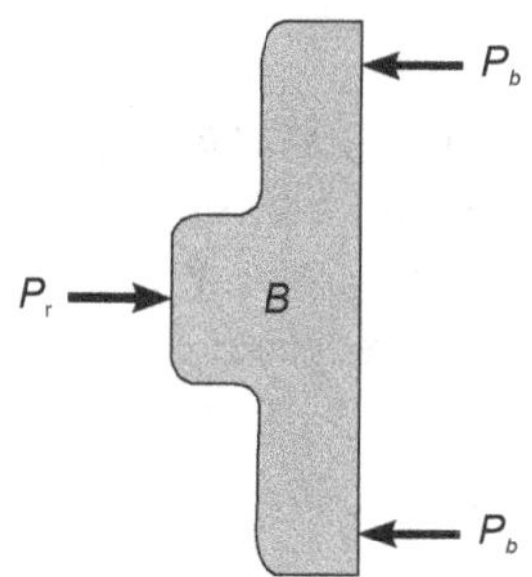

$$\delta_{D/B} = \frac{1}{4}(0.002\,\text{m}) = 0.0005\,\text{m} \tag{3}$$

But $\delta_{D/B} = \delta_D - \delta_B$ where and represent the displacements of D and B. If we assume that casting A is held in a fixed position while the nuts at D and H are being tightened, these displacements are equal to the deformations of th bolts and of the rod, respectively. We have, therefore,

$$\delta_{D/B} = \delta_b - \delta_r \tag{4}$$

Substituting from (1), (2),and (3) into (4), we obtain

$$0.0005\,\text{m} = 7.368 \times 10^{-9} P_b + 4.210 \times 10^{-9} P_r \tag{5}$$

Free body: Casting B

$$\xrightarrow{+}\Sigma F = 0: \qquad\qquad P_r - 2P_b = 0, \quad P_r = 2P_b \tag{6}$$

Forces in bolts and rod

Substituting for P_r from (6) into (5), we have

$$0.0005\,\text{m} = 7.368 \times 10^{-9} P_b + 4.210 \times 10^{-9}(2P_b)$$

$$P_b = 31.67\text{kN}$$

$$P_r = 2P_b = 2(31.67\,\text{kN}) = 63.34\,\text{kN}$$

Stress in rod

$$\sigma_r = \frac{P_r}{A_r} = \frac{63.34\,\text{kN}}{\frac{1}{4}\pi(0.036\,\text{m})^2} \qquad\qquad \sigma_r = 62.2\,\text{MPa}$$

PROBLEM 3.19

Determine the values of the stress in portions AC and CB of steel bar shown in Figure, when the temperature of the bar is $-50°C$, knowing that a close fit exists at both of the rigid supports when the temperature is $+25°C$. Use the values $E = 200$ GPa and $\alpha = 12 \times 10^{-6}/°C$ for steel.

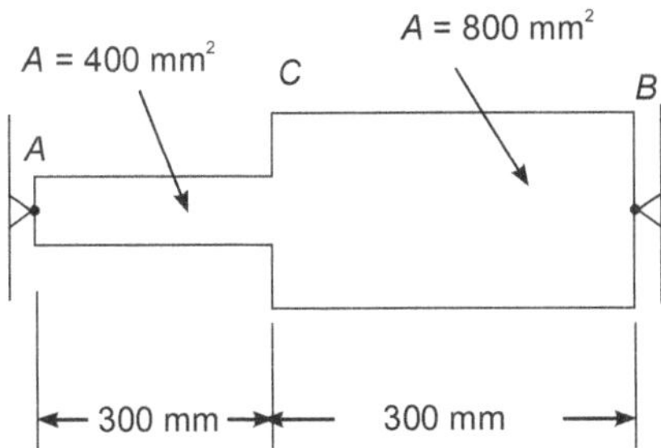

We first determine the reactions at the supports. Since the problem is statically indeterminate, we detach the bar from its support at B and let it undergo the temperature change.

$$\Delta T = (-50^\circ C) - (25^\circ C) = -75^\circ C$$

The corresponding deformation (Figure) is

$$\delta_r = \alpha(\Delta T)L = (12 \times 10^{-6}\,/\,^\circ C)(-75^\circ C)(0.6\,\text{m})$$
$$= -540 \times 10^{-6}\,\text{m}$$

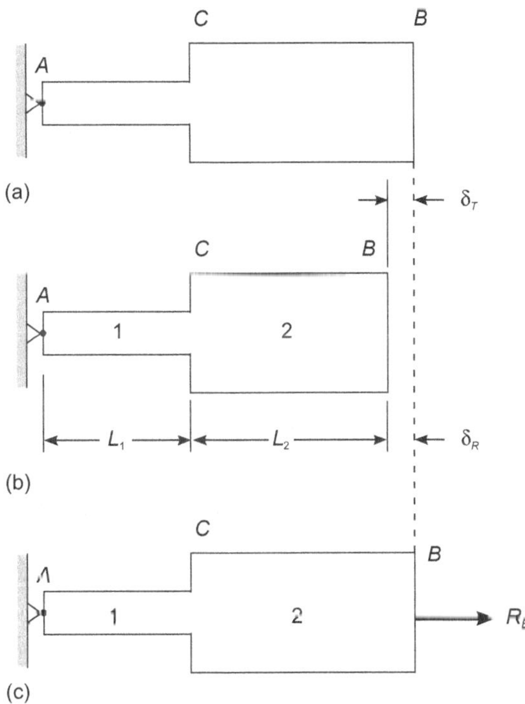

Applying now the unknown force R_B at end B (Figure), we express the corresponding deformation δ_R. Substituting

$$L_1 = L_2 = 0.3\,\mathrm{m}$$

$$A_1 = 400 \times 10^{-6}\,\mathrm{m}^2 \qquad A_2 = 800 \times 10^{-6}\,\mathrm{m}^2$$

$$P_1 = P_2 = R_B \qquad E = 200 \times 10^9\,\mathrm{Pa}$$

$$\delta_R = \frac{P_1 L_1}{A_1 E} + \frac{P_2 L_2}{A_2 E}$$

$$= \frac{R_B}{200 \times 10^9\,\mathrm{Pa}} \left(\frac{0.3\,\mathrm{m}}{400 \times 10^{-6}\,\mathrm{m}^2} + \frac{0.3\,\mathrm{m}}{800 \times 10^{-6}\,\mathrm{m}^2} \right)$$

$$= (5.625 \times 10^{-9}\,\mathrm{m}\,/\,\mathrm{N}) R_B$$

Expressing that the total deformation of the bar must be zero as a result of the imposed constraints, we write

$$\delta = \delta_T + \delta_R = 0$$

$$\delta = -540 \times 10^6\,\mathrm{m} + (5.625 \times 10^{-9}\,\mathrm{m/N}) R_B = 0$$

from which we obtain

$$R_B = 96.0 \times 10^3\,\mathrm{N} = 96.0\,\mathrm{kN}$$

The reaction at A is equal and opposite.

Noting that the forces in the two portions of the bar are $P_1 = P_2 = 96.0\,\mathrm{kN}$, we obtain the following values of the stress in portions AC and CB of the bar:

$$\sigma_1 = \frac{P_1}{A_1} = \frac{96.0 \times 10^3\,\mathrm{N}}{400 \times 10^{-6}\,\mathrm{m}^2} = 240\,\mathrm{MPa}$$

$$\sigma_2 = \frac{P_2}{A_2} = \frac{96.0 \times 10^3\,\mathrm{N}}{800 \times 10^{-6}\,\mathrm{m}^2} = 120\,\mathrm{MPa}$$

We cannot emphasise too strongly the fact that, while the total deformation of the bar must be zero, the deformations of the portions AC and CB are not zero. A solution of the problem based on the assumption that these deformations are zero would therefore be wrong. Neither can the values of the strain in AC or CB be assumed equal to zero. To amplify this point, we shall now determine the strain in ε_{AC} portion AC of the bar. The strain ε_{AC} may be divided into two component parts; one is the thermal strain ϵ_T produced in the unrestrained bar by the temperature change ΔT (Figure). Therefore,

$$\varepsilon_T = \alpha \Delta T = (12 \times 10^{-6}\,/\,^\circ\mathrm{C})(-75^\circ\mathrm{C})$$

$$= -900 \times 10^{-6} = -900\mu$$

The other component of ε_{AC} is associated with the stress due to the force R_B applied to the bar (Figure). From Hooke's law, we express this component of the strain as

$$\frac{\sigma_1}{E} = \frac{240\,\text{MPa}}{200\,\text{GPa}} = 1200 \times 10^{-6} = 1200\,\mu$$

Adding the two components of the strain in *AC*, we obtain

$$\varepsilon_{AC} = \varepsilon_T + \frac{\sigma_1}{E} = -900\mu + 1200\mu$$
$$= +300\mu$$

A similar computation yields the strain in portion *CB* of the bar:

$$\varepsilon_{CB} = \varepsilon_T + \frac{\sigma_2}{E} = -900\mu + 600\mu$$
$$= -300\mu$$

The deformations δ_{AC} and δ_{CB} of the two portions of the bar are expressed respectively as

$$\delta_{AC} = \varepsilon_{AC}(AC) = (+300\,\mu)(0.3\,\text{mm}) = +90\,\mu m$$
$$\delta_{CB} = \varepsilon_{CB}(CB) = (-300\,\mu)(0.3\,\text{m}) = -90\,\mu m$$

We thus check that, while the sum $\delta = \delta_{AC} + \delta_{CB}$ of the two deformations is zero, neither of the deformations is zero.

PROBLEM 3.20

The 10-mm-diameter rod *CE* and the 15-mm-diameter rod *DF* are attached to the rigid bar *ABCD* as shown. Knowing that the rods are made of aluminum and using $E = 70\,\text{GPa}$, determine (a) the force in each rod caused by the loading shown, (b) the corresponding deflection at point *A*.

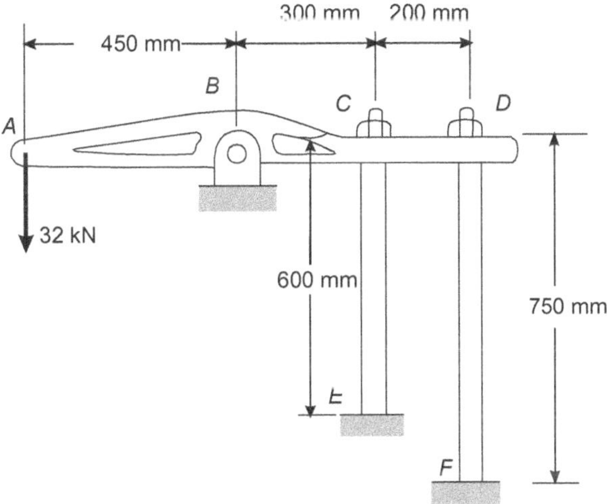

Statics   Considering the free body of bar *ABCD*, we note that the reaction at *B* and the forces exerted by the rods are indeterminate. However, using statics we may write

$$\Sigma M_B = 0: \qquad (32\,\text{kN})(0.45\,\text{m}) - F_{CE}(0.3\,\text{m}) - F_{DF}(0.5\,\text{m}) = 0$$

$$0.3F_{CE} + 0.5F_{DF} = 14.4 \times 10^3 \tag{1}$$

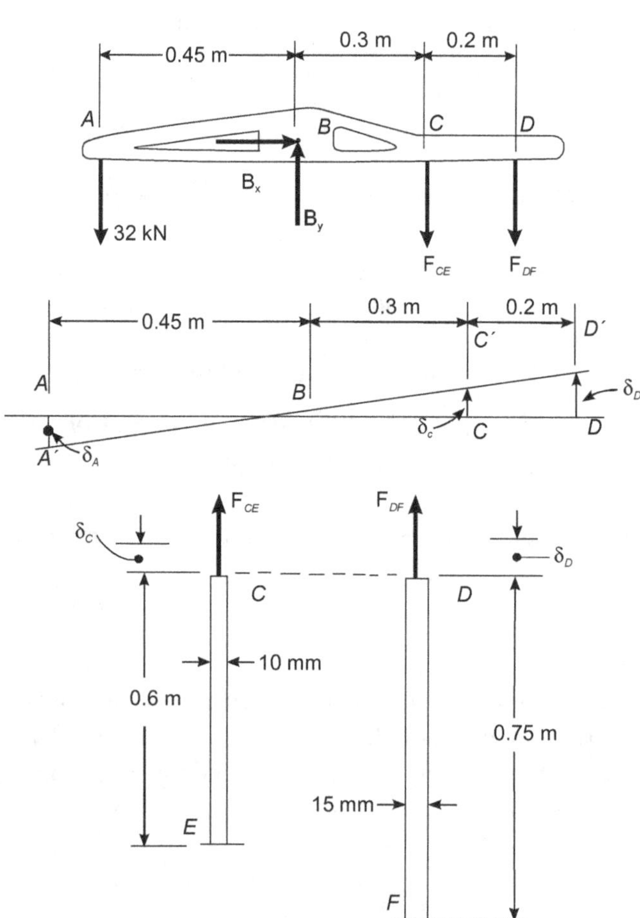

Geometry   After application of the 32-kN load, the position of the bar is $A'B'C'D'$. From the similar triangles BAA', BCC' and BDD' we have

$$\frac{\delta_C}{0.3\,m} = \frac{\delta_D}{0.5\,m} \qquad\qquad \delta_c = 0.6\delta_D \tag{2}$$

and

$$\frac{\delta_A}{0.45\,m} = \frac{\delta_D}{0.5\,m} \qquad\qquad \delta_A = 0.9\delta_D \tag{3}$$

Deformations: Using equation for deformation, we have

$$\delta_C = \frac{F_{CE}L_{CE}}{A_{CE}E} \qquad \delta_D = \frac{F_{DF}L_{DF}}{A_{DF}E}$$

Substituting for δ_C and δ_D into (2),we write

$$\delta_C = 0.6\delta_D, \delta_C = \frac{F_{CE}L_{CE}}{A_{CE}E} = 0.6\frac{F_{DF}L_{DF}}{A_{DF}E}$$

$$F_{CE} = 0.6\frac{L_{DF}A_{CE}}{L_{CE}A_{DF}}F_{DF} = 0.6\left(\frac{0.75\,\mathrm{m}}{0.60\,\mathrm{m}}\right)\left[\frac{\frac{1}{4}\pi(0.010\,\mathrm{m})^2}{\frac{1}{4}\pi(0.015\,\mathrm{m})^2}\right]F_{DF} \qquad F_{CE} = 0.333F_{DF}$$

Force in each rod: Substituting for F_{CE} into (1), we have

$$0.3(0.000F_{DF})+0.5F_{DF} = 14.4\times10^3 \qquad\qquad F_{DF} = 24\,\mathrm{kN}$$
$$F_{CE} = 0.333F_{DF} = 0.333(24\,\mathrm{kN}) \qquad\qquad F_{CE} = 8\,\mathrm{kN}$$

Deflections   The deflection of point D is

$$\delta_D = \frac{F_{DF}L_{DF}}{A_{DF}E} = \frac{(24\,\mathrm{kN})(0.75\,\mathrm{m})}{\frac{1}{4}\pi(0.015\,\mathrm{m})^2(70\,\mathrm{GPa})} \qquad \delta_D = 1.455\,\mathrm{mm}$$

Using (3) ,we write

$$\delta_A = 0.9\delta_D = 0.9(1.455\,\mathrm{mm}) \qquad\qquad \delta_A = 1.310\,\mathrm{mm}$$

PROBLEM 3.21

The rigid bar *CDE* is attached to a pin support at *E* and rests on the 30-mm-diameter brass cylinder *BD*. A 22-mm-diameter steel rod *AC* passes through a hole in the bar and is secured by a nut: which is snugly fitted when the temperature of the entire assembly is 20°C. The temperature of the brass cylinder is then raised to 50°C while the steel rod remain at 20°C. Assuming that no stresses were present before the temperature change, determine the stress in the cylinder.

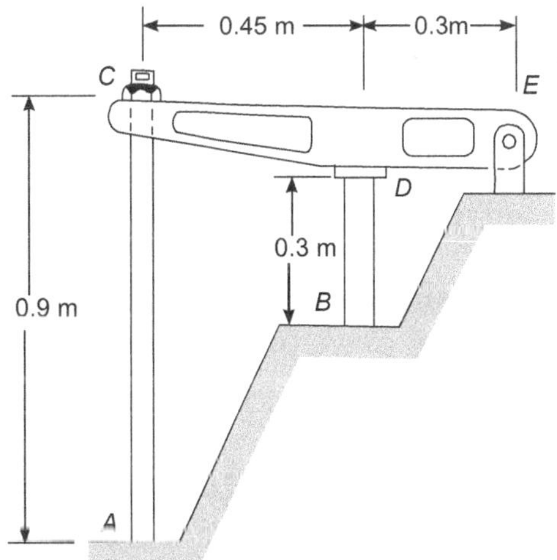

Rod AC: Steel   Cylinder BD: Brass

$E = 200\,GPa$ $E = 105\,GPa$

$\alpha = 12 \times 10^{-6}\,/\,°C$ $\alpha = 18.8 \times 10^{-6}\,/\,°C$

Statics: Considering the free-body of the entire assembly, we write

$$+\;\Sigma M_E = 0: \qquad R_A(0.75\,\text{m}) - R_B(0.3\,\text{m}) = 0 \qquad R_A = 0.4 R_B \tag{1}$$

Deformations We shall use the method of superposition, considering R_B as redundant. With the support at B removed, the temperature rise of the cylinder causes point B move down through δ_T. The reaction R_B must cause a deflection δ_1 of the same magnitude as δ_T so that the final deflection of point B will be zero (Figure).

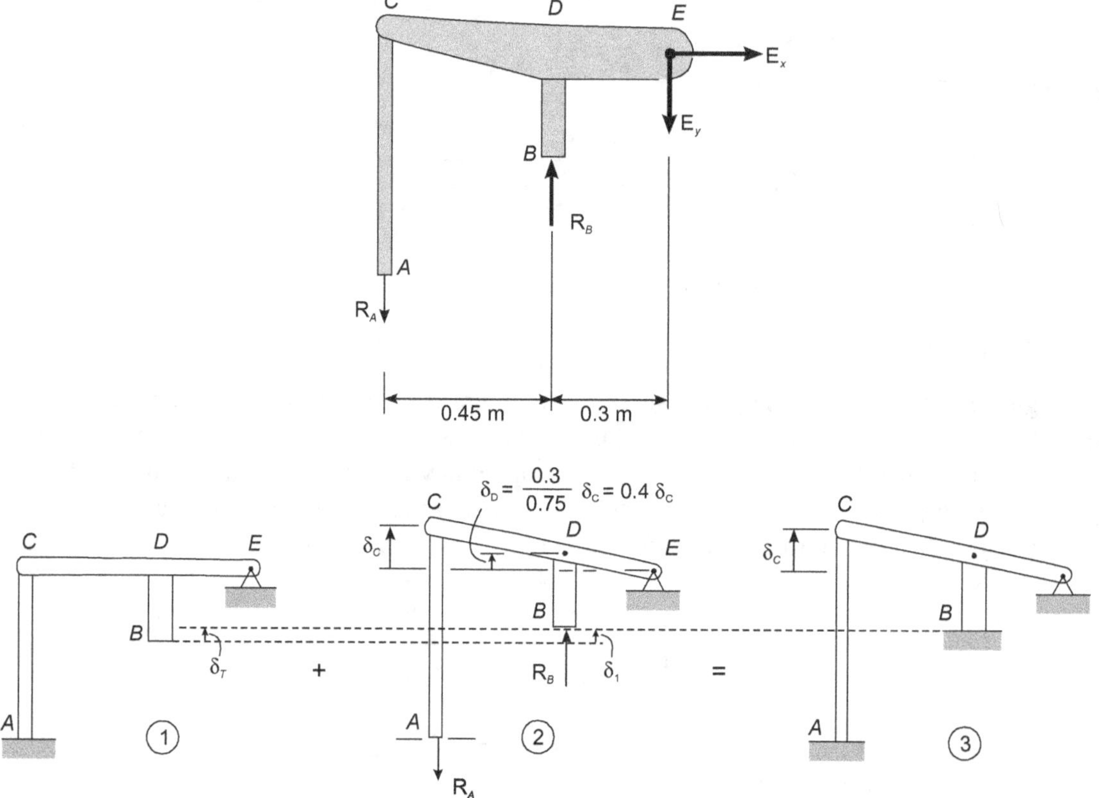

Deflection δ_T Due to a temperature rise $50°C - 20°C = 30°C$, the length of the brass cylinder increases by δ_T.

$$\delta_T = L(\Delta T)\alpha = (0.3\,\text{m})(30°C)(18.8 \times 10^{-6}\,/\,°C) = 169.2 \times 10^{-6}\,\text{m} \downarrow$$

Deflection δ_1 From Figure we note that $\delta_D = 0.4\,\delta_C$ and that $\delta_1 = \delta_D + \delta_{B/D}$

$$\delta_C = \frac{R_A L}{AE} = \frac{R_A(0.9\,m)}{\frac{1}{4}\pi(0.22\,m)^2(200\,GPa)} = 11.84 \times 10^{-9}\,R_A \uparrow$$

$$\delta_D = 0.40\delta_c = 0.4(11.84 \times 10^{-9}\,R_A) = 4.74 \times 10^{-9}\,R_A \uparrow$$

$$\delta_{B/D} = \frac{R_B L}{AE} = \frac{RB(0.3\,m)}{\frac{1}{4}\pi(0.03\,m)^2(105\,GPa)} = 4.04 \times 10^{-9}\,R_B$$

We recall from (1) that $R_A = 0.4\,R_B$ and write

$$\delta_1 = \delta_D + \delta_{B/D} = [4.74(0.4R_B) + 4.04R_B]10^{-9} = 5.94 \times 10^{-9}\,R_B \uparrow$$

But

$$\delta_T = \delta_1: \qquad 169.2 \times 10^{-6}\,m = 5.94 \times 10^{-9}\,R_B \qquad R_B = 28.5\,kN$$

Stress in cylinder:

$$\sigma_B = \frac{R_B}{A} = \frac{28.5\,kN}{\frac{1}{4}\pi(0.03)^2} \qquad \sigma_B = 40.3\,MPa$$

PROBLEM 3.22

An 800-mm long cylindrical rod of cross-sectional area $A_r = 45$ mm^2 is placed inside a tube of the same length and of cross-sectional area $A_t = 60$ mm^2. The ends of the rod and tube are attached to a rigid support on one side, and to a rigid plate on the other, as shown in the longitudinal section of Figure. The rod and tube are both assumed to be elastoplastic, with moduli of elasticity $E_r = 200$ GPa and $E_t = 100$ GPa, and yield strengths $(\sigma_r)_y = 200$ MPa and $(\sigma_t)_y = 250$ MPa. Draw the load-deflection diagram of the rod-tube assembly when a load P is applied to the plate as shown.

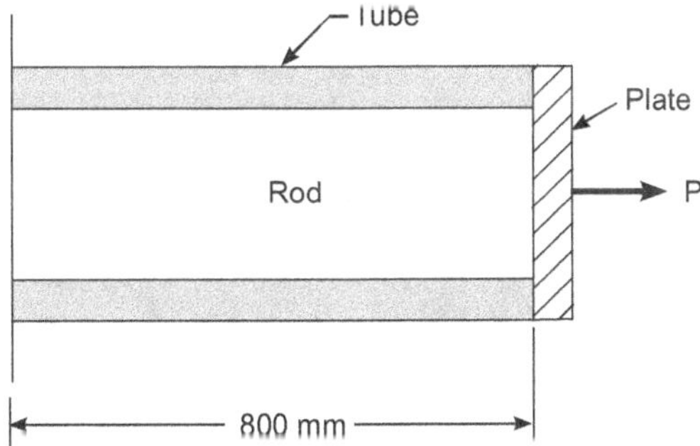

We first determine the internal force and the elongation of the rod as it begins to yield:

$$(P_r)_Y = (\sigma_r)_Y\,A_r = (200\,MPa)(45\,mm^2) = 9\,kN$$

$$(\delta_r)_Y = (\varepsilon_r)_Y\,L = \frac{(\sigma_r)_Y}{E_r'}\,L = \frac{200\,MPa}{200\,Gpa}(800\,mm) = 0.8\,mm$$

Since the material is elastoplastic, the force-elongation diagram of the rod alone consists of an oblique straight line and of a horizontal straight line, as shown in Figure. Following the same procedure for the tube, we have

$$(P_t)_Y = (\sigma_t)_Y \, A_t = (250\,\text{MPa})(60\,\text{mm}^2) = 15\,\text{kN}$$

$$(\delta_t)_Y = (\varepsilon_t)_Y \, L = \frac{(\sigma_t)_Y}{E_t} L = \frac{250\,\text{MPa}}{100\,\text{GPa}}(800\,\text{mm}) = 2\,\text{mm}$$

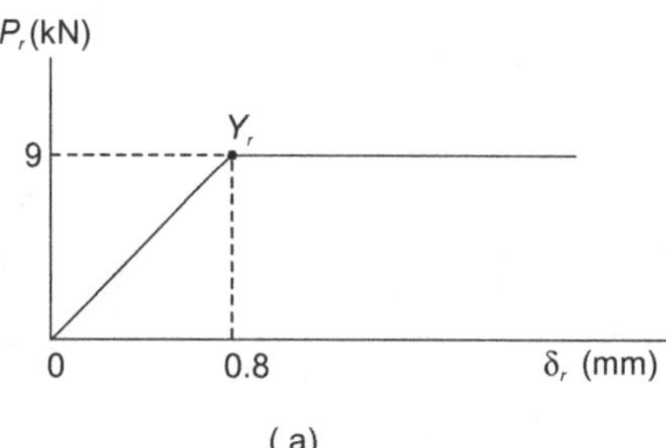

(a)

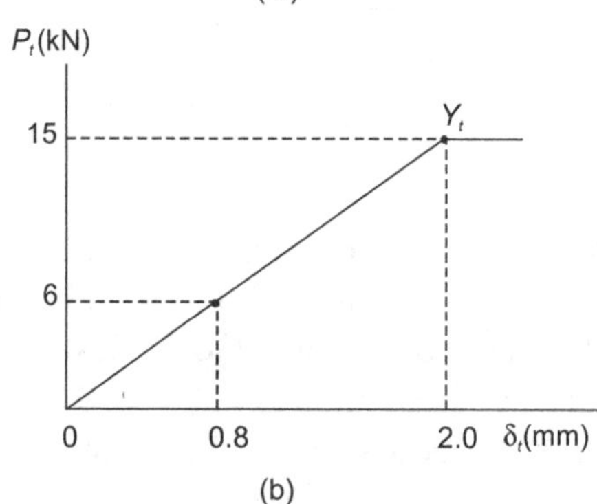

(b)

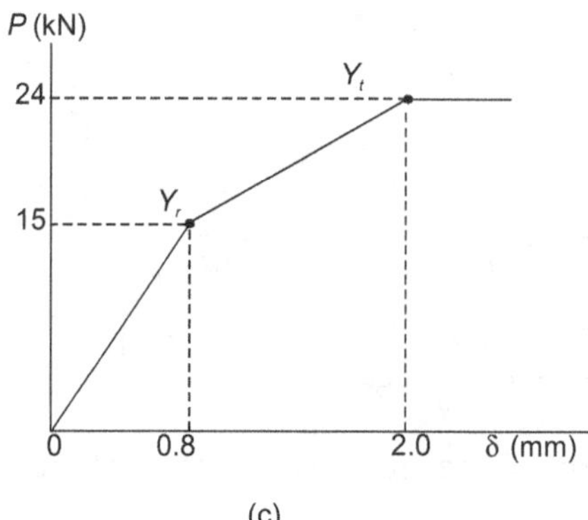

(c)

The load=deflection diagram of the tube alone is shown in Figure above. Observing that the load and deflection of the rod–tube combination are, respectively,

$$P = P_r + P_t \qquad\qquad \delta = \delta_r = \delta_t$$

we draw the required load-deflection diagram by adding the ordinates of the diagrams obtained for the rod and for the tube (Figure.). Points Y_r and Y_t correspond to the onset of yield in the rod and in the tube, respectively.

REVIEW QUESTIONS

SHORT QUESTIONS

1. Define 'Volumetric strain'.

2. Define Poisson's Ratio.

3. Define Resistance, proof resilience and modulus of resilience.

4. Calculate the force required for punching a hole of 10 mm diameter through a mild steel plate of 5 mm thick. Take the maximum shear strength of mild steel as 300 N/mm². Also find the compressive stress on the punch.

5. Define the terms proof resilience and modulus of resilience.

6. State Hooke's Law.

7. Define bulk modulus.

8. State the principle of superposition.

9. A rod of diameter 30 mm and length 400 mm was found to elongate 0.35mm when it was subjected to a load of 65 kN. Compute the modulus of elasticity of the material of this rod.

10. What is strain energy and write its unit in S.I. system?

11. Define the term: Elastic limit.

12. Give the relation between Young's modulus and bulk modulus

13. Define "Shear Stress".

14. Distinguish between rigid and deformable bodies.

15. A short bar of length 100 mm tapers uniformly from a diameter 40 mm to a diameter of 30 mm and carries an axial compressive load of 200 kN. Find the change in length of the bar.

16. Compare Strength, Stiffness and Stability.

17. State the relation between the modulus of elasticity, rigidity modulus, bulk modulus and Poisson's ratio.

LARGE QUESTIONS

1. Two vertical rods one of steel and other of copper are each rigidly fixed at the top and 600 mm apart. Diameters and lengths of the rods are 25 mm and 5 mm respectively. A cross bar fixed to the rods at the lower end carries a load of 7 kN such that the cross bar remains horizontal even after loading. Find the steps in each rod and the position of the load on the cross bar. Assume the modulus of elasticity for steel and copper as 200 kN/mm² and 100 kN/mm² respectively.

2. A cast iron flat 300 mm long and 30 mm (thickness) × 60 mm (width) uniform cross section, is acted upon by the following forces:

   30 kN tensile in the direction of the length

   360 kN tensile in the direction of the width

   240 kN tensile in the direction of the thickness.

   Calculate the direct strain, net strain in each direction and change in volume of the flat. Assume the modulus of elasticity and Poisson's ratio for cast iron as 140 kN/mm² and 0.25 respectively.

3. A steel bar 4 cm × 4 cm in section, 3 meters long is subjected to an axial pull of 128 kN. Taking E = 20 × 10¹⁰ N/m². Calculate the alteration in the length of the bar. Calculate also the amount of energy stored in the bar during extension.

4. A steel wire 6 mm diameter is used for lifting a load of 1.5 kN at its lowest end, the length of the wire hanging vertically being 160 meters. Taking the unit weigth of steel = 78 kN/m³ and E = 2 × 10⁵ N/mm², calculate the elongation of the wire.

5. i. In an experiment, a bar of 30 mm diameter is subjected to a pull of 60 kN. The measured extension on gauge length of 200 mm is 0.09 mm and the change in diameter is 0.0039 mm. Calculate the Pisson's ratio and the values of the three modulii.

   ii. An alloy specimen has modulus of elasticity of 120 GPa and modulus of rigidity of 45 GPa. Determine the Poisson's ratio of the material.

6. i. An alloy circular bar ABCD 3 m long is subjected to a tensile force of 50 kN as shown in figure. If the stress in the middle portion BC is not to exceed 150 MPa, then what should be its diameter? Also find the length of the middle portion, if the total extension of the bar should not exceed by 3 mm. Take E = 100 GPa.

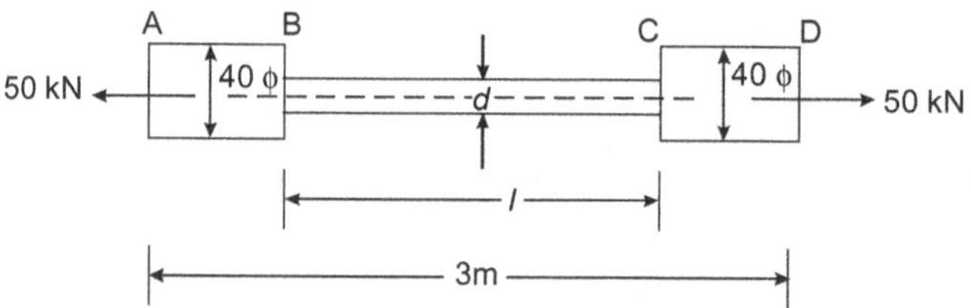

   ii. A circular bar rigidly fixed at its both ends uniformly tapers from 75 mm at one end to 50 mm at the other end. If its temperature is raised through 26 K, what will be the maximum stress developed in the bar. Take E as 200 GPa and a as 12 × 10⁻⁶/K for the bar material.

7. i. An aluminum cylinder of diameter 60 mm located inside a steel cylinder of internal diameter 60 mm and wall thickness 15 mm. The assembly is subjected to a compressive force of 200 kN. Determine the forces and stresses developed in steel and aluminum as 50 GPa.

   ii. Steel railroad 10 m long is laid with a clearance of 3 mm at a temperature of 15°C. At what temperature will the rails just touch? . What stress would be induced in the rails at that temperature if there were no initial clearance?. Assume coefficient of expansion, a = 11.7 × 10⁻⁶/°C and modulus of elasticity, E = 200 GPa.

8. i. A steel bar 250 mm long of section 50×50 mm is subjected to stresses of 100 N/mm² tensile along the axis and 30 N/mm² compressive along the sides. The increase in volume was observed to be 50 mm³. Determine the value of Poisson's ratio, modulus of rigidity and bulk modulus. Take E = 200 GPa.

   ii. A vertical rod 2 m long, fixed at the upper end, is 1500 mm² in area for 1 m and 2000 mm² in area for 1 m. A collar is attached to the free end. Through what height can a load of 100 kg fall on the collar to cause a maximum stress of 50 MPa? Take E = 200 GPa.

9. i. The ultimate stress, for a hollow steel column which carries an axial load of 1.9 MN is 480 N/mm². If the external diameter of the column is 200 mm, determine the internal diameter. Take factor of safety = 4.

   ii. Draw the stress-strain curve for mild steel subjected to tension and indicate the salient points.

10. i. Derive an expression for volumetric strain for a rectangular bar which is subjected to three mutually perpendicular tensile stresses.

   ii. A test element is subjected to three mutually perpendicular unequal stresses. Find the change in volume of the element, if the algebraic sum of these stresses is equal to zero.

11. The load P is applied on the bars as shown in Figure. Find the safe load P if the stresses in brass and steel are not to exceed 60 N/mm² and 120 N/mm² respectively. E for steel = 200 kN/mm²·, E for brass = 100 kN/mm². The copper rods are 40 mm × 40 mm in section and the steel rod is 50 mm × 50 mm in section. Length of steel rod is 250 mm and copper rod is 150 mm.

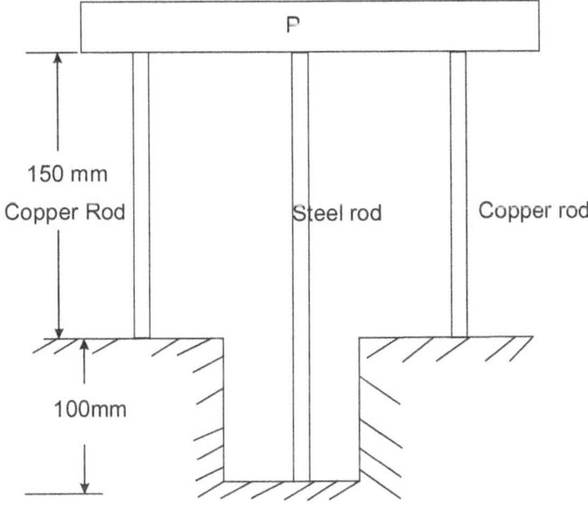

12. A steel bar ABCD 4 m long is subjected to forces as shown in Figure. Find the elongation of the bar. Take E for the steel as 200 GPa.

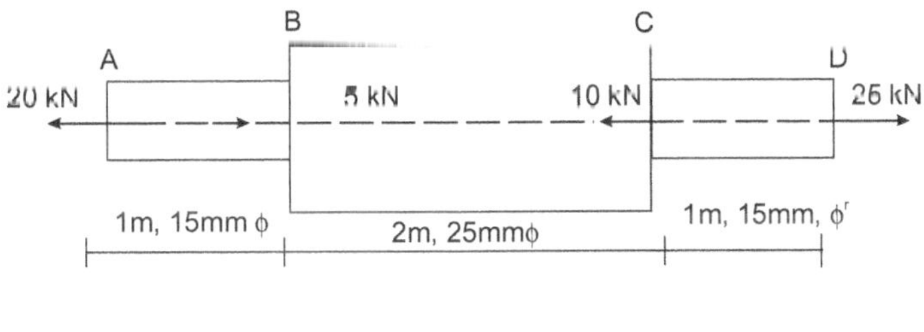

4

GENERAL THEORY OF ELASTICITY

DIFFERENTIAL EQUATION OF EQUILIBRIUM

In simple confingurations discussed earlier, concept of average stress across a section was developed and can work well for simple configurations. However for complex bodies, the stress distribution varies across sections as we saw in stress concentration problems. Therefore we need to consider the stress at a point in an elastic body (Figure 4.1). Let us consider now the variation of the stress as we change the position of the point. For this purpose the conditions of equilibrium of a small rectangular parallelepiped with the sides $\delta x, \delta y, \delta z$ (Figure 4.2) must be studied. The components of stresses acting on the sides of this small element and their positive directions are indicated in the Figure 4.2. Here we take into account the small changes of the components of stress due to the small increases $\delta x, \delta y, \delta z$ of the coordinates. Thus designating the mid-points of the sides of the element by, 1,2,3,4,5,6 as in Figure 4.2, we distinguish between the value of σ_x at point 1, and its value at point 2, writing these $(\sigma_x)_1$ and $(\sigma_x)_2$, respectively. The symbol σ_x itself denotes, of course, the value of this stress component at the point x, y, z. In calculating the forces acting on the element we consider the sides as very small, and the force is obtained by multiplying the stress at the centroid of a side by the corresponding area of this side.

If we let X, Y, Z called body force components per unit volume of the element, then the equation of equilibrium obtained by summing all the forces acting on the element in the x direction is

$$\left[(\sigma_x)_1 - (\sigma_x)_2\right]\delta y\delta z + \left[(\tau_{xy})_3 - (\tau_{xy})_4\right]\delta x\delta z$$
$$\left[(\tau_{xz})_5 - (\tau_{xz})_6\right]\delta x\delta y + X\delta x\delta y\delta z = 0 \tag{4.1}$$

Design for strength-basics of theory of elasticity. The points to be considered are

1. Material behaviour (Ductile or brittle)
2. Effect of loads
3. Internal resistance to external loads
4. Linear elastic deformations and following aspects are not discussed in this text.
   * Yielding

* ❀ Plastic flow

* ❀ Disproportionate increase in deformation

* ❀ These phases are absent for brittle materials

Effect of Loads

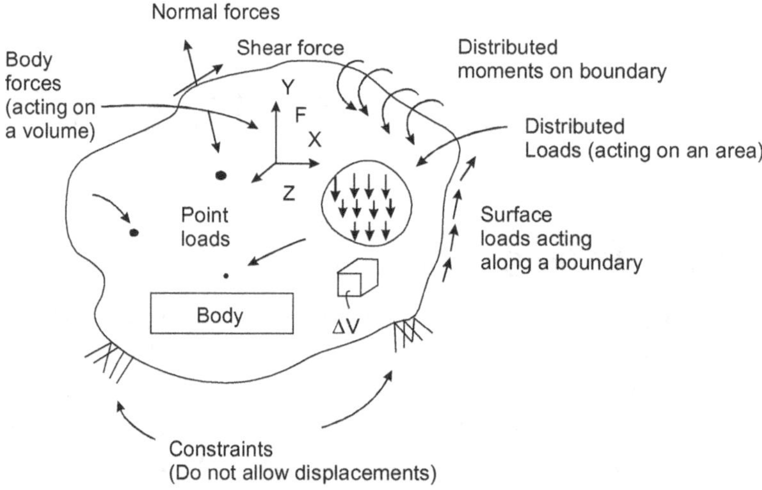

Figure 4.1 Loads acting on a body

A machine member, in general, is subjected to different types of loads.

Any point or location within the body resists the external loads by developing stresses which cause strains. Recall,

$$\text{Stress}, \sigma = \frac{\text{Force}}{\text{Area}}(\text{N/mm}^2)$$

$$\text{Strain}, \in = \frac{\text{Elongation}}{\text{Original length}}\left(\frac{\text{mm}}{\text{mm}}\right)$$

The two other equations of equilibrium are obtained in the same manner. After dividing by $\delta x, \delta y, \delta z$ and proceeding to the limit by shrinking the element down to the point x, y, z, we find,

$$\frac{\partial \sigma_x}{\partial x} + \frac{\partial \tau_{xy}}{\partial y} + \frac{\partial \tau_{xz}}{\partial z} + X = 0$$

$$\frac{\partial \sigma_y}{\partial y} + \frac{\partial \tau_{xy}}{\partial x} + \frac{\partial \tau_{yz}}{\partial z} + Y = 0 \qquad (4.2)$$

$$\frac{\partial \sigma_z}{\partial z} + \frac{\partial \tau_{xz}}{\partial x} + \frac{\partial \tau_{yz}}{\partial y} + Z = 0$$

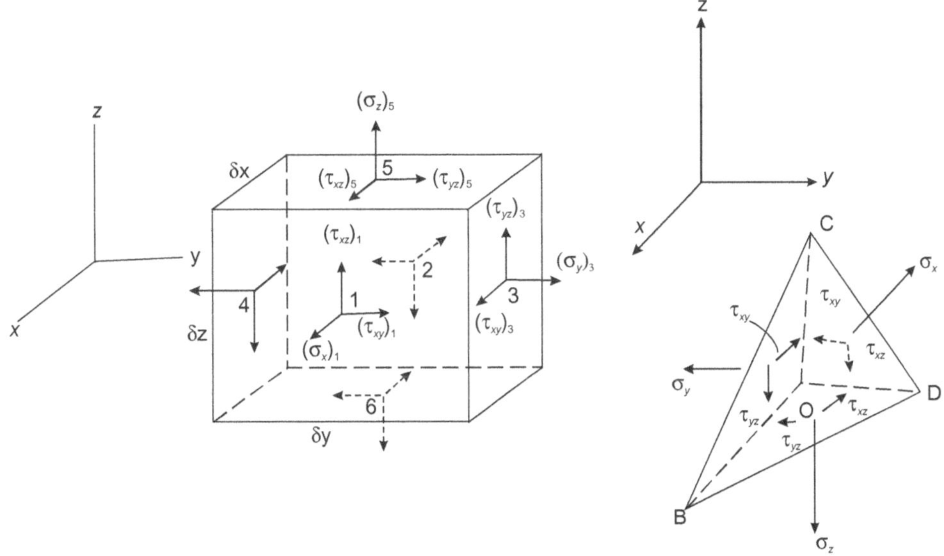

Figure 4.2

Equations (4.2) must be satisfied at all points throughout the volume of the body. The stresses vary over the volume of the body, and when we arrive at the surface they must be such as to be in equilibrium with the external forces on the surface of the body. (See tetrahedran Figure 4.2).

$$\overline{X} = \sigma_x l + \tau_{xy} m + \tau_{xz} n$$

$$\overline{Y} = \sigma_y m + \tau_{yz} n + \tau_{xy} l \qquad (4.3)$$

$$\overline{Z} = \sigma_z n + \tau_{xz} l + \tau_{yz} m$$

in which l, m, n are the direction cosines of the external normal to the surface of the body at the point under consideration, $\overline{X}, \overline{Y}, \overline{Z}$ are components of body force shown in the reference directions. If the problem is to determine the state of stress in a body subjected to the action of given forces, it is necessary to solve equation (4.2) and the solution must be such as to satisfy the boundary conditions, equation (4.3). These three equations, containing six components of stress, namely $\sigma_x, \ldots \tau_{yz}$, are not sufficient for the determination of these components. The problem is a statically indeterminate one, and in order to obtain the solution we must proceed as we did in the case of two-dimensional problems, i.e., the elastic deformations of the body must also be considered. Note that stress is a Tensor defined by its six components along the three directions.

CONDITIONS OF COMPATIBILITY

These conditions generate additional equations needed which consider deformation and strain. It should be noted that the six components of strain at each point are completely determined by the three functions u, v, w, representing the components of displacement, along the directions x, y, z.

Hence, the components of strain cannot be taken arbitrarily as functions of x, y,z but are subject to relations that follow.

Note that strain is also a Tensor defined by six components along the three direction.

$$\frac{\partial^2 \epsilon_x}{\partial y^2} = \frac{\partial^3 u}{\partial x \partial y^2} \qquad \frac{\partial^2 \epsilon_y}{\partial x^2} = \frac{\partial^3 v}{\partial x^2 \partial y} \qquad \frac{\partial^2 \gamma_{xy}}{\partial x \partial y} = \frac{\partial^3 u}{\partial x \partial y^2} + \frac{\partial^3 v}{\partial x^2 \partial y}$$

From which

$$\frac{\partial^2 \epsilon_x}{\partial y^2} + \frac{\partial^2 \epsilon_y}{\partial x^2} = \frac{\partial^2 \gamma_{xy}}{\partial x \partial y} \qquad\qquad (4.4)$$

Two more relations of the same kind can be obtained by cyclical interchange of the letters x, y, z.

From the derivatives

$$\frac{\partial^2 \epsilon_x}{\partial y \partial z} = \frac{\partial^3 u}{\partial x \partial y \partial z} \qquad \frac{\partial \gamma_{yz}}{\partial x} = \frac{\partial^2 v}{\partial x \partial z} + \frac{\partial^2 w}{\partial x \partial y}$$

$$\frac{\partial \gamma_{xz}}{\partial y} = \frac{\partial^2 u}{\partial y \partial z} + \frac{\partial^2 w}{\partial x \partial y} \qquad \frac{\partial \gamma_{xy}}{\partial z} = \frac{\partial^2 u}{\partial y \partial z} + \frac{\partial^2 v}{\partial x \partial z}$$

We find that

$$2\frac{\partial^2 \epsilon_x}{\partial y \partial z} = \frac{\partial}{\partial x}\left(-\frac{\partial \gamma_{yz}}{\partial x} + \frac{\partial \gamma_{xz}}{\partial y} + \frac{\partial \gamma_{xy}}{\partial z} \right) \qquad\qquad (4.5)$$

Two more relations of the kind equation (4.5) can be obtained by interchange of the letters x,y,z. We thus arrive at the following six differential relations between the components of strain, which must be satisfied by virtue of the above equations.

$$\frac{\partial^2 \epsilon_x}{\partial y^2} + \frac{\partial^2 \epsilon_y}{\partial x^2} = \frac{\partial^2 \gamma_{xy}}{\partial x \partial y} \qquad 2\frac{\partial^2 \epsilon_x}{\partial y \partial z} = \frac{\partial}{\partial x}\left(-\frac{\partial \gamma_{yz}}{\partial x} + \frac{\partial \gamma_{xz}}{\partial y} + \frac{\partial \gamma_{xy}}{\partial z} \right)$$

$$\frac{\partial^2 \epsilon_y}{\partial z^2} + \frac{\partial^2 \epsilon_z}{\partial y^2} = \frac{\partial^2 \gamma_{yz}}{\partial y \partial z} \qquad 2\frac{\partial^2 \epsilon_y}{\partial x \partial z} = \frac{\partial}{\partial y}\left(\frac{\partial \gamma_{yz}}{\partial x} - \frac{\partial \gamma_{xz}}{\partial y} + \frac{\partial \gamma_{xy}}{\partial z} \right) \qquad (4.6)$$

$$\frac{\partial^2 \epsilon_z}{\partial x^2} + \frac{\partial^2 \epsilon_x}{\partial z^2} = \frac{\partial^2 \gamma_{xz}}{\partial x \partial z} \qquad 2\frac{\partial^2 \epsilon_z}{\partial x \partial y} = \frac{\partial}{\partial y}\left(\frac{\partial \gamma_{yz}}{\partial x} + \frac{\partial \gamma_{xz}}{\partial y} - \frac{\partial \gamma_{xy}}{\partial z} \right)$$

These differential relations are called the conditions of compatibility (3No's)

Hooke's law (also called constitutive relations between stress and strain)

$$\in_x = \frac{1}{E}(\sigma_x - v\,\overline{\sigma_y + \sigma_x}),$$

$$\in_y = \frac{1}{E}(\sigma_y - v\,\overline{\sigma_x + \sigma_z}),$$ (4.7)

$$\in_z = \frac{1}{E}(\sigma_z - v\,\overline{\sigma_x + \sigma_y}),$$

By using Hooke's law equations (4.6) can be transformed into relations between the components of stress. Take, for instance, the conditions

$$\frac{\partial^2 \in_y}{\partial z^2} + \frac{\partial^2 \in_z}{\partial y^2} = \frac{\partial^2 \gamma_{yz}}{\partial y\,\partial z}$$ (4.8) [2nd equation set (4.6)]

From stress–strain, relations using the notation $\theta = \sigma_x + \sigma_y + \sigma_z$ and $e = \in_x + \in_y + \in_z$, we get

$$\in_y = \frac{1}{E}\left[(1+v)\sigma_v - v\theta\right]$$

$$\in_z = \frac{1}{E}\left[(1+v)\sigma_z - v\theta\right]$$ (4.9)

$$\gamma_{yz} = \frac{2(1+v)\tau_{yz}}{E} \qquad \left(e = \frac{1-2v_{nm}}{E}\theta\right)$$

Substituting these expressions in equation (4.8), we obtain

$$(l+v)\left(\frac{\partial^2 \sigma_y}{\partial z^2} + \frac{\partial^2 \sigma_z}{\partial y^2}\right) - v\left(\frac{\partial^2 \theta}{\partial z^2} + \frac{\partial^2 \theta}{\partial v^2}\right) = 2(1+v)\frac{\partial^2 \tau_{yz}}{\partial y\partial z}$$ (4.10)

The right side of this equation can be transformed by using the equations of equilibrium equation (4.2). From these equations we find

$$\frac{\partial \tau_{yz}}{\partial y} = -\frac{\partial \sigma_z}{\partial z} - \frac{\partial \tau_{xz}}{\partial x} - Z$$

$$\frac{\partial \tau_{yz}}{\partial z} = -\frac{\partial \sigma_y}{\partial y} - \frac{\partial \tau_{xy}}{\partial x} - Y$$ (4.11)

Differentiating the first of these equations with respect to z and the second with respect to y, and adding them together, we find

$$2\frac{\partial^2 \tau_{yz}}{\partial y\partial z} = -\frac{\partial^2 \sigma_z}{\partial z^2} - \frac{\partial^2 \sigma_y}{\partial y^2} - \frac{\partial}{\partial x}\left(\frac{\partial \tau_{xz}}{\partial z} + \frac{\partial \tau_{xy}}{\partial y}\right) - \frac{\partial Z}{\partial z} - \frac{\partial Y}{\partial y}$$ (4.12)

Or, by using the first of eqs. (4.2)

$$2\frac{\partial^2 \tau_{yz}}{\partial y \partial z} = -\frac{\partial^2 \sigma_x}{\partial x^2} - \frac{\partial^2 \sigma_y}{\partial y^2} - \frac{\partial^2 \sigma_z}{\partial z^2} + \frac{\partial X}{\partial x} - \frac{\partial Y}{\partial y} - \frac{\partial Z}{\partial z} \tag{4.13}$$

Substituting this in equation (4.10) and using, to simplify the writing, the symbol

$$\nabla^2 = \frac{\partial^2}{\partial x^2} + \frac{\partial^2}{\partial y^2} + \frac{\partial^2}{\partial z^2} \qquad (\nabla^2 = \text{Laplacian operator})$$

We find

$$(1+v)\left(\nabla^2\theta - \nabla^2\sigma_x - \frac{\partial^2\theta}{\partial x^2}\right) - v\left(\nabla^2\theta - \frac{\partial^2\theta}{\partial x^2}\right) = (1+v)\left(\frac{\partial X}{\partial x} - \frac{\partial Y}{\partial y} - \frac{\partial Z}{\partial z}\right) \tag{4.14}$$

Two analogous equations can be obtained from the two other conditions of compatibility of the type (4.8).

Adding together all three equations of the type (4.14) we find

$$(1-v)\nabla^2\theta = -(1+v)\left(\frac{\partial X}{\partial x} + \frac{\partial Y}{\partial y} + \frac{\partial Z}{\partial z}\right) \tag{4.15}$$

Substituting this expression for $\nabla^2\theta$ in equation 4.14,

$$\nabla^2\sigma_x + \frac{1}{1+v}\frac{\partial^2\theta}{\partial x^2} = -\frac{v}{1-v}\left(\frac{\partial X}{\partial x} + \frac{\partial Y}{\partial y} + \frac{\partial Z}{\partial z}\right) - 2\frac{\partial X}{\partial x} \tag{4.16}$$

We can obtain three equations of this kind, corresponding to the first three of equation (4.6). In the same manner the remaining three conditions (4.6) can be transformed into equations of the following kind:

$$\nabla^2\tau_{yz} + \frac{1}{1+v}\frac{\partial^2\theta}{\partial y \partial z} = -\left(\frac{\partial Z}{\partial y} + \frac{\partial Y}{\partial z}\right) \tag{4.17}$$

If there are no body forces or if the body forces are constant, equation (4.16) and (4.17) become

$$(1+v)\nabla^2\sigma_x + \frac{\partial^2\theta}{\partial x^2} = 0 \qquad (1+v)\nabla^2\tau_{yz} + \frac{\partial^2\theta}{\partial y \partial z} = 0$$

$$(1+v)\nabla^2\sigma_y + \frac{\partial^2\theta}{\partial y^2} = 0 \qquad (1+v)\nabla^2\tau_{xz} + \frac{\partial^2\theta}{\partial x \partial z} = 0 \tag{4.18}$$

$$(1+v)\nabla^2\sigma_z + \frac{\partial^2\theta}{\partial z^2} = 0 \qquad (1+v)\nabla^2\tau_{xy} + \frac{\partial^2\theta}{\partial x \partial y} = 0$$

We see that in addition to the equations of equilibrium (4.2) and the boundary conditions (4.3), the stress components in an isotropic body must satify the six conditions of compatibility (4.6) and (4.7) or the six conditions (4.18). This system of equations is generally sufficient for determining the stress components without ambiguity.

$$\gamma_{xy} = \frac{\partial u}{\partial y} + \frac{\partial v}{\partial x}; \gamma_{xz} = \frac{\partial u}{\partial z} + \frac{\partial w}{\partial x}$$

$$\gamma_{yz} = \frac{\partial v}{\partial y} + \frac{\partial w}{\partial y} \tag{4.19}$$

$$\epsilon_x = \frac{\partial u}{\partial x}; \epsilon_y = \frac{\partial v}{\partial y}; \epsilon_z = \frac{\partial w}{\partial z}$$

DETERMINATION OF DISPLACEMENTS

When the components of stress are found from the previous equations, the components of strain can be calculated by using Hooke's law. Then equations (4.19) are used for the determination of the displacements u, v, w. Differentiating equation (4.19) with respect to x, y, z we can obtain 18 equations containing 18 second derivatives of u, v, w, from which all these derivatives can be determined. For u, for instance, we obtain

$$\frac{\partial^2 u}{\partial x^2} = \frac{\partial \epsilon_x}{\partial x} \qquad \frac{\partial^2 u}{\partial y^2} = \frac{\partial \gamma_{xy}}{\partial y} - \frac{\partial \epsilon_y}{\partial x} \qquad \frac{\partial^2 u}{\partial z^2} = \frac{\partial \gamma_{xz}}{\partial z} - \frac{\partial \epsilon_z}{\partial x}$$

$$\frac{\partial^2 u}{\partial x \partial y} = \frac{\partial \epsilon_x}{\partial y} \qquad \frac{\partial^2 u}{\partial x \partial z} = \frac{\partial \epsilon_x}{\partial z} \qquad \frac{\partial^2 u}{\partial y \partial z} = \frac{1}{2}\left(\frac{\partial \gamma_{xz}}{\partial y} + \frac{\partial \gamma_{xy}}{\partial z} - \frac{\partial \gamma_{yz}}{\partial x} \right) \tag{4.20}$$

The second derivatives for the two other components of displacement v and w can be obtained by cyclical interchange in equation (4.20) of the letters x, y, z.

Now u, v, w can be obtained by double integration of these second derivatives. The introduction of arbitrary constants of integration will result in adding to the values of u, v, w linear functions in x, y, z, as it is evident that such functions can be added to u, v, w without affecting such equations as (4.20). To have the strain components (4.19) unchanged by such an addition, the additional linear functions must have the form

$$\left. \begin{array}{l} u' = a + hy - cz \\ v' = d - bx + ez \\ w' = f + cx - ey \end{array} \right\} \tag{4.21}$$

This means that the displacements are not entirely determined by the stresses and strains.

EQUATIONS OF EQUILIBRIUM IN TERMS OF DISPLACEMENTS

One method of solution of the problems of elasticity is to eliminate the stress components from equation (4.2) and (4.3) by using Hooke's law and to express the strain components in terms of displacements using equation (4.19), we arrive at three equations of equilibrium containing only the three unknown functions u, v, w.

$$\sigma_x = \lambda e + 2G \frac{\partial u}{\partial x} \tag{4.22}$$

$$\tau_{xy} = G\gamma_{xy} = G\left(\frac{\partial u}{\partial y} + \frac{\partial v}{\partial x}\right)$$

$$\tau_{xz} = G\gamma_{xy} = G\left(\frac{\partial w}{\partial x} + \frac{\partial u}{\partial z}\right) \tag{4.23}$$

where
$$\lambda = vE / (1+v)(1-2v)$$

We get,

$$(\lambda + G)\frac{\partial e}{\partial x} + G\left(\frac{\partial^2 u}{\partial x^2} + \frac{\partial^2 u}{\partial y^2} + \frac{\partial^2 u}{\partial z^2}\right) + X = 0 \tag{4.24}$$

The two other equations can be transformed in the same manner. Then, using the symbol ∇^2 the equations of equilibrium (2.2) become

$$(\lambda + G)\frac{\partial e}{\partial x} + G\nabla^2 u + X = 0$$

$$(\lambda + G)\frac{\partial e}{\partial y} + G\nabla^2 v + Y = 0 \tag{4.25}$$

$$(\lambda + G)\frac{\partial e}{\partial z} + G\nabla^2 w + Z = 0$$

and, when there are no body forces,

$$(\lambda + G)\frac{\partial e}{\partial x} + G\nabla^2 u = 0$$

$$(\lambda + G)\frac{\partial e}{\partial y} + G\nabla^2 v = 0 \tag{4.26}$$

$$(\lambda + G)\frac{\partial e}{\partial z} + G\nabla^2 w = 0$$

GENERAL SOLUTION FOR THE DISPLACEMENTS

It is easily verified by substitution that the differential equations (4.26) of equilibrium in terms of displacement are satisfied by

$$u = \phi_1 - \alpha \frac{\partial}{\partial x}(\phi_0 + x\phi_1 + y\phi_2 + z\phi_3)$$

$$v = \phi_2 - \alpha \frac{\partial}{\partial y}(\phi_0 + x\phi_1 + y\phi_2 + z\phi_3)$$

$$w = \phi_3 - \alpha \frac{\partial}{\partial z}(\phi_0 + x\phi_1 + y\phi_2 + z\phi_3)$$

where $4\alpha = 1/(1-v)$ and the four function $\phi_0, \phi_1, \phi_2, \phi_3$ are harmonic, i.e., (They satisfy Lapalace equation)

$$\nabla^2\phi_0 = 0; \quad \nabla^2\phi_1 = 0; \quad \nabla^2\phi_2 = 0; \quad \nabla^2\phi_3 = 0$$

It can be shown that this solution is general, even when ϕ_0 is omitted.

PRINCIPAL STRESSES, PRINCIPAL DIRECTIONS AND STRESS INVARIANTS HAVING 6 COMPONENTS OF STRESS IN THREE DIRECTIONS

The stress tensor $[\sigma]$ can be transformed in a different set of mutually perpendicullar planes, where the stress vector has the same directions as the normal to the plane. As a consequence, the shearing stress vanishes. The stress in those parincipl planes are the principal stresses and their normals are the principal directions of the stress state.

Let us consider a principal plane. The stress acting on it has only the normal component σ, so that the components of the stres vector are $X_x = l\sigma, Y_y = m\sigma, Z_z = n\sigma$. Substituting these values in expression 4.3, we get the homogenous system of linear equations, (l, m, n are direction cosines)

$$\begin{pmatrix} \sigma_x - \sigma & \tau_{xy} & \tau_{xz} \\ \tau_{xy} & \sigma_{y-\sigma} & \tau_{yz} \\ \tau_{xz} & \tau_{yz} & \sigma_{z-\sigma} \end{pmatrix} \begin{Bmatrix} l \\ m \\ n \end{Bmatrix} = \begin{Bmatrix} 0 \\ 0 \\ 0 \end{Bmatrix} \tag{4.27}$$

Such a system of equations has the trivial solution $l = m = n = 0$, and has other non-zero solutions only, if the determinant of the system matrix, $[C]$, vanishes. The direction cosines l, m, and n cannot be zero simultaneously, since they are the components of a unit vector. Thus, the second posssibility (zero determinant) must yield, as expressed by the condition.

$$\begin{vmatrix} \sigma_x - \sigma & \tau_{xy} & \tau_{xz} \\ \tau_{xy} & \sigma_y - \sigma & \tau_{yz} \\ \tau_{xz} & \tau_{yz} & \sigma_z - \sigma \end{vmatrix} = -\sigma^3 + I_1\sigma^2 - I_2\sigma + I_3 = 0 \tag{4.28}$$

In this expression the quantities, I_1, I_2, and I_3 take the values,

$$I_1 = \sigma_x + \sigma_y + \sigma_z$$

$$I_2 = \begin{vmatrix} \sigma_x & \tau_{xy} \\ \tau_{xy} & \sigma_y \end{vmatrix} + \begin{vmatrix} \sigma_x & \tau_{xy} \\ \tau_{xy} & \sigma_y \end{vmatrix} + \begin{vmatrix} \sigma_y & \tau_{yz} \\ \tau_{yz} & \sigma_z \end{vmatrix}$$

$$= \sigma_x \sigma_y + \sigma_x \sigma_z + \sigma_y \sigma_z - \tau_{xy}^2 - \tau_{xz}^2 - \tau_{yz}^2$$

$$I_3 = \begin{vmatrix} \sigma_x & \tau_{xy} & \tau_{xz} \\ \tau_{xy} & \sigma_y & \tau_{yz} \\ \tau_{xz} & \tau_{yz} & \sigma_z \end{vmatrix} = \sigma_x \sigma_y \sigma_z + 2\tau_{xy}\tau_{xz}\tau_{yz} - \sigma\tau_{yz}^2 - \sigma\tau_{xz}^2 - \sigma\tau_{xy}^2$$

The roots of equation (4.28) are the stresses, which satisfy equation (4.27), with non-simultaneous zero direction cosines l, m and n. They represent the normal stresses in faces, where the shearing stress is zero, which means that they are principal stresses. The direction cosines of the normals to these facets—the principal directions—may be computed by substituting σ in expression 4.27 for one of the roots of equation (4.28) and considering the supplementary condition $l^2 + m^2 + n^2 = 1$, since, with that substitution, equations (4.27) become linearly dependent $(|C| = 0)$. Usually the principal stresses are denoted by σ_1, σ_2, and σ_3 with $\sigma_1 > \sigma_2 > \sigma_3$.

The roots of equation (4.28) must not vary when the reference system is rotated, since they represent the principal stresses, which are intrinsic values of stress state and therefore must not depend on the particular reference system used to describe the stress tensor. For this reason, equation (4.28) designated as the characteristic equation of the stress tensor. The roots of this equation will be independent of the reference system if the coefficients I_1, I_2, I_3 are insensitive to coordinate changes. These coefficients are therefore invariants of the stress tensor.

SUMMARY

In the most general case of a loaded three-dimensional solid, a small volume Δv scooped out from the solid, will have the stress picture as shown below (enlarged)

- Equations of equilibrium in x, y, z directions are obtained by considering $\Delta v \rightarrow 2$ a point.
- As both normal and shear stresses (three normal, 3 shear) vary with locations, these are different on opposite faces by incremental values.
- Therefore, three equations (all partial differential equations) of equilibrium equation (4.2).
- Similarly by considering three displacements u, v and w as continuous variables (no discontinuity exists as material is considered homogeneous) and the strains (3-normal, 3-shear),6 equations (compatibility of displacement and strains) are obtained.

Result is total 9 equations but 15 unknowns (6 stresses, 6 strains, 3 displacements)

Balance 6 equations are obtained from Hooke's Law $(\sigma \, \alpha \in)$

Relating 6 stresses with 6 strains through elastic constants;

✺ It may appear that there are $6 \times 6 = 36$ elastic constants, to form the above relations.

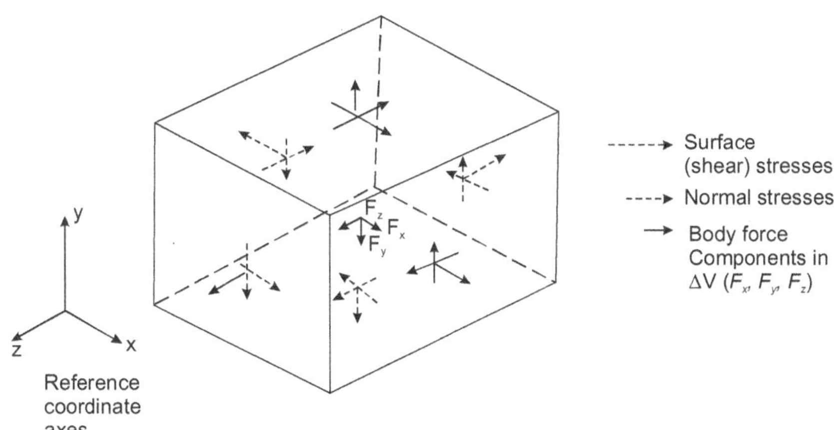

Figure 4.3 Free body diagram of small element Δv

✺ By considerations of (1) complementary nature of shear stresses, (2) Normal and shear stresses are related only to normal and shear strains, (3) isotropy of the material.

The 36 constants, reduce finally to 2 elastic constants only. These are

✺ Young's modulus $\left(E - \text{units} \dfrac{N}{\text{mm}^2} \right)$

✺ Poisson's ratio (ν – dimensionless)

✺ Very familiar and used extensively in almost all designs and followed in design and data hand books with values given for most commonly used materials ($v = 0.3$). To recapulate the stress–strain relation are,

$$
\text{Normal stresses} \atop \text{(Tensile or compressive)}
\left\{
\begin{aligned}
\sigma_{xx} &= \frac{E}{(1+v)(1-2v)}\left\{(1-v)\in_{xx} + v(\in_{yy} + \in_{zz})\right\} \\
\sigma_{yy} &= \frac{E}{(1+v)(1-2v)}\left\{(1-v)\in_{yy} + v(\in_{zz} + \in_{xx})\right\} \\
\sigma_{zz} &= \frac{E}{(1+v)(1-2v)}\left\{(1-v)\in_{zz} + v(\in_{xx} + \in_{yy})\right\}
\end{aligned}
\right\}
\text{Normal strain} \atop \text{(elongation or compressions)}
$$

$$\text{Shear stresses} \quad \left\{ \begin{array}{l} \sigma_{xy}(=\tau_{xy})=G\gamma_{xy}=\dfrac{E}{2(1+v)}\gamma_{xy} \\[2ex] \sigma_{yz}(=\tau_{yz})=G\gamma_{yz}=\dfrac{E}{2(1+v)}\gamma_{yz} \\[2ex] \sigma_{zx}(=\tau_{zx})=G\gamma_{zx}=\dfrac{E}{2(1+v)}\gamma_{zx} \end{array} \right\} \quad \begin{array}{l} \text{Shear strains} \\ \text{(Rotation or} \\ \text{distortions)} \end{array}$$

$$G = \text{Shear or rigidity modulus} = \frac{E}{2(1+v)}$$

The cause–effect relation is

- Normal stresses produce change in volume

- Shear stresses cause change in shape

Rigorous theory of elasticity solutions to relatively simple engineering problems in 3-D even based on assumptions like

   i. Small deformations, small strains

   ii. Linear elastic behaviour

   iii. Homogeneous material

   iv. Isotropy

   v. Regular and well-defined geometries of body shapes and cross sections

   vi. Simple loadings

   vii. Idealized constraints

are quite laborious and complex and need a computer to numerically handle the 15 partial differential equations (3-equil; 6-compatibility; 6-stress–strain) in 15 unknowns (3 – displ; 6-stresses; 6-strains). Therefore, without much affecting the physical principles, the rigorousness of theory of elasticity is relaxed (with complete knowledge of implications of such relaxations and limitations they impose on validity of solutions), so that design engineers can use simplified equations for design calculations. The simplified treatment is presented commonly under strength of materials.

INTERESTING POINTS TO PONDER

ANISOTROPY, ISOTROPY AND ORTHOTROPY

In 3D Theory of Elasticity we explained that stress or strain at a point is described by a column vector of 6 quantites namely 3 normal stresses (and strains) and 3 shear stresses (and strains). When it came to stress-strain relations, in the most general case, it is described by a 6*6 matrix of elastic constants. Due to the symmetry of the matrix elements about the diagonal, the number of elastic constants can be reduced to 21. Thic can further be reduced to only 2 namely E, v if the material is isotropic, that is if the material's elastic properties are same in all directions.

However, due to certain manufacturing processes such as rolling of sheet metals, there exists a certain orientation of the crystals in the material and, the elastic properties of the metal become direction dependent. Such materials are called Anisotropic.

As a special case, like in plywoods, the properties are different along the ply and across the ply. Only in two specified perpendicular directions the properties vary. Such materials are called orthotropic materials. Even a plate stiffened along the length by longitudinal stiffeners can be considered to be orthotropic, since E can be construcd as different along the direction of stiffeners and its perpendicular dircction.

In case of fibre reinforced composites, the fibres are randomly oriented and we have an anisotropic behaviour of the material.

5

PRINCIPAL STRESSES AND STRAINS

In the earlier chapter 3, we saw how a normal force at a particular section can produce components of direct normal and shear stresses in a different plane. To recaptulate,

STRESSES IN A TENSILE MEMBER

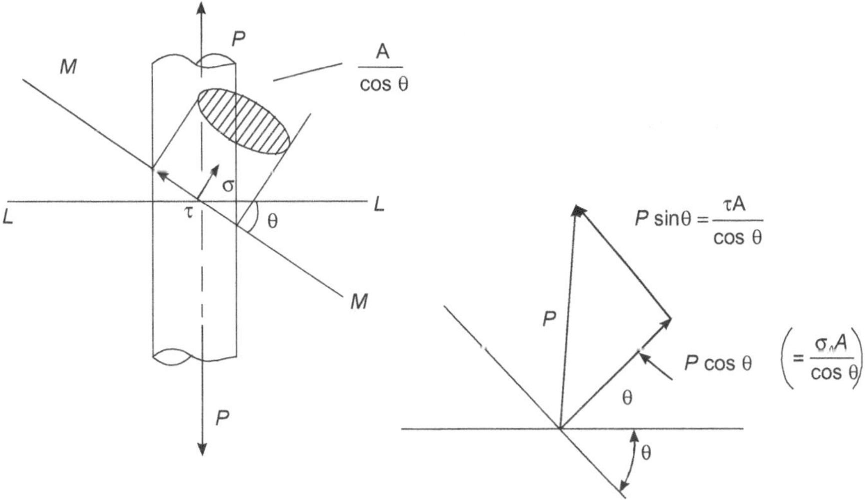

Figure 5.1 Stress in a tensile member

When a bar of uniform sectional area A is subjected to an axial load P, then the stress acting on a cross-section is given by LL normal to the axis p/A. Considering another section given by the plane MM inclined at θ to LL, the area cut by the plane is $\dfrac{A}{\cos\theta}$. Let the normal stress across MM be σ_n.

Resolving perpendicular to MM,

$$\sigma_n \cdot \frac{A}{\cos\theta} = P\cos\theta$$

$$\therefore \sigma_n = \frac{P}{A}\cos^2\theta$$

Further there will be a shearing stress of τ acting parallel to *MM* and resolving in this direction

$$\tau . \frac{A}{\cos\theta} = P \sin\theta$$

$$\tau = \frac{P}{A} \sin\theta \cos\theta$$

This means that when a rod is subjected to pure tension, both tensile and shearing stresses are produced. In a material under direct compression, the corresponding stresses would be compressive and shearing. It is possible that the shearing stress produced may be more important than the applied stress. The greatest shearing stress may be calculated as follows:

Since $\tau = \dfrac{P}{2A} \sin\theta \cos\theta = \dfrac{P}{A} \dfrac{\sin 2\theta}{2}$

The greatest value is when $\sin 2\theta = 1$ of $\theta = 45°$

$\therefore$ The greatest shearing stress produced $\tau_{max} = \dfrac{P}{2A}$

Let us consider different cases when stress combinations occur in planes of section of loaded material or component.

TWO MUTUALLY PERPENDICULAR NORMAL STRESSES

At any point in a material where stress is acting, it is possible to assume that the point consists of a very small triangular block, such that the stresses act across the faces of the block.

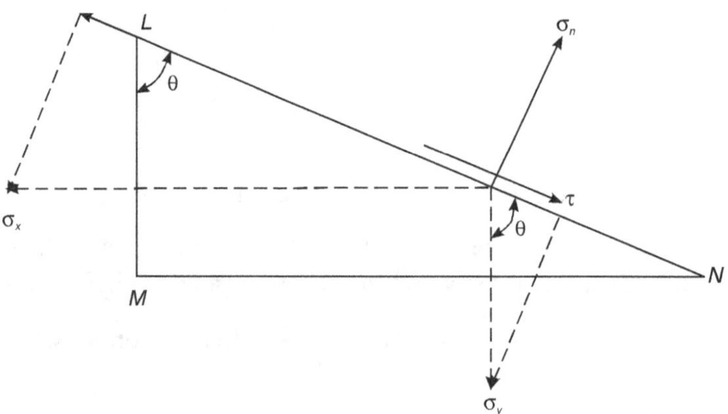

Figure 5.2

Consider that direct stresses σ_x and σ_y act across the faces *LM* and *MN* and that the block has unit depth perpendicular to *LMN*. Let the stresses τ and σ_n act on the same plane at an angle θ to *LM*.

Resolving normal to *LN*:

$$\sigma_n \times LN = \sigma_x \times LM \cos\theta + \sigma_y \, MN \sin\theta$$

$$\sigma_n = \sigma_x \frac{LM}{LN}\cos\theta + \sigma_y \frac{MN}{LN}\sin\theta$$

$$= \sigma_x \times \cos^2\theta + \sigma_y \sin^2\theta$$

$$= \frac{\sigma_x}{2}\times 2\cos^2\theta + \frac{\sigma_y}{2}\times 2\sin^2\theta$$

$$= \frac{\sigma_x}{2}(1-\sin^2\theta+\cos^2\theta)+\frac{\sigma_y}{2}(1-\cos^2\theta+\sin^2\theta)$$

$$= \frac{\sigma_x+\sigma_y}{2}+\sigma_x\left[\frac{\cos^2\theta-\sin^2\theta}{2}\right]-\sigma_y\left[\frac{\cos^2\theta-\sin^2\theta}{2}\right]$$

$$= \frac{\sigma_x+\sigma_y}{2}+\frac{\sigma_x-\sigma_y}{2}\cos 2\theta$$

When $\theta = 0$, then $\sigma_n = \dfrac{\sigma_x+\sigma_y}{2}+\dfrac{\sigma_x-\sigma_y}{2}=\sigma_x$

and when $\theta = \dfrac{\pi}{2}$, then $\sigma_n = \dfrac{\sigma_x+\sigma_y}{2}-\dfrac{\sigma_x-\sigma_y}{2}=\sigma_y$

Resolving parallel to *LN*

$$\tau \times LN = \sigma_x \times LM \sin\theta - \sigma_y \times MN \cos\theta$$

$$\tau = \sigma_x \times \frac{LM}{LN}\sin\theta - \sigma_y \times \frac{MN}{LN}\cos\theta$$

$$= \sigma_x \cos\theta \sin\theta - \sigma_y \sin\theta \cos\theta$$

$$= (\sigma_x-\sigma_y)\sin\theta \cos\theta$$

$$= \frac{\sigma_x-\sigma_y}{2}\sin 2\theta$$

The maximum value of τ occurs when $2\theta = \dfrac{\pi}{2}$ or $\theta = \dfrac{\pi}{4}$ and then

$$\tau_{max} = \frac{\sigma_x-\sigma_y}{2}$$

The resultant stress $\sigma_r = \sqrt{\sigma_n^2+\tau^2}$

$\tan\phi = \dfrac{\tau}{\sigma_n}$ where ϕ is the angle which the resultant stress makes with the normal to the plane

and is called obliquity.

TWO-DIMENSIONAL STRESS SYSTEM—GENEAL

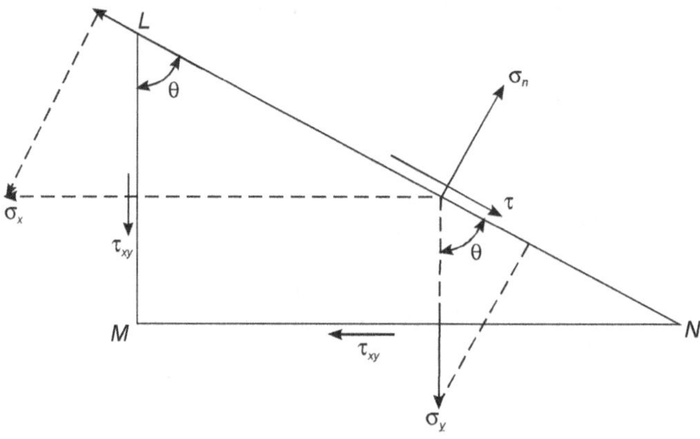

Figure 5.3

Consider the stress acting on a small triangular block (taken as a point in the limiting case). A general stress system will consist of normal and shearing stresses acting across the faces of the block. Consider some arbitray plane *LN* at angle θ to the plane of the stress.

Resolving normal to *LN*

$$\sigma_n \times LN = \tau_{xy} LM \sin\theta + \sigma_x LM \cos\theta + \tau_{xy} MN \cos\theta + \sigma_y MN \sin\theta$$

$$\sigma_n = \tau_{xy} \frac{LM}{LN} \sin\theta + \sigma_x \frac{LM}{LN} \cos\theta + \tau_{xy} \frac{MN}{LN} \cos\theta + \sigma_y \frac{MN}{LN} \sin\theta$$

$$= \tau_{xy} \cos\theta \sin\theta + \sigma_x \cos^2\theta + \tau_{xy} \sin\theta \cos\theta + \sigma_y \sin^2\theta$$

$$= 2\tau_{xy} \sin\theta \cos\theta + \sigma_x \cos^2\theta + \sigma_y \sin^2\theta$$

$$\sigma_n = \tau_{xy} \sin 2\theta + \frac{\sigma_x + \sigma_y}{2} + \frac{\sigma_x - \sigma_y}{2} \cos 2\theta$$

Resolving along *LN*

$$\tau \times LN = \tau_{xy} \times MN \sin\theta + \sigma_x \times LM \sin\theta - \tau_{xy} + LM \cos\theta - \sigma_y \times MN \cos\theta$$

$$\tau = \tau_{xy} \frac{MN}{LN} \sin\theta + \sigma_x \frac{LM}{LN} \sin\theta - \tau_{xy} \frac{LM}{LN} \cos\theta - \sigma_y \frac{MN}{LN} \cos\theta$$

$$= \tau_{xy} \sin\theta \sin\theta + \sigma_x \cos\theta \sin\theta - \tau_{xy} \cos\theta \cos\theta - \sigma_y \sin\theta \cos\theta$$

$$= \tau_{xy} \sin^2\theta + \sigma_x \sin\theta \cos\theta \cdot \tau_{xy} \cos^2\theta - \sigma_y \sin\theta \cos\theta$$

$$= \tau_{xy} (\sin^2\theta - \cos^2\theta) + \left(\frac{\sigma_x - \sigma_y}{2} \right) \sin 2\theta$$

$$\tau = \left[\frac{\sigma_x - \sigma_y}{2} \right] \sin 2\theta - \tau_{xy} \cos 2\theta$$

PRINCIPAL PLANE AND PRINCIPAL STRESSES

A body is subjected to stresses in one plane or in different planes. We can always find three mutually perpendicular planes along which the stresses at a certain body can be resolved completely into stresses normal to these planes.

These planes which pass through the point in such a manner that the resultant stress across them is totally a normal stress are known as "principal planes" and normal stresses across these planes are termed as "principal stresses". The plane carrying the maximum normal stress is called the major principal plane and the corresponding stress the major principal stress. The plane carrying the minimum normal stress is known as minor principal plane and the corresponding stress as minor principal stress.

In the two-dimensional stress system, let us find out the principal stresses. For this, the maximum and minimum values σ_n of must be obtained.

$$\sigma_n = \tau_{xy} \sin 2\theta + \left(\frac{\sigma_x + \sigma_y}{2}\right) + \left(\frac{\sigma_x - \sigma_y}{2}\right)\cos 2\theta$$

$$\frac{d\sigma_n}{d\theta} = 2\tau_{xy} \cos 2\theta - 2\frac{(\sigma_x - \sigma_y)}{2}\sin 2\theta = 0$$

$$\left(\frac{\sigma_x - \sigma_y}{2}\right)\sin 2\theta - \tau_{xy} \cos 2\theta = 0$$

σ_n is maximum or minimum at this angle θ which defines the orientation of principal planes where there is no shear stress acting. Conversely principal stresses can be computed by considering planes with zero shear stress.

Also

$$\left(\frac{\sigma_x - \sigma_y}{2}\right)\sin 2\theta = \tau_{xy} \cos 2\theta$$

$$\text{or} \tan 2\theta = \frac{2\tau_{xy}}{\sigma_x - \sigma_y}$$

Let us now examine, where exactly maximum shear stress exists.

Differentiating the expression for τ w.r.t θ

$$\frac{d\tau}{d\theta} = \left(\frac{\sigma_x - \sigma_y}{2}\right)2\cos 2\theta + \tau_{xy} \sin 2\theta \times 2 = 0$$

$$(\sigma_x - \sigma_y)\cos 2\theta = -2\tau_{xy} \sin 2\theta$$

$$\tan 2\theta = \frac{-(\sigma_x - \sigma_y)}{2\tau_{xy}}$$

Maximum τ occurs on this plane, the inclination of which is given by the above equation.

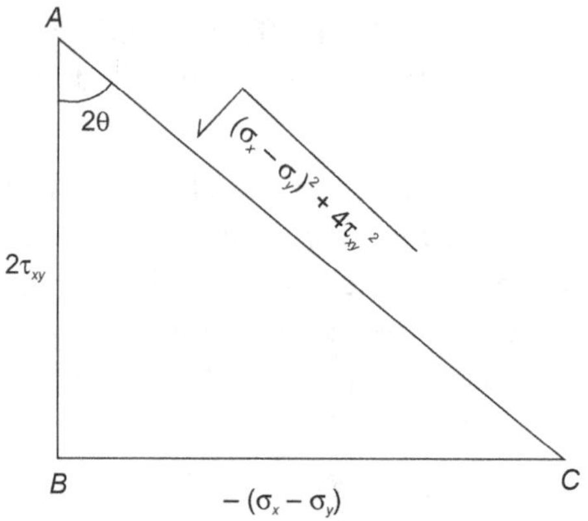

Figure 5.4

$$\tan 2\theta = \frac{-(\sigma_x - \sigma_y)}{2\tau_{xy}}$$

$$\sin 2\theta = \frac{-(\sigma_x - \sigma_y)}{\sqrt{(\sigma_x - \sigma_y)^2 + 4\tau_{xy}^2}}$$

$$\cos 2\theta = \frac{2\tau_{xy}}{\sqrt{(\sigma_x - \sigma_y)^2 + 4\tau_{xy}^2}}$$

$$\tau = \left(\frac{\sigma_x - \sigma_y}{2}\right)\sin 2\theta - \tau_{xy}\cos 2\theta$$

τ will obtain maximum value if $\sin 2\theta$ and $\cos 2\theta$ are duly substituted the values corresponding to

$$\tan 2\theta = \frac{-(\sigma_x - \sigma_y)}{2\tau_{xy}}$$

Substituting,

$$\tau_{max} = \left(\frac{\sigma_x - \sigma_y}{2}\right) - \left(\frac{\sigma_x - \sigma_y}{\sqrt{(\sigma_x - \sigma_y)^2 + \sqrt{4\tau_{xy}^2}}}\right)$$

$$= -\tau_{xy}\frac{2\tau_{xy}}{\sqrt{(\sigma_x - \sigma_y)^2 + \sqrt{4\tau_{xy}^2}}}$$

$$= \frac{-\dfrac{(\sigma_x - \sigma_y)^2}{2} - 2\tau_{xy}^2}{\sqrt{(\sigma_x - \sigma_y)^2 + 4\tau_{xy}^2}}$$

$$= -\frac{1}{2} \frac{(\sigma_x - \sigma_y)^2 + 4\tau_{xy}^2}{\sqrt{(\sigma_x - \sigma_y)^2 + 4\tau_{xy}^2}}$$

$$\tau_{max} = -\frac{1}{2}\left[\sqrt{(\sigma_x - \sigma_y)^2 + 4\tau_{xy}^2}\right]$$

Similarly consider the plane where principal stresses occur, the inclination is given by

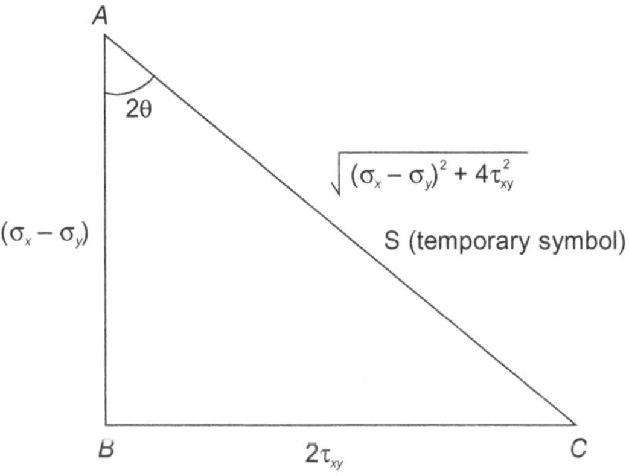

Figure 5.5

$$\tan 2\theta = \frac{2\tau_{xy}}{(\sigma_x - \sigma_y)}$$

$$\sin 2\theta = \frac{2\tau_{xy}}{\sqrt{(\sigma_x - \sigma_y)^2 + 4\tau_{xy}^2}}$$

$$\cos 2\theta = \frac{(\sigma_x - \sigma_y)}{\sqrt{(\sigma_x - \sigma_y)^2 + 4\tau_{xy}^2}}$$

Substituting these values in σ_n derived earlier, $S^2 = (\sigma_x - \sigma_y)^2 + 4\tau_{xy}^2$

$$\sigma_n = \tau_{xy} \cdot \frac{2\tau_{xy}}{S} + \frac{\sigma_x + \sigma_y}{2} + \frac{\sigma_x - \sigma_y}{2} \cdot \frac{(\sigma_x - \sigma_y)}{S}$$

$$= \frac{\sigma_x + \sigma_y}{2} + \frac{(\sigma_x - \sigma_y)^2}{2S} + \frac{2\tau_{xy}^2}{S}$$

$$-\frac{1}{2}\left\{(\sigma_x + \sigma_y) + \frac{(\sigma_x - \sigma_y)^2 + 4\tau_{xy}^2}{S}\right\} - \frac{1}{2}\left[(\sigma_x + \sigma_y) + S\right]$$

$$= \frac{1}{2}\left[(\sigma_x + \sigma_y)\right] + \frac{1}{2}\sqrt{(\sigma_x - \sigma_y)^2 + 4\tau_{xy}^2}$$

In such a plane the shear stresses are zero.

If $\tau = 0$, $\tan 2\theta = \dfrac{2\tau_{xy}}{(\sigma_x - \sigma_y)}$

Note $\;\;$ $\mathrm{Tan}\, 2\theta$ is +ve in the 1st quadrant and IIIrd quadrant. Therefore, $\sin 2\theta$ and $\cos 2\theta$ in third quadrant are negative. If we substitute the –ve values for $\sin 2\theta$ and $\cos 2\theta$, we get

$$\sigma_n = \frac{\sigma_x + \sigma_y}{2} - \frac{1}{2}\sqrt{(\sigma_x - \sigma_y)^2 + 4\tau_{xy}^2}$$

The two different signs $+$ and $-$ in the expressions for normal stresses indicate major principal and minor principal stresses. These are given the designation

So,

$$\sigma_{1,2} = \left(\frac{\sigma_x + \sigma_y}{2}\right) \pm \sqrt{\left(\frac{\sigma_x + \sigma_y}{2}\right)^2 + \tau_{xy}^2}$$

If we add, σ_1 and σ_2

$$\sigma_1 + \sigma_2 = \sigma_x + \sigma_y$$

If we subtract

$$\sigma_1 - \sigma_2 = 2\sqrt{\left(\frac{\sigma_x - \sigma_y}{2}\right) + \tau_{xy}^2}$$

$$= \sqrt{(\sigma_x - \sigma_y)^2 + 4\tau_{xy}^2}$$

$$= 2\tau_{max}$$

$$\text{or } \tau_{max} = \left(\frac{\sigma_1 - \sigma_2}{2}\right) \qquad \left(\tan 2\theta = \frac{2\tau_{xy}}{\sigma_x - \sigma_y}\right)$$

GRAPHICAL METHODS

Mohr's Circle

A German scientist Otto Mohr devised a graphical method for finding out the normal and shear stresses on any interface of an element when it is subjected to two perpendicular stresses (see Figure 5.6(a). Method is explained as follows.

Mohr's circle construction for like stresses

Steps of construction (see Figure 5.6(b))

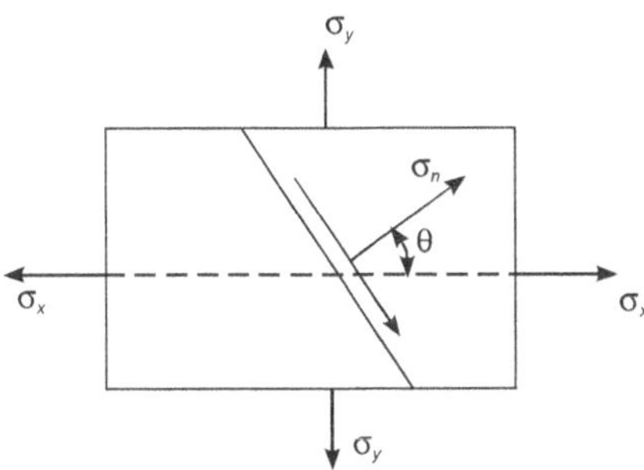

Figure 5.6(a)   Normal and shear stresses on an oblique plane

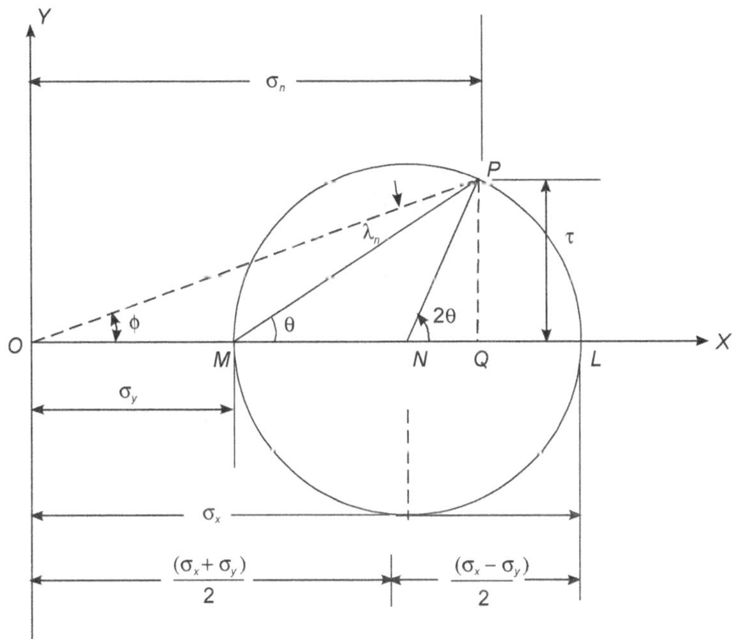

Figure 5.6(b)   Normal and shear stress on an oblique plane—Mohr's circle

1. Using some suitable scale, measure OL and OM equal σ_x to σ_y and respectively on the axis OX.

2. Bisect LM at N.

3. With N as centre and NL or NM radius, draw a circle.

4. At the centre N draw a line NP at an angle 2θ, in the same direction as the normal to the plane makes with the direction of σ_x. In Figure 5.6(a) which represents the stress system, the normal to the plane makes an angle θ, with the direction of σ_x. In Figure 5.6(a) which represents the

stress system, the normal to the plane makes an angle θ, with the direction of σ_x in the anticlockwise direction. The line *NP* therefore, is drawn in the anticlockwise direction.

5.   From *P*, drop a perpendicular *PQ* on the axis *OX*. *PQ* will represent τ and $OQ\sigma_x$.
Now from stress diagram,

$$NP = NL = \frac{\sigma_x - \sigma_y}{2}$$

$$PQ = NP\sin 2\theta = \frac{\sigma_x - \sigma_y}{2}\sin 2\theta = \tau$$

Similarly, $OQ = ON + NQ = \dfrac{\sigma_x + \sigma_y}{2} + \dfrac{\sigma_x - \sigma_y}{2}\cos 2\theta = \sigma_x$

Also from stress circle, τ is maximum when $2\theta = 90°$ or $\theta = 45°$. Recall 45° plane to have the maximum shear stress when the tensile test in UTM was discussed.

$$\tau_{max} = \frac{\sigma_x - \sigma_y}{2}$$

Mohr's circle construction for unlike stresses

See Figure 5.7(a). In case σ_x and σ_y are not like, the same procedure will be followed except that σ_x and σ_y will be measured to the opposite sides of the origin. The construction is given in Figure 5.7(b). It may be noted that the direction of σ_y will depend upon its position with respect to the point O. If it is to the right of *O*, the direction of σ_n will be the same as that of σ_x.

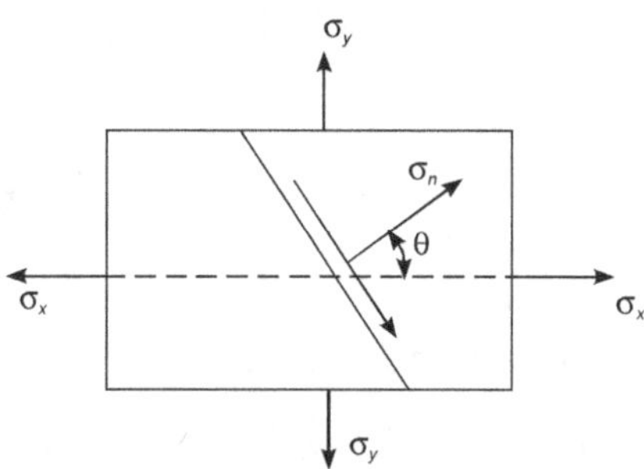

Figure 5.7(a)

Procedure is practically same except σ_x and σ_y will be measured on opposite sides of origin. Direction of σ_n will depend upon its position with respect to 'O'. If right, σ_n will have same direction as clockwise shear + anticlockwise shear − , + values above O_x,− values below O_x

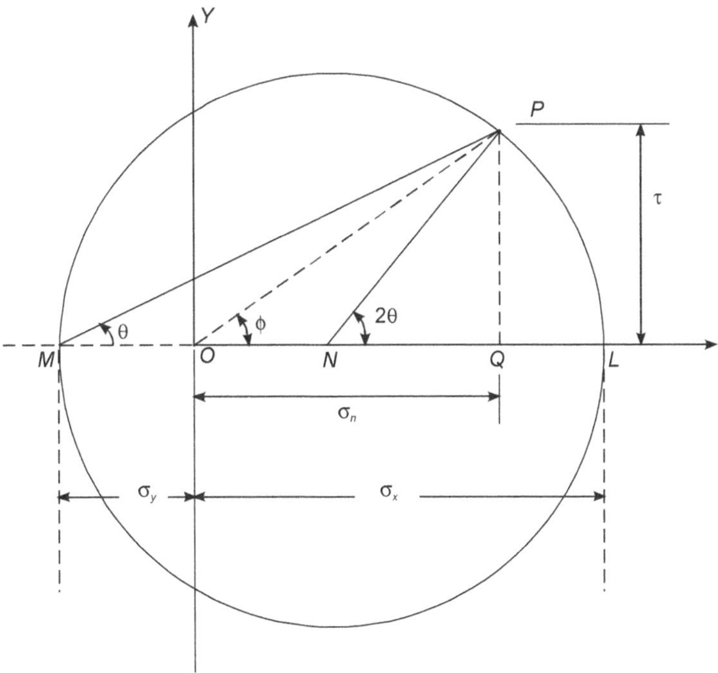

Figure 5.7(b)

Mohr's circle construction for two perpendicular direct stresses with state of simple shear

See Figure 5.8(a) and 5.8(b). Following steps of construction are followed if the material is subjected to direct stresses σ_x and σ_y along with a state of simple shear. Refer Figure 5.8(a).

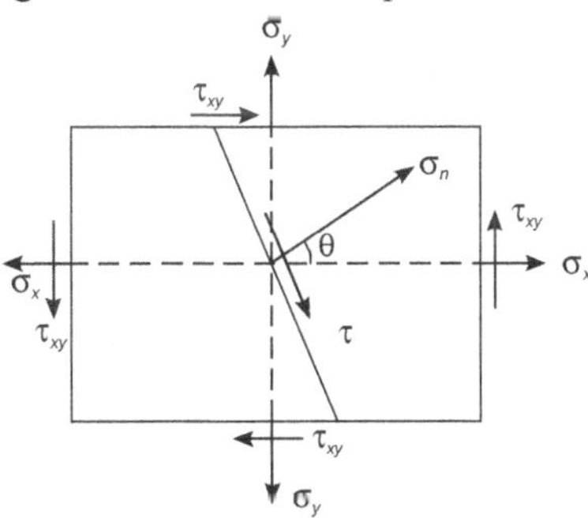

Figure 5.8(a)

Figure 5.8(b)

STEPS

See Figure 5.8(b)

1. Using some suitable scale, measure $OL = \sigma_x$ and $OM = \sigma_y$ along the axis OX.

2. At L draw LT perpendicular to OX and equal to τ_{xy}. LT has been drawn downward (as per sign conventions and adopted) because τ_{xy}. is acting up with respect to the plane across which σ_x is acting tending to rotate it in the anticlockwise direction and is negative.

3. Similarly, make MS perpendicular to OX and equal to the τ_{xy}. but above OX.

4. Join ST to cut the axis in N.

5. With N as centre and NS or NT as radius, draw a circle.

6. At N, make NP at angle 2θ with NT in the anticlockwise direction.

7. Draw PQ perpendicular to the axis. PQ will give τ while OQ will give σ_n and OP will give σ_r.

Proof

Let the radius of the stress circle b R.

Then $R = \sqrt{NL^2 + LT^2} = \sqrt{\left(\dfrac{\sigma_x - \sigma_y}{2}\right)^2 + \tau_{xy}^2}$

Also, $R\cos\beta = NL = \dfrac{\sigma_x - \sigma_y}{2}$

$$R\sin\beta = LT = \tau_{xy}$$

Now,

$$OQ = ON + NQ = ON + R\cos(2\theta - \beta)$$
$$= ON + R\cos 2\theta \cos\beta + R\sin 2\theta \sin\beta$$
$$= \frac{\sigma_x + \sigma_y}{2} + \frac{\sigma_x - \sigma_y}{2}\cos 2\theta + \tau_{xy}\sin 2\theta$$
$$= \sigma_n$$

Similarly,

$$PQ = R\sin(2\theta - \beta) = R\sin 2\theta \cos\beta - R\cos 2\theta \sin\beta$$
$$= \frac{\sigma_x - \sigma_y}{2}\sin 2\theta - \tau_{xy}\cos 2\theta$$

The following conclusions are drawn from Mohr's circle.

i.   When P coincides with V, σ_n attains the maximum value.
$$\sigma_{n(\max)} = OV = ON + NV$$
$$= \frac{\sigma_x + \sigma_y}{2} + \sqrt{\left[\frac{\sigma_x - \sigma_y}{2}\right]^2 + \tau_{xy}^2}$$

$\sigma_{n(\max)}$ or (σ_1) *is known as major principal stress.*

$$\tau = 0; \ \sigma_{r(\max)} = \sigma_{n(\max)}$$
$$\tan 2\theta = \tan\beta = \frac{\tau_{xy}}{\left[\dfrac{\sigma_x - \sigma_y}{2}\right]} = \frac{2\tau_{xy}}{\sigma_x - \sigma_y}$$

ii.   When P coincides with U, σ_n attains minimum value
$$\sigma_{n(\min)} = OU = ON - NU$$
$$= \frac{\sigma_x + \sigma_y}{2} - \sqrt{\left[\frac{\sigma_x - \sigma_y}{2}\right]^2 + \tau_{xy}^2}$$

$\sigma_{n(min)}$ or (σ_2) *is known as minor principal stress.*

$$\tau = 0, \sigma_{r(min)} = \sigma_{n(min)} \quad \theta = 90° + \beta/2$$

iii. When $2\theta = \beta + 90°$, τ attains the maximum value , τ_{max}

$$\tau_{max} = \sqrt{\left[\frac{\sigma_x - \sigma_y}{2}\right]^2 + \tau_{xy}^2} = \frac{\sigma_1 - \sigma_2}{2}$$

When $2\theta = \beta + 270°$

$$\tau_{max} = -\sqrt{\left[\frac{\sigma_x - \sigma_y}{2}\right]^2 + \tau_{xy}^2}$$

PROBLEM 5.1

When a certain thin-walled tube is subjected to internal pressure and torque, stresses in the tube wall are (a) 120 MN/m² (tensile) (b) 60 MN/m² (tensile) in a direction at right angles to (a), (c) complementary shear stress of 90 MN/m² in the direction of (a) and (b). (i) Calculate the normal and tangential stresses on the two planes which are equally inclined to (a) and (b) (ii) What are the results if due to an end thrust; (b) is compressive and (a) and (c) being unchange.

Solution

$$\sigma_x = 120 \text{ MN/m}^2 \text{(tensile)}$$

$$\sigma_y = 60 \text{ MN/m}^2 \text{(tensile)}$$

$$\tau_{xy} = 90 \text{ MN/m}^2 \text{(shear)}$$

i. $\sigma_n = ?\ \tau = ?$

When $\theta = 45°$

$$\sigma_n = \frac{(\sigma_x + \sigma_y)}{2} + \frac{(\sigma_x - \sigma_y)}{2}\cos 2\theta + \tau_x \sin 2\theta$$

$$= \frac{120 + 60}{2} + \frac{120 - 60}{2}\cos 90° + 90\sin 90° = 90 + 0 + 90$$

$$\sigma_n = 180 \text{ MN/m}^2 \text{(tensile)}$$

$$\tau = \frac{\sigma_x - \sigma_y}{2}\sin 2\theta - \tau_{xy}\cos 2\theta = \frac{120 - 60}{2}\sin 90° + 90\cos 90°$$

$$= 30 \text{ MN/m}^2 \text{(shear)}$$

Similarly, when $\theta = 135°$

$$\sigma_n = \frac{120+60}{2} + \frac{120-60}{2}\cos 270° + 90\sin 270°$$

$$\sigma_n = 90+0-90 = 0$$

$$\tau = \frac{120-60}{2}\sin 270° - 90\cos 270° = -30-0$$

$$\tau = -30 \text{ MN/m}^2$$

ii.   In the case with end thrust:

$$\sigma_x = 120 \text{ MN/m}^2$$

$$\sigma_y = -60 \text{ MN/m}^2$$

$$\tau_{xy} = 90 \text{ MN/m}^2$$

When $\theta = 45°$

$$\sigma_n = \frac{120-60}{2} + \frac{120+60}{2}\cos 90° + 90\sin 90°$$

$$= 30+0+90 = 120 \text{ MN/m}^2$$

$$\tau = \frac{120-(-60)}{2}\sin 90° - 90\cos 90°$$

$$= 90-0 = 90 \text{ MN/m}^2$$

When $\theta = 135°$

$$\sigma_n = \frac{120-60}{2} + \frac{120+60}{2}\cos 270° + 90\sin 270°$$

$$\sigma_n = 30+0-90 = -60 \text{ MN/m}^2$$

$$\tau = \frac{120-(-60)}{2}\sin 270° - 90\cos 270°$$

$$\tau = -90-0 = -90 \text{ MN/m}^2$$

iii.   When End thrust exists

$$\sigma_x - 120 \text{ MN/m}^2 \quad \sigma_y = -60\text{MN/m}^2 \quad \tau_{xy} = 90 \text{ MN/m}^2$$

$$\theta = 45° \sigma_n - \frac{120-60}{2} + \frac{120+60}{2}\cos 90° + 90\sin 90° - 30+0+90$$

$$= 120 \text{ MN/m}^2$$

$$\tau = \frac{120-(-60)}{2}\sin 90° - 90\cos 90° = 90-0$$

$$= 90 \text{ MN/m}^2$$

$$\theta = 135°\sigma_n = \frac{120-60}{2} + \frac{120+60}{2}\cos 270° + 90\sin 270°$$

$$= 30 + 0 - 90 - 60 \text{ MN/m}^2$$

$$\tau = \frac{120-(-60)}{2}\sin 270° - 90\cos 270°$$

$$= -90 - 0 = -90 \text{ MN/m}^2$$

PROBLEM 5.2

Two mutually perpendicular planes of an element of material are subjected to direct stresses of 10.5 MN/m² (tensil) and 3.5 MN/m² (comp) and shear stress of 7 MN/m².

Find graphically or otherwise,

(i) The magnitude and direction of principal stresses σ_1 and σ_2 (ii) Magnitude of the normal and shear stresses on a plane on which the shear stress is maximum.

Analytical method

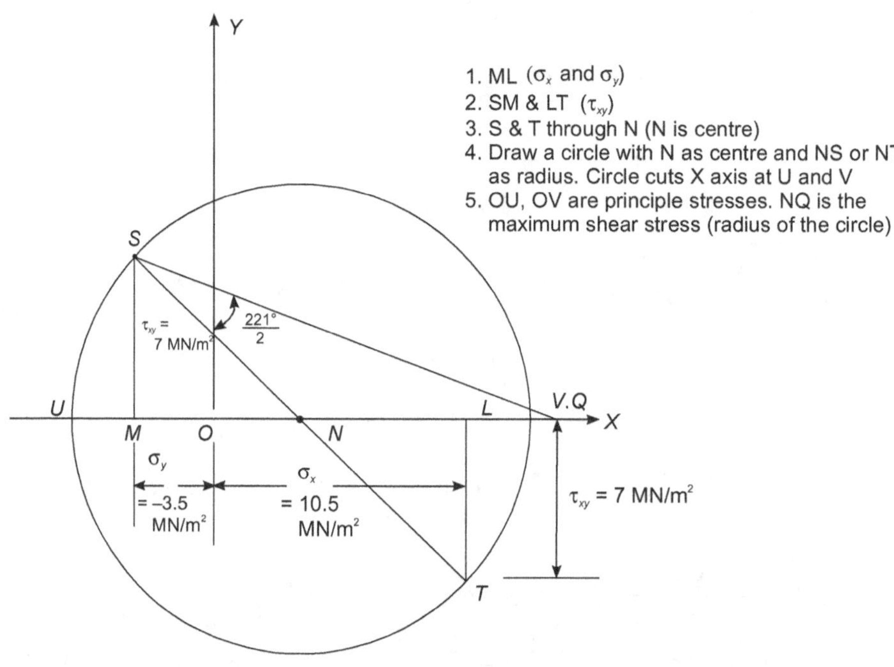

Figure

$$\sigma_x = 10.5 \text{ MN/m}^2 \text{ (tensile)}, \sigma_y = 3.5 \text{ MN/m}^2 \text{ (comp)} \tau_{xy} = 7 \text{ MN/m}^2$$

i. Principal stresses σ_1 and σ_2

we know that

$$\sigma = \left(\frac{\sigma_x + \sigma_y}{2}\right) \pm \sqrt{\left(\frac{\sigma_x - \sigma_y}{2}\right)^2 + \tau_{xy}^2}$$

$$= \frac{10.5 + (-3.5)}{2} = \pm \sqrt{\left(\frac{10.5 - (-3.5)}{2}\right)^2 + 7^2}$$

$$= 3.5 \pm \sqrt{7^2 + 7^2} = 3.5 \pm 9.9$$

or $\sigma_1 = 13.41 \text{ MN/m}^2$, $\sigma_2 = -6.4 \text{ MN/m}^2$

Hence, maximum principal stress,

$$\sigma_1 = 13.4 \text{ MN/m}^2 \text{ (tensile)}$$

Minimum principal stress

$$\sigma_2 = 6.4 \text{ MN/m}^2 \text{ (comp.)}$$

Directions of principal stresses θ_1, θ_2

$$\tan 2\theta = \frac{2\tau_{xy}}{\sigma_x - \sigma_y} = \frac{2 \times 7}{10.5 - (-3.5)}$$

$$2\theta = 45° \text{ or } 225°$$

$$\theta_1 = 22°30' \text{ and } \theta_2 = 112°30'$$

ii. Magnitudes of normal (σ_n) and shear stresses (τ)

$$\sigma_n = \frac{\sigma_x + \sigma_y}{2} + \frac{\sigma_x - \sigma_y}{2} \cos 2\theta + \tau_{xy} \sin 2\theta$$

$$\sigma_n = \frac{10.5 + (-3.5)}{2} + \frac{10.5 - (-3.5)}{2} \cos 45° + 7 \sin 45°$$

$$\sigma_n = 3.5 + 7 \cos 45° + 7 \sin 45°$$

$$- 13.4 \text{ MN/m}^2 \text{ (tensile)}$$

$$\tau = \frac{\sigma_x - \sigma_y}{2} \sin 2\theta - \tau_{xy} \cos 2\theta$$

$$= \frac{10.5 - (-3.5)}{2} \sin 45° - 7 \cos 45°$$

$$- 0$$

Maximum shear stress

$$= \frac{\sigma_1 - \sigma_2}{2} = \frac{13.4 - (-6.4)}{2} = \frac{19.8}{2}$$

$$= 9.9 \text{ MN/m}^2$$

Normal stress in plane where maximum shear exists $= \dfrac{\sigma_x + \sigma_y}{2} = \dfrac{\sigma_1 + \sigma_2}{2} = \dfrac{7}{2} = 3.5 \text{ MN/m}^2$

PROBLEM 5.3

For the state of plane stress shown in Figure determine

(a) the principal planes, (b) the principal stresses, (c) the maximum shearing stress and the corresponding normal stress.

(a) *Principal planes*   Following the usual sign convention, we write the stress components as

$$\sigma_x = +50 \text{ MPa} \quad \sigma_y = -10 \text{ MPa} \quad \tau_{xy} = +40 \text{ MPa}$$

Substituting into equation for $\tan 2\theta$,, we have

$$\tan 2\theta_p = \frac{2\tau_{xy}}{\sigma_x - \sigma_y} = \frac{2(+40)}{50 - (-10)} = \frac{80}{60} = 1.33$$

$$2\theta_p = \tan^{-1} 1.33 = 53.1° \text{ and } 180° + 53.1° = 233.1°$$

$$\theta_p = 26.6° \text{ and } 116.6°$$

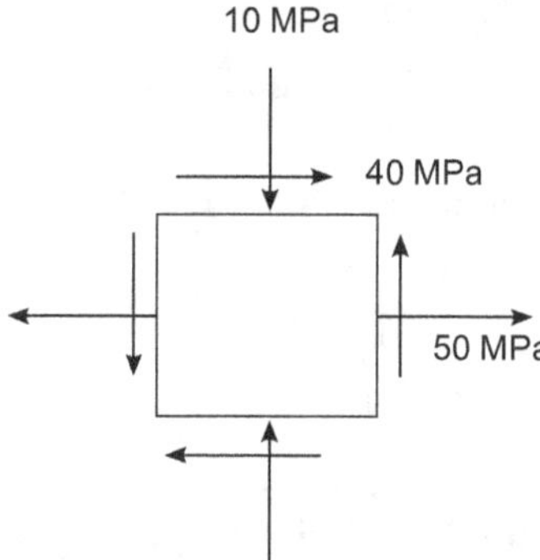

(b) *Principal stresses*   Formula yields

$$\sigma_{max, min} = \frac{\sigma_x + \sigma_y}{2} \pm \sqrt{\left(\frac{\sigma_x - \sigma_y}{2}\right)^2 + \tau_{xy}^2} = 20 \pm \sqrt{(30)^2 + (40)^2}$$

$$\sigma_{max} = 20 + 50 = 70 \text{ MPa} \qquad \sigma_{min} = 20 - 50 = -30 \text{ MPa}$$

The principal planes and principal stresses are sketched in Figure.

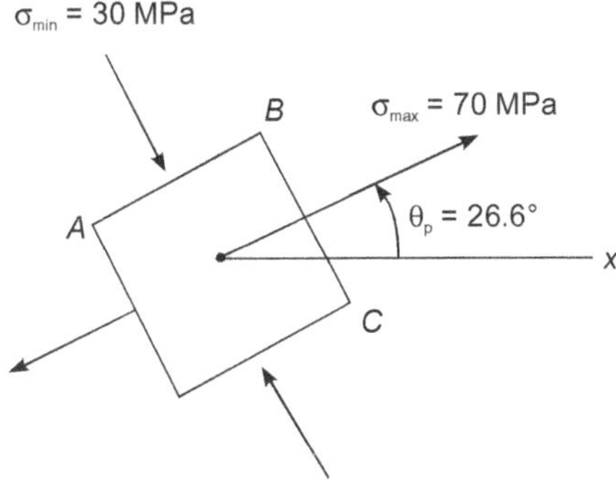

Making $\theta = 26.6°$ in equation, we check that the normal stress exerted on face BC of the element is the maximum stress:

$$\sigma_{x'} = \frac{50-10}{2} + \frac{50+10}{2}\cos 53.1° + 40\sin 53.1°$$

$$= 20 + 30\cos 53.1° + 40\sin 53.1° = 70 \text{ MPa} = \sigma_{max}$$

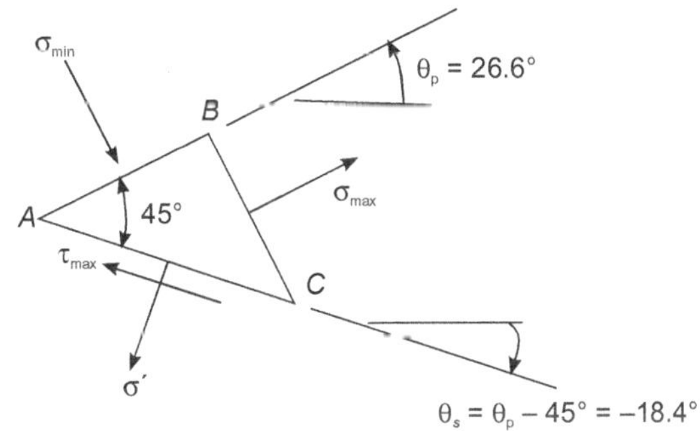

(c) ***Maximum shearing stress***   Formula τ_{max} yields

$$\tau_{max} = \sqrt{\left(\frac{\sigma_x - \sigma_y}{2}\right)^2 + \tau_{xy}^2} = \sqrt{(30)^2 + (40)^2} = 50 \text{ MPa}$$

Since σ_{max} and σ_{min} have opposite signs, the value obtained for τ_{max} actually represents the maximum value of the shearing stress at the point considered. The orientation of the planes of maximum shearing stress and the sense of the shearing stresses are best determined by passing a section along the diagonal plane AC of the element of Figure. Since the faces AB and BC of the element are contained in the principal planes, the diagonal plane AC must be one of the planes of maximum shearing stress (Figure). Furthermore, the equilibrium conditions for the prismatic element

ABC require that the shearing stress exerted on *AC* be directed as shown. The cubic element corresponding to the maximum shearing stress is shown in Figure. The normal stress on each of the four faces of the element is given by:

$$\sigma' = \sigma_{ave} = \frac{\sigma_x + \sigma_y}{2} = \frac{50 - 10}{2} = 20 \text{ MPa}$$

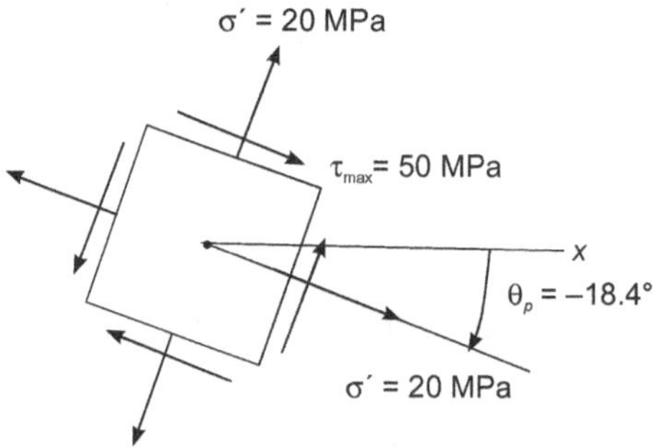

PROBLEM 5.4

A single horizontal force *P* of magnitude 900 N is applied to end *D* of lever *ABD*. Knowing that portion *AB* of the lever has a diameter of 36 mm, determine (a) the normal and shearing stresses on an element located at point *H* and having sides parallel to the *x* and *y* axes, (b) the principal planes and the principal stresses at point *H*.

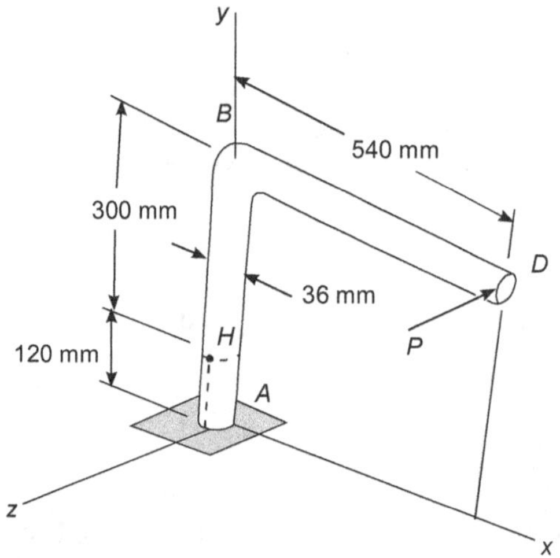

Solution

We replace the force *P* by an equivalent force-couple system at the centre *C* of the transverse section containing point *H* :

$$P = 900 \text{ N} \qquad T = (900 \text{ N})(0.540 \text{ m}) = 486 \text{ N.m}$$

$$M_x = (900\ \text{N})(0.300\ \text{m}) = 270\ \text{N.m}$$

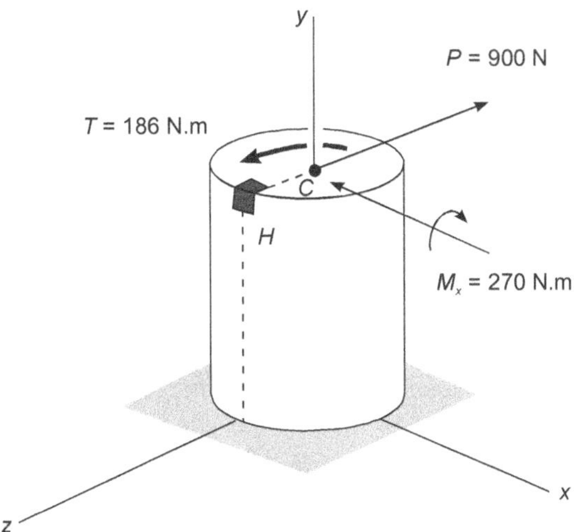

(a) **Stresses** $\sigma_x, \sigma_y, \tau_{xy}$ **at Point H** Using the sign convention shown in Figure, we determine the sense and the sign of each stress component by carefully examining the sketch of the force-couple system at point C:

$$\sigma_x = 0 \qquad \sigma_y = +\frac{Mc}{I} = +\frac{(270\,\text{N.m})(0.018\,\text{m})}{\dfrac{1}{4}\pi(0.018\,\text{m})^4}$$

$$\sigma_y = 58.9\ \text{MPa}$$

$$\tau_{xy} = \frac{Tc}{J} = +\frac{(486\,\text{N.m})(0.018\,\text{m})}{\dfrac{1}{2}\pi(0.018\,\text{m})^4}$$

$$\tau_{xy} = 53.1\ \text{MPa}$$

We note that the shearing force P does not cause any shearing stress at point H.

(b) **Principal planes and principal stresses** Substituting the values of the stress components into equation, we determine the orientation of the principal planes

$$\tan 2\theta_p = \frac{2\tau_{xy}}{\sigma_x - \sigma_y} = \frac{2(53.1)}{0 - 58.9} = -1.80$$

$$2\theta_p - \tan^{-1}(-1.8) - 61.0^\circ \quad \text{and}$$

$$180^\circ - 61.0^\circ = +119^\circ$$

$$\theta_p = -30.5^\circ \quad \text{and} \quad +59.5^\circ$$

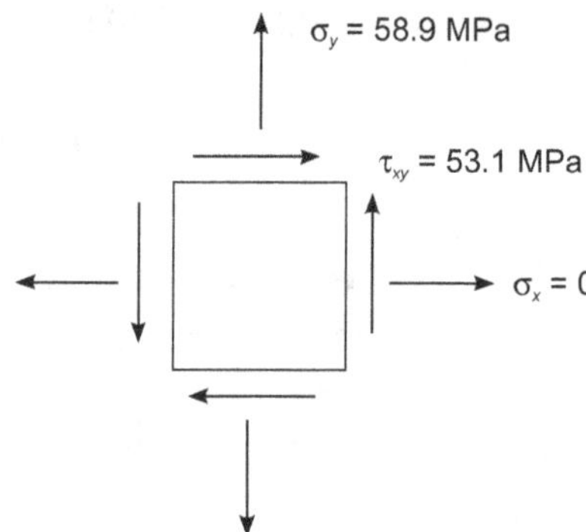

Substituting into equation, we determine the magnitudes of the principal stresses:

$$\sigma_{max,\,min} = \frac{\sigma_x + \sigma_y}{2} \pm \sqrt{\left(\frac{\sigma_x - \sigma_y}{2}\right)^2 + \tau_{xy}^2}$$

$$= \frac{0+58.9}{2} \pm \sqrt{\left(\frac{0-58.9}{2}\right)^2 + (53.1)^2} = +29.45 \pm 60.72$$

$$\sigma_{max} = +90.2 \text{ MPa}$$

$$\sigma_{min} = -32.3 \text{ MPa}$$

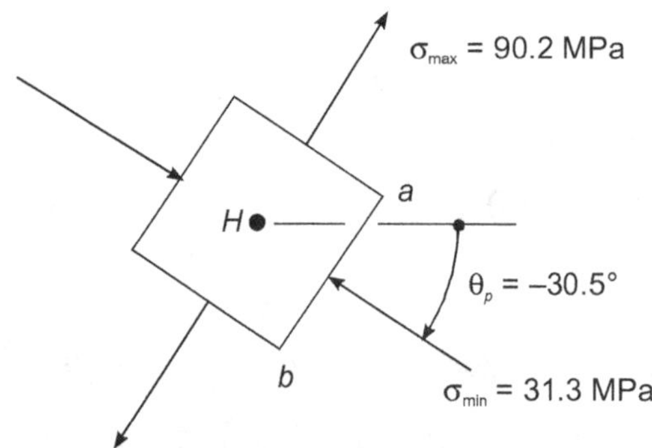

Considering face *ab* of the element shown, we make $\theta_p = -30.5°$ in equation and find $\sigma_{x'} = -31.3$ MPa. We conclude that the principal stresses are as shown.

PROGRAM 5.5

For the state of plane stress shown, determine (a) the principal planes and the principal stresses, (b) the stress components exerted on the element obtained by rotating the given element counterclockwise through 30°.

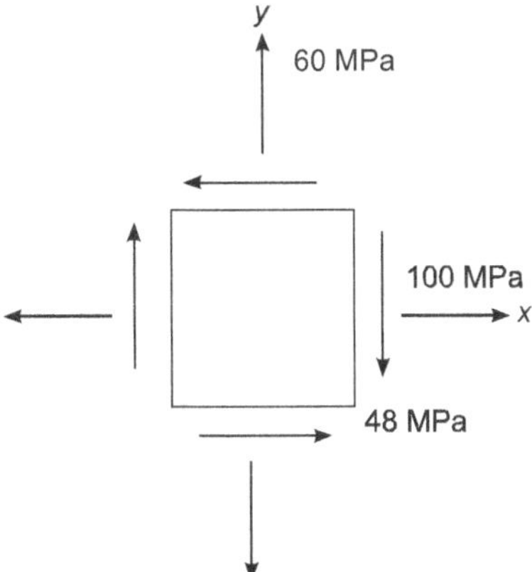

Construction of Mohr's circle   We note that on a face perpendicular to the x axis, the normal stress is tensile and the shearing stress tends to rotate the element clockwise; thus we plot X at a point 100 units to the right of the vertical axis and 48 units above the horizontal axis. In a similar fashion, we examine the stress components on the upper face and plot point Y(60, –48). Joining points X and Y by a straight line, we define the centre C of Mohr's circle. The abscissa of C, which represents $\sigma_{ave,}$ and the radius R of the circle may be measured directly or calculated as follows:

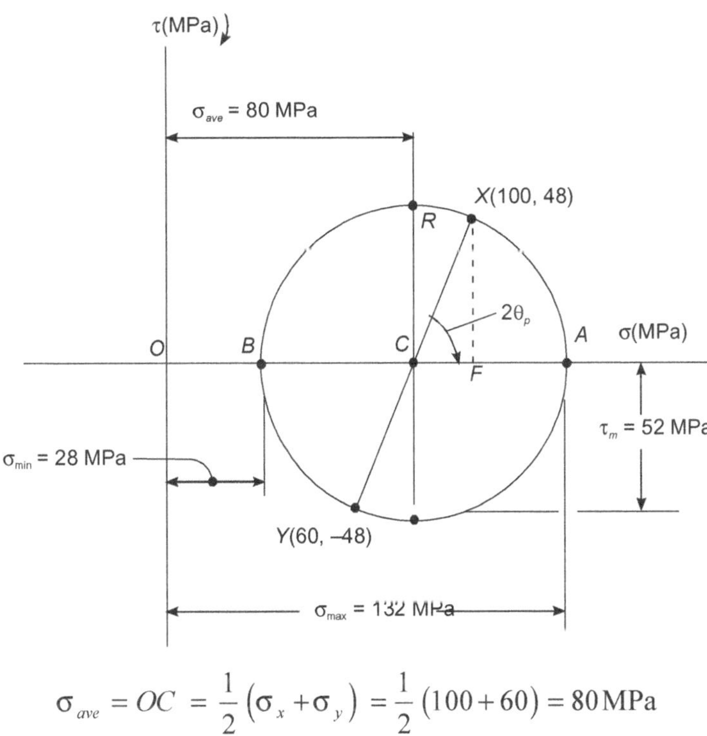

$$\sigma_{ave} = OC = \frac{1}{2}\left(\sigma_x + \sigma_y\right) = \frac{1}{2}\left(100 + 60\right) = 80\,\text{MPa}$$

$$R = \sqrt{(CF)^2 + (FX)^2} = \sqrt{(20)^2 + (48)^2} = 52 \text{ MPa}$$

(a) *Principal planes and principal stresses* We rotate the diameter XY clockwise through $2\theta_p$ until it coincides with the diameter AB. We have

$$\tan 2\theta_p = \frac{XF}{CF} = \frac{48}{20} = 2.4 \qquad 2\theta_p = 67.4° \qquad \theta_p = 33.7°$$

The principal stresses are represented by the abscissas of points A and B:

$$\sigma_{max} = OA = OC + CA = 80 + 52 \qquad \sigma_{max} = +132 \text{ MPa}$$
$$\sigma_{min} = OB = OC - BC = 80 - 52 \qquad \sigma_{min} = +28 \text{ MPa}$$

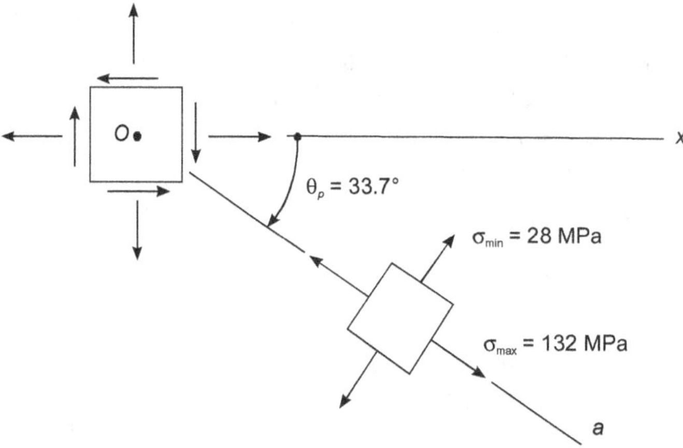

Since the rotation which brings XY into AB is clockwise, the rotation which brings Ox into the axis Oa corresponding to σ_{max} is also clockwise; we obtain the orientation shown for the principal planes.

(b) *Stress components on element rotated* 30°. Points X' and Y' on Mohr's circle which correspond to the stress components on the rotated element are obtained by rotating XY counter clockwise through $2\theta = 60°$. We find

$$\phi = 180° - 60° - 67.4° \qquad\qquad \phi = 52.06°$$
$$\sigma_{x'} = OK = OC - KC = 80 - 52 \cos 52.6° \qquad \sigma_{x'} = +48.4 \text{ MPa}$$
$$\sigma_{y'} = OL = OC + CL = 80 + 52 \cos 52.6° \qquad \sigma_{y'} = +111.6 \text{ MPa}$$
$$\tau_{x'y'} = KX' = 52 \sin 52.6° \qquad\qquad \tau_{x'y'} = 41.3 \text{ MPa}$$

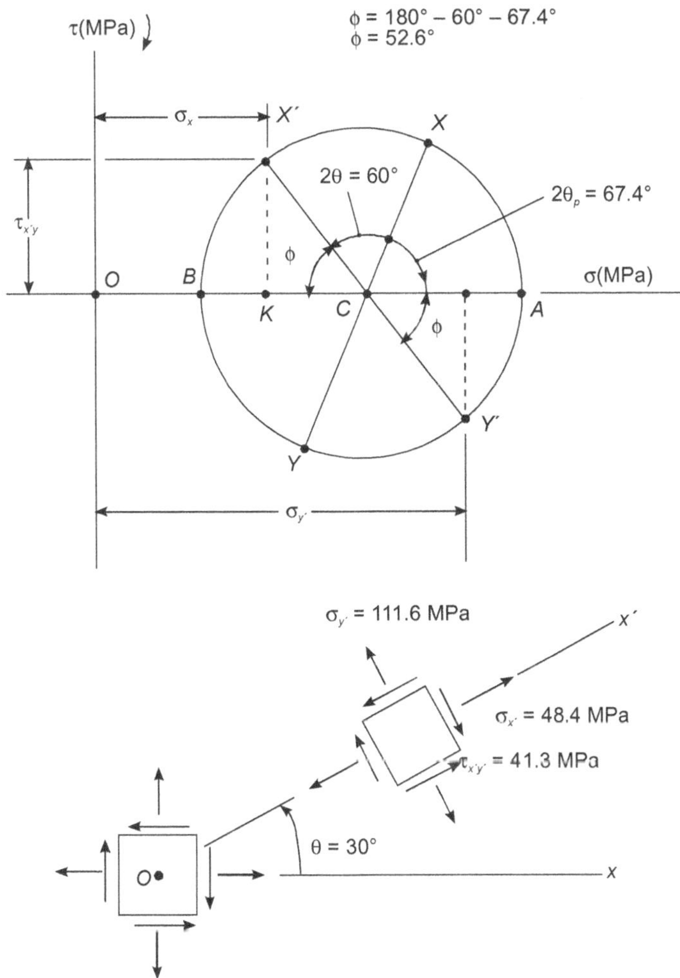

Since X' is located above the horizontal axis, the shearing stress on the face perpendicular to Ox' tends to rotate the element clockwise.

PROBLEM 5.6

A state of planes stress consists of a tensile stress $\sigma_0 = 80$ MPa exerted on vertical surfaces and of unknown shearing stresses. Determine (a) the magnitude of the shearing stress τ_0 for which the largest normal stress is 100 MPa, (b) the corresponding maximum shearing stress.

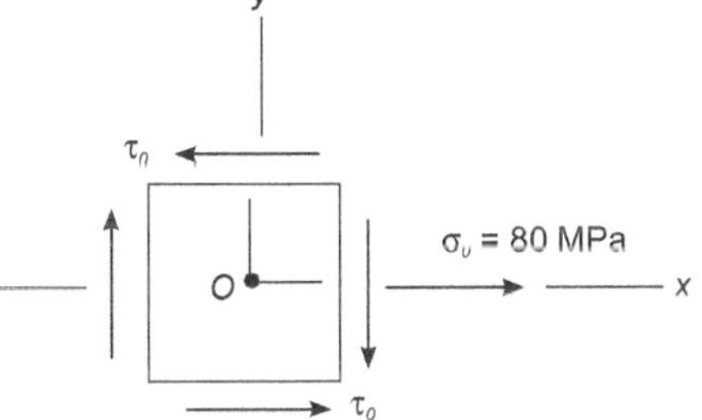

Construction of Mohr's circle   We shall assume that the shearing stresses act in the senses shown. Thus, the shearing stress τ_0 on a face perpendicular to the x axis tends to rotate the element clockwise and we plot the point X of coordinates 80 MPa and τ_0 above the horizontal axis. Considering a horizontal face of the element, we observe that $\sigma_y = 0$ and that τ_0 tends to rotate the element counterclockwise; thus we plot point Y at a distance τ_0 below O.

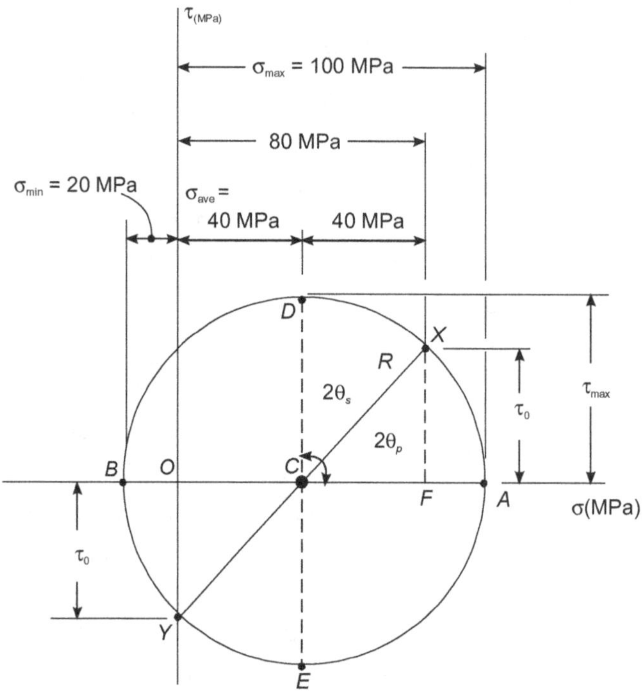

We note that the abscissa of the centre C of Mohr's circle is

$$\sigma_{ave} = \frac{1}{2}(\sigma_x + \sigma_y) = \frac{1}{2}(80 + 0) = 40 \text{ MPa}$$

The radius R of the circle is determined by observing that the maximum normal stress, $\sigma_{max} = 100$ MPa, is represented by the abscissa at point A and writing

$$\sigma_{max} = \sigma_{ave} + R$$
$$100\,\text{MPa} = 40\,\text{MPa} + R$$
$$R = 60\,\text{MPa}$$

(a) *Shearing stress* τ_0   Considering the right triangle CFX, we find

$$\cos 2\theta_p = \frac{CF}{CX} = \frac{CF}{R} = \frac{40 \text{ MPa}}{60 \text{ MPa}}$$

$$2\theta_p = 48.2° \downarrow$$

$$\theta_p = 24.1° \downarrow$$

$$\tau_0 = FX = R\sin 2\theta_p = (60 \text{ MPa})\sin 48.2°$$

$$\tau_0 = 44.7 \text{ MPa}$$

(b) *Maximum shearing stress* The coordinates of point D of Mohr's circle represent the maximum shearing stress and the corresponding normal stress.

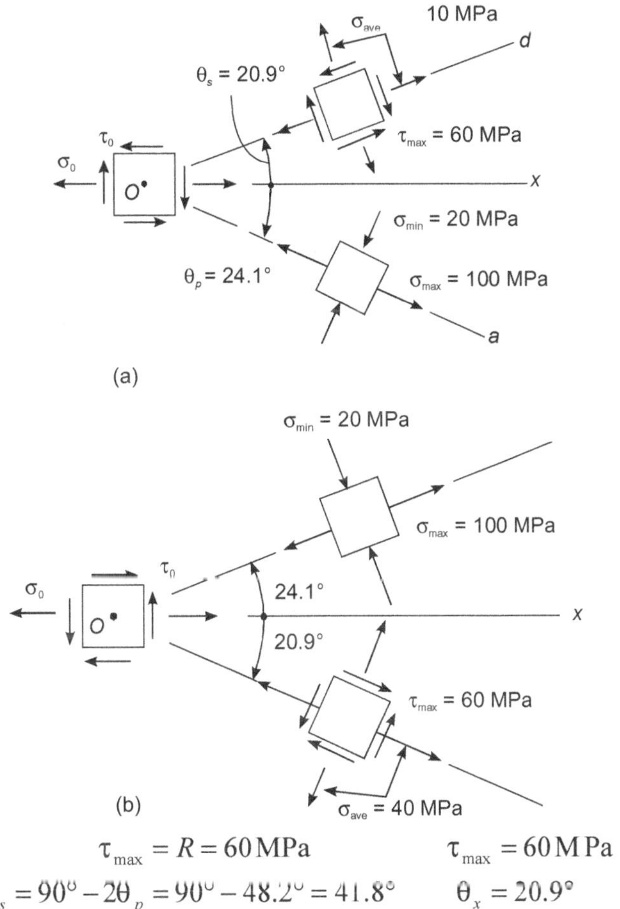

$$\tau_{max} = R = 60\,\text{MPa} \qquad\qquad \tau_{max} = 60\,\text{MPa}$$

$$2\theta_s = 90° - 2\theta_p = 90° - 48.2° = 41.8° \qquad \theta_x = 20.9°$$

The maximum shearing stress is exerted on an element which is oriented as shown in Figure a. (The element upon which the principal stresses are exerted is also shown.)

Note If our original assumption regarding the sense of τ_0 was reversed, we would obtain the same circle and the same answers, but the orientation of the elements would be as shown in Figure b.

PROBLEM 5.7

For the state of plane stress shown in Figure, determine the three principal planes and principal stresses, (b) the maximum shearing stress.

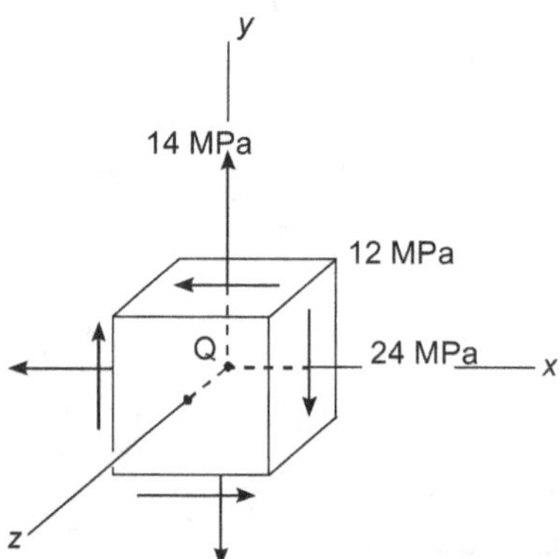

(a) *Principal planes and principal stresses*   We construct Mohr's circle for the transformation of stress in the *xy* plane (Figure). Point *X* is plotted 24 units to the right of the τ axis and 12 units above the σ axis (since the corresponding shearing stress tends to rotate the element clockwise). Point *Y* is plotted 14 units to the right of the τ axis and 12 units below the σ axis. Drawing the line *XY,* we obtain the centre *C* of Mohr's circle for the *xy* plane: its abscissa is

$$\sigma_{ave} = \frac{\sigma_x + \sigma_y}{2} = \frac{24 + 14}{2} = 19 \text{ MPa}$$

Since the sides of the right triangle *CFX* are *CF* = 24 − 19 = 5 MPa and *FX* = 12 MPa, the radius of the circle is

$$R = CX = \sqrt{(5)^2 + (12)^2} = 13 \text{ MPa}$$

The principal stresses in the plane of stress are

$$\sigma_a = OA = OC + CA = 19 + 13 = 32 \text{ MPa}$$
$$\sigma_b = OB = OC - BC = 19 - 13 = 6 \text{ MPa}$$

Since the faces of the element which are perpendicular to the z axis are free of stress, these faces define one of the principal planes, and the corresponding principal stress is $\sigma_z = 0$. The other two principal planes are defined by points A and B on Mohr's circle. The angle θ_p through which the element should be rotated about the z axis to bring its faces to coincide with these planes (Figure) is half the angle ACX. We have

$$\tan 2\theta_p = \frac{FX}{CF} = \frac{12}{5}$$
$$2\theta_p = 67.4° \quad \theta_p = 33.7°$$

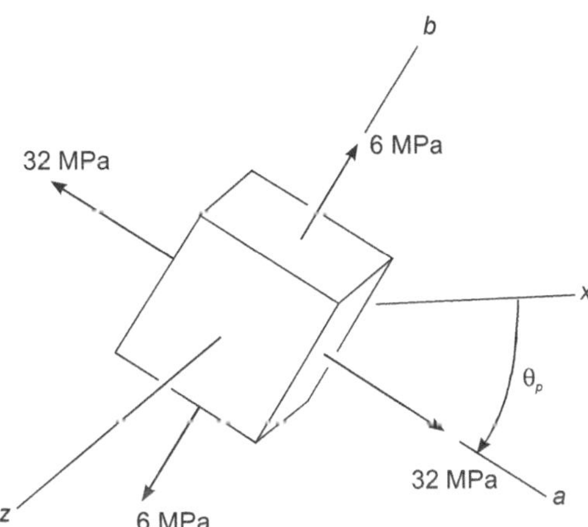

(b) *Maximum shearing stress*   We now draw the circles of diameter OB and OA, which correspond respectively to rotations of the element about the a and b axes (Figure). We note that the maximum shearing stress is equal to the radius of the circle of diameter OA. We thus have

$$\tau_{max} = \frac{1}{2}\sigma_a = \frac{1}{2}(32 \text{ MPa}) = 16 \text{ MPa}$$

Since points D' and E', which define the planes of maximum shearing stress, are located at the ends of the vertical diameter of the circle corresponding to a rotation about the b axis, the faces of the element of Figure may be brought to coincide with the planes of maximum shearing stress through a rotation of 45° about the b axis.

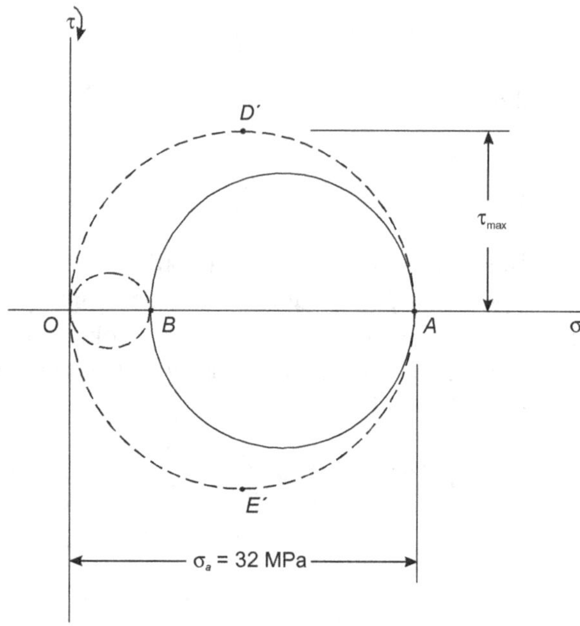

Figure

CONCLUSION

Mohr's circle construction has been used in this chapter to arrive at the principal stresses and maximum shear stress. The same construction can also be used for strains. In a plane problem, if $\in_x, \in_y$ and γ_{xy} are known, we can determine $\in_1, \in_2$ and γ_{12} by Mohr's circle construction. Similarly in the case of moments of inertia of plane areas (Chapter 10), if we know I_{xx}, I_{yy} and I_{xy}, we can use Mohr's circle construction to determine, the principal moments of inertia I_1 and I_2 and maximum product of inertia I_{12}.

Mohr's circle construction can also be used for a body subjected to three-dimensional state of stress from which we can determine the three principal stresses and three maximum shearing stresses. This will be very useful when we discuss *Theories of Failure* in Chapter 6.

REVIEW QUESTIONS

SHORT QUESTIONS

1. Define the terms: Principal planes and principal stresses.

2. What are the uses of a Mohr's circle?

3. Draw a Mohr's circle for the given shear stress q.

LARGE QUESTIONS

1. The state of stress (in N/mm²) acting at a certain point of the strained material is shown in Figure. Compute

    i.   The magnitude and nature of principal stresses and

    ii. The orientation of principal planes

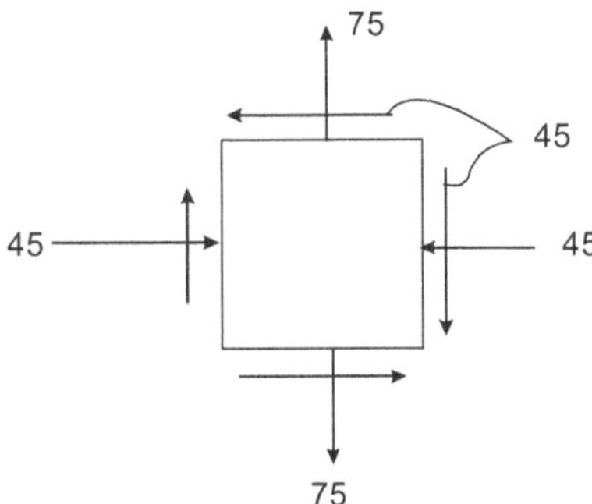

2. A short metallic column of 500 mm² cross sectional area carries an axial load 100 kN. For a plane inclined at 60° with the direction of load, calculate normal stress, tangential stress, resultant stress, maximum shear stress and obliquity of the resultant stress.

3. i. A point in a strained material is subjected to two mutually perpendicular tensile stress of 200 MPa and 100 MPa. Determine the intensities of normal, shear and resultant stresses on a plane inclined at 30° with the axis of the minor tensile stress.

    ii. A point is subjected to a tensile stress of 250 MPa in the horizontal direction and another tensile stress of 100 MPa in the vertical direction. The point is also subjected to a simple shear stress of 25 MPa, such that when it is associated with the major tensile stress, it tends to rotate the element in the clockwise direction. What is the magnitude of the normal and shear stresses on a section inclined at an angle of 20° with the major tensile stress?

4. At a certain point in a strained material, the stresses on two planes, at right angles to each other are 20 MPa and 10 MPa both tensile. They are accompanied by a shear stress of magnitude of 10 MPa. Find graphically or otherwise, the location of principal planes and evaluate the principal stresses.

5. A point in a strained material is subjected to stresses shown in Figure. Using Mohr's circle method, determine the normal and tangential stresses across the oblique plane. Check the answer analytically.

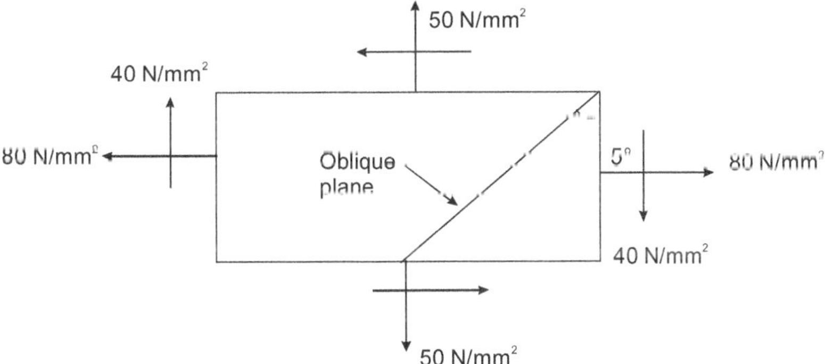

6. A point in a strained material is subjected to mutually perpendicular stress of 600 N/mm² (compressive). It is also subjected to a shear stress of 100 N/mm². Draw Mohr's circle, and find the principal stresses and maximum shear.

7. At a point in a strained material the stresses on the vertical plane are 50 MPa and 40 MPa anticlockwise in effect. On the horizontal plane the normal stress is 30 MPa compressive stress. Determine

 i. principal stresses and their planes

 ii. maximum shear plane and the stresses on it

 iii. the stresses on an inclined plane at 30° anticlockwise from VP and the resultant stress on it and

 iv. stresses on an inclined plane at 40° anticlockwise from horizontal plane and the resultant stress on it.

6

THEORIES OF FAILURE

INTRODUCTION

Due to a large number of examples of compound stresses met within engineering practice, the cause of 'failure' (inability of a part or an assembly to meet its functional requirements) or permanent set under such conditions has attracted considerable attention. Certain theories have been advanced to explain the cause of failure and many of the theories have received considerable experimental investigation. No great uniformity of opinion has been reached, and there is still greater room for further experimental investigation and refinement of these theories.

The principal theories are:

1. Maximum principal stress theory (Rankine's Theory)
2. Maximum shear stress or stress difference theory (Guset Theory/Tresca Criterion)
3. Strain energy theory (Haigh's Theory)
4. Shear strain energy theory (Mises–Henky Theory/Vonmises Criterion)
5. Maximum principal strain theory (St. Venant's Theory)

In all above theories:

σ_{et}, σ_{ec} = Tensile stress at the elastic limit in simple tension and compression respectively

$\sigma_1, \sigma_2, \sigma_3$ = Principal stresses in any complex system (such that $\sigma_1 > \sigma_2 > \sigma_3$)

It may be assumed that the loading is gradual or static (and there is no cyclic or impact loading).

MAXIMUM PRINCIPAL STRESS THEORY

❋ This is the simplest and the oldest theory of failure associated with Rankine.

❋ According to this theory failure will occur when the maximum principal tensile stress (σ_1) in the complex system reaches the value of the maximum stress at the elastic (σ_{et}) in simple tension or the minimum principal stress (that is, the maximum principal compressive stress) reaches the elastic limit stress (σ_{ec}) in simple compression.

$$\sigma_1 = \sigma_{et} \text{ (in simple tension)} \tag{6.1}$$

$$|\sigma_3| = \sigma_{ec} \text{ (in simple compression)}$$

$$|\sigma_3| \text{ means numerical value of } \sigma_3$$

❋ If the maximum principal stress is the design criterion, the maximum principal stress must not exceed the working σ for the material. Hence

$$\sigma_1 \leq \sigma$$

❋ This theory disregards the presence/effect of other principal stresses and of the shearing stresses on other planes through the element. For brittle materials which do not fail by yielding, but fail by brittle fracture, the maximum principal stress theory is considered to be reasonably satisfactory and approximately correct for ordinary cast iron and brittle materials.

The maximum principal stress theory is contradicted in the following cases.

1. On a mild steel specimen when simple tension test is carried out sliding occurs approximately 45° to the axis of the specimen; this shows that the failure in this case is due to maximum shear stress rather than the direct tensile stress.

2. It has been found that a material which is even though weak in simple compression yet can sustain hydrostatic pressure far in excess of the elastic limit in simple compression.

MAXIMUM SHEAR STRESS/STRESS DIFFERENCE THEORY/GUSET'S OR TRESCA THEORY (MSS THEORY)

❋ This theory implies that failure will occur when the maximum shear stres τ_{max} in the complex system reaches the value of the maximum shear stress in simple tension at the elastic limit, i.e.

$$\tau_{max} = \frac{\sigma_1 - \sigma_3}{2} = \frac{\sigma_{et}}{2} \text{ in simple tension} \tag{6.2}$$

$$\sigma_1 - \sigma_3 = \sigma_{et}$$

In actual design σ_{et} in the above equation is replaced by the safe stress.

This theory gives good correlation with results of experiments on ductile materials and has been found to give satisfactory results for them.

Following points are worth noting:

i. This theory does not give accurate results for the state of stress of pure shear in which the maximum amount of shear is developed (i.e., torsion test).

ii. This theory is not applicable in the case where the state of stress consists of triaxial tensile stresses of nearly equal magnitude reducing the shearing stress to a small magnitude, so that failure would be by brittle fracture rather than by yielding.

iii. It does not give as close results as found by experiments on ductile materials. However it gives safe results.

STRAIN ENERGY THEORY (HAIGH'S THEORY)

❀ This theory which has a thermodynamic analogy and a logical basis is due to Haigh. It states that the failure of a material occurs when the total strain energy in the material reaches the total strain energy of the material at the elastic limit in simple tension.

In a three-dimensional stress system, the strain energy per unit volume is given by:

$$U = \frac{1}{2E}\left[(\sigma_1^2 + \sigma_2^2 + \sigma_3^2) - \frac{2}{\nu}(\sigma_1\sigma_2 + \sigma_2\sigma_3 + \sigma_3\sigma_1)\right] \quad \text{(where } \sigma_1, \sigma_2 \text{ and } \sigma_3 \text{ are of the same sign)}$$

Hence at the point of failure

$$U = \frac{1}{2E}\left[\sigma_1^2 + \sigma_2^2 + \sigma_3^2 - \frac{2}{\nu}(\sigma_1\sigma_2 + \sigma_2\sigma_3 + \sigma_3\sigma_1)\right] = \frac{\sigma_e^2}{2E}$$

$$(\sigma_1^2 + \sigma_2^2 + \sigma_3^2) - \frac{2}{\nu}(\sigma_1\sigma_2 + \sigma_2\sigma_3 + \sigma_3\sigma_1) = \sigma_e^2$$

In actual design σ_e^2 in the above equation is replaced by the allowable stress obtained by dividing σ_e^2 by F.O.S (F.O.S = Factor of safety).

Taking two-dimensional case $(\sigma_3 = 0)$, the equation reduces to

$$\sigma_1^2 + \sigma_2^2 - \frac{2}{\nu}\cdot\sigma_1\sigma_2 = \sigma_e^2$$

If σ is the working stress in the material, the design criterion may be stated as follows:

$$(\sigma_1^2 + \sigma_2^2 - \frac{2}{\nu}\sigma_1\sigma_2) \le \sigma^2$$

The following points are to be noted.

i. The results of this theory are similar to the experimental results for ductile materials (i.e., the materials which fail by general yielding) for which $\sigma_{et} = \sigma_{ec}$ approximately.

ii. The theory does not apply to materials for which σ_{et} is quite different from σ_{ec}.

iii. The theory does not give results exactly equal to the experimental results even for ductile materials, even though the results are close to the experimental.

SHEAR STRAIN ENERGY THEORY (DE THEORY)

This theory is also called distortion energy theory or Mises-Henky theory. According to this theory "the elastic failure occurs where the shear strain energy per unit volume in the stressed material reaches a value equal to the shear strain energy per unit volume at the elastic limit point in the simple tension test".

Shear strain energy due to the principal stresses σ_1, σ_2 and σ_3 per unit volume of the stress material,

$$U_s = \frac{1}{12G}\left[(\sigma_1 - \sigma_2)^2 + (\sigma_2 - \sigma_3)^2 + (\sigma_3 - \sigma_1)^2\right] \qquad (6.4)$$

But for the simple tension test at the elastic limit point where there is only one principal stress, i.e., σ_{et} we have the shear strain energy per unit volume given by

$$U_s = \frac{1}{12G}\left[(\sigma_{et} - 0)^2(0-0)^2 + (0 - \sigma_{et})^2\right] \qquad \left|\begin{array}{l} \therefore \sigma_1 = \sigma_{et} \\ \sigma_2 = 0 \\ \sigma_3 = 0 \end{array}\right.$$

Equating the two energies, we get

$$(\sigma_1 - \sigma_2)^2 + (\sigma_2 - \sigma_3)^2 + (\sigma_3 - \sigma_1)^2 = 2\sigma_{et}^{\ 2}$$

In actual design σ_{et} in equation (1) is replaced by safe equivalent stress in σ_t simple tension.

Best results are obtained for ductile materials with this theory. This above theory has been found to give best results for ductile material for which $\sigma_{et} = \sigma_{ec}$ approximately.

The following points are worth noting.

i. The theory does not agree with the exerimental results for the material for which σ_{et} is quite different from σ_{ec}.

ii. The theory gives $\sigma_{et} = 0$ for hydrostatic pressure or tension, which means that the material will never fail under any hydrostatic pressure or tension and this is obviously not correct.

Actually when three equal tensions are applied in three principal directions, brittle fracture occurs and as such maximum principal stress theory will give reliable results in this case.

iii. This theory is regarded as one to which most of the ductile materials conform under the action of various types of loading.

MAXIMUM PRINCIPAL STRAIN THEORY

This theory is associated with St. Venant. The theory states that the failure of a material occurs when the principal tensile strain in the material reaches the strain at the elastic limit in simple tension or when the minimum principal strain (i.e., maximum principal compressive strain) reaches the elastic limit strain in simple compression. Principal strain in the direction of principal stress σ_1,

$$e_1 = \frac{1}{E}\left(\sigma_1 - \frac{1}{v}(\sigma_2 + \sigma_3)\right) \qquad (6.5)$$

Principal strain in the direction of the principal stress σ_3,

$$e_3 = \frac{1}{E}\left[\sigma_3 - \frac{1}{\nu}(\sigma_1 + \sigma_2)\right]$$

The conditions to cause failure according to the maximum principal strain theory are

$$e_1 > \frac{\sigma_{et}}{E} \quad (e_1 \text{ must be} + \text{ve})$$

$$|e_3| > \frac{\sigma_{ec}}{E} \quad (e_3 \text{ must be} - \text{ve})$$

$$\frac{1}{E}\left[\sigma_1 - \frac{1}{\nu}(\sigma_2 + \sigma_3)\right] > \frac{\sigma_{et}}{E} \text{ and } \frac{1}{E}\left[\sigma_3 - \frac{1}{\nu}(\sigma_1 + \sigma_2)\right] > \frac{\sigma_{ec}}{E}$$

and   or

$$\sigma_1 - \frac{1}{\nu}[\sigma_2 + \sigma_3] > \sigma_{et}$$

$$\sigma_3 - \frac{1}{\nu}[\sigma_1 + \sigma_2] > \sigma_{ec}$$

To prevent failure,

$$\sigma_1 - \frac{1}{\nu}(\sigma_2 + \sigma_3) < \sigma_{et}$$

$$\sigma_3 - \frac{1}{\nu}(\sigma_1 + \sigma_2) < \sigma_{ec}$$

At the point of elastic failure

$$\sigma_1 - \frac{1}{\nu}(\sigma_2 + \sigma_3) = \sigma_{et}$$

and

$$\left|\sigma_3 - \frac{1}{\nu}(\sigma_1 + \sigma_2)\right| = \sigma_{ec}$$

For design purposes

$$\sigma_1 - \frac{1}{\nu}(\sigma_2 + \sigma_3) = \sigma_t$$

$$\left|\sigma_3 - \frac{1}{\nu}(\sigma_1 + \sigma_2)\right| = \sigma_c$$

(where σ_t and σ_c are the safe stresses)

The following are important points,

i.   The theory overestimates the behaviour of ductile materials

ii. The theory does not fit well with the experimental results except for brittle materials for biaxial tension-compression state of stress (for which it is sometimes recommended) and is not much used in practice.

GRAPHICAL REPRESENTATION OF VARIOUS THEORIES OF FAILURE FOR A 2-DIMENSIONAL STRESS SYSTEM

All the theories of failure discussed in this chapter can be graphically represented in a simple diagram as shown in Figure 6.1.

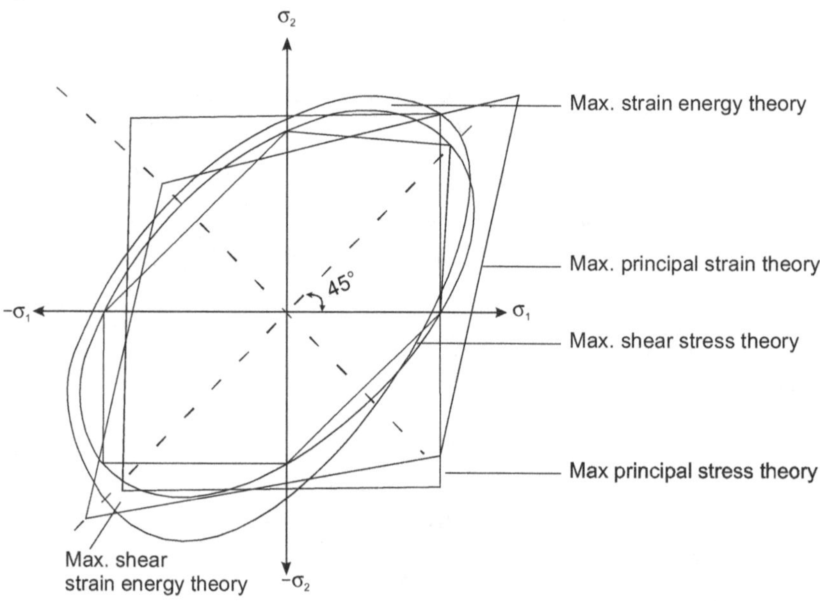

Figure 6.1 Graphical representation of theories of failure

SUMMARY—THEORIES OF FAILURE

To predict when failures occur under combined loading knowing S_{yp} or S_u

1. Maximum principal stress theory (Rankine)

    For ductile materials $\sigma_1 = \sigma_{yp}$ for failure

    $\sigma_1 \leq (\sigma_{yp})/FS$ for design

    For brittle materials $\sigma_1 = \sigma_u$ for failure

    $\sigma_1 = (\sigma_u)/FS$ for design

2. Maximum shear stress theory(Tresca) (MSS theory)

    $\tau_{max} = \tau_{yp} = \sigma_{yp}/2$

    $(\sigma_1 - \sigma_3)/2 = (\sigma_{yp})/2$ if $\sigma_1 > \sigma_2 > \sigma_3$

$\sigma_1 - \sigma_3 = \sigma_{yp}$ for failure

$(\sigma_1 - \sigma_3)/FS \le \sigma_{yp}/FS$ for design

3. Maximum principal strain theory (St. Venant)

$\sigma_1 - v(\sigma_2 + \sigma_3) = \sigma_{yp}$ for failure; $\sigma_1 > \sigma_2 > \sigma_3$

4. Maximum strain energy theory (Haigh)

$$\sigma_1^2 + \sigma_2^2 + \sigma_3^2 - 2\mu(\sigma_1\sigma_2 + \sigma_2\sigma_3 + \sigma_3\sigma_1) = \sigma_{yp}^2$$

5. Maximum shear strain energy theory (Vonmises) (DE theory)

$$(\sigma_1 - \sigma_2)^2 + (\sigma_2 - \sigma_3)^2 + (\sigma_3 - \sigma_1)^2 = 2\sigma yp^2 \text{ for failure}$$

PROBLEM 6.1

The state of plane stress shown occurs at a critical point of a steel machine component. As a result of several tensile tests, it has been found that the tensile yield strength is S_y = 250 MPa for the grade of steel used. Determine the factor of safety with respect to yield, using (a) the maximum-shearing-stress criterion, and (b) the maximum-distortion energy criterion.

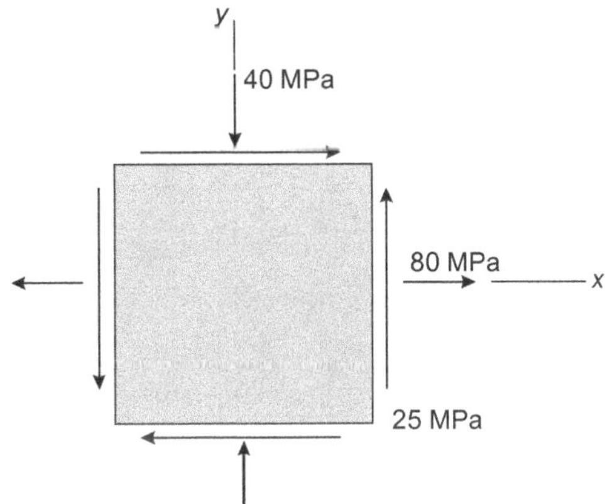

Mohr's circle We construct Mohr's circle for the given state of stress and find

$$s_{ave} = OC = \frac{1}{2}(s_x + s_y) = \frac{1}{2}(80 - 40) = 20\,\text{MPa}$$

$$\tau_m = R = \sqrt{(CF)^2 + (FX)^2} = \sqrt{(60)^2 + (25)^2} = 65\,\text{MPa}$$

(a) *Maximum-shearing-stress criterion* Since for the grade of steel used the tensile strength is S_y = 250 MPa, the corresponding shearing stress yields is

$$t_Y = \frac{1}{2}s_y = \frac{1}{2}(250\,\text{MPa}) = 125\,\text{MPa}$$

For $t_m = 65\,\text{MPa}$ $F.S. = \dfrac{t_y}{t_m} = \dfrac{125\,\text{MPa}}{65\,\text{MPa}}$ $F.S. = 1.92$

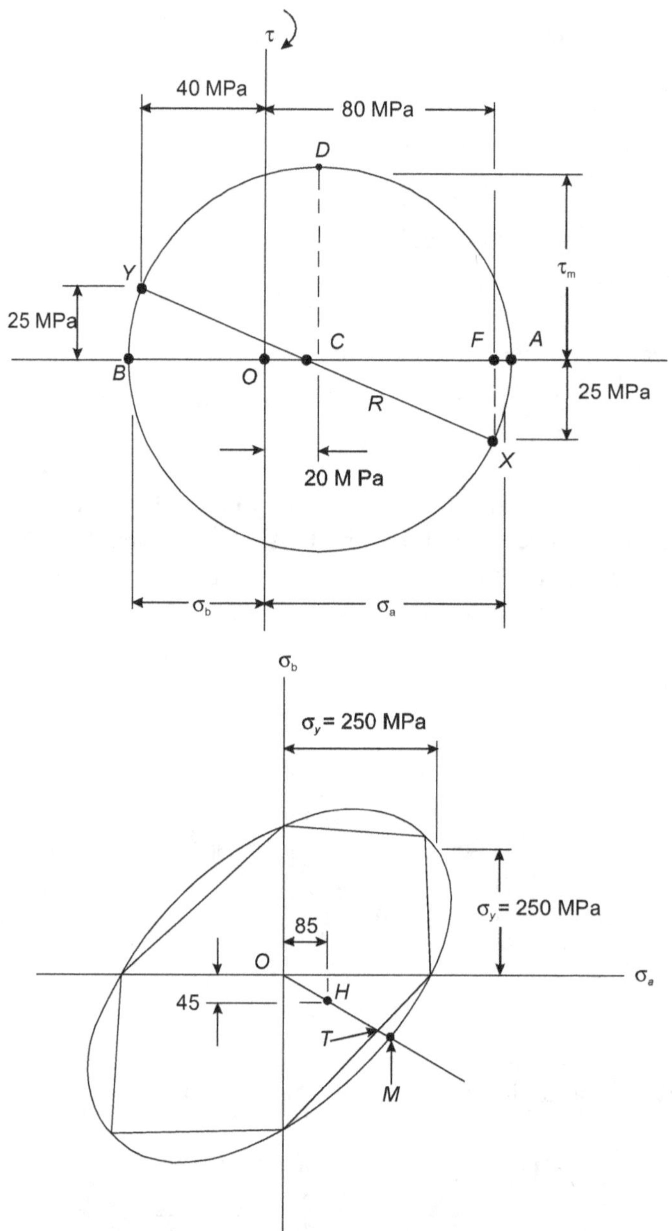

Principal stresses

$$s_a = OC + CA = 20 + 65 = 20 + 65 = +85\,\text{MPa}$$

$$s_b = OC - BC = 20 - 65 = -45\,\text{MPa}$$

(b) *Maximum-distortion-energy criterion* Introducing a factor of safety into equation (6.4) we write

$$\sigma_a^2 - \sigma_a\sigma_b + \sigma_b^2 = \left(\frac{\sigma_y}{F.S.}\right)^2$$

For $\sigma_a = +85\,\text{MPa}, \sigma_b = -45\,\text{MPa},$ and $\sigma_y = 250\,\text{MPa},$ we have

$$(85)^2 - (85)(-45) + (45)^2 = \left(\frac{250}{F.S.}\right)^2$$

$$114.3 = \frac{250}{F.S.} \qquad F.S. = 2.19$$

Comment. For a ductile material $\sigma_y = 250\,\text{MPa}$, with we have drawn the hexagon associated with the maximum-shearing-stress criterion and the ellipse associated with the maximum-distortion-energy criterion. The given state of plane stress is represented by point H of coordinates $S_a = 85$ MPa and $S_b = -45$ MPa We note that the straight line drawn through points O and H intersects the hexagon at the point T and the ellipse at point M. For each criterion, the value obtained for *F.S.* may be verified by measuring the line segments indicated in the Figure and computing their ratios:

(a) $F.S = \dfrac{OT}{OH} = 1.92$ (b) $F.S. = \dfrac{OM}{OH} = 2.19$

PROBLEM 6.2

This example illustrates the use of a failure theory to determine the strength of a mechanical element or component. The example may also clear up any confusion existing between the phrases 'strength of a machine part', 'strength of a material', and 'strength of a part at a point'.

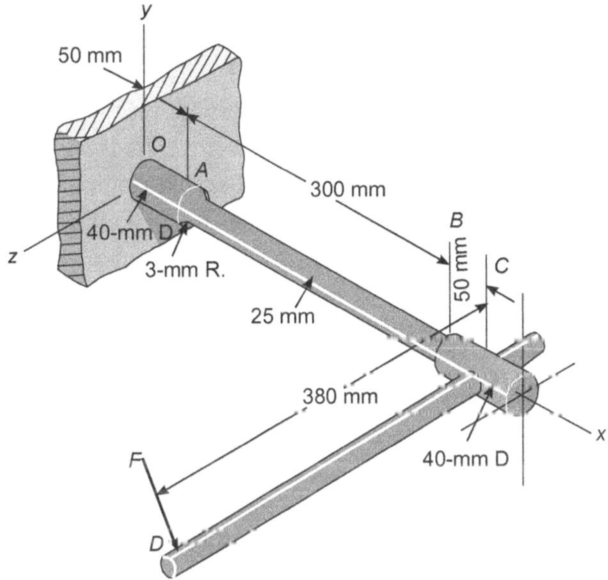

A certain force F applied at D near the end of the 380-mm lever shown in Fig. which is quite similar to a socket wrench, results in certain stresses in the cantilevered bar $OABC$. This bar $(OABC)$ is of AISI 1035 steel, forged and heat-treated so that it has a minimum (ASTM) yield strength of 560 MPa. We presume that this component would be of no value after yielding. Thus the force F required to initiate yielding can be regarded as the strength of the component part. Find this force.

Solution

We will assume that lever DC is strong enough and hence not a part of the problem. A 1035 steel, heat-treated, will have a reduction in area of 50 percent or more and hence is ductile material at normal temperatures. This also means that stress concentration at shoulder A need not be considered. A stress element at A on the top surface will be subjected to a tensile bending stress and a torsional stress. This point, on the 25-mm diameter section is the weakest section, and governs the strength of the assembly. The two stresses are

$$\sigma_x = \frac{M}{I/c} = \frac{32M}{\pi d^3} = \frac{32(0.355\,F)}{\pi(0.025^3)} = 231.424\,F$$

$$\tau_{zx} = \frac{Tr}{J} = \frac{16T}{\pi d^3} = \frac{16(0.38\,F)}{\pi(0.025^3)} = 123.860\,F$$

Employing the distortion-energy theory, we find, from equation (6.4), that

$$\sigma' = (\sigma_x^2 + 3\tau_{zx}^2)^{1/2} = [(231.424F)^2 + 3(123.860F)^2]^{1/2} = 315.564F$$

Equating the von Mises stress σ'_{eq} to S_y, we solve for F and get

$$F = \frac{S_y}{315.564} = \frac{560 \times 10^6}{315.564} = 1.77\,\text{kN}$$

In this example the strength of the material at point A is $S_y = 560$ MPa. The strength of the assembly or component is F = 1.8 kN.

Let us see how to apply the MSS theory. For a point undergoing plane stress with only one non-zero normal stress and one shear stress, the two non-zero principal stresses σ_A and σ_B will have opposite signs and hence from the MSS theory equation (6.2) gives

$$\sigma_A - \sigma_B = 2\left[\left(\frac{\sigma x}{2}\right)^2 + \tau_{zx}^2\right]^{1/2} = (\sigma_x^2 + 4\tau_{zx}^2)^{1/2}$$

$$(\sigma_x^2 + 4\tau_{zx}^2)^{1/2} = S_y$$

$$[(231424F)^2 + 4(123860\,F)^2]^{1/2} = 339002\,F = 560 \times 10^6$$

$$\therefore F = 1.65\,\text{kN}$$

which is about 7 percent less than found for the DE theory. As stated earlier, the MSS theory is more conservative than the DE theory.

PROBLEM 6.3

The cantilevered tube shown in Figure is to be made of 2014 aluminum alloy treated to obtain a specified minimum yield strength of 276 MPa. We wish to select a stock size tube of 42mm σD and thick 5 mm or 42 mm σD and thick 4mm using a design factor of safety FS = 4. The bending load is F = 1.75 kN, the axial tension is P = 9.0 kN, and the torsion is T = 72 N. m. What is the realized factor of safety?

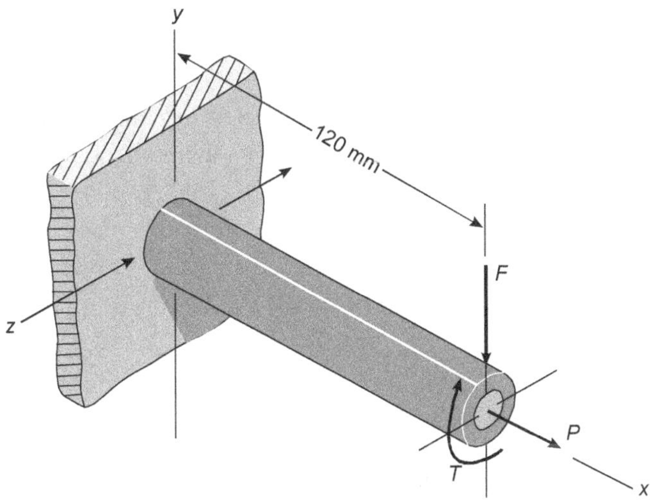

Solution

Since the maximum bending moment is M = 120 F, the normal stress, for an element on the top surface of the tube at the origin, is

$$\sigma_x = \frac{P}{A} + \frac{Mc}{I} = \frac{9}{A} + \frac{120(1.75)(d_0/2)}{I} = \frac{9}{A} + \frac{105d_0}{I} \tag{1}$$

where

d_0 is diameter

$\dfrac{P}{A}$ is direct stress

$\dfrac{Mc}{I}$ is bending stress

Total longitudinal stress = direct stress + Bending stress

If millimeters are used for the area properties, the stress is in gigapascals.

The torsional stress at the same pont is

$$\tau_{zx} = \frac{Tr}{J} = \frac{72(d_0/2)}{J} = \frac{36d_0}{J} \tag{2}$$

For accuracy, we choose the distortion-energy theory as the design basis. The von Mises stress is

$$\sigma' = (\sigma_x^2 + 3\tau_{zx}^2)^{1/2} \tag{3}$$

On the basis of the given factor of safety of 4 the goal for σ' is

$$\sigma' \leq \frac{S_y}{F.S} = \frac{0.276}{4} = 0.0690 \text{ GPa} \tag{4}$$

where we have used gigapascals in this relation to agree with eqs. (1) and (2).

We can compute these for the two tube sizes. $A = 42 \times 5$-mm tube is satisfactory. The von Mises stress is found to be $\sigma' = 0.06043$ GPa for this size. Thus the realized factor of safety is

$$n = \frac{S_y}{\sigma'} = \frac{0.276}{0.06043} = 4.57$$

For the next smaller size, a 42- $\times$ 4-mm tube $\sigma' = 0.07105$ GPa , giving a factor of safety of

$$n = \frac{S_y}{\sigma'} = \frac{0.276}{0.07105} = 3.88$$

INTERESTING POINTS TO PONDER

VONMISES THEORY

Quite often for ductile materials, the most commonly used failure theory is Vonmises theory or maximum shear strain energy theory. The advantage is that all the principal stresses σ_1, σ_2, σ_3, are used unlike maximum shear stress theory wherein only two at a time is taken for calculation leaving the third one, viz., σ_1 and σ_2, σ_3 and σ_1.

VonMises theory also gives the concept of equivalent stress σ_{eq} also known as VonMises stress. σ_{eq} is simply termed as the single value of stress which can replace the existing state of stress to asses the safety of the design. Moreover when fatigue analysis or creep analysis is done, the material properties are plotted with σ vs n or σ vs time as the case is. In such plots also, even if the body is subjected to a 3-dimensional state of stress, we can determine σ_{eq} from VonMises theory and σ_{eq} in such plots to determine no. of cycles failure (fatigue analysis) or time to rupture (creep analysis). The calculations made in this way are quite acceptable for design. All standard finite element analysis program also give the print out or plot of VonMises stresses in various elements (Chapter 18) and can be used for design check if safe or otherwise.

7

EXPERIMENTAL METHODS FOR DETERMINING STRAINS AND STRESSES

A structural member or a machine part may be of such a form or may be loaded in such a way that the direct use of formulas for the calculation of stresses and strains produced in it is impractical. One can then resort either to numerical techniques such as the finite-element method or to experimental methods. The experimental methods can be applied to the actual member in some cases, or to a model thereof, or to a more conventionalized specimen. Which choice is made depends upon the results desired, the accuracy needed, and the costs associated with the several methods. There has been a tremendous increase in the use of numerical methods in recent years, but comparable increases have also been seen in the use of experimental techniques. Indeed, many investigations make use of both numerical methods and experimental techniques to cross-feed information from one to the other for increased accuracy, cost effectiveness and establishing validation of results obtained using numerical methods.

MEASUREMENT OF STRAIN

The most direct way of determining the stresses produced in a component is to measure the accompanying strains. Measurement is comparatively easy when the stress is fairly uniform over a considerable length of the part in question, but becomes difficult when the stress is localized or varies greatly with position. Short gauge lengths and great precision require stable gauge elements and stable electronic amplification. In an isotropic material under uniaxial stress one strain measurement is all that is needed. On a free surface under biaxial stress conditions two measured orthogonal strains will give the stresses in those same directions. Three measured surface strains will allow the determination of the stresses in all directions at that position. At a free edge in a member which is thin compared to its other dimensions, one might measure the change in thickness or the through-thickness strain by a gauge to determine the stress parallel to the edge such as in the bottom of the groove at the edge of a plate .

The most important instruments and techniques used for strain measurement are discussed as follows.

Mechanical measurement   A direct measurement of strain can be made with an Invar tape over a gauge length of several metres or with a pair of dividers over a reasonable fraction of a metre. For shorter gauge lengths, mechanical amplification can be used, but friction is a problem and vibration can make them difficult to mount and to read. Optical magnification using mirrors still requires mechanical levers or rollers and is an improvement but still not satisfactory for most applications. In a laboratory setting, however, such mechanical and optical magnification can be used successfully.

Brittle coatings   Surface coatings which are brittle and expected to crack at strain levels well within the elastic limit of most structural materials, provide a means of locating points of maximum strain and the directions of principal strains. Under well-controlled environmental conditions and with suitable calibration such coatings can yield reasonably acceptable results. However, we can use other methods to fine-tune as now the location is established by brittle coatings.

Electrical strain and displacement gauges   The evolution of electrical gauges has led to a variety of configurations where changes in resistance, capacitance, or inductance can be related to strain and displacement with proper instrumentation

Electric resistance strain gauge   For the electrical resistance strain gauges, the gauge lengths vary from less than 0.01 inch to several inches . The gauge grid material may be metallic or a semiconductor and can be obtained in alloys which are designed to provide minimum output due to temperature strains alone and comparatively large outputs due to stress-induced strains. The semiconductor strain gauges have the largest changes in resistance for a given strain but are generally very sensitive to temperature.

Capacitance strain gauge   Capacitance strain gauges are larger and more massive than bonded electric resistance strain gauges and are more widely used for applications beyond the upper temperature limits of the bonded resistance strain gauges.

Inductance strain gauges   The change in air gap in a magnetic circuit can create a large change in inductance depending upon the design of the rest of the magnetic circuit. The large change in inductance is accompanied by a large change in force across the gap, and so the very sensitive inductance strain gauges can be used only on more massive structures. They have been used as overload indicators on presses with no electronic amplification necessary. The linear relationship between core motion and output voltage of a linear differential transformer makes possible accurate measurement of displacements over a wide range of gauge lengths and under a wide variety of conditions.

Of the above, the electrical resistance strain gauge is most commonly employed by design, production and testing establishment as they are convenient to handle and provide accurate results needed to validate and verify design and analytical calculation. They are also relatively cheaper and require standard instrumentation available commercially. Almost all mechanical measurements or instrumentation laboratories of engineering/technical institutions have this facility. Hence, we are discussing this topic in more depth compared to other methods of strain measurements listed herein.

The electrical resistance strain gauge discussed in this chapter is basically a piece of very thin foil or fine wire which exhibits a change in resistance proportional to the mechanical strain imposed on it . In order to handle such a delicate filament, it is either mounted on or bonded to some type of carrier material and is known as the bonded strain gauge .

The strain gauge is used widely by stress analysts in the experimental determination of stresses and in checking the results obtained from analytical or numerical techniques. The electrical resistance strain gauge is more often preferred and used.

Resistance Strain Gauges—Theoretical background

Consider a conductor of uniform cross-sectional area A and length L, made of a material with resistivity ρ . The resistance R of such a conductor is given by

$$R = \frac{\rho L}{A} \tag{7.1}$$

If this conductor is now stretched or compressed, its resistance will change because of dimensional changes (length and cross-sectional area) and because of a fundamental property of materials called $p\bar{i}$-$\bar{e}z\bar{o}$- resistance (pronounced $p\bar{i}$-$\bar{e}z\bar{o}$- resistance), which indicates a dependence of resistivity ρ on the mechanical strain. To find how a change dR in R depends on the basic parameters, we differentiate equation.(7.1) to get

$$dR = \frac{A(\rho\, dL + L d\rho) - \rho\, L\, dA}{A^2} \tag{7.2}$$

Since volume $V = AL$, $dV = A\, dL + L dA$

Also

dV = change in volume

       = new volume after deformation − old volume

       = product of sides after deformation − old volume

$dV = L(1+\varepsilon)(1-\varepsilon V)^2 - AL$

$$dV = L(1+\varepsilon)\, A(1-\varepsilon v)^2 - AL \tag{7.3}$$

Where ε = unit strain and v = Poisson's ratio. Since ε is small, $(1-v\varepsilon)^2 \approx 1- 2v\varepsilon$ and equation. (7.3) becomes

$$dV = AL\varepsilon(1-2v) = A\, dL + L dA \tag{7.4}$$

and since $\varepsilon = dL/L$,

$$A\, dL(1-2v) = A\, dL + L dA \tag{7.5}$$

$$- 2v\, AdL = LdA \tag{7.6}$$

Substituting in equation (7.2) yields ,

$$dR = \frac{\rho\, AdL + LAd\rho + 2v\rho\, AdL}{A^2} \tag{7.7}$$

and thus

$$dR = \frac{\rho dL(1 + 2v)}{A} + \frac{Ld\rho}{A} \tag{7.8}$$

Dividing by equation (7.1) gives

$$\frac{dR}{R} = \frac{dL}{L}(1 + 2v) + \frac{d\rho}{\rho} \tag{7.9}$$

and finally

$$\text{Gauge factor} \triangleq \frac{dR/R}{dL/L} = \underbrace{1}_{\substack{\text{resistance}\\\text{change due}\\\text{to length}\\\text{change}}} + \underbrace{2v}_{\substack{\text{resistance}\\\text{change due}\\\text{to are change}}} + \underbrace{\frac{d\rho/\rho}{dL/L}}_{\substack{\text{resistance change due}\\\text{to piezoresistance}\\\text{effect}}} \tag{7.10}$$

Thus if the gauge factor is known, measurement of dR/R allows measurement of the strain $dL/L = \varepsilon$. This is principle of the resistance strain gauge.

The basic principle of the resistance strain gauge is implemented in several different ways :

1. Unbonded metal-wire gauge

2. Bonded metal-wire gauge

3. Bonded metal-foil gauge Figure (7.1)

4. Vacuum-deposited thin-metal-film gauge

5. Sputter-deposited thin-metal-film gauge

6. Bonded semiconductor gauge

7. Diffused semiconductor gauge

The bonded metal-wire gauge (totally largely superseded by the bonded metal-foil construction) has been applied to both stress analysis and transducers. A grid of fine wire is cemented to the specimen surface, where strain is to be measured (Figure 7.1). Embedded in a matrix of cement, the wires cannot buckle and thus faithfully follow both the tension and the compression strains of the specimen.

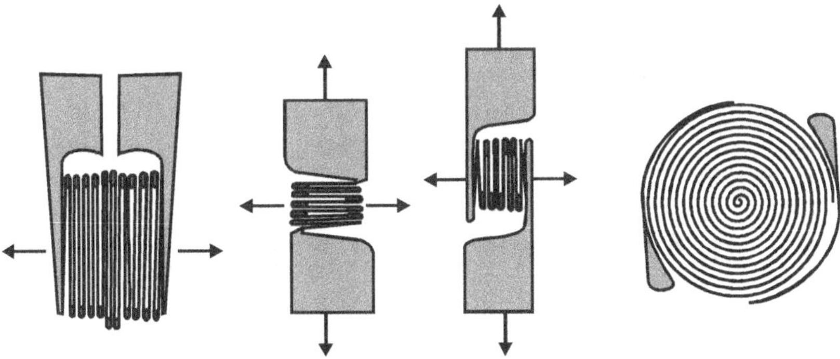

Figure 7.1   Foil strain gauges

Bonding Techniques   The proper functioning of a strain gauge is completely dependent on the bond which holds it to the structure undergoing test. If the bond does not faithfully transmit the strain from the test piece to the wire or foil of the gauge, the results obtained cannot be accurate. Usually, the manufacturer of the strain gauges will recommend cements which are compatible with their use and will provide guidelines for their proper installation .

Strain-gauge Construction

Since the foil used in a strain gauge must be very fine or thin to have a sufficiently high electrical resistance (usually between 60 and 350 ohms.), it is difficult to handle. For example, the foil used in gauges is often about 0.1 mil in thickness. Some use has been made of wire filaments in strain gauges, but this type of gauge is seldom used except in special or high-temperature applications. In order to handle this delicate foil, it must be provided with a carrier medium or backing material, usually a piece of paper, plastic, or epoxy. The cement provides so much lateral resistance to the foil that it can be shortened significantly without buckling; then compressive as well tensile strains can be measured. Lead wires or connection terminals are often provided on foil gauges, as illustrated in the typical foil gauge shown in Figure 7.2.

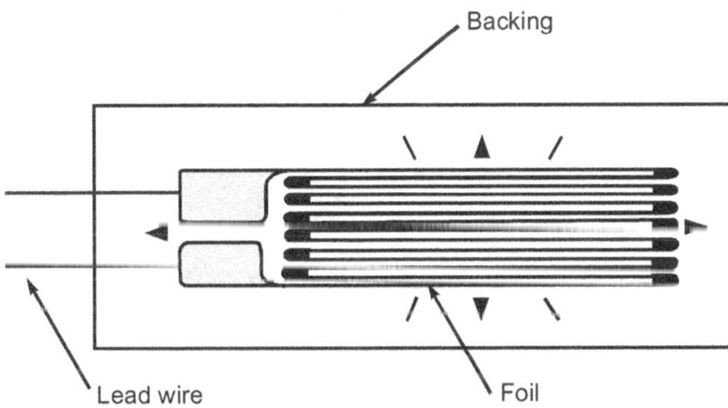

Figure 7.2 Typical construction of a foil strain gauge

Transverse Sensitivity

Because of its construction, a portion of the foil in each gauge lies in the transverse direction and will respond to transverse strain. Therefore the gauge factor K of a gauge * is always slightly smaller than the gauge factor of the material of which it is fabricated. One of the desirable features of foil-type gauges is their low transverse sensitivity.

STRAIN-GAUGE CIRCUITRY AND INSTRUMENTATION

WHEATSTONE BRIDGE

The circuitry and associated instrumentation block diagram are shown in Fig. 7.3(a) and 7.3(b) respectively. In the potentiometer circuit it is necessary to block the DC component of the output voltage with a capacitor before feeding the signal to the input of an amplifier. The same effect can be achieved by suppressing the DC component of the signal by connecting two potentiometer circuits in parallel and taking the output signal from corresponding points in the two branches of the resulting network, as shown in Figure 7.3. This circuit arrangement is generally referred to as a Wheatstone bridge, and represents one of the most precise methods known for measuring (or comparing) resistances. Advantages of the Wheatstone bridge over the potentiometer circuit are (1) much greater flexibility in circuit arrangements for signal augmentation, temperature compensation, and cancellation or separation of variables, (2) capacity for accurately indicating combined static and dynamic strains, and (3) virtually complete freedom from error due to resistive changes in the conductors connecting the supply voltage to the network.

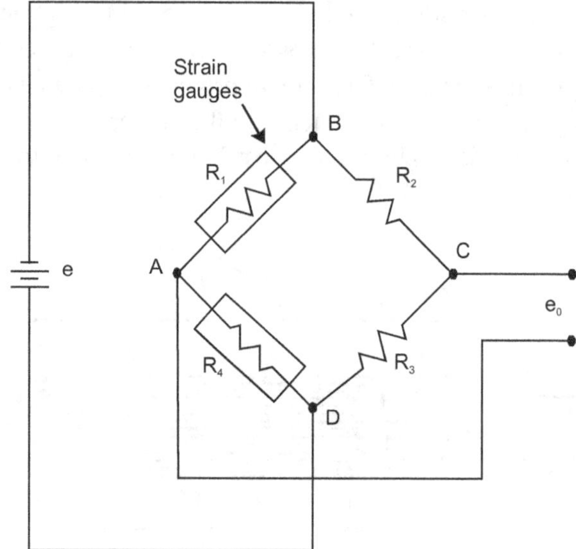

(a) Wheatstone bridge circuit for static and dynamic strain measurement

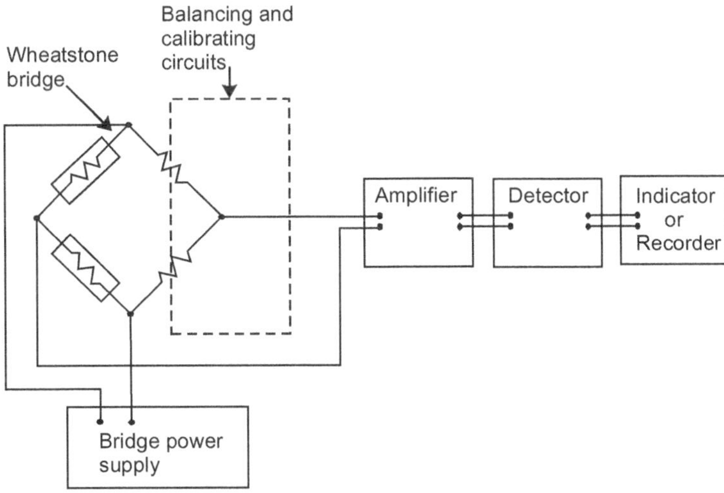

(b) Block diagram of basic elements of strain-gauge instrumentation system

Figure 7.3

Temperature Effects

When a bonded strain gauge is used in measurements, any change in resistance in the strain-gauge measurement system is interpreted as resulting from a strain. If thermal expansion is not induced, then this change will result from a mechanical strain. However, if thermal expansion is induced, then there will be a change in resistance resulting from the mechanical strain, and in addition, there will be a change in resistance resulting from the response of the strain gauge to changes in temperature. The strain indication which results from such a temperature effect is known as an " apparent strain". In the measurement of static strains, the effects of temperature represent the largest potential source of error and require some form of temperature compensation, which is shown in Figure 7.4.

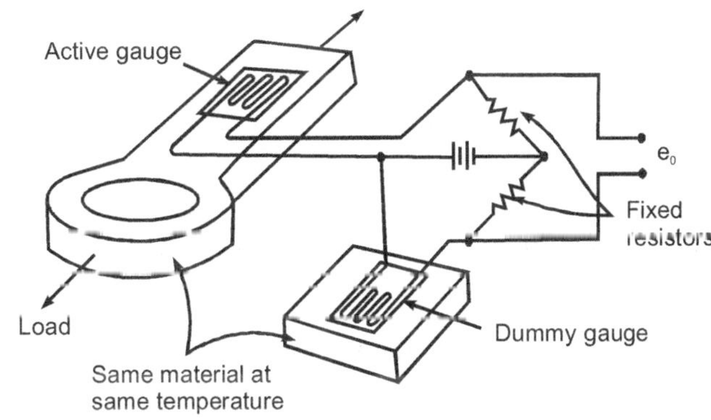

Figure 7.4 Strain gauge temperature compensation

Gauge selection is dependent on the space limitation and steepness of strain gradient in any region. The strain gauge indicates the average strain over the length of the gauge; in a region of steep strain gradient, this indicated value may be much less than the maximum strain. The shorter the gauge used in such a region, the closer is the gauge indication to the maximum strain (Figure 7.5). A single gauge can be used in only the very limited case where a stress exists in one direction only, and that direction must be known. If stresses exist in several directions, or if the direction of a singly existing stress is unknown, a strain-gauge rosette consisting of three or more gauges must be employed.

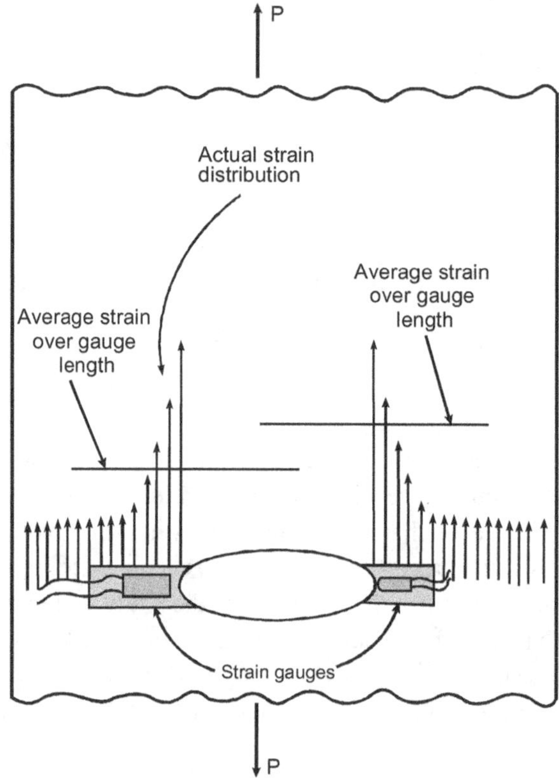

Figure 7.5   Effect of gauge length on zones of steep strain gradients

The right side strain gauge indicates a higher strain and extremely small length of strain gauge should be employed to measure peak strain. This aspect is very important while validating results of finite element analysis (FEA) over very small elements used in severe stress-gradient zones. The gauge length should be compatible with element dimensions if validation results are used for acceptance of FEA results

In addition to single-element gauges,gauge combination called rosettes (Figure.7.6) are available in many configurations for specific stress-analysis or transducer applications. While individual gauges conceivably could be cemented down in the same patterns, precise relative orientation of the several gauges is critical in most of these applications, and this is much more easily obtained in rosette manufacture than by the user with single gauges. One rosette commonly used in stress analysis solves the problem of a surface stress whose magnitude and direction are totally unknown. Theory shows that measurements with a 3-gauge rosette allow calculation of all the desired information. Since such measurements are intended to define stress at a point, ideally the three gauges should be superimposed on that point.

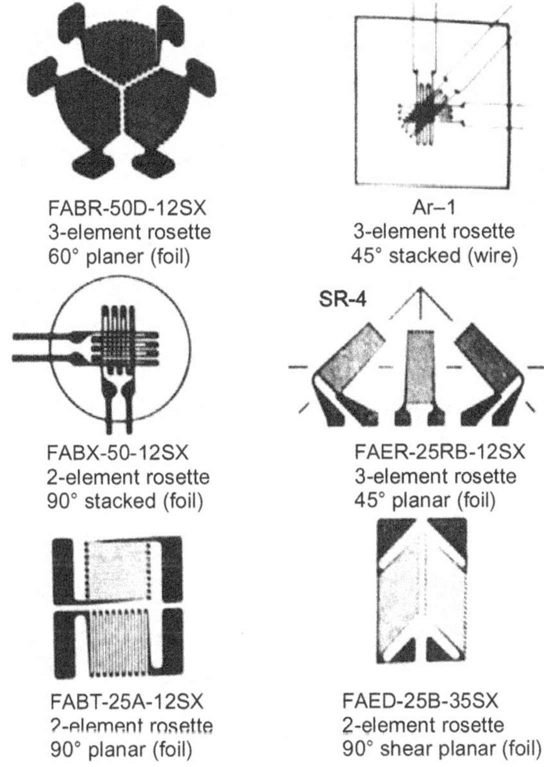

Figure 7.6 Strain-gauge rosettes

Interferometric strain gauges   A simple strain gauge with a short length and high sensitivity can be created by several methods. In one, a diffraction grating is deposited at the desired location and in the desired direction and the change in grating pitch under strain is measured. With a metallic grid these strain gauges can be used at elevated temperatures. Another method, also usable at high temperatures, makes use of the interference of light reflected from the inclined surfaces of two very closely spaced indentations in the surface of a metallic specimen.

Moiré techniques   All Moiré techniques can be explained by optical interference.

Geometric moiré   Geometric moiré techniques use grids of alternate equally wide bands of relatively transparent or light-coloured material and opaque or dark-coloured material in order to observe the relative motion of two such grids. The most common technique uses an alternate transparent and opaque grid to produce photographically a matching grid on the flat surface of the specimen.

Then the full-field relative motion is observed between the reproduction and the original when the specimen is loaded. Similarly, the original may be used with a projector to reproduce the photographic image on the specimen and then produce interference with the projected image after loading. White light can be used, as the interference is due merely to geometric blocking of the light as it passes through or is reflected from the grids.

Moiré interferometry   Interferometry provides a means of producing both specimen gratings and reference gratings. Virtual reference gratings of more than 100,000 lines per inch have been utilized. Moire interferometry provides contour maps of in-plane displacements, and with the fine pitches attainable, differentiation to obtain strains from this experimental process is comparable to that used in the finite- element method of numerical analysis where displacement fields are generally the initial output .

Holographic and laser speckle interferometry   The rapid evolution of holographic and laser speckle interferometry is related to the development of high-power lasers and to the development of digital computer enhancement of the resulting images. Various techniques are used to measure the several displacement components to diffuse reflecting surfaces.

X-ray diffraction   X-ray diffraction makes possible the determination of changes in interatomic distance and thus the measurement of elastic strain. The method has the particular advantages that it can be used at points of high stress concentration and to determine residual stresses without cutting the object of investigation.

PHOTOELASTIC ANALYSIS

When we discussed general theory of elasticity for isotropic materials, the equations for plane problems can be solved using stress function approach. This approach provides the basic principle that irrespective of the used material, the stress distribution is same for a given loading. Photoelastic analysis is based on this principle. A model made of perspex loaded as in the main components is used in most photoelastic analysis studies and provides information on stress distribution in the actual component.

When a beam of polarized light passes through an elastically stressed transparent isotropic material, the beam may be treated as having been decomposed into two rays polarized in the planes of the principal stresses in the material. In birefringent materials the indexes of refraction of the material encountered by these two rays will depend upon the principal stresses. Therefore, interference patterns will develop which are proportional to the differences in the principal stresses.

Two-dimensional analysis   With suitable optical elements—polarizers and wave plates of specific relative retardation—both the principle stress differences and the directions of principle stresses can be determined at every point in a two-dimensional specimen. Many suitable photoelastic plastics are available. The material properties that must be considered are transparency, sensitivity (relative index of refraction change with stress), optical and mechanical creep , modulus of elasticity , ease of machining, cost, and stability (freedom from stresses developing with time).

Three-dimensional analysis   Several photoelastic techniques are used to determine stresses in three-dimensional specimens . If information is desired at a single point only, the optical polarizers, wave plates, and photoelastically sensitive material can be embedded in a transparent model and two-dimensional techniques used. A modification of this technique, stress freezing, is possible in some biphase materials. By heating, loading, cooling, and unloading, it is possible to lock permanently into the specimen, on a molecular level, stresses proportional to those present under load. Since equilibrium exists at a molecular level, the specimen can be cut into two-dimensional slices and all secondary principal stress differences determined. The secondary principal stresses at a point are defined as the largest and smallest normal stresses in the plane of the slice; these in general will not correspond with the principal stresses at that same point in the three-dimensional structure. If desired, the specimen can be cut into cubes and the three principal stress differences determined. The individual principal stresses at a given point cannot be determined from photoelastic data taken at that point alone since the addition of a hydrostatic stress to any cube of material would not be revealed by differences in the indexes of refraction. Mathematical integration techniques, which start at a point where the hydrostatic stress component is known, can be used with photoelastic data to determine all individual principal stresses .

Photoelastic coating   Photoelastic coatings have been sprayed , bonded in the form of thin sheets, or cast directly in place on the surface of models or structures to determine the two-dimensional surface strains. The surface is made reflective before bonding the plastics in place so the effective thickness of the photoelastic plastic is doubled and all two-dimensional techniques can be applied with suitable instrumentation .

ANALOGIES

Certain problems in elasticity like stress concentration problem, Torsion of non circular section are described by equations that cannot be easily solved but they are mathematically identical with the equations that describe some other physical phenomenon which can be investigated experimentally.

Membrane analogy   This is especially useful in determining the torsion properties of bars having noncircular sections. If in a thin flat plate holes are cut having the outlines of various sections and over each of these holes a soap film (or other membrane) is stretched and slightly distended by pressure from one side, the volumes of the bubbles thus formed are proportional to the torsional rigidities of the corresponding sections and the slope of a bubble surface at any point is proportional to the shear stress caused at that point of the corresponding section by a given twist per unit length of bar. By cutting in the plate, one hole the shape of the section to be studied and another hole that is circular, the torsional properties of the irregular section can be determined by comparing the bubble formed on the hole of that shape with the bubble formed on the circular hole since the torsional properties of the circular section are known .

Electrical analogy for isopachic lines   **Isopachic lines** are lines along which the sums of the principle stresses are equal in a two-dimensional plane stress problem . The voltage at any point on a uniform two-dimensional conducting surface is governed by the same form of equation as is the principle stress sum. Teledeltos paper is a uniform layer of graphite particles on a paper backing and makes an excellent material from which to construct the electrical analog. The paper is cut to a geometric outline corresponding to the shape of the two-dimensional structure or part, and boundary potentials are applied by an adjustable power supply. The required boundary potentials are obtained from a photoelastic study of the part where the principal stress sums can be found from the principal stress difference on the boundaries.

8

BENDING STRESSES IN BEAMS

PURE BENDING

Prismatic bars are subjected to equal and opposite couples. These are basically treated as one-dimensiional structures. In chapter 3, we considered bars subjected to axial tension or compression. Here in this chapter, the bars or beams are subjected to lateral or transverse loads causing bending.

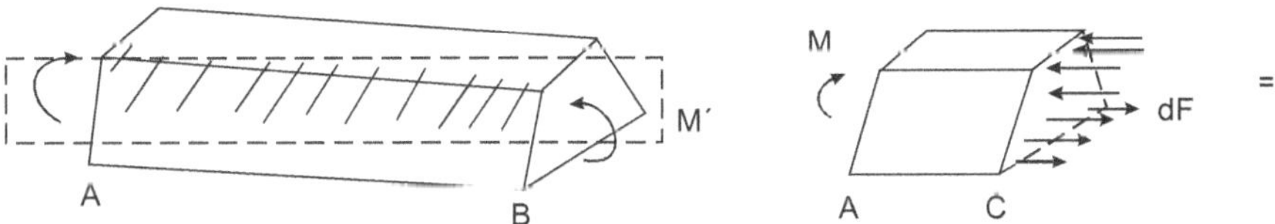

Figure 8.1

    The conditions of equilibrium of portion AC of the member require that the elementary forces exerted on AC by the other portion be equivalent to the couple M (Figure 8.1). Thus the internal forces in any cross section of a member in pure bending are equivalent to a couple. The moment of that couple is known as bending moment in the section.

    Assign +ve sign when the member is bent as shown in Figure 8.2 and − ve sign when the senses of M and M' are reversed.

    An example for pure bending is shown in Figure 8.2 (simply supported beam)

    Take a section in between B and C, say at E

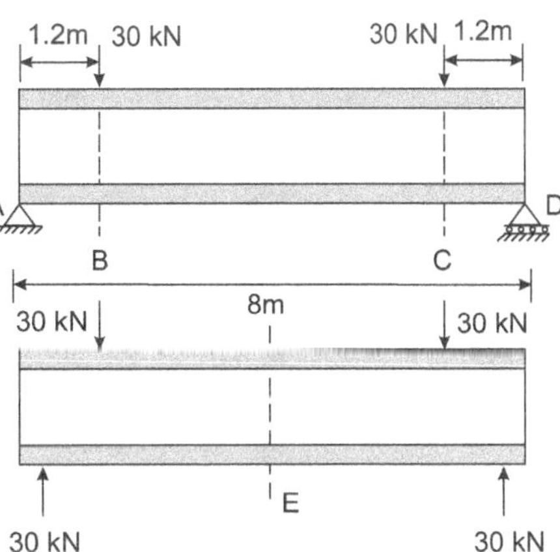

Figure 8.2

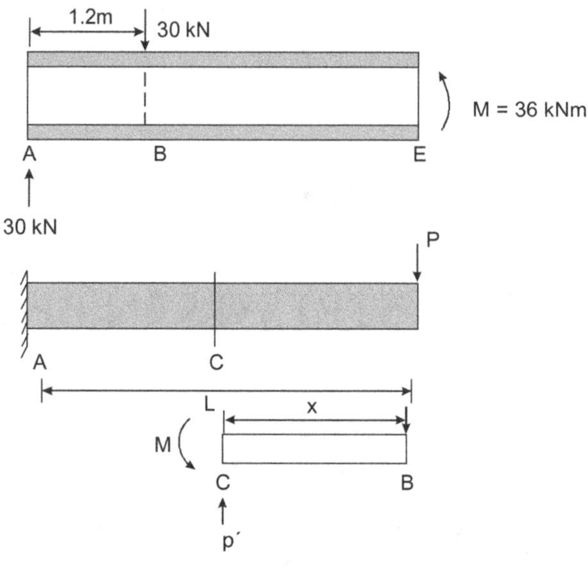

Figure 8.3

The internal forces at section E must be equivalent to a couple of 36 kNm. For the cantilever beam (Figure 8.3),

$$M = Px$$

Bending moment and shear force diagrams are most important for this chapter and to recapitulate, bending moment and shear force distribution in beams, a few examples are given for illustration. The bending moment and shear force diagram are necessary to compute the stress in the beam and carryout the design process. Recall the following relations.

$$w = \frac{dV}{dx} \tag{8.1}$$

$$V = \frac{dM}{dx} \tag{8.2}$$

where,

   V—shear force

   w—load intensity and

   M—bending moment

PROBLEM 8.1

Draw the shear and bending moment diagrams for the cantilever beam AB. The distributed load of 48 kN/m extends over 2 m of the beam and the 40 kN load is applied at E.

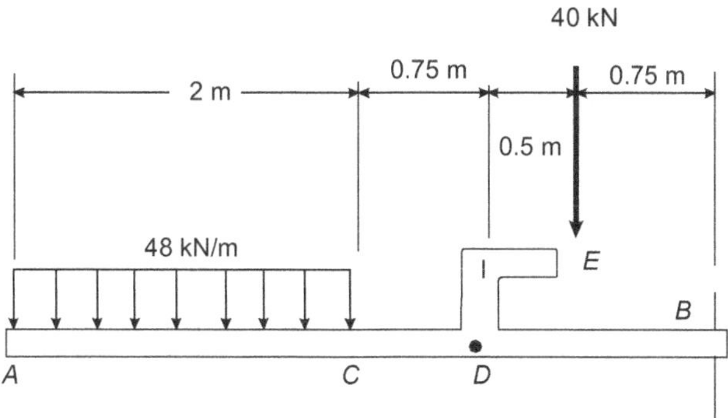

Solution

The 40 kN load is replaced by an equivalent force-couple system acting on the beam at point D. The reaction at B is determined by considering the entire beam as a free body.

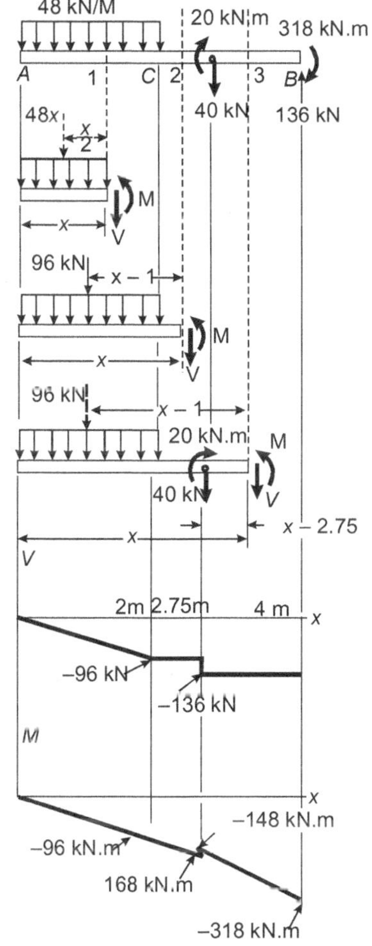

From A to C. We determine the internal forces at a distance x from point A by considering the portion of beam to the left of section 1. That part of the distributed load acting on the free body is replaced by its resultant, and we write

$$+\uparrow \Sigma F_y = 0: \quad -48x - V = 0 \quad V = -48x \text{ kN}$$

$$+\,\mathord{\rotatebox[origin=c]{0}{$\circlearrowleft$}}\, \Sigma M_1 = 0: \quad 48x(\tfrac{1}{2}x) + M = 0 \; M = -24x^2 \text{ kNm}$$

Since the free-body diagram shown may be used for all values of x smaller than 2 m, the expressions obtained for V and M are valid in the region $0 < x < 2$ m.

From C to D Considering the portion of beam to the left of section 2 and again replacing the distributed load by its resultant, we obtain

$$+\uparrow \Sigma F_y = 0: \quad -96 - V = 0 \quad V = -96 \text{ kN}$$

$$+\,\mathord{\rotatebox[origin=c]{0}{$\circlearrowleft$}}\, \Sigma M_2 = 0: \quad 96(x-1) + M = 0 \; M = 96 - 96x \text{ kN.m}$$

These expressions are valid in the region 2 m $< x < 2.75$ m.

From D to B Using the portion of beam to the left of section 3, we obtain for the region $2.75m < x < 4$ m.

$$V = -136 \text{ kN} \quad M = 226 - 136x \text{ kN.m}$$

The shear and bending-moment diagrams for the entire beam may now be plotted. We note that the couple of moment 20 kNm applied at point D introduces a discontinuity into the bending-moment diagram just as a point load of 40 kN introduces a discontinuity in the shear force diagram.

PROBLEM 8.2

Draw the shear and bending-moment diagrams for the beam and loading shown.

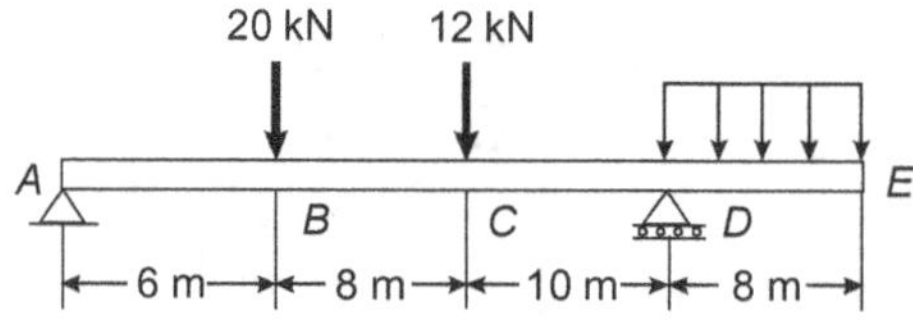

Solution

Considering the entire beam as a free body, we determine the reactions:

$$+\,\mathord{\rotatebox[origin=c]{0}{$\circlearrowleft$}}\, \Sigma M_A = 0: \quad D(24\text{m}) - (20\text{kN})(6\text{m}) - (12\text{kN})(14\text{m}) - (12kN)(28\text{m}) = 0$$

$$D = +26\text{kN} \qquad\qquad D = 26\text{kN} \uparrow$$

$$+\uparrow \Sigma F_y = 0: \quad A_y - 20\text{kN} - 12\text{kN} + 26\text{kN} - 12\text{kN} = 0$$

$$A_y = +18\text{kN} \qquad\qquad A_y = 18 \text{ kN} \uparrow$$

$$\xrightarrow{+} \Sigma F_x = 0: \quad A_y = 0 \qquad\qquad A_x = 0$$

We also note that at both A and E the bending moment is zero; thus two points (indicated by dots) are obtained on the bending-moment diagram.

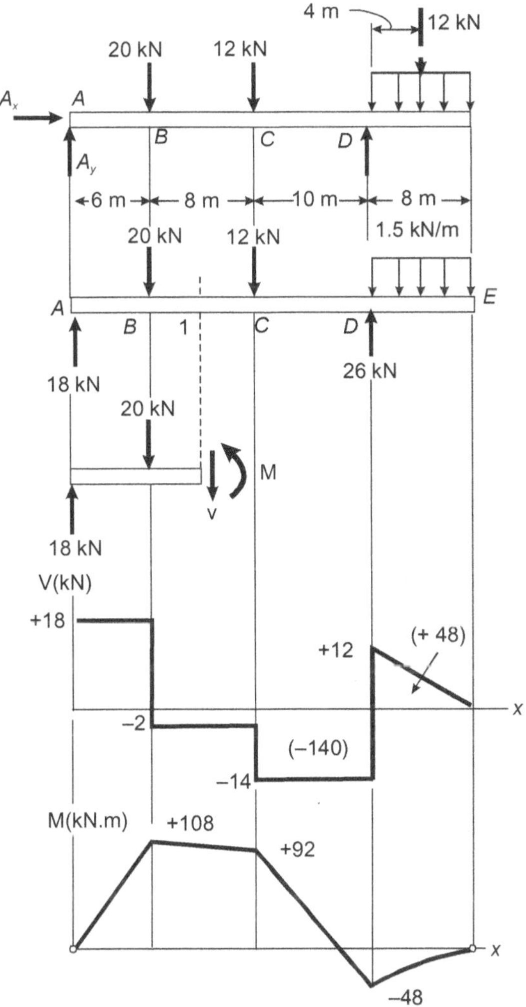

Shear diagram Since $dV/dx = -w$, we find that between concentrated loads and reactions the slope of the shear diagram is zero (i.e., the shear is constant). The shear at any point is determined by dividing the beam into two parts and considering either part as a free body. For example, using the portion of beam to the left of section 1, we obtain the shear between B and C:

$$+\uparrow \Sigma F_y = 0: \quad +18\,\text{kN} - 20\,\text{kN} - V = 0 \qquad V = -2\,\text{kN}$$

We also find that the shear is + 12 kN just to the right of D and zero at end E. Since the slope $dV/dx = -w$ is constant between D and E, the shear diagram between theses two points is a straight line.

Bending-moment diagram   We recall that the area under the shear curve between two points is equal to the change in bending moment between the same two points. For convenience, the area of each portion of the shear diagram is computed and is indicated in parentheses on the diagram. Since the bending moment M_A at the left end is known to be zero, we write

$$M_B - M_A = +108 \qquad M_B = +108 \ \text{kNm}$$
$$M_C = M_B = -16 \qquad M_C = +92 \ \text{kNm}$$
$$M_D - M_C = -140 \qquad M_D = -48 \ \text{kNm}$$
$$M_E - M_D = +48 \qquad M_L = 0$$

Since M_E is known to be zero, a check of the computations is obtained.

Between the concentrated loads and reactions the shear is constant; thus, the slope dM/dx is constant and the bending-moment diagram is drawn by connecting the known points with straight lines. Between D and E where the shear diagram is an oblique straight line, the bending-moment diagram is a parabola.

From the V and M diagrams we note that $V_{\text{max}} = 18\,\text{kN}$ and $M_{\text{max}} = 108\,\text{kNm}$.

PROBLEM 8.3

Draw the shear and bending-moment diagrams for the beam and loading shown and determine the location and magnitude of the maximum bending moment.

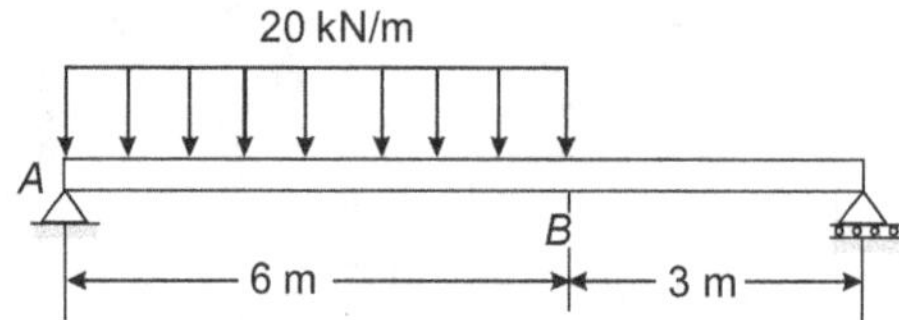

Solution

Considering the entire beam as a free body, we obtain the reactions

$$R_A = 80 \ \text{kN} \uparrow \qquad R_C = 40 \ \text{kN} \uparrow$$

Shear diagram   The shear just to the right of A is $V_L = +80\,\text{kN}$. Since the change in shear between two points is equal to minus the area under the load curve between the same two points, we obtain V_B by writing

$$V_B - V_A = -(20\text{kN}/\text{m})\,(6 \ \text{m}) = -120\,\text{kN}$$
$$V_B = -120 + V_A = -120 + 80 = -40\,\text{kN}$$

The slope $\dfrac{dv}{dx} = -w$ being constant between A and B, the shear diagram between theses two points is represented by a straight line. Between B and C, the area under the load curve is zero; therefore,

$$V_C - V_B = 0 \qquad V_C = V_B = -40 \text{ kN}$$

and the shear is constant between B and C.

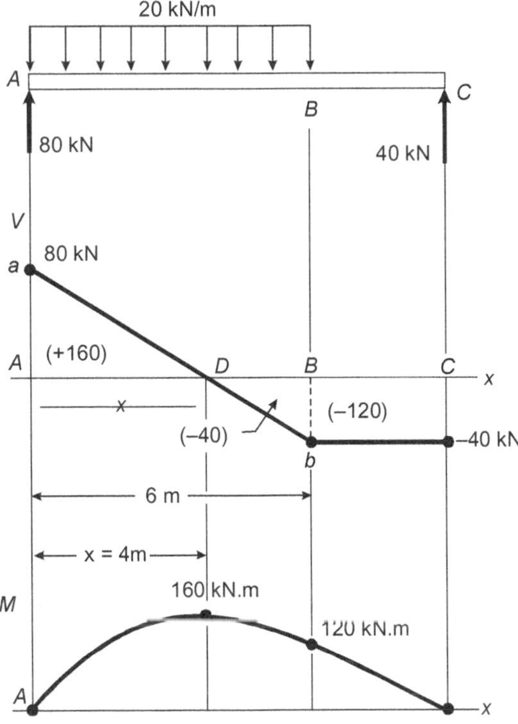

Bending-moment Diagram   We note that the bending moment at each end of the beam is zero. In order to determine the maximum bending moment, we locate the section D of the beam where $V = 0$. We write

$$V_D - V_A = -wx$$
$$0 - 80\,\text{kN} = -(20\,\text{kN/m})x$$

and, solving for x: $\qquad x = 4\text{m}$

The maximum bending moment occurs at point D, where we have $dM/dx = V = 0$. The areas of the various portions of the shear diagram are computed and are given (in parentheses) on the diagram. Since the area of the shear diagram between two points is equal to the change in bending moment between the same two points, we write

$$M_D - M_A = +160 \text{ kNm} \qquad M_D = +160 \text{ kNm}$$
$$M_B - M_D = +40 \text{ kNm} \qquad M_B = +120 \text{ kNm}$$
$$M_C - M_B = -120 \text{ kNm} \qquad M_C = 0$$

The bending-moment diagram consists of an arc of parabola followed by a segment of straight line; the slope of the parabola at A is equal to the value of V at that point.

The maximum bending moment is $M_{max} = M_D = +160$ kNm

PROBLEM 8.4

Consider a simple beam with a uniformly increasing load intensity from an end, as shown in Figure. The total applied load is W.(a) Construct shear and moment diagrams with the aid of the integration process. (b) Derive expressions for V and M.

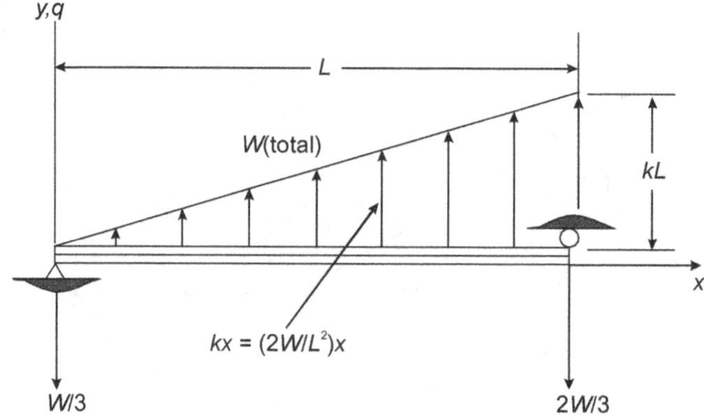

Figure (a)

Solution

(a) Since the total load $W = kL^2/2$, $k = 2W/L^2$. For the given load distribution, the downward reactions are $W/3$ and $2W/3$, as shown in Figure. Therefore, the shear diagram given in Figure (b) begins and ends as shown. Since the rate of applied load is smaller on the left end than on the right, the shear diagram is concave upward. The point of zero shear occurs where the reaction on the left is balanced by applied load; that is,

$$\frac{W}{3} = \frac{1}{2}x_1\frac{2W}{L^2}x_1 \text{ hence, } x_1 = \frac{L}{\sqrt{3}}$$

At x_1, the bending moment is maximum;therefore,

$$M_{max} = M\left(\frac{L}{\sqrt{3}}\right) = -\frac{W}{3}\frac{L}{\sqrt{3}} + \frac{1}{2}\frac{L}{\sqrt{3}}\frac{2W}{L^2}\frac{L}{\sqrt{3}}\left(\frac{1}{3}\frac{L}{\sqrt{3}}\right) = -\frac{2WL}{9\sqrt{3}}$$

By following the rules, the moment diagram has the shape shown in Figure (c). Although the shear and bending moment diagrams could be sketched qualitatively, it was necessary to supplement the results analytically for determining the critical values.

(a) Applying equation 8.1 and integrating twice, one has

$$\frac{d^2M}{dx^2} = q = +kx = +\frac{2W}{L^2}x \,(\text{Load intensity})$$

$$\frac{dM}{dx} = \frac{kx^2}{2} + C_1 \text{ and } M = \frac{kx^3}{6} + C_1 x + C_2$$

However, the boundary conditions require that the moments at $x - 0$ and $x = L$ be zero; that is, $M(0) = 0$ and $M(L) = 0$. Therefore, since

$$M(0) = 0, \ C_2 = 0$$

And, similarly, since $M(L) = 0$,

$$\frac{kL^3}{6} + C_1 L = 0 \text{ or } C_1 = -\frac{kL^2}{6}$$

With these constants,

$$V = \frac{dM}{dx} = \frac{kx^2}{2} - \frac{kL^2}{6} = \frac{Wx^2}{L^2} - \frac{W}{3}$$

$$M = \frac{kx^2}{6} - \frac{kL^2 x}{6} = \frac{Wx^3}{3L^2} - \frac{Wx}{3}$$

These results agree with those found earlier.

Distribution of shearing stresses depend upon P′ (shear force) and normal stress distribution can be obtained from M and is the same as if the beam were in pure bending.

We shall use methods of statics here to derive relations which must be satisfied by the stresses experienced at any cross section by a prismatic bar in pure bending.

Denoting σ and τ the normal and shear stresses, we shall express the resultant of system of internal forces as equivalent to the couple M. Statics tells that couple M actually consists of two equal and opposite forces. The sum of the components of these forces in any direction is therefore zero. Moreover the moment of the couple is the same about any axis perpendicular to its plane and is zero about any axis contained in that plane.

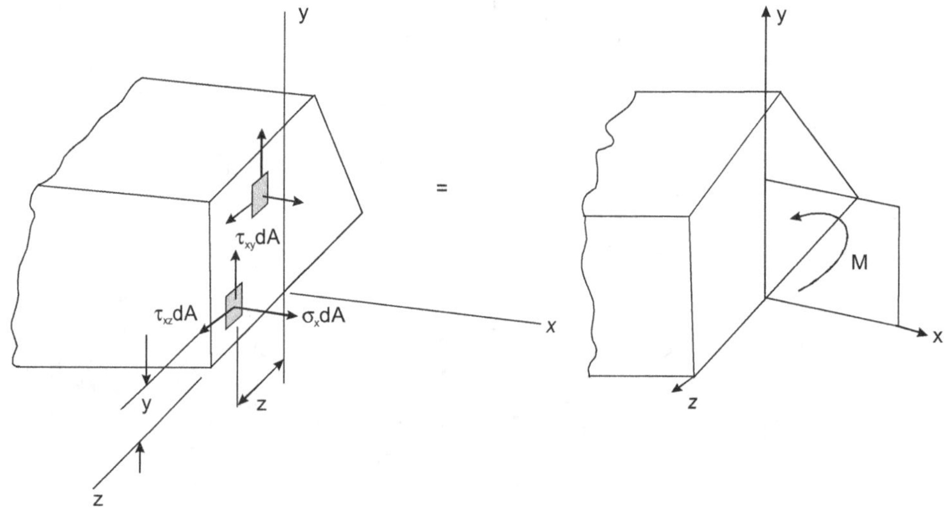

Figure 8.4 Stresses in pure bending

Let us select the coordinate axes.

Equivalence of the elementary forces and of the couple M = Sum of the components of forces and moment of forces = Corresponding components and moments of couple M.

X components

$$\int \sigma_x \, dA = 0$$

Moments about y axis

$$\int z\sigma_x \, dA = 0$$

1. Become trivial if bar is symmetric with respect to plane containing M and if y axis is chosen in that plane.

    Moments about z axis

    $$\int -y\sigma_x \, dA = M$$

2. Minus sign is due to $\sigma x \geq 0$ tensile stress leads to a –ve moment (clockwise) $\sigma_x dA$ about z axis.

    (Three more equations we can write for shearing stresses. They will be zero as we shall see later)

This is an example of a statically indeterminate problem as statics alone cannot help to derive the stress distribution. The deformation needs to be considered to arrive at the stress distribution.

DEFORMATION OF A SYMMETRIC MEMBER IN PURE BENDING

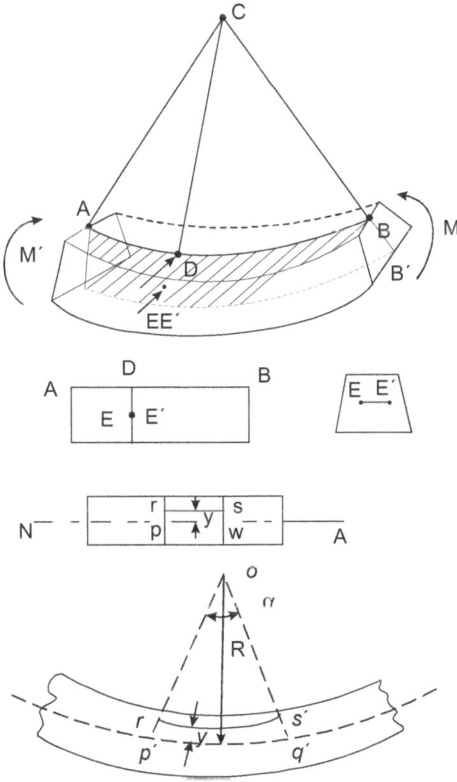

Figure 8.5

Assumptions in the theory of bending

1. Material is homogeneous
2. Material obeys Hooke's law (Max stress within yield stress)
3. E is same for compression and tension
4. Plane transverse sections before bending remain plane after bending.
5. No resultant push or pull on the cross section of the beam. Net axial force is zero.
6. Loads applied in the plane of bending. (Transverse or lateral loads)
7. Transverse section is symmetrical about a line paving through the centre of gravity in the plane of bending
8. The radius of curvature of the beam before bending is very large in comparison to its dimensions in the transverse direction.

Treat the short portion of the bent beam as a part of the are of a circle (Figure 8.5). It follows that at the outer radii, the material will be in tension and at the inner radii in compression. At some radius there will be no stress. This layer of the material is the neutral layer or neutral axis.

$$p'q' = R\alpha$$

$$r's' = (R - y)\alpha$$

Initially $pq = rsp'q' = pq$

The strain in $rs = \dfrac{rs - r's'}{rs}$ but $rs = pq = p'q$

$\therefore$ Strain $= \dfrac{p'q' - r's'}{rs}$

$p'q' = R\alpha$

$r's' = (R - y)\alpha$

Strain $= \dfrac{R\alpha - (R - y)\alpha}{R\alpha} = \dfrac{y}{R}$

If the stress in $rs = \sigma$, Young's modulus $= E$

Then strain $= \dfrac{\sigma}{E}$

$$\frac{\sigma}{E} = \frac{y}{R} \text{ or } \frac{\sigma}{y} = \frac{E}{R}$$

If a transverse section is viewed, let a strip of area δA lie at a distance y from neutral axis (Figure 8.6) be considered.

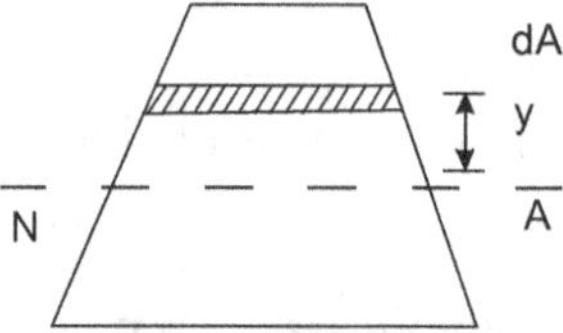

Figure 8.6

The normal force $\perp$ to plane of section $= \dfrac{E}{R}.y.\delta A$ (Force = stress × area)

The moment of this force about neutral axis $= \dfrac{E}{R}.y.\delta Ay$

$$= \frac{E}{R} y^2 \delta A$$

This is the resisting moment of the material caused by the stress and the total resisting moment is $\Sigma \frac{E}{R} y^2 dA$ or $\frac{E}{R} \Sigma y^2 dA$

$\Sigma y^2 dA$ is the geometric or section or area moment of inertia about the neutral axis I_{NA} or I.

$\therefore$ Resisting moment $M = \frac{E}{R} I$

Since the resisting moment balances the applied bending moment

$$M = \frac{E}{R} I \text{ or } \frac{M}{I} = \frac{E}{R}$$

$$\therefore \frac{M}{I} = \frac{\sigma}{y} = \frac{E}{R} \qquad (8.3)$$

This equation is known as the bending equation.

POSITION OF NEUTRAL AXIS

Consider the cross section of a beam, there will be no resultant force on the section for condition of equilibrium. The force acting on small area δA at a distance y from neutral axis is

$$\delta F = \sigma \delta A = \frac{E}{R} y \delta A$$

Total normal force $= \Sigma \frac{E}{R} y \delta A$

$\therefore$ for zero resultant force $\Sigma y \delta A = 0$

$\Sigma y \delta A$ or $\int y dA$ is the moment of the sectional area about the neutral axis, and this moment is zero, the axis must pass through the centre of the area.

Hence neutral axis or neutral layer passes through the centre of area.

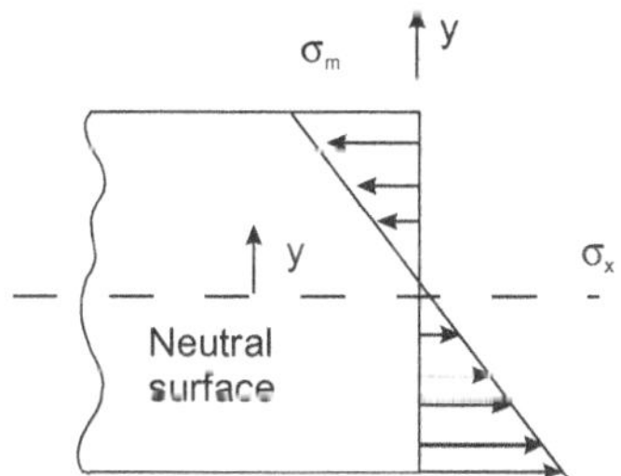

SECTION MODULUS

$$\frac{M}{I} = \frac{\sigma}{y}$$

$\sigma = \dfrac{My}{I}$ (Normal stress varies linearly with distance from NA

$$= \frac{M}{z}$$

$$\frac{I}{y} = z \text{ is called section modulus.} \tag{8.4}$$

This is of more importance in design than MI. The maximum stress σ corresponding to maximum y is given by $\dfrac{M}{z}$.

Calculation of section moduli for circular and rectangular cross sections

Rectangular section:

$$I = \frac{bh^3}{12}$$

$$y_{max} = \frac{h}{2}$$

$$z = \frac{I}{y_{max}} = \frac{bh^2}{6}$$

$$= \frac{1}{6}Ah \text{ (since } A = bh)$$

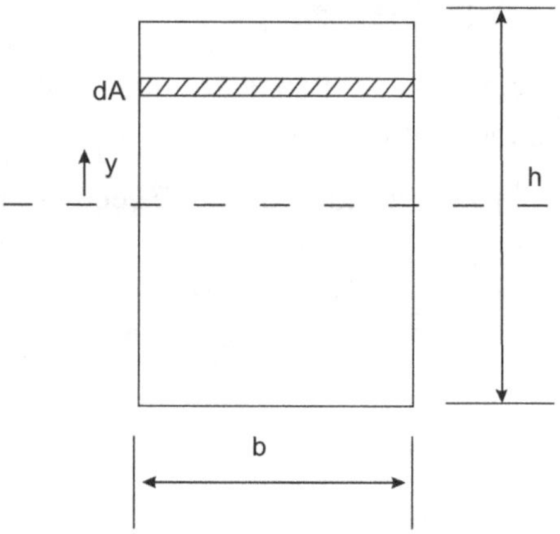

If two beams have same 'A' beam with a larger depth will have larger Z, thus will be more effective in resisting bending. (However larger *h* may result in lateral instability of the beam which puts a restriction on choosing appropriate depth *h*)

Circular section

$$I = \frac{\pi d^4}{64} \text{ for a circular section}$$

$$y_{max} = \frac{d}{2}$$

$$\therefore \text{Section modulus} = \frac{I}{y_{max}} = \frac{\pi d^4 / 64}{d / 2} = \frac{\pi d^3}{32}$$

For a hollow circular section like in the case of hollow pipes or tubes

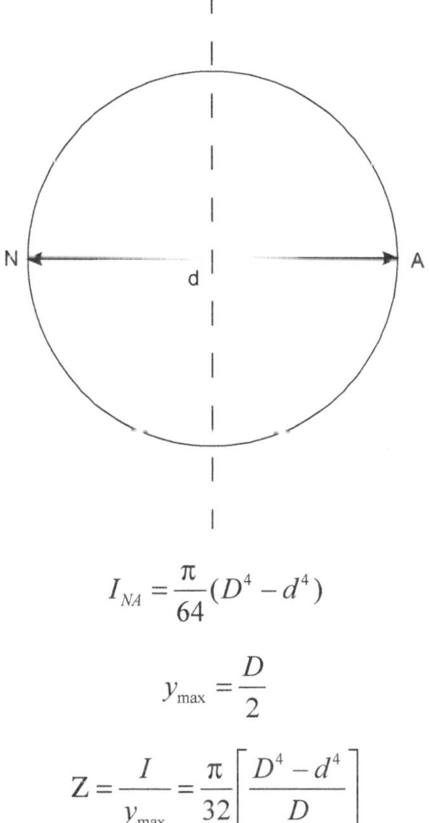

$$I_{NA} = \frac{\pi}{64}(D^4 - d^4)$$

$$y_{max} = \frac{D}{2}$$

$$Z = \frac{I}{y_{max}} = \frac{\pi}{32}\left[\frac{D^4 - d^4}{D}\right]$$

PROBLEM 8.5

A 250 mm (depth) × 150 mm (width) rectangular beam is subjected to a maximum bending moment of 750 kNm. Determine.

i.   The maximum stress in the beam

ii. If $E = 200$ GN/m², find the radius of the portion where bending is maximum.

iii. The value of the longitudival stress at a distance of 65 mm from top surface of the beam.

Solution

Width $= b = 150$ mm $= 0.15$ m

Depth $= d = 250$ mm $= 0.25$ m

Maximum bending moment $= M = 750$ kNm

$E = 200$ GN /m

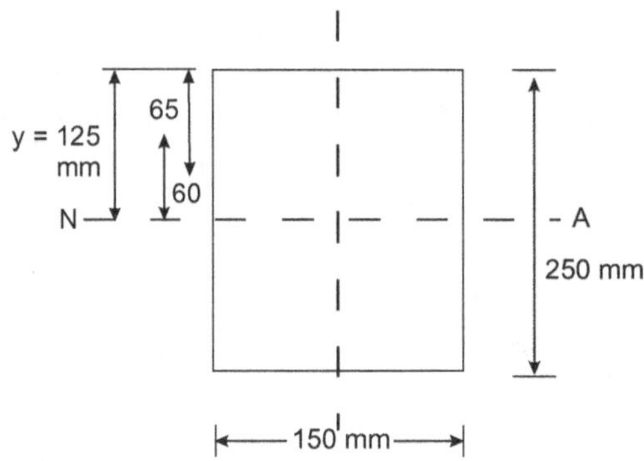

i.   *Max stress in the beam*

Moment of inertia of the section about NA $= \dfrac{bd^3}{12} = \dfrac{0.15 \times 0.253}{12} = 0.0001953\,\text{m}^4$

Distance of neutral axis from the top surface of the beam $= y = \dfrac{d}{2} = 0.125\,\text{m}$

$$\frac{M}{I} = \frac{\sigma}{y} \therefore \sigma = \frac{My}{I}$$

$$= \frac{750 \times 10^3 \times 0.125}{0.0001953}$$

Maximum stress in the beam $= 4.8 \times 10^8\,\dfrac{\text{N}}{\text{m}^2} = 480\,\dfrac{\text{MN}}{\text{m}^2}$

ii. Radius of curvature R

$$\frac{M}{I} = \frac{E}{R}$$

$$R = \frac{EI}{M} = \frac{200 \times 10^9 \times 0.0001953}{750 \times 10^3}$$

$$= 52.08\,\text{m}$$

iii. Longitudinal stress at a distance of 65 mm from top surface of the beam

$$\frac{M}{I} = \frac{\sigma}{y} = \frac{E}{R}$$

$$\sigma = \frac{My_1}{I} = \frac{750 \times 10^3 \times (60 \times 10^{-3}) \times 10^{-6}}{0.0001953} \frac{MN}{m^2}$$

$$= 230.4 \ MN/m^2$$

PROBLEM 8.6

A symmetrical section 200 mm deep has a moment of inertia of $2.26 \times 10^{-5} m^4$ about its neutral axis. Determine the largest span over which when simply supported, the beam would carry a uniformly distributed load of 4 KN/m run without the stress due to bending exceeding 125 MN/m².

Solution

Depth of the symmetrical section $= d = 200$ mm

M.I about neutral axis $I = 2.26 \times 10^{-5} m^4$

$$\text{u.d.l} = w = \frac{4kN}{m} \ Nm.$$

Maximum bending stress $\sigma = 125 \ MN/m^2$

Longest span l.

$$\frac{M}{I} = \frac{\sigma}{y}$$

$$M = \frac{\sigma I}{y} = \frac{125 \times 10^6 \times 2.26 \times 10^{-5}}{0.2 / 2}$$

$$= 28.25 \ kNm$$

Maximum bending moment due to u.d.l.is

$$= \frac{wl^2}{8} = \frac{4l^2}{8} = 0.5l^2 \ kNm$$

$$0.5l^2 = 28.25$$

$$l - 7.516 \, m$$

PROBLEM 8.7

Composite beams

A beam simply supported at ends and having a cross section as shown in figure is loaded with a u.d.l over whole of its span. If the beam is 8 m long, find the u.d.l if maximum permissible bending stress in tension is limited to 30 MN/m^2 and in compression to 45 MN/m^2. What are the actual maximum bending stresses set up in the section.

Solution

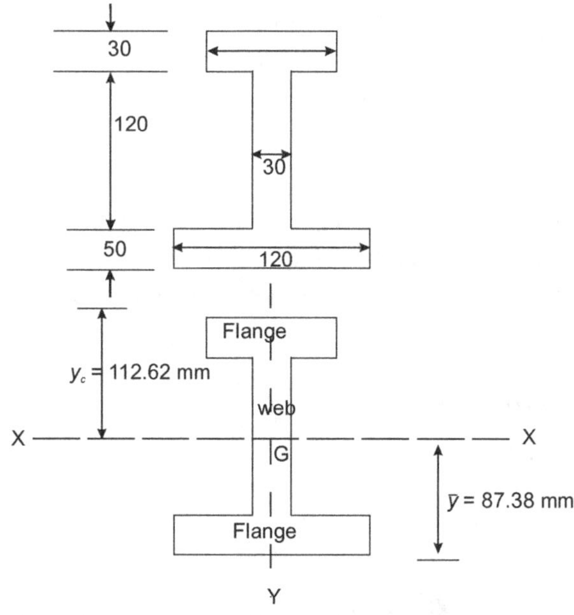

Component	Area 'A' mm^2	Centroidal distance from bottom edge, 'y' mm	Ay mm^3
Top Flange	100 × 30 = 3000	200-30/2 = 185mm	555000
Web	120 × 30 = 3600	50+120/2 = 110mm	396000
Bottom Flange	120 × 50 – 6000	0+50/2 = 25mm	150000
Total	$\sum A = 12,600$		$\sum Ay = 1101000$

Distance of the centroidal axis from bottom flange $= y = \dfrac{\sum Ay}{\sum A} = \dfrac{1101000}{12,600} = 87.38$ mm

$I_{xx} = I$ top flange $+ I$ web $+ I$ bottom flange

$$= \left[\frac{100 \times 30^3}{12} + 100 \times 30 \times (112.62 - 15)^2 \right] +$$

$$\left[\frac{30 \times 120^3}{12} + 30 \times 120 \times (110 - 87.38)^2 \right] + \left[\frac{120 \times 50^3}{12} + 120 \times 50 \times (87.38 - 25)^2 \right]$$

$$= (225000 + 28588993) + (4320000 + 1841992) + (1250000 + 23347586)$$

$$= 28813993 + 6161992 + 24597586 = 5957 \times 10^4 \, \text{mm}^4 = 5957 \times 10^{-8} \, \text{m}^4$$

Maximum bending moment $= M = \dfrac{wl^2}{8} = \dfrac{w8^2}{8} = 8w$

(where $w = udl$)

For Tension side of the I.Section

$$\frac{M}{I} = \frac{\sigma_t}{y_t}$$

$$M = I \times \frac{\sigma_t}{y_t} = \frac{5957 \times 10^{-8} \times 30 \times 10^6}{87.38 \times 10^{-3}}$$

$$= 20452 \text{ Nm} = 20.452 \text{ kNm}$$

For compression side of the I section

$$\frac{M}{I} = \frac{\sigma_c}{y_c}$$

$$M = \frac{\sigma_c I}{y_c} = \frac{5957 \times 10^{-8} \times 45 \times 10^6}{112.62 \times 10^{-3}}$$

$$= 23802.6 \text{ Nm} = 23.8026 \text{ kNm}$$

$\therefore$ Moment of resistance $= 20.452$ kNm (min.of the above two values)

$$8w = 20.452 \text{ kNm}$$

$$w = 2.556 \text{ kN/m}$$

Max σ at top $= \dfrac{M}{I} y_c = \dfrac{20.452 \times 10^3 \times 112.62 \times 10^{-3}}{5957 \times 10^{-8}} = 38.6$ MN/m^2 (Compressive)

Max σ at bottom $\dfrac{M}{I} y_t = 30$ MN/m^2 (tensile)

PROBLEM 8.8

The rectangular tube shown is extruded from an aluminum alloy for which $\sigma_y = 150$ MPa, and $E = 70$ GPa. Neglecting the effect of fillets, determine (a) the bending moment M for which the factor of safety will 3.00 (b) the corresponding radius of curvature of the tube.

Moment of inertia Considering the cross-sectional area of the tube as the difference between the two rectangles shown and recalling the formula for the centroidal moment of inertia of a rectangle, we write

$$I = \frac{1}{12}(0.080)(0.120)^3 - \frac{1}{12}(0.064)(0.104)^3 \quad I = 5.52 \times 10^{-6} \text{ m}^4$$

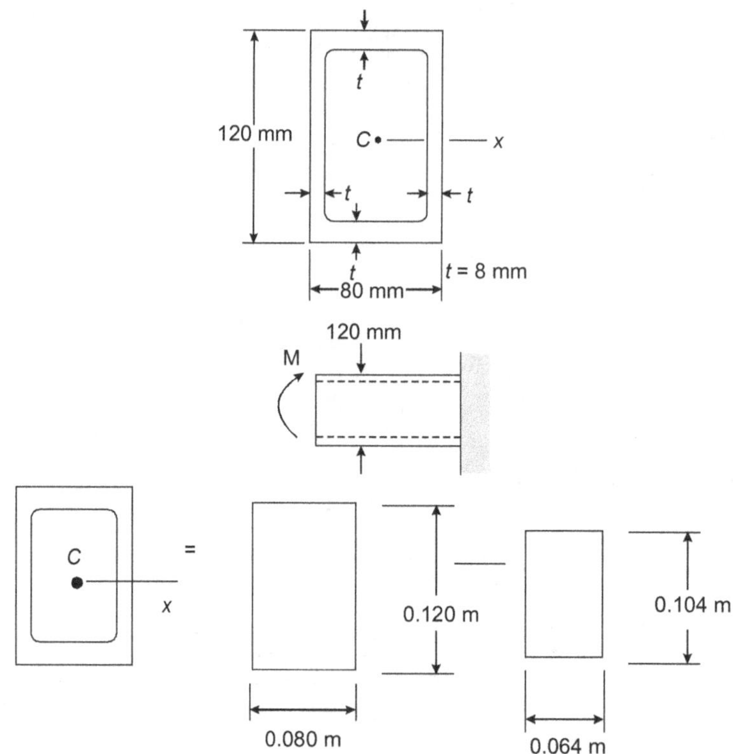

Allowable stress   For a factor of safety of 3.00 and an ultimate stress of 300 MPa, we have

$$\sigma_{all} = \frac{\sigma_U}{F.S.} = \frac{300 \text{ MPa}}{3.00} = 100 \text{MPa}$$

Since $\sigma_{all} < \sigma_y$, the tube remains in the elastic range.

a. *Bending moment* with $c = \frac{1}{2}(0.120\,\text{m}) = 0.060\,\text{mm}$, we write

$$\sigma_{all} = \frac{Mc}{I}$$

$$M = \frac{I}{c}\sigma_{all} = \frac{5.52}{0.060\,\text{m}} \times 10^{-6}\,\text{m}^4(100\text{ MPa})$$

$$M = 9.20\,\text{kNm}$$

b. *Radius of curvature*   Recalling that $E = 70$ GPa, we substitute this value and the values obtained for I and M in to equation (8.3) and find

$$\frac{1}{R} = \frac{M}{EI} = \frac{9.20 \text{ kN.m}}{(70\,\text{GPa})(5.52 \times 10^{-6}\,\text{m}^4)} = 23.8 \times 10^{-3}\,\text{m}^{-1}$$

$$R = 42\,\text{m}$$

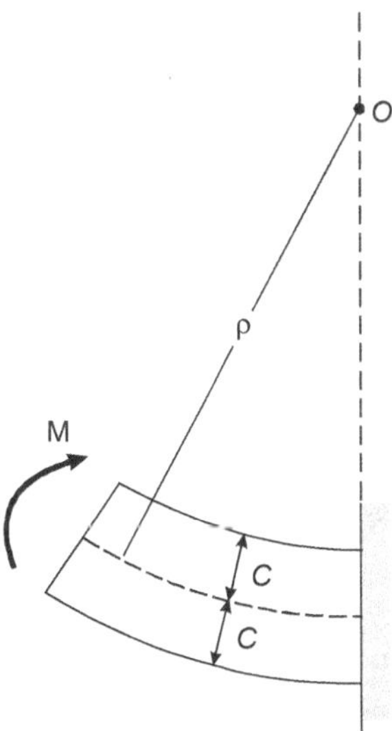

Alternative solution Since we know that the maximum stress is $\sigma_{all} = 100$ MPa, we may determine the maximum strain $\in$.

$$\varepsilon_m = \sigma_{all} / E = 100 \, \text{MPa}/70 \, \text{GPa} = 1429 \, \mu$$

$$\varepsilon_m = \frac{c}{R};$$

$$R = \frac{c}{\varepsilon_m} = \frac{0.060 \, \text{m}}{1429 \, \mu} = 42.0 \, \text{m}$$

PROBLEM 8.9

A cast-iron machine part is acted up by the 3-kNm couple shown. Knowing that $E = 165$ GPa and neglecting the effect of fillets, determine (a) the maximum tensile and compressive stresses in the casting. (b) the radius of curvature of the casting.

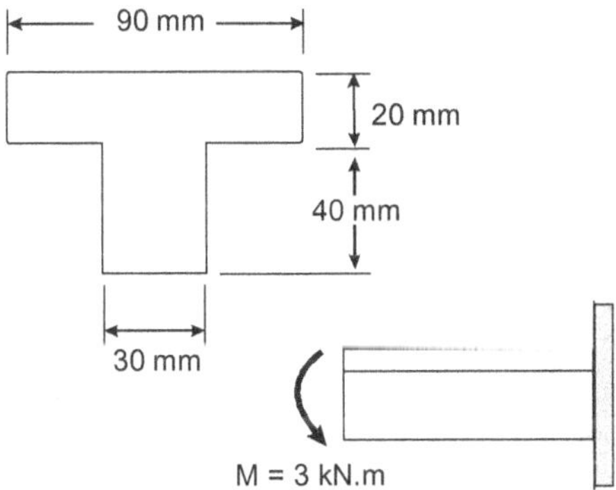

Centroid We divide the T-shaped cross section into the two rectangles shown and write

	Area, mm²	*y*, mm	*yA*, mm³
1.	$(20)(90) = 1800$	50	90×10^3
2.	$(40)(30) = 1200$	20	24×10^3
	$\sum A = 3000$		$\sum \bar{y}A = 114 \times 10^3$

$$\bar{Y} \sum A = \sum \bar{y}A$$

$$\bar{Y}(3000) = 114 \times 10^3$$

$$\bar{Y} = 38 \text{ mm}$$

Centroidal moment of inertia   The parallel-axis theorem is used to determine the moment of inertia of each rectangle with respect to the axis x′ which passes through the centroid of the composite section . Adding the moments of inertia of the rectangles, we write

$$I_{x'} = \sum (I + Ad^2) = \sum \left(\frac{1}{12}bh^3 + Ad^2 \right)$$

$$= \frac{1}{2}(90)(20)^3 + (90 \times 20)(12)^2 + \frac{1}{2}(30)(40)^3 + (30 \times 40)(18)^2$$

$$= 868 \times 10^3 \text{ mm}^4$$

$$I = 868 \times 10^{-9} \text{ m}^4$$

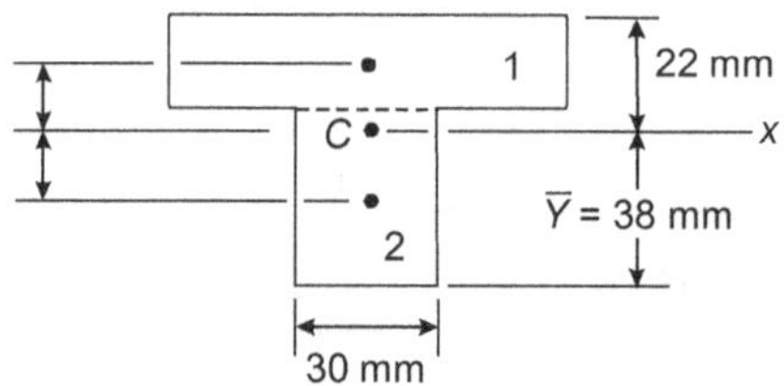

a. *Maximum tensile stress*   Since the applied couple bends the casting downward, the centre of curvature is located below the cross section. The maximum tensile stress occurs at point A, which is farthest from the centre of curvature.

$$\sigma_A = \frac{Mc_A}{I} = \frac{(3 \text{ kN.m})(0.022 \text{ m})}{868 \times 10^{-9} \text{ m}^4} \qquad \sigma_A = +76.0 \text{ MPa}$$

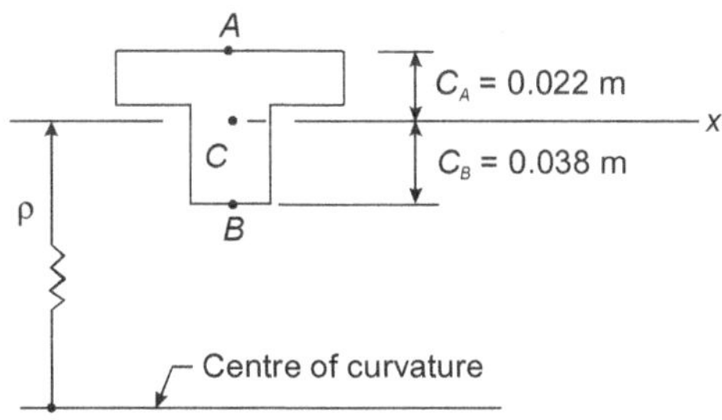

Maximum compressive stress This occurs at point B; we have

$$\sigma_B = \frac{Mc_B}{I} = \frac{(3 \text{ kNm})(0.038\,\text{m})}{868 \times 10^{-9}\,\text{m}^4} \qquad \sigma_B = -131.3\,\text{MPa}$$

b. *Radius of curvature*   From equation (8.3), have

$$\frac{1}{\rho} = \frac{M}{EI} = \frac{3 \text{ kNm}}{(165 \text{ GPa})(868 \times 10^{-9}\,\text{m}^4)}$$

$$= 20.95 \times 10^{-3}\,\text{m}^{-1}$$

$$R = 47.7\,\text{m}$$

REVIEW QUESTIONS

SHORT QUESTIONS

1. A hollow circular bar having external diameter twice the inner diameter is used as a beam. If the bar is subjected to a bending moment of 40 kNm and the allowable bending stress in the beam is limited to 100 MN/m², find the inner diameter of the bar.

2. Write down any four types of beams.

3. Write the expansion for section modulus.

4. Mention the various types of beams with respect to support conditions.

5. What is meant by shear centre?

6. State the assumptions made in the theory of simple bending?

7. What are the assumptions made in theory of simple bending?

8. Give the relations between load, shear force and bending moment.

9. Mention and sketch any two types of supports and cording for the beams.

10. Sketch the bending stress as well as shear stress distribution for a beam of rectangular cross section.

11. What do you understand by the term 'Point of contraflexure'?

12. What is the value of bending moment corresponding to a point having a zero shear force?

13. A cantilever beam of length 6 m carries a uniformly distributed load of 3 kN/m over its entire length; draw the shear force and bending moment diagram.

14. Sketch the bending stress distribution diagrams for the cantilever and simply supported beams of rectangular cross-section under uniform loading.

15. Define shear force and bending moment.

LARGE QUESTIONS

1. Draw the shear force and bending moment diagrams for the beam shown in Figure. Also determine the maximum bending moment and location of point of contra flexure.

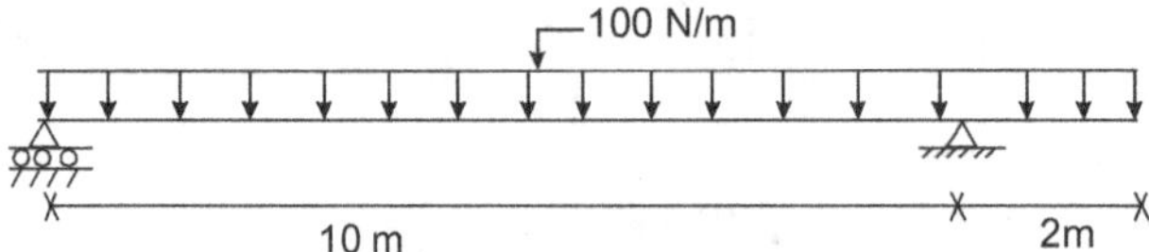

2. A cast iron pipe 300 mm internal diameter, metal thickness 15 mm, is supported at two points 6 m apart. Find the maximum bending stress in the metal of the pipe when it is running full of water. Assume the specific weight of cast iron and water as 72 kN/m² and 10 kN/m³ respectively.

3. A simply supported beam of span 6 meters carries a point load of 1 kN at 1 meter from the left end and another point load of 4 kN at 1 meter from the right end. It carries a uniformly distributed load of 2 kN/m run over a distance of 2 meters at midspan of the beam. Find the SF and BM values. Draw SFD and BMD.

4. A simply supported beam of 4 m span is carrying loads as shown in figure. Draw the shear force and bending moment diagrams for the beam.

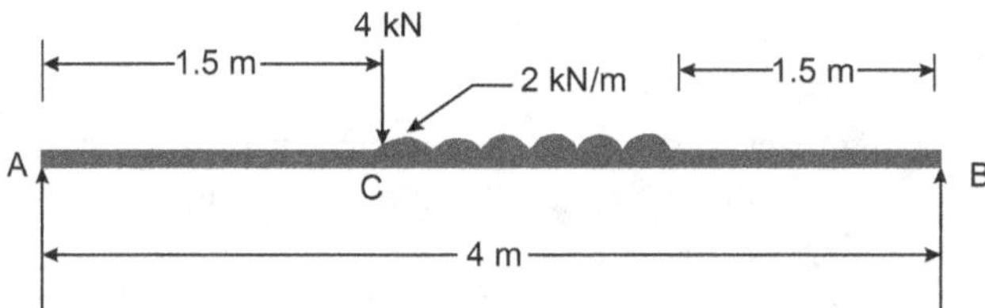

5. A horizontal beam 10 m long is carrying a uniformly distributed load of 1 kN/m. The beam is supported on two supports 6 m apart. Find the position of the supports, so that bending moment on the beam is as small as possible. Also draw the shear force and bending moment diagrams.

6. A beam ABCD, simply supported at A and D. Where length AB = 2 m; BC = 1 m; CD 2.5 m, carries a point load of 10 kN at B and a concentrated load of 20 kN/m run between CD.

   i. Draw the shear force and bending moment diagram for the above loading.

   ii. If the width of the beam is 150 mm and depth of the beam is 450 mm, calculate the maximum bending stress for the beam. Take E = 210 GPa.

7. Draw the shear force diagram and the bending moment diagram for the following beam and mark all the salient points.

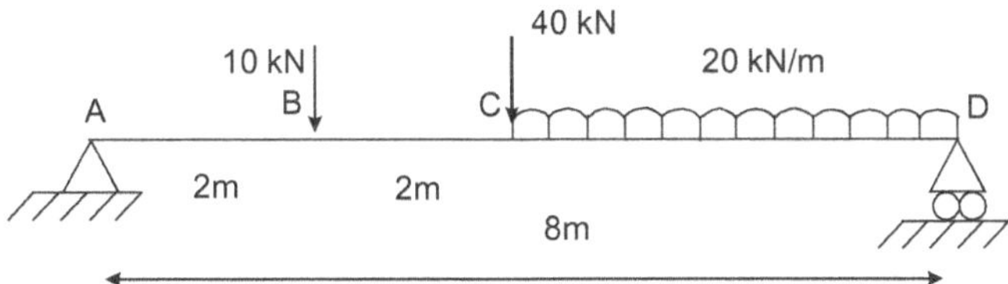

8. Determine the maximum tensile and compressive stresses in a simply supported beam of span 4 m subjected to uniformly distributed load of 5 kN/m throughout its span. The cross-section is as shown in figure in mm.

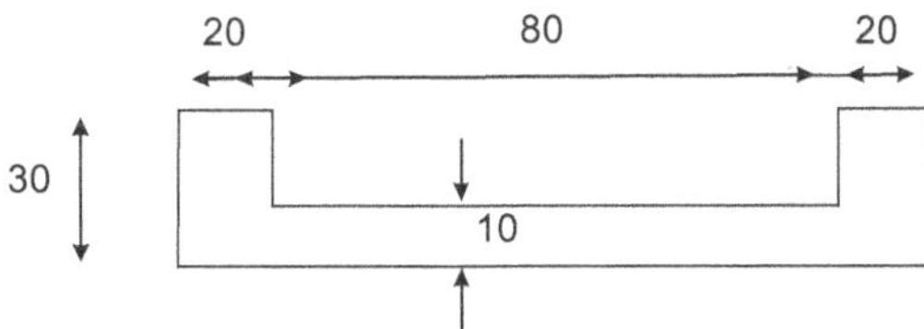

9. A square beam 20 mm x 20 mm in section and 2m long is supported at the ends. Ther beam fails when a point load of 400 N is applied at the centre of beam. What uniformly distributed load per meter length will break a cantilever of same material 40 mm wide, 60 mm deep and 3 m long?

9

SHEARING STRESSES IN BEAMS

INTRODUCTION

While dealing with pure bending of beams, due to transverse or lateral loading, two diagrams namely the Shear Force Diagram (SFD) and the Bending Moment Diagram (BMD) were necessary to study the distribution of BM and SF.

Bending Moments cause longitudinal or fibre stresses, that vary linearly in a section passing through zero value at the neutral axis. However we did not dwell much on the shear forces. In this chapter, we discuss about shear forces and shear stresses.

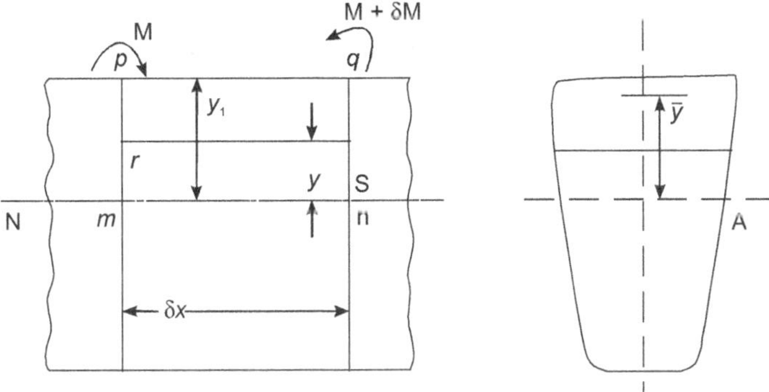

Figure 9.1   Bending moment in an element of a beam

SHEAR STRESSES IN BEAMS

We considered the longitudinal stresses due to bending moments in a section. Let us now consider the shear stresses caused by shear forces in sections of the beam.

Refer Figure 9.1, consider a short slice of a beam of length δx with a variation of bending moment over its length from M to $(M + \delta M.)$ Take a layer 'rs' at a distance y from the neutral axis; let the width of this layer be b and hence its area $(\delta x \times b)$. Between the successsive faces of the slice, there will be an excess force, since the stresses at the face 'pr' will be less than that on the face 'qs'. This excess force will be balanced by the shearing stress acting along the layer 'rs'.

Then the total force on face $'pr' = \sum\limits_{y}^{y_1} \dfrac{My}{I}(\delta y \times b)$

$$= \int\limits_{y}^{y_1} \dfrac{M}{I} Yb \; dy$$

and total force on 'sq' $= \int\limits_{y}^{y_1} \dfrac{(M+\delta M)}{I} \cdot (y.b)\,dy$

or the excess force between *sq* and *rp*

$$= \int\limits_{y}^{y1} \dfrac{\delta M}{I}(y.b)dy$$

$$= \dfrac{\delta M}{I}\int\limits_{y}^{y1}(y.b)dy$$

But $\int\limits_{y}^{y1} y^b \, dy$ is the moment of area of the face *'pr'* about neutral axis.

or $\int\limits_{y}^{y1} y.b.dy = A\overline{y}$ ($\overline{y}$ is the distance of the centroidal axis from the neutral axis)

This excess force is balanced by the shearing stress τ acting along *'rs'*. The force due to this stress is $t'\,(b.dx)$

$$\therefore \tau \times (b \cdot \delta x) = \dfrac{\delta M}{I}\, A\overline{y}$$

$$\tau = \dfrac{\delta M}{\delta x} \times \dfrac{1}{Ib} \times A\overline{y} \tag{9.1}$$

In the limit as $\delta x \to 0$

$\dfrac{\delta M}{\delta x} = S,$ the shearing force (see equation 9.3)

$$\therefore \tau = \dfrac{S.A\overline{y}}{Ib} \tag{9.2}$$

where, b is the width of the beam at the point where the shearing force is considered and $A\overline{y}$ is the moment of the sectional area above that point about the neutral axis.

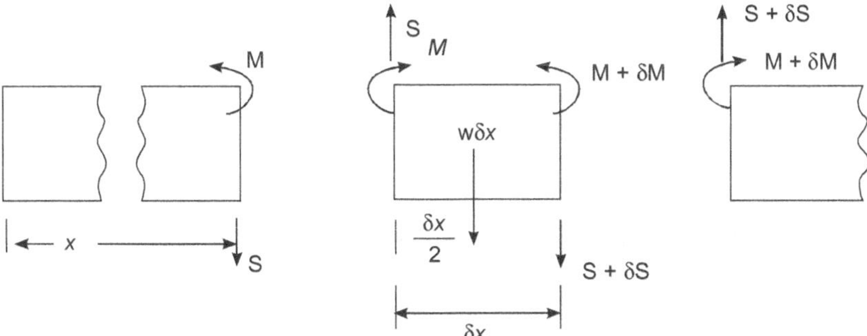

where,

w—Transverse load per unit length, S—Shear force

δx—Length of element of beam along the length

Figure 9.2 Elemental length of beam subjected to S & M

1. Vertical forces

$$(S + \delta S) - S = w\delta x$$

$$\delta s = w\delta x$$

$$\frac{\delta s}{\delta x} = w \quad \frac{ds}{dx} = w$$

2. Coluples

$$M - (\delta M + M) = -S\delta x + wd \times \left(\frac{\delta x}{2}\right)$$

$$\delta M = S\delta x$$

$$S = \frac{\delta M}{\delta x}$$

$$\text{as } \delta x \to 0 \quad S = \frac{dM}{dx} \quad \text{(rate of change of M)}$$

$$(9.3)$$

Figure 9.3a and b provides illustration as to how the shear stress cause sliding between layers in horizontal and vertical direction.

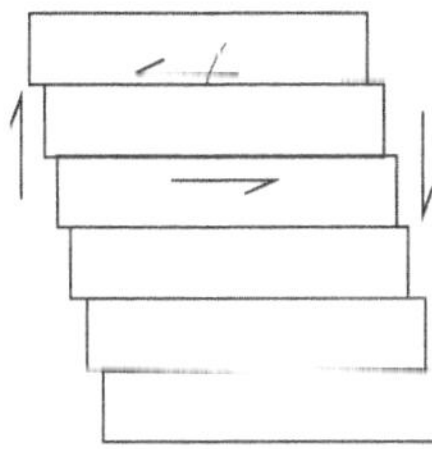

Figure 9.3(a) Sliding between layers

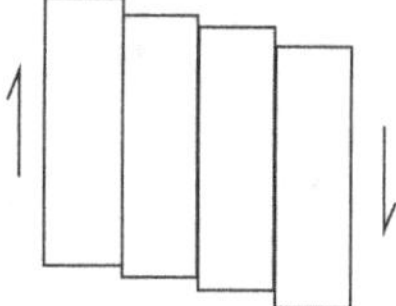

Figure 9.3(b) Sliding between vertical segments

VARIATION OF SHEAR STRESS IN BEAM CROSS-SECTION

Rectangular cross section Let b be the width and d the depth of the rectangular section. Let τ be the shear stress in a layer at a distance y from NA, where a particular section is subjected to shear force S.

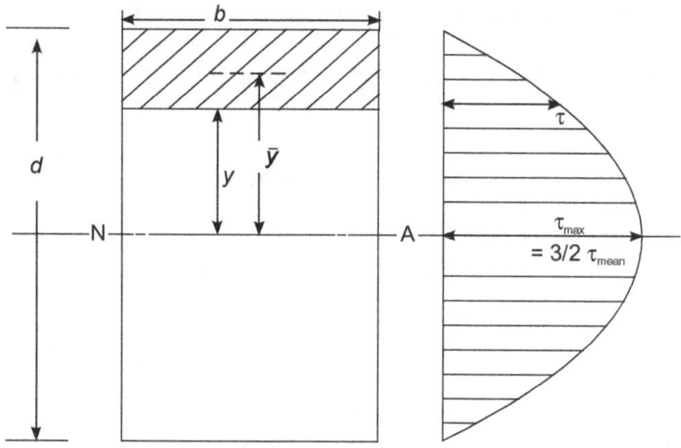

Figure 9.4

Shaded area,

$$A = b\left(\frac{d}{2} - y\right),\ \bar{y} = \frac{1}{2}\left(\frac{d}{2} + y\right),\ I = \frac{bd^3}{12}$$

$$A\bar{y} = \frac{b}{2}\left(\frac{d}{2} - y\right)\left(\frac{d}{2} + y\right) = \frac{b}{2}\left(\frac{d^2}{4} - y^2\right) \qquad \bar{y} = y + \left(\frac{d}{2} - y\right)\frac{1}{2}$$

$$= \frac{y}{2} + \frac{d}{4}$$

$$\tau = \frac{SA\bar{y}}{I.b} = \frac{S \times \dfrac{b}{2}\left(\dfrac{d^2}{4} - y^2\right)}{\dfrac{bd^3}{12} \times b} \qquad = \frac{1}{2}\left(\frac{d}{2} + y\right)$$

$$\tau = \frac{12 S \times b}{b \times bd^3 \times 2}\left(\frac{d^2}{4} - y^2\right)$$

Figure 9.4 shows the shear stress distribution (parabolic) in the rectangular section.

The maximum value of τ is at the neutral axis where $y = 0$

or

$$\tau_{max} / y = 0, \text{ i.e, at } NA = \frac{12s}{bd^3} \times \frac{d^2}{8} = \frac{3}{2}\frac{S}{bd} = \frac{3}{2}\tau_{av} \tag{9.4}$$

where τ_{av} = average shear stress $= \dfrac{S}{bd}$

When we use theory of elasticity solution (chapter 4) for a beam assuming a stress function, we get a solution simultaneously for bending and shear stresses. We can see here a close link between theory of elasticity and strength of materials approach in modeling problems.

Solid circular section

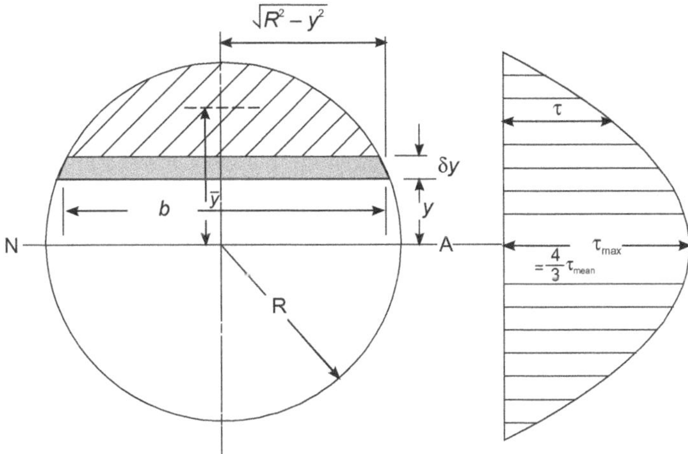

Figure 9.5

Figure 9.5 shows a solid circular section of radius R. Consider an elementary strip δy of width at a distance y from NA (Neutral axis).

Let the width of the strip be b, then

$$b = 2\sqrt{R^2 - y^2}$$

Area of the elementary strip

$$\delta a - b \times \delta y = 2\sqrt{R^2 - y^2} \times \delta y$$

Moment of elemental area about $NA = da' \, y$

$$= 2\sqrt{R^2 - y^2} \times y\delta y$$

For shaded area, total moment $= \int\limits_{y}^{R} 2y\sqrt{R^2 - y^2}\, dy = \int\limits_{y}^{R} b \cdot y \cdot dy$

Now

$$b = 2\sqrt{R^2 - y^2} \qquad \therefore b^2 = 4(R^2 - y^2)$$

Differentiating both sides, we get

$$2b \cdot db = -4 \times 2y\, dy \text{ or } y\, dy = -\frac{b\, db}{4}$$

When $y = y$, $b = b$

and when $y = R$, $b = 0$

$$\therefore \int\limits_{y}^{R} b.y.dy = \int\limits_{b}^{0} -b \times \frac{b \cdot db}{4} = \int\limits_{0}^{b} \frac{b^2 \cdot db}{4} = \left[\frac{b^3}{12}\right]_{0}^{b} = \frac{b^3}{12}$$

Now, $\tau = \dfrac{S}{Ib} \times$ Moment of shaded area $\left(A\bar{y}\right)$

$$\tau = \frac{S}{Ib} \times \frac{b^3}{12} \text{ or } \tau = \frac{S}{12T} b^2$$

$$= \frac{S}{12 \times \dfrac{\pi R^4}{4}}[4(R^2 - y^2)] = \frac{4}{3}\frac{S}{\pi R^4}(R^2 - y^2)$$

$$\tau = \frac{4}{3}\frac{S}{\pi R^4}(R^2 - y^2)$$

i.e., $\hphantom{xxxxxxxxxxxxxxxxxxxxxxxxxxxxxxxxxxxxxxxxxxxxxxxxxxxx}$ (9.5)

At $y = R$, and at $y = 0$, τ will be the maximum

Hence

$$\tau_{max} = \frac{4}{3}\frac{S}{\pi R^2}$$

$$\tau_{mean} = \frac{S}{\pi R^2}$$

$$\tau_{max} = \frac{4}{3}\tau_{mean}$$

Symmetrical I-Section

Shearing stresses in flanges

$$\tau = \frac{S}{IB} \times A\overline{y}$$

$$A\overline{y} = B\left[\frac{D}{2} - y\right] \times \frac{1}{2} \times \left[\frac{D}{2} + y\right]$$

$$= \frac{B}{2}\left[\frac{D^2}{4} - y^2\right]$$

$$\tau = \frac{S}{IB} \times \frac{B}{2}\left(\frac{D^2}{4} - y^2\right) = \frac{S}{2I}\left(\frac{D^2}{4} - y^2\right) \tag{9.6}$$

Shearing stress in web

$$\tau = \frac{S}{Ib} \times A\overline{y}$$

$$A\overline{y} = B \cdot \frac{(D-d)}{2} \times \left(\frac{D+d}{4}\right) + \left(\frac{d}{2} - y\right)b \times \frac{1}{2}\left(\frac{d}{2} + y\right)$$

$$= \frac{B}{8}(D^2 - d^2) + \frac{b}{2}\left(\frac{d^2}{4} - y^2\right)$$

$$\tau = \frac{S}{8Ib}\left[B(D^2 - d^2) + b(d^2 - 4y^2)\right] \tag{9.7}$$

The distribution of shear stress caused by these separate effects is shown in Figure 9.6 below.

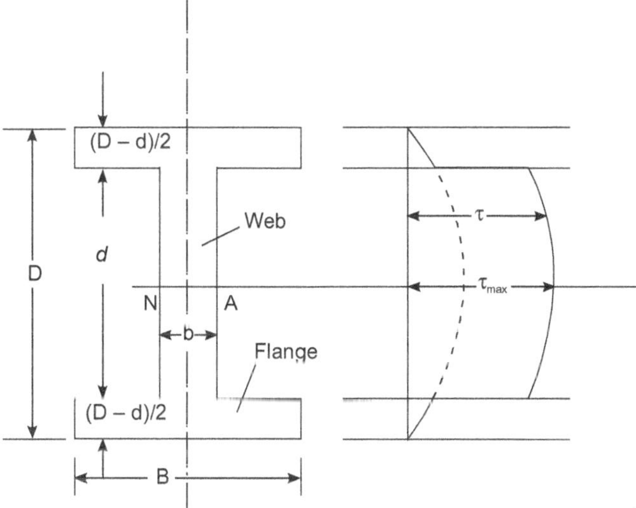

Figure 9.6

The value of τ will be maximum at $y = 0$

$$\tau_{max} = \frac{S}{8Ib}\left[B(D^2 - d^2) + bd^2\right]$$

It may be noted that at the inner surface of the flange, owing to the change in width of the section, the numerical value of the shearing stress changes from

$$\frac{S}{8Ib}(B(D^2 - d^2)) \text{ to } \frac{S}{8I} \times \frac{B}{b} \times (D^2 - d^2)$$

(In designn of built up beams, this is important)

Note   While bending stresses are maximum (or minimum) at the extreme layers and zero at the neutral layer of the beam, the shear stress distributions shows the opposite trend, i.e., zero at the extreme layers and maximum at the neutral layer. That is, never are both reaching maximum at the same location!

Shear Centre and Shear Flow Concepts

Shear centre is the centroid of shear forces in the cross section of a flexural member. In beams, cross sections that are symmetrical about horizontal and vertical axes, shear centre is the same as the centroid of the sectional area.

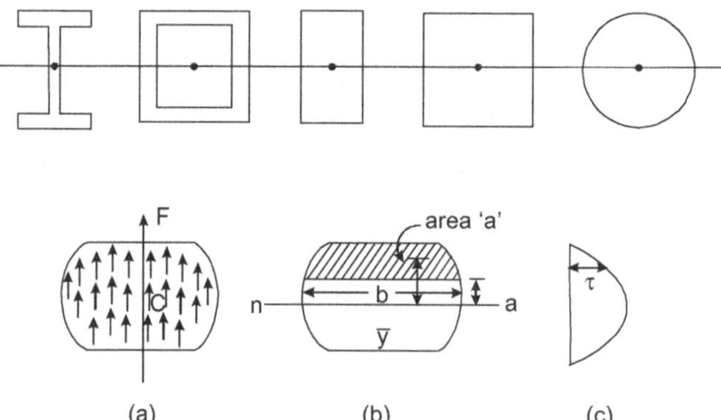

Figure 9.7   Shows the variables involved in equation 9.4 and 9.5 for τ_{max}

We have already seen that a shear force F across a beam section produces different values of shear stress on different fibres depending upon their distance y from the neutral axis. Thus,

$$\tau = F \cdot a\bar{y} / Ib$$

On a beam section G their resultant is F and it passes through the shear centre C. This is the way the beam resists the external force. Every section of the beam gets acted on by a force F and resists it by a combined shear force F equal and opposite and in the same line.

Normally we expect the load system and the reaction system to lie in vertical plane passing through the line of shear centres in a beam (Figure 9.8(c)). In this manner, we will have no induced torsion. On the other hand, if the load line is offset from the shear centre (Fig. 9.8(d)), say '*e*', we will have an induced torsion equal to *F.e*. We must ensure that the load line passes through *C* to avoid torsional shear stress.

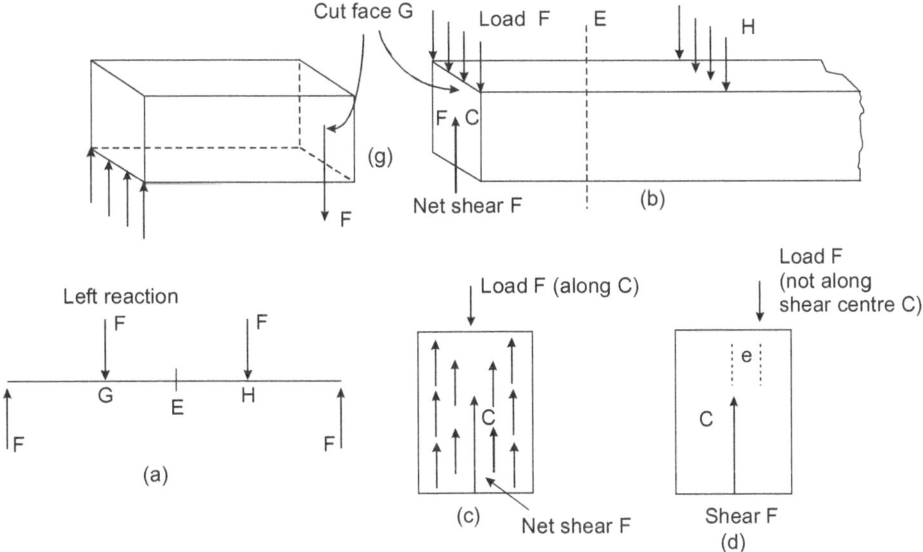

Figure 9.8 Importance of shear centre (a) is schematic loading diagram in a beam (b) and (c) show the vertical shear stresses.

Shear centre is not a problem in symmetrical sections about *X* and *Y* axes. In such sections the shear center *C* would be on the *Y* axis. So, if the load is along the *Y* axis, no torsion is experienced as there is no eccentricity between *Y* and *C*.

PROBLEM 9.1

Beam *AB* is made of three planks glued together and is subjected, in its plane of symmetry, to the loading shown. Knowing that the width of each glued joint is 20 mm, determine the average shearing stress in each joint at section *n-n* of the beam. The location of the centroid of the section is given in the sketch and the centroidal moment of inertia is known to be $I = 8.63 \times 10^{-6}\, m^4$

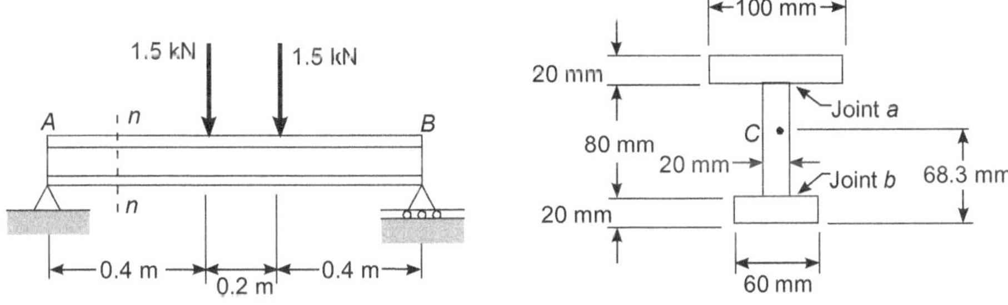

Solution

Since the beam and loading are both symmetric with respect to the centre of the beam, we have $A = B = 1.5\,\text{kN}\uparrow$. Considering the portion of the beam of the left of section *n-n* as a free body, we write

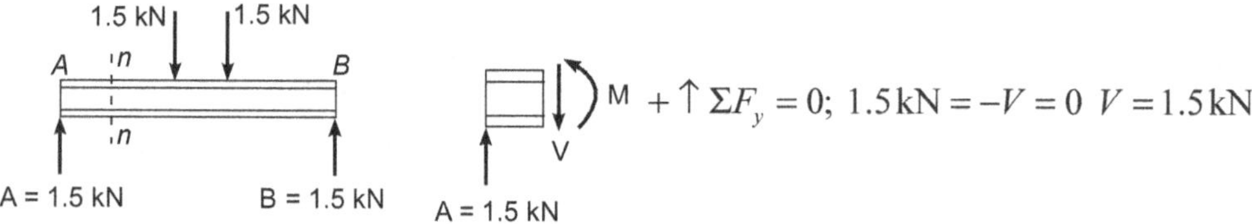

We pass the section *a-a* through the glued joint and separate the cross-sectional area into two parts. We choose to determine Q by computing the first moment with respect to the neutral axis of the area above sectio *-a*

$$Q = A\bar{y}_1 = [(0.100\,\text{m})(0.020\,\text{m})](0.0417\,\text{m}) = 83.4\times10^{-6}\,\text{m}^3$$

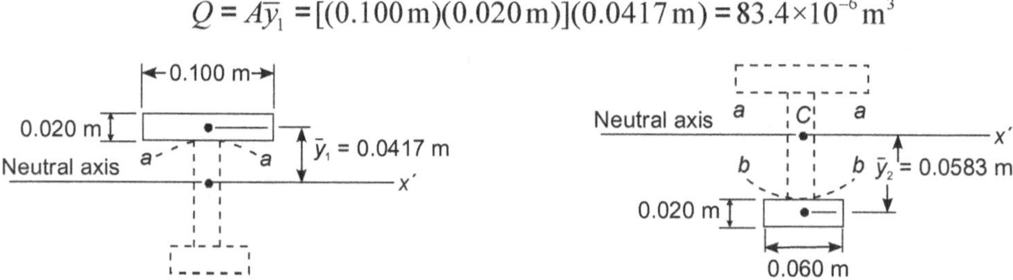

Recalling that the width of the glued joint is $t = 0.020$ m, we use Equation (9.2) to determine the average shearing stress in the joint.

$$\tau_{ave} = \frac{VQ}{It} = \frac{(1500\,\text{N})(83.4\times10^{-6}\,\text{m}^3)}{(8.63\times10^{-6}\,\text{m}^4)(0.020\,\text{m})}$$

We now pass section *b-b* and compute Q by using the area below the section.

$$Q = A\bar{y}_2[(0.060\,\text{m})(0.020\,\text{m})](0.0583\,\text{m}) = 70.0\times10^{-6}\,\text{m}^3$$

$$\tau_{ave} = \frac{VQ}{It} = \frac{(1500\,\text{N})(70.0\times10^{-6}\,\text{m}^3)}{(8.63\times10^{-6}\,\text{m}^4)(0.020\,\text{m})}$$

PROBLEM 9.2

A machine part has a T-shaped cross section and is acted upon in its plane of symmetry by the single force shown. Determine (a) the maximum compressive stress at section *n-n*, (b) the maximum shearing stress.

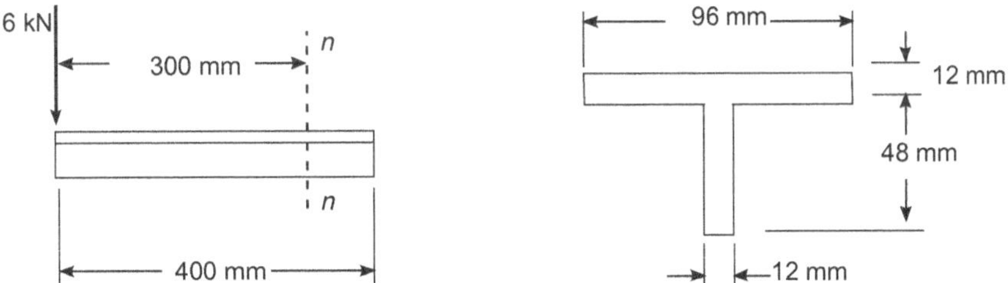

Solution

Neutral axis: The neutral axis passes through the centroid C of the cross section. Using the axis b-b as a reference axis and choosing the positive sense downward, we write

$$\bar{Y} = \frac{\Sigma A \bar{y}}{\Sigma A} \frac{(96\,\text{min})(12\,\text{mm})(-6\,\text{mm}) + (48\,\text{mm})(12\,\text{mm})(24\,\text{mm})}{(96\,\text{mm})(12\,\text{mm}) + (48\,\text{mm})(12\,\text{mm})}$$

$$= \frac{6912\,\text{mm}^3}{1728\,\text{mm}^2} = 4.00\,\text{mm}$$

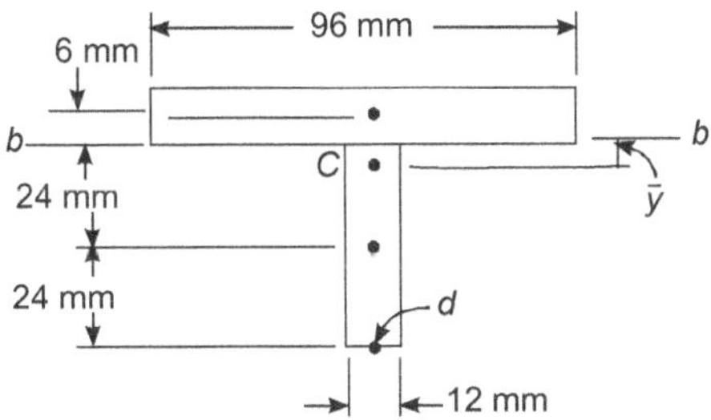

Centroidal moment of inertia using the parallel-axis theorem:

$$I = \frac{1}{12}(96\,\text{mm})(12\,\text{mm})^3 + (96\,\text{mm})(12\,\text{mm})(6\,\text{mm} + 4\,\text{mm})^2$$

$$+ \frac{1}{12}(12\,\text{mm})(48\,\text{mm})^3 + (12\,\text{mm})(48\,\text{mm})(24\,\text{mm} - 4\,\text{mm})^2$$

$$I = 470 \times 10^3\,\text{mm}^4 = 470 \times 10^{-9}\,\text{m}^4$$

(a) Maximum compressive stress: At section n-n the bending moment is $M = (6\,\text{kN})(0.3\,\text{m}) = 1800\,\text{N.m}$. The maximum compressive stress occurs at point d where $c = 48\,\text{mm} - 4\,\text{mm}$. We write

$$S_m = \frac{Mc}{I} = \frac{(1800\,\text{N.m})(0.044\,\text{m})}{470 \times 10^{-9}\,\text{m}^4} \qquad S_m = 168.5\,\text{MPa}$$

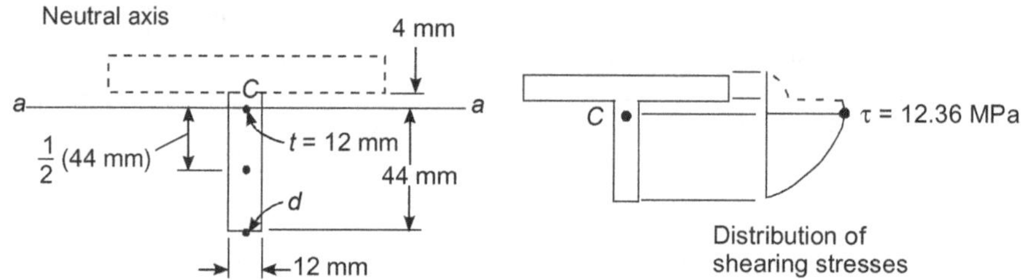

(b) Maximum shearing stress: The maximum value of Q occurs at the neutral axis. Since in this cross section the width t is minimum at the neutral axis, the maximum shearing stress will occur there. Passing section a-a at the neutral axis, we separate the cross-sectional area into two portions. Choosing the area below section a-a, we have.

$$Q = [(44\,\text{mm})(12\,\text{mm})]\frac{44\,\text{mm}}{2} = 11.62'\ 10^3\,\text{mm}^3$$

$$= 11.62'\ 10^{-6}\,\text{m}^3$$

Nothing that $V = 6$ kN and $t = 12$ mm, we write

$$\tau_m = \frac{VQ}{It} = \frac{(6000\,\text{N})(11.62\times10^{-6}\,\text{m}^3)}{(470\times10^{-9}\,\text{m}^4)(0.012\,\text{m})} \qquad t_m = 12.36\,\text{MPa}$$

REVIEW QUESTIONS

SHORT QUESTIONS

1.   Draw the shear stress distribution across a beam section which is

   (a) A rectangular section and

   (b) I-section

2.   Sketch the shear stress distribution diagram across the depth of a T– section.

LARGE QUESTIONS

1.   A timer beam 150 mm x 250 mm in cross section is simply supported at its ends and has a span of 3.5 m. The maximum safe allowable stress in bending is 7500 kN/m². Find the maximum safe UDL which the beam can carry. What is the maximum shear stress in the beam for the UDL calculated?

2.   A 300 mm by 125 mm rolled steel joist of I-section, flanges 12.5 mm thick and web 8.25 mm thick, is subjected to a bending moment of 3 kNm and a shear force of 100 kN at a perpendicular section.

   i.   Calculate the values of the maximum principal stress at neutral axis, top of web and outer edge of flange.

ii.   Draw the shear stress distribution diagram across the section.

3.   An R.S. Joist 55 cm by 19 cm having flange and web thickness 1.5 cm and 0.99 cm respectively, is used as a beam. If at a section, it is subjected to shear force of 100 kN, find the greatest intensity of shear stress in the beam taking,

   i.   Web vertical.

   ii.   Web horizontal. Show the variation of shear in both cases.

4.   The shear force acting on a beam section is 100 kN. The cross section is shown in Figure. I = 3142210 mm⁴. Obtain shear stress distribution across the section.

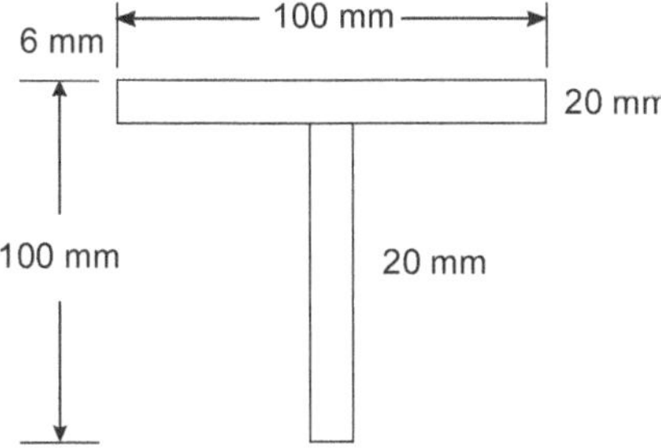

5.   A beam of triangular cross section having base width of 100 mm and height of 150 mm is subjected to a shear force of 15 KN. Find the value of maximum shear stress, and sketch the shear stress distribution along the depth of beam.

10

PROPERTIES OF PLANE AREAS

The engineers must be familiar with this subject, because of their importance in connection with the analysis of bending and torsion. Certain relations between the moments of inertia and product of inertia of a plane area (here called a section) are indicated in this chapter. Moment of Inertia of a section is an indication of the measure of resistance, the section offers to bending and torsion. The equations are given with reference to Figure 10.1, and the notation is as follows: A is the area of the section; X and Y are rectangular axes in the plane of the section intersecting at any point 0; Z is a polar axis through 0; U and V are rectangular axes through 0 inclined at the point 0; l and m are rectangular axes parallel to X and Y, respectively, and intersecting at G, the centroid of the area; $\bar{y}$ and $\bar{x}$ are the distances of X and Y from l and m, respectively; r is the distance from 0 to dA. The concepts and method of evaluating properties of plane areas are illustrated with several examples.

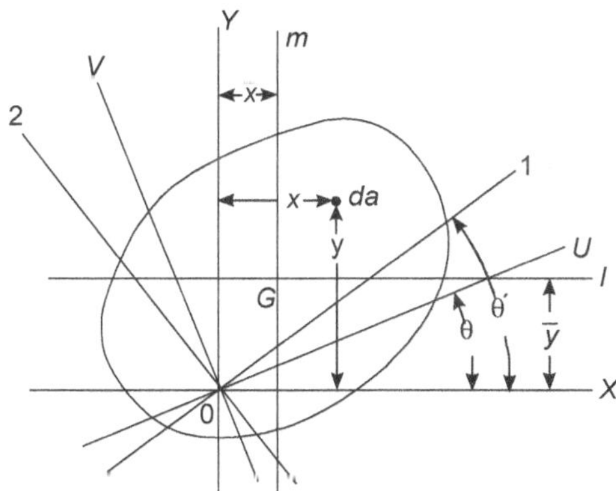

Figure 10.1   Area showing coordinates

PROBLEM 10.1

For the triangular area of Figure 1, determine (a) the first moment Q_x of the area with respect to the x axis, (b) the ordinate $\bar{y}$ of the centroid of the area.

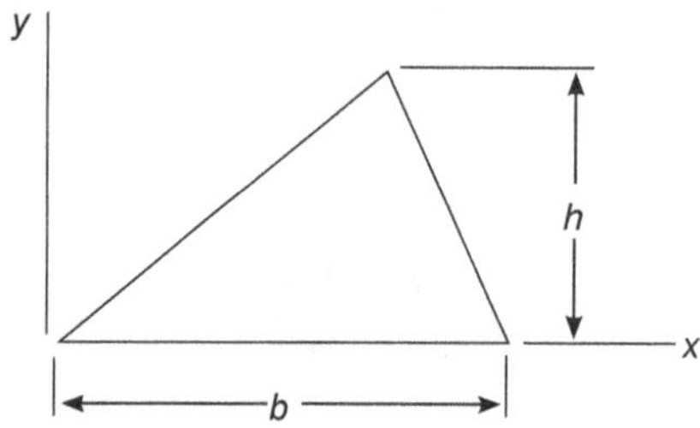

Figure (a)

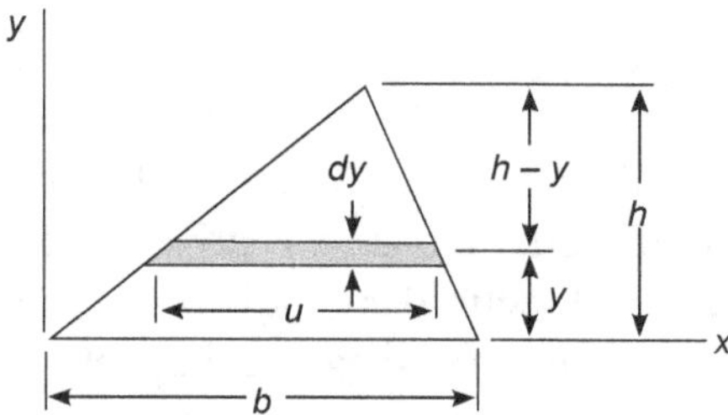

Figure (b)   Elemental area of a section

(a) *First moment Q_x*   We select as an element of area a, horizontal strip of length u and thickness dy, and note that all the points within the element are at the same distance y from the x axis (Figure 2). From similar triangles, we have

$$\frac{u}{b} = \frac{h-y}{h}$$

$$u = b\frac{h-y}{h}$$

and

$$dA = udy = b\frac{h-y}{h}dy$$

The first moment of the area with respect to the x axis is

$$Q_x = \int_A ydA = \int_0^h yb\frac{h-y}{h}dy = \frac{b}{h}\int_0^h \left(hy - y^2\right)dy$$

$$= \frac{b}{h}\left[h\frac{y^2}{2} - \frac{y^3}{3}\right]_0^h$$

$$Q_x = \frac{1}{6}bh^2 \tag{10.1}$$

(b) *Ordinate of centroid*   Observing that $A = \dfrac{1}{2}bh$

and $Q_x = A\overline{y}$, we have $\dfrac{1}{6}bh^2 = \dfrac{1}{2}(bh)\overline{y}$

$$\overline{y} = \frac{1}{3}h \tag{10.2}$$

DETERMINATION OF THE FIRST MOMENT AND CENTROID OF A COMPOSITE AREA

Centroid is a point in a body, which represents the single point through which the net downward weight of the body, running the weights of all individual elements that constitute the total body weight, passes through.

Consider an area A, such as the trapezoidal area shown in Figure 10.2, which may be divided into small geometric shapes. As we saw in the preceding section, the first moment Qx of the area with respect to the x axis is represented by the integral $\int y dA$, which extends over the entire area A. Dividing A into its component parts A1, A2, A3, we write

$$Q_x = \int_A y dA = \int_{A_1} y dA = \int_{A_2} y dA = \int_{A_3} y dA$$

or,

$$Q_x = A_1\overline{y}_1 + A_2\overline{y}_2 + A_3\overline{y}_3$$

where $\overline{y}_1$, $\overline{y}_2$ and $\overline{y}_3$ represent the ordinates of the centroids of the component areas. Extending this result to an arbitrary number of component areas, and noting that a similar expression may be obtained for Q_y, we write

$$Q_x = \sum_i A_i\overline{y}_i$$

$$Q_y = \sum_i A_i\overline{x}_i$$

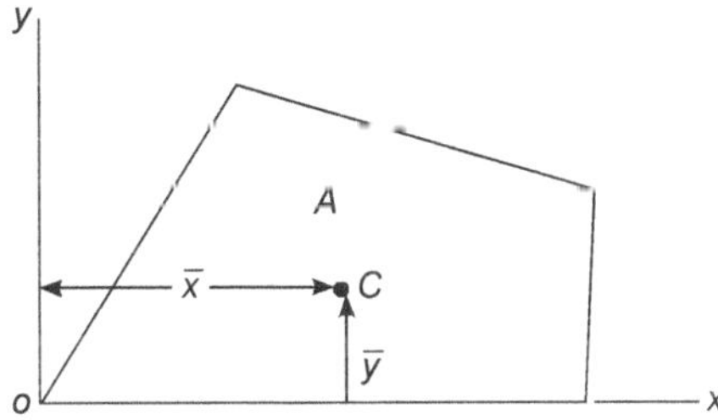

Figure 10.2   Composite area

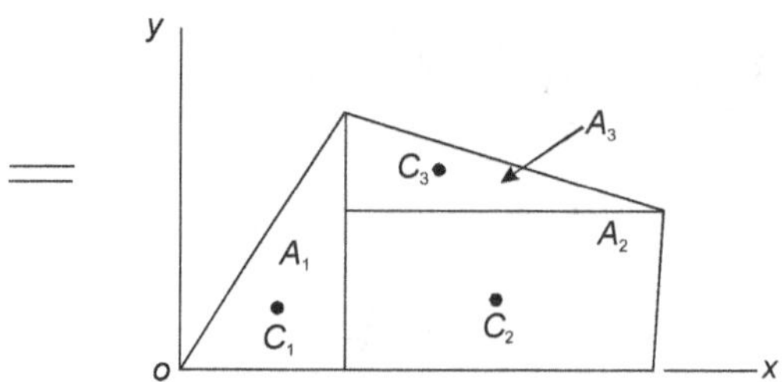

Figure 10.3

To obtain the coordinates $\bar{X}$ and $\bar{Y}$ of the centroid C of the composite area A, we substitute $Q_x = A\bar{Y}$ and $Q_y = A\bar{X}$ into above equations. We obtain,

$$A\bar{Y} = \sum_i A_i \bar{y}_i \quad A\bar{X} = \sum_i A_i \bar{x}_i$$

Solving for $\bar{X}$ and $\bar{Y}$ and recalling that the area A is the sum of the component areas A_i, we write

$$\bar{X} = \frac{\sum_i A_i \bar{x}_i}{\sum_i A_i} \quad \bar{Y} = \frac{\sum_i A_i \bar{y}_i}{\sum_i A_i} \tag{10.3}$$

PROBLEM 10.2

Locate the centroid C of the area A shown in Figure (A 'T' section)

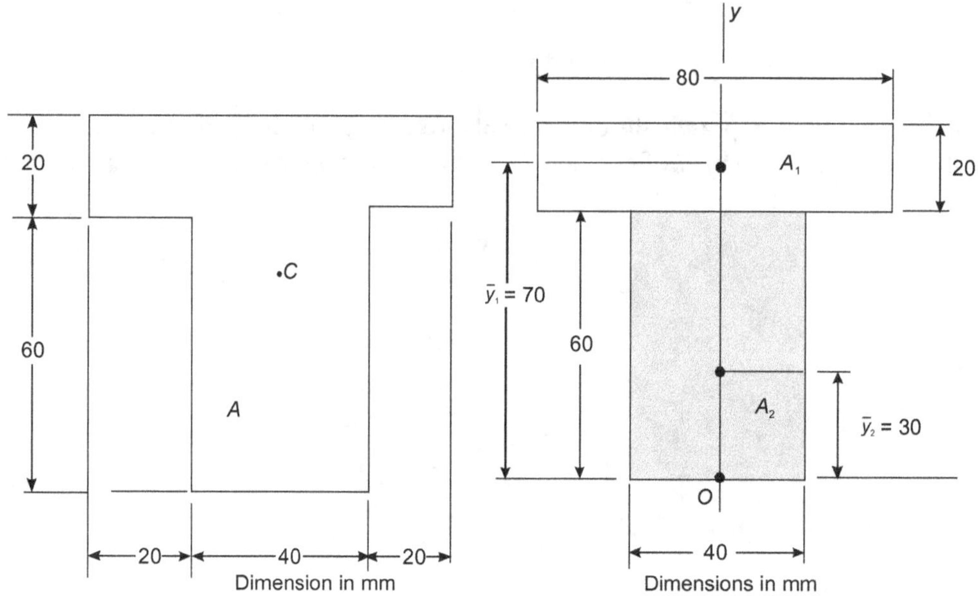

Figure 1   **Figure 2**

Selecting the coordinate axes shown in Figure 2, we note that centroid C must be located on the y axis, since this axis is an axis of symmetry; thus, $\overline{X} = 0$.

Dividing A into its component parts A_1 and A_2, we use the second of equation (10.3) to determine the ordinate $\overline{Y}$ of the centroid. The actual computation is best carried out in tabular form.

	Area (mm²)	$\overline{y}_i$ (mm)	$A_i\overline{y}_i$ (mm)³
A_1	$(20)(80) = 1600$	70	112×10^3
A_2	$(40)(60) = 2400$	30	72×10^3
	$\sum_i A_i = 4000$		$\sum_i A_i y_i = 184 \times 10^3$

$$\overline{Y} = \frac{\sum_i A_i \overline{y}_i}{\sum Ai} = \frac{184 \times 10^3\,\text{mm}^3}{4 \times 10^3\,\text{mm}^3} = 46\,\text{mm}$$

PROBLEM 10.3

Referring to the area A of problem 2, we consider the horizontal x' axis through its centroid C. (Such an axis is called a centroidal axis.). Denoting by A' the portion of A located above that axis (Figure 3), determine the first moment of A' with respect to the x' axis.

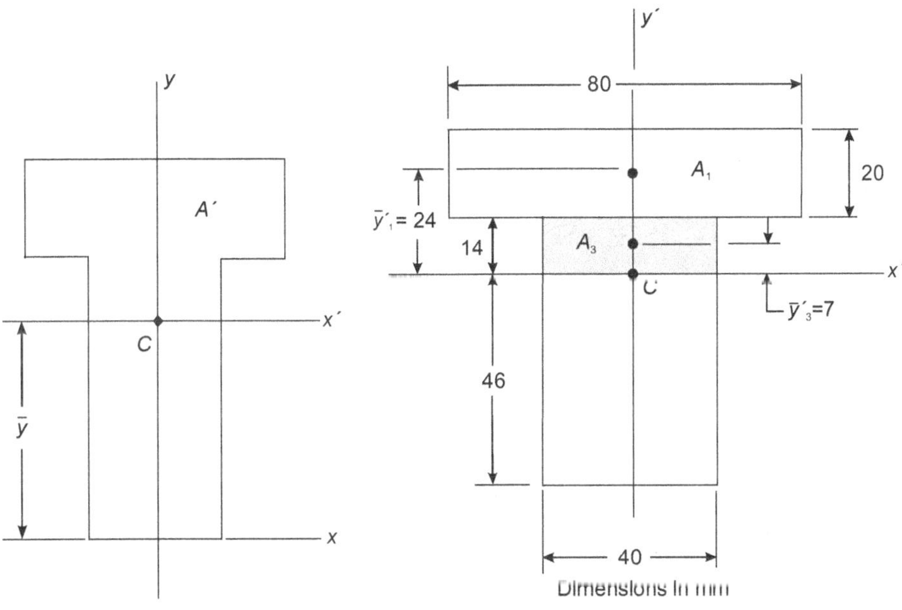

Figure 1   **Figure 2**

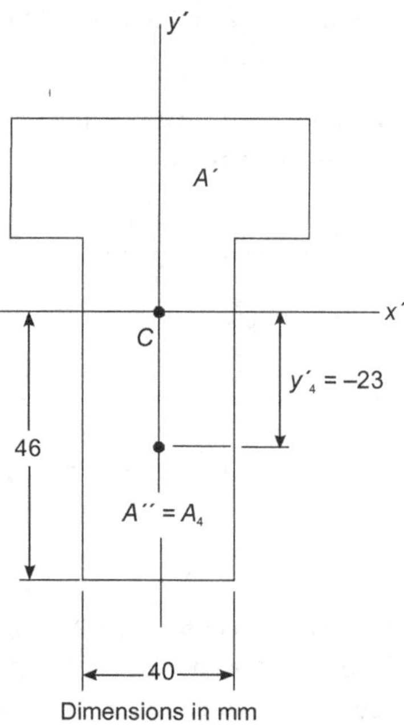

Figure 3

Solution

We divide the area A' into its components A_1 and A_3 (Figure 2). Recalling from Example that C is located 46 mm above the lower edge of A, we determine the ordinates $\bar{y}'_1$ and $\bar{y}'_3$ of A_1 and A_3 and express the first moment $Q'_{x'}$ of A' with respect x' to as follows:

$$Q'_{x'} = A_1\bar{y}'_1 + A_3\bar{y}'_3$$
$$= (20 \times 80)(24) + (14 \times 40)(7) = 42.3 \times 10^3 \text{ mm}^3$$

Alternate solution   We first note that since the centroid C of A is located on the x' axis, the first moment $Q'_{x'}$ of the entire area A with respect to that axis is zero:

$$Q_{x'} = A\bar{y}' = A(0) = 0$$

Denoting be A'' the portion of A located below the x' axis and by $Q''_{x'}$ its first moment with respect to that axis, we have therefore

$$Q'_{x'} = Q'_{x'} + Q''_{x'} = 0 \text{ or } Q'_{x'} = -Q''_{x'}$$

which shows that the first moments of A' and A'' have the same magnitude and opposite signs. Referring to Figure 3, we write

$$Q''_{x'} = A_4\bar{y}'_4 = (40 \times 46)(-23) = -42.3 \times 10^3 \text{ mm}^3$$

and

$$Q_{x'}' = -Q_{x'}'' = +42.3 \times 10^3 \text{ mm}^3$$

SECOND MOMENT, OR MOMENT OF INERTIA, OF AN AREA; RADIUS OF GYRATION

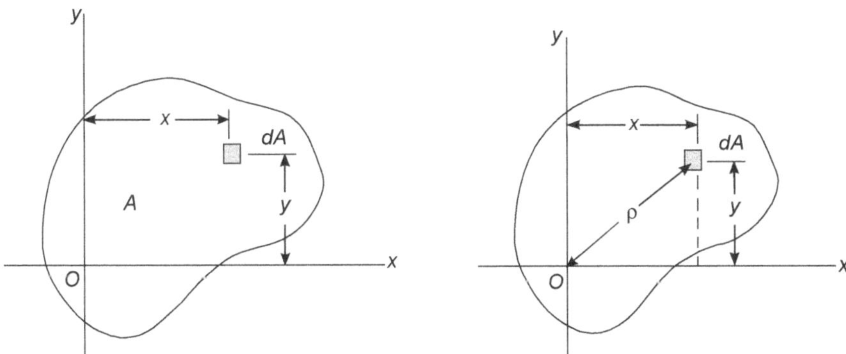

Figure 10.4   Figure 10.5

Consider again an area A located in the xy plane (Figure 10.4) and the element of area dA of coordinates x and y. The second moment, or moment of inertia, of the area A with respect to the x axis, and the second moment, or moment of inertia, of A with respect to the y axis are defined respectively as

$$I_x = \int_A y^2 dA \quad I_y = \int_A x^2 dA \tag{10.4}$$

These integrals are referred to as rectangular moments of inertia, since they are computed from the rectangular coordinates of the element dA. While each integral is actually a double integral, it is possible in many applications to select elements of area dA in the shape of thin horizontal or vertical strips, and thus reduce the computations to integrations in a single variable. This is illustrated in problem 10.4.

We now define the polar moment of inertia of the area A with respect to point O (Figure 10.5) as the integral

$$J_O = \int_A \rho^2 dA \tag{10.5}$$

where ρ is the radial distance from O to the element dA. While this integral is again a double integral, it is possible in the case of a circular area to select elements of area dA in the shape of thin circular rings, and thus reduce the computation of J_O to a single integration (*see* problem 10.4).

We note from Equations (10.4) and (10.5) that the moments of inertia of an area are positive quantities. If SI units are used, moments of inertia are expressed in m^4 or mm^4. Remember that we are dealing with plane areas and moment of inertia is the property of the section. This is not to be confused with mass moment of inertia encountered in problems of dynamics.

An important relation may be established between the polar moment of inertia J_0 of a given area and the rectangular moments of inertia I_x and I_y of the same area. Noting that $\rho^2 = x^2 + y^2$, we write

$$J_O = \int_A \rho^2 \, dA = \int_A (x^2 + y^2) \, dA = \int_A y^2 \, dA + \int_A x^2 \, dA$$

or

$$J_O = I_x + I_y \tag{10.6}$$

The radius of gyration of an area A with respect to the x axis is defined as the quantity r_x, which satisfies the relation

$$I_x = r_x^2 A \tag{10.7}$$

where I_x is the moment of inertia of A with respect to the x axis. Solving equation (10.7) for r_x, we have

$$r_x = \sqrt{\frac{I_x}{A}} \tag{10.8}$$

In a similar way, we may define the radii of gyration with respect to the y axis and the origin O. We write

$$I_y = r_y^2 A \qquad r_y = \sqrt{\frac{I_y}{A}} \tag{10.9}$$

$$J_O = r_O^2 A \qquad r_O = \sqrt{\frac{J_O}{A}} \tag{10.10}$$

Substituting for J_0, I_x and I_y in terms of the corresponding radii of gyration in equation (10.6), we observe that

$$r_O^2 = r_x^2 + r_y^2 \tag{10.11}$$

To summarize,

$$I_x = \int dA.y^2$$

$$I_y = \int dA.x^2$$

$$J_z = \int dA.r^2$$

$$H_{xy} = \int dA.xy \text{ (This is called product of inertia)}$$

$$k_x = \sqrt{\frac{I_x}{A}}$$

$$k_y = \sqrt{\frac{I_y}{A}} \qquad\qquad (10.12)$$

PROBLEM 10.4

For the rectangular area of Figure, determine (a) the moment of inertia I_x of the area with respect to the centroidal x axis, (b) the corresponding radius of gyration r_x.

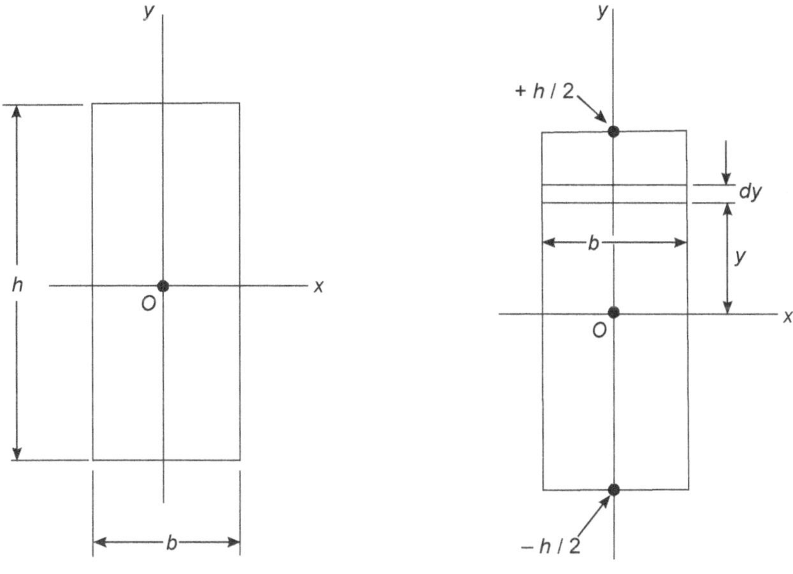

Figure 1   **Figure 2**

(a) *Moment of inertia* I_x We select as an element of area a horizontal strip of length b and thickness dy (Figure 2). Since all the points within the strip are at the same distance y from the x axis, the moment of inertia of the strip with respect to that axis is

$$dI_x = y^2 dA = y^2 (bdy)$$

Integrating from $y = -h/2$ to $y = +h/2$, we write

$$I_x = \int_A y^2 dA = \int_{-h/2}^{+h/2} y^2 (bdy) = \frac{1}{3} b \left[y^3 \right]_{-h/2}^{+h/2}$$

$$= \frac{1}{3} b \left(\frac{h^3}{8} + \frac{h^3}{8} \right)$$

or

$$I_x = \frac{1}{12} bh^2$$

(b) *Radius of gyration r_x* From equation (10.7), we have

$$I_x = r_x^2 A \qquad \frac{1}{12} bh^3 = r_x^2 (bh)$$

And, solving for r_x,

$$r_x = \frac{h}{\sqrt{12}}$$

The following equations and statements hold true:

$$J_z = I_x + I_y = I_u + I_v = I_1 + I_2$$

$$I_x = I_l + A\overline{y}^2$$

$$I_u = I_x \cos^2\theta + I_y \sin^2\theta - H_{xy} \sin 2\theta$$

$$\theta' = \frac{1}{2} \arctan \frac{2H_{xy}}{I_y - I_x} \tag{10.3}$$

$$I_{1,2} = \frac{1}{2}\left(I_x + I_y\right) \pm \sqrt{\frac{1}{4}\left(I_y - I_x\right)^2 + H_{xy}^2} \quad \left\{ \begin{array}{l} (+ \text{ for } I_1 \text{ the max}) \\ (- \text{ for } I_2 \text{ the min}) \end{array} \right\}$$

$$H_{xy} = H_{lm} + A\overline{x}\,\overline{y}$$

$$H_{uv} = H_{xy} \cos 2\theta - \frac{1}{2}\left(I_y - I_x\right)\sin 2\theta$$

$$H_{12} = 0$$

Which of the axes, 1 or 2, is the axis of maximum moment of inertia and which is the axis of minimum moment of inertia must be ascertained by calculation, unless the shape of the section is such as to make this obvious.

Any axis of symmetry is one of the principal axes for every point thereon.

The product of inertia is zero for any pair of axes, one of which is an axis of symmetry.

If the moment of inertia for one of the principal axes through a point is equal to that for any other axis through that point, it follows that the moments of inertia for all axes through that point are equal. (This refers to axes in the plane of section). Thus the moment of inertia of a square, an equilateral triangle, or any section having two or more axes of identical symmetry is the same for any central axis.

PROBLEM 10.5

For the circular area of Figure 1, determine (a) the polar moment of inertia J_o,

(b) the rectangular moments of inertia I_x and I_y.

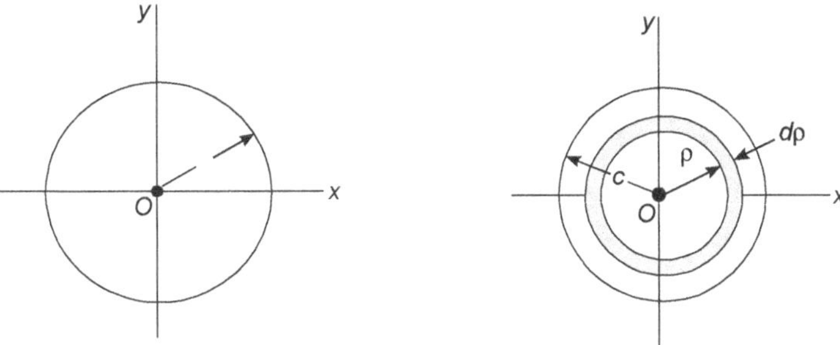

Figure 1   **Figure 2**

(a) *Polar moment of inertia*   We select as an element of area ring of radius ρ and thickness $d\rho$ (Figue 2). Since all the points within the ring are at the same distance ρ from the origin O, the polar moment of inertia of the ring is

$$dJ_O = \rho^2 dA = \rho^2 (2\pi\rho \, d\rho)$$

Integrating in ρ from 0 to c, we write

$$J_O = \int_A \rho^2 dA = \int_0^c \rho^2 (2\pi\rho \, d\rho) = 2\pi \int_0^c \rho^3 d\rho$$

$$J_O = \frac{1}{2}\pi c^4$$

(b) *Rectangular moments of inertia*   Because of the symmetry of the circular area, we have $Ix = I_y$. Recalling equation (10.6), we write

$$J_O = I_x + I_y = 2I_x \quad \frac{1}{2}\pi c^4 = 2I_x$$

and thus,

$$I_x = I_y = \frac{1}{4}\pi c^4$$

The results obtained in the preceding two examples and the moments of inertia of other common geometric shapes, are listed in table at the end of this chapter and can be used in design calculations involving such section.

PARALLEL-AXIS THEOREM

Consider the moment of inertia I_x of an area A with respect to an arbitrary x axis (Figure 10.6). Denoting by y the distance from an element of area dA to that axis , we recall that

$$I_x = \int_A y^2 dA$$

Let us now draw the centroidal x' axis, i.e., the axis parallel to the x axis which passes through the centroid C of the area. Denoting by y' the distance from the element dA to that axis, we write $y = y' + d$, where d is the distance between the two axes. Substituting for y in the integral representing I_x, we write

$$I_x = \int_A y^2 dA = \int_A (y' + d)^2 dA$$

$$I_x = \int_A y'^2 dA + 2d \int_A y' dA + d^2 \int_A dA \tag{10.14}$$

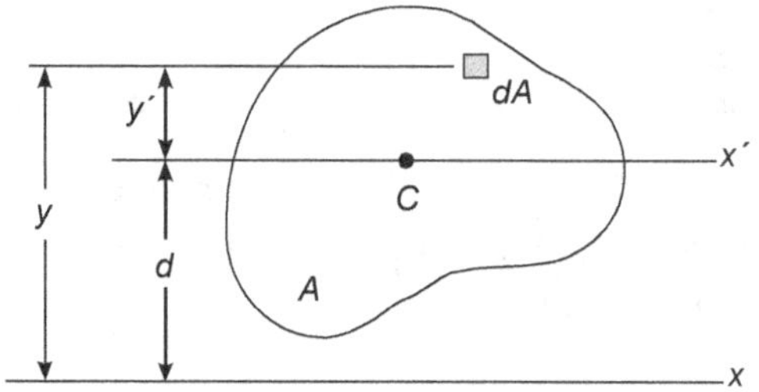

Figure 10.6

The first integral in equation (10.14) represents the moment of inertia $\overline{I}'_x$ of the area with respect to the centroidal x' axis. The second integral represents the first moment $Q_{x'}$ of the area with respect to the x' axis and is equal to zero, since the centroid C of the area is located on that axis. Indeed, we recall that

$$Q_{x'} = A\overline{y}' = A(0) = 0$$

Finally, we observe that the last integral in equation (10.14) is equal to the total area A. We have, therefore,

$$I_x = \overline{I}_{x'} + Ad^2 \tag{10.15}$$

This formula expresses that the moment of inertia I_x of an area with respect to an arbitrary x axis is equal to the moment of inertia $\overline{I}_{x'}$ of the area with respect to the centroidal x' axis parallel to the x axis, plus the product Ad^2 of the area A and the square of the distance d between the two axes. This result is known as the parallel-axis theorem. It makes it possible to determine the moment of inertia of an area with respect to a given axis, when its moment of inertia with respect to a centroidal axis of the same direction is known. Conversely, it makes it possible to determine the moment of inertia $\overline{I}_{x'}$ of an area A with respect to a centroidal x' axis when the moment of inertia I_x of A with respect to a parallel axis is known, by subtracting from I_x the product Ad^2. We should note that the parallel-axis theorem may be used only if one of the two axes involved is a centroidal axis.

A similar formula may be derived, which relates the polar moment of inertia J_o of an area with respect to an arbitrary point O and the polar moment $\overline{J}_C$ of of the same area with respect to its centroid C. Denoting by d the distance between O and C, we write

$$J_O = \overline{J}_C + Ad^2 \tag{10.16}$$

DETERMINATION OF THE MOMENT OF INERTIA OF A COMPOSITE AREA

Consider a composite area A made of several component parts A_1, A_2, and so forth. Since the integral representing the moment of inertia of A may be subdivided into integrals extending over A_1, A_2 and so forth, the moment of inertia of A with respect to a given axis will be obtained by adding the moments of inertia of the areas A_1, A_2 and so forth, with respect to the same axis. The moment of inertia of an area made of several of the common shapes shown in the Table 10.1 of this chapter can be obtained from the formulas given in that table. Before adding the moments of inertia of the component areas, however, the parallel-axis theorem should be used to transfer each moment of inertia to the desired axis.

Table 10.1   Properties of sections

Form of section	Area and distances from centroid to extremities	Moments and products of inertia and radii of gyration about central axes
1. Square	$A = a^2$ $y_1 = y_2 = \dfrac{a}{2}$ $y_3 = 0.707 a \cos\left(\dfrac{\pi}{4} - \alpha\right)$	$I_1 = I_2 = I_3 = \dfrac{1}{12} a^4$ $r_1 = r_2 = r_3 = 0.2887 a$
2. Rectangle	$A = bd$ $y_1 = \dfrac{d}{2}$ $y_2 = \dfrac{b}{2}$	$I_1 = \dfrac{1}{12} bd^3$ $I_2 = \dfrac{1}{12} db^3$ $I_1 > I_2$ if $d > b$ $r_1 = 0.2887 d$ $r_2 = 0.2887 b$
3. Hollow rectangle	$A = bd - b_i d_i$ $y_1 = \dfrac{d}{2}$ $y_2 = \dfrac{b}{2}$	$I_1 = \dfrac{bd^3 - b_i d_i^3}{12}$ $I_2 = \dfrac{db^3 - d_i b_i^3}{12}$ $r_1 = \left(\dfrac{I_1}{A}\right)^{1/2}$ $r_2 = \left(\dfrac{I_2}{A}\right)^{1/2}$

(Contd.,)

Table 10.1   (*Continues*)

Form of section	Area and distances from centroid to extremities	Moments and products of inertia and radii of gyration about central axes
4. Tee section	$A = tb + t_w d$ $y_1 = \dfrac{bt^2 + t_w d(2t + d)}{2(tb + t_w d)}$ $y_2 = \dfrac{b}{2}$	$I_1 = \dfrac{b}{3}(d+t)^3 - \dfrac{d^3}{3}(b - t_w) - A(d + t - y_1)^2$ $I_2 = \dfrac{tb^3}{12} + \dfrac{dt_w^3}{12}$ $r_1 = \left(\dfrac{I_1}{A}\right)^{1/2}$ $r_2 = \left(\dfrac{I_2}{A}\right)^{1/2}$
5. Channel section	$A = tb + 2t_w d$ $y_1 = \dfrac{bt^2 + 2t_w d(2t + d)}{2(tb + 2t_w d)}$ $y_2 = \dfrac{b}{2}$	$I_1 = \dfrac{b}{3}(d+t)^3 - \dfrac{d^3}{3}(b - 2t_w) - A(d + t - y_1)^2$ $I_2 = \dfrac{(d+t)b^3}{12} - \dfrac{d(b - 2t_w)^3}{12}$ $r_1 = \left(\dfrac{I_1}{A}\right)^{1/2}$ $r_2 = \left(\dfrac{I_2}{A}\right)^{1/2}$
6. Wide-flange beam with equal flanges	$A = 2bt + t_w d$ $y_1 = \dfrac{d}{2} + t$ $y_2 = \dfrac{b}{2}$	$I_1 = \dfrac{b(d + 2t)^3}{12} - \dfrac{(b - t_w)d^3}{12}$ $I_2 = \dfrac{b^3 t}{6} + \dfrac{t_w^3 d}{12}$ $r_1 = \left(\dfrac{I_1}{A}\right)^{1/2}$ $r_2 = \left(\dfrac{I_2}{A}\right)^{1/2}$

(Contd.,)

Table 10.1 (Continues)

Form of section	Area and distances from centroid to extremities	Moments and products of inertia and radii of gyration about central axes
7. Equal-legged angle 	$A = t(2a - t)$ $y_{1a} = \dfrac{0.7071(a^2 + at - t^2)}{2a - t}$ $y_{1b} = \dfrac{0.7071a^2}{2a - t}$ $y_2 = 0.7071a$	$I_1 = \dfrac{a^4 - b^4}{12} - \dfrac{0.5ta^2b^2}{a+b}$ $I_2 = \dfrac{a^4 - b^4}{12}$ where, $b = a - t$ $r_1 = \left(\dfrac{I_1}{A}\right)^{\frac{1}{2}}$ $r_2 = \left(\dfrac{I_2}{A}\right)^{\frac{1}{2}}$

Notation: A = Area (length)^2; y = Distance to extreme fibre (length); I = Moment of inertia (length)^4; r = Radius of gyration (length)

11

TORSION OF CIRCULAR SHAFTS

SHAFT

We consider shafts of cylindrical cross section—hollow or solid made of mild steel, alloy steel and copper alloys. Any rotating machine like compressor, turbines, pumps, blowers, fans, generators essentially is designed with a rotating shaft. Such safts are subjected to the following loads.

1. Torsional load
2. Bending load
3. Axial load
4. Combination of above three loads

The shafts are designed on the basis of strength and rigidity. Rigidity is needed to minimize deflection so that problems of rubbing against stationary members during operation is avoided.

The following values are usually adopted for the design of shaft:

$$\sigma = 112 \frac{MN}{m^2}, \text{ the maximum permissible tensile/ compressive stress}$$

$$\tau = 56 \frac{MN}{m^2}, \text{ maximum permissible shear stress, usually half of the stress value in tension.}$$

The ultimate tensile stress for commercial steel shafting may be $315 \frac{MN}{m^2}$ for hot rolled and turned low carbon steel and $490 \frac{MN}{m^2}$ cold finished low carbon steel, for corresponding stresses at the elastic limit would be about $160 \frac{MN}{m^2}$ and $315 \frac{MN}{m^2}$ respectively. In shafts with keyways, allowable stresses are 75% of the values given (due to stress concentration effects produced by the keyways).

TORSION OF SHAFTS

To transmit energy by rotation, it is necessary to apply a turning force. In the case of a shaft, if the force is applied tangentially in the shaft surface and in the plane of transverse cross-section, the torque or twisting moment may be calculated by multiplying the force with the radius of the shaft. If the shaft is subjected to two opposing turning moments it is said to be in pure torsion and it will exhibit the tendency of shearing off at every cross-section which is perpendicular to longitudiual axis.

CIRCULAR SHAFTS

TORSION EQUATION

Assumptions: We shall now derive the torsion equation just as we derived the bending equation for a beam.

The torsion equation is based on following assumptions.

  i. The material of the shaft is uniform throughout.

  ii. The shaft which is circular in section remains circular after twisting caused by torques.

  iii. A plane section of shaft normal to its axis before loading remains plane after the torques have been applied.

  iv. The twist along the length of the shaft is uniform throughout.

  v. The distance between any two normal cross-sections remain the same after the application of torque (Not true for noncircular section).

Note   Noncircular sections 'warping', which caused by the axial displacement of two normal sections and as such this is not considered in this text. Theory of elasticity solution and membrane anology techniques used enable to determine torsional shear stresses in non circular shafts.

  vi. Maximum shear stress induced in the shaft due to application of torque does not exceed its elastic limit value.

Let

T = Maximum twisting torque (N.mm)

D = Diameter of the shaft (mm)

Ip = Polar moment of inertia (mm^4)

τ = Shear stress (N/mm^2)

G = Modulus of rigidity (N/mm^2)

θ = Angle of twist (radians) and

l = Length of the shaft (mm)

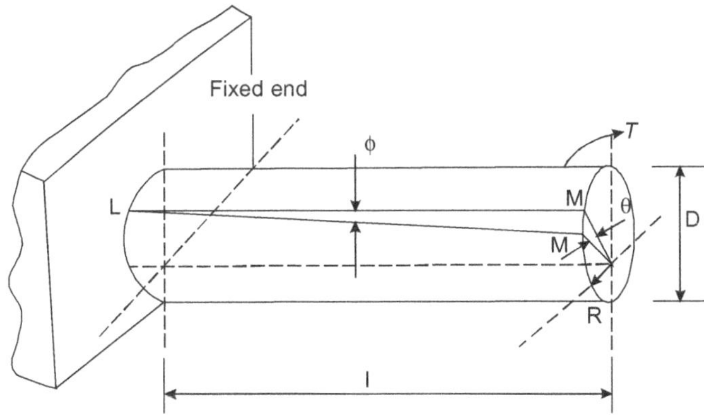

Figure 11.1

In Figure 11.1 is shown a shaft fixed at one end and torque being applied at the other end. If a line LM is drawn on the shaft it will be distorted to LM′, thus cross-section will be twisted through angle θ and surface by angle ϕ.

$$\text{Here shear, strain (Rotation of the line LM on the shaft surface) } \phi = \frac{MM'}{l}$$

$$\text{Also } \phi = \frac{\tau}{G} \left(\text{Shear strain} = \frac{\text{Shear stress}}{\text{Rigidity modulus}} \right)$$

$$\therefore \frac{MM'}{l} = \frac{\tau}{G}$$

or

$$\frac{R\theta}{l} = \frac{\tau}{G} \; [\because MM' = R \times \theta \; R \text{ being radius of the shaft}]$$

$$\therefore \frac{\tau}{R} = \frac{G\theta}{l} \tag{1}$$

Consider an elementary ring of thickness dx at a radius x and let the shear stress at this radius be τ_x.

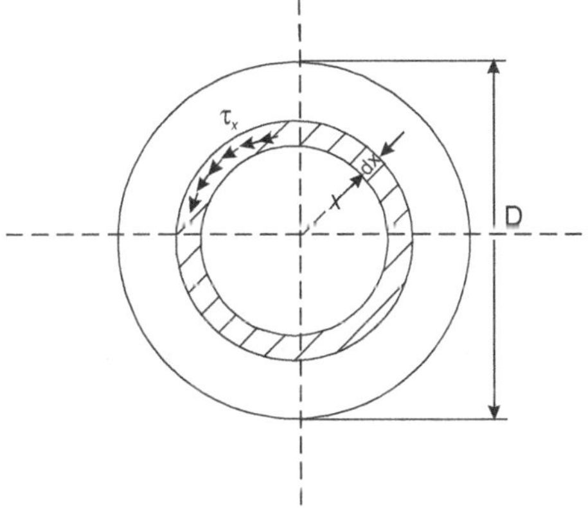

Figure 11.2

The turning force on the elementary ring $= \tau_x \cdot 2\pi x \cdot dx$

Turning moment due to this turning force, $dT = \tau_x \cdot 2\pi x \cdot dx \times x$

$$\text{Total turning moment, } T = \int_0^R \tau_x \cdot 2\pi \, x dx \times x$$

$$= 2\pi \int_0^R \tau_x \cdot x^2 \cdot dx = 2\pi \int_0^R \frac{\tau \cdot x}{R} \cdot x^2 \cdot dx \left[\because \frac{\tau}{R} = \frac{\tau_x}{x} \text{ or } \tau_x = \frac{\tau \cdot x}{R} \right]$$

$$= 2\pi \frac{\tau}{R} \int_0^R x^3 dx$$

$$= 2\pi \frac{\tau}{R} \cdot \frac{R^4}{4}$$

$$T = \tau \cdot \frac{\pi R^3}{2} = \tau \cdot \frac{\pi}{16} D^3 \qquad \text{(strength of solid shaft)}$$

$$\text{or } T = \frac{\tau}{R} \cdot \frac{\pi R^4}{2} = \frac{\tau}{R} I_p \left(\because \text{For a circular shaft, } I_p = \frac{\pi}{32} D^4 = \frac{\pi}{2} R^4 \right)$$

$$\frac{T}{I_p} = \frac{\tau}{R} \qquad\qquad (2)$$

From equations (1) and (2) we have

$$\frac{T}{I_p} = \frac{\tau}{R} = \frac{G\theta}{l} \qquad\qquad (11.1)$$

This is called torsion equation

Note

From the relation, $\dfrac{T}{I_p} = \dfrac{\tau}{R}$,

We have $T = \tau \times \dfrac{I_p}{R}$

For a given shaft, I_p and R are constants $\dfrac{I_p}{R}$ and is thus a constant and is known as polar modulus of the shaft section.

Thus $T = \tau \times z_p$

For a shaft of given material τ, the maximum permissible shear stress is fixed and thus the greatest twisting moment that the shaft can withstand is proportional to the polar modulus of the shaft. Polar modulus of the section is thus measure of strength of shaft in torsion.

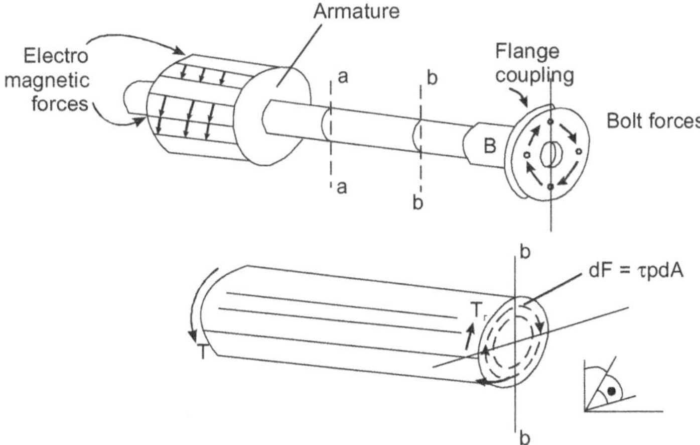

Figure 11.3   Shaft subjected to torque

HOLLOW CIRCULAR SHAFTS

Consider a hollow circular shaft subject to a torque T.

Let

R = Outer radius of the shaft

r = Inner radius of the shaft, and

τ = shear stress at radius R.

Following the same procedure, we have

dT = Turning moment on the elementary ring

$$= \tau_x \cdot 2\pi x \cdot dx \cdot x$$

Integrating both sides, we get

$$\int dT = \int_r^R \tau_x \cdot 2\pi x \cdot dx \cdot x$$

But $\dfrac{\tau x}{x} = \dfrac{\tau}{R}$ or $\tau_x = \dfrac{\tau}{R} \cdot x$

$$\therefore \int dT = \int_r^R \frac{\tau x}{R} \cdot 2\pi x \cdot dx \cdot x = \frac{2\pi \tau}{R} \int x^3 dx$$

$$T = \frac{2\pi\tau}{R}\left[\frac{x^4}{4}\right]_r^R = \frac{\pi}{2} \cdot \frac{\tau}{R}(R^4 - r^4)$$

or

$$T = \frac{\pi}{16}\tau\left[\frac{D^4 - d^4}{D}\right] \quad \text{(strength of hollow shaft)}$$

or

But

$$I_p = \frac{\pi}{32}(D^4 - d^4) = \frac{\pi}{2}(R^4 - r^4)$$

$$\therefore T = \frac{\tau}{R} \cdot I_p$$

$$\frac{T}{I_p} = \frac{\tau}{R} = \frac{G\theta}{l} \quad \text{(Torsion equation)}$$

TORSIONAL RIGIDITY

From the relation $\dfrac{T}{I_p} = \dfrac{G\theta}{l}$

We have $\theta = \dfrac{Tl}{GI_p}$

Since G, I_p, l are constants for a given shaft, θ the angle of twist is directly proportional to T (the twisting moment).

$$\frac{T}{\theta} = \frac{GI_p}{l}\left(\frac{GI_p}{l} \text{ is called torsional rigidity}\right) \tag{11.2}$$

The quantity GI_p is known as torsional rigidity and is represented by k or μ. From the above relation, we have

$$k = \frac{GI_p}{l} = \frac{T}{\theta} \quad \text{(Torsional spring constant)} \tag{11.3}$$

POWER TRANSMITTED BY SHAFT

Consider a force F newtons acting tangentially on the shaft of radius R. If the shaft due to this turning moment (F × R) starts rotating at N r.p.m then

Work supplied to the shaft / sec $= F \times$ distance moved/sec

$$= F \times 2\pi\frac{RN}{60} \text{ Nm/s}$$

$$P = F \times \frac{2\pi RN}{60} \text{ watts} = \frac{T \times 2\pi RN}{60 \times 1000} \text{ kw}$$

$$P = \frac{2\pi NT}{60 \times 1000}\,\text{kw} \tag{11.4}$$

where T is the mean or average torque in Nm.

IMPORTANCE OF ANGLE OF TWIST AND VARIOUS STRESSES IN SHAFT

Angle of Twist

In many problems such as torsion of shafts of accurate milling and drilling machines the angle of twist (θ) is required to be restricted. In such applications, the shaft are designed on the basis of limiting angle of twist and checked for shearing stresses. In torsional vibration problems, the angle of twist is also needed. These considerations make it necessary to find suitable expression for angle of twist.

A shaft for which angle of twist is significant, should always be designed or checked for angle of twist in addition to the design stresses in shafts.

Stresses in Shafts

In a shaft the following significant stresses occur.

i. A maximum shear stress occurs on the cross-section of the shaft at its outer surface.

ii. The maximum longitudinal shear stress occurs at the surface of the shaft on the longitudinal planes passing through the longitudinal axis of the shaft.

iii. The maximum tensile stress (i.e., major principal stress) occurs at planes 45° to the maximum shearing stress planes at the surface of the shaft. This stress is equal to the maximum shear stress on the cross-section of the shaft.

iv. The maximum compressive stress (i.e., or min principal stress) occurs on the planes at 45° to the longitudinal and the cross-sectional planes at the surface of the shaft. This stress is equal to the maximum shear stress on the cross-section.

- ✦ These are critical stresses as they govern the failure of the shaft. These stresses develop simultaneously and hence should be considered simultaneously for design purposes.

- ✦ For most engineering materials, fortunately the shear strength is the smallest as compared to the tensile and compressive stresses and in such cases only the maximum shear stress on the cross-section of the shaft is the significant stress for design.

- ✦ For materials for which tensile and compressive strengths are lower than the shear strength, the shaft design should be done for the lowest strength.

MODULUS OF RUPTURE

A modulus of rupture corresponding to the modulus of rupture in bending, may be defined as follows:

"The maximum fictitious shear stress calculated by the torsion formula by using the experimentally maximum torque (i.e., ultimate torque) required to rupture a shaft".

$$\text{Mathematically,} \ \tau_r = \frac{T_u R}{I_p} \tag{11.5}$$

where

τ_r = Modulus of rupture in torsion (also called computing ultimate twisting strength).

T_u = Ultimate torque at failure

R = Outer radius of the shaft

The torsion formula $\dfrac{T}{I_p} = \dfrac{\tau}{R}$ is not applicable beyond the limit of proportionality.

COMPARISON OF SOLID AND HOLLOW SHAFTS

(a) Comparison by Strength

In this case it is assumed that both shafts have same length, same material, same weight and hence the same maximum shear stress.

Let

D_s = Diameter of solid shaft

d_H = Internal diameter of hollow shaft

D_H = External diameter of hollow shaft

A_s = Cross-sectional area of solid shaft

A_H = Cross-sectional area of hollow shaft

T_s = Torque transmitted by the solid shaft

T_H = Torque transmitted by the hollow shaft

Now $\quad T_s = \tau . \dfrac{\pi}{16} D_s^3$

$$T_H = \tau . \frac{\pi}{16} \left[\frac{D_H^4 - d_H^4}{D_H} \right]$$

$$\therefore \frac{\text{Strength of hollow shaft}}{\text{Strength of solid shaft}} = \frac{T_H}{T_s} = \frac{\tau \, \dfrac{\pi}{16} \, \dfrac{D_H^4 - d_H^4}{D_H}}{\tau \cdot \dfrac{\pi}{16} D_s^3}$$

or

$$\frac{T_H}{T_s} = \frac{D_H{}^4 - d^4{}_H}{D_H . D_s{}^3} \tag{1}$$

Let

$$\frac{D_H}{d_H} = n$$

$\therefore D_H = n d_H$. Substituting in equation (1) we get

$$\frac{T_H}{T_s} = \frac{n^4 d_H^4 - d_H^4}{n d_H D_s^3} = \frac{d_H^4 (n^4 - 1)}{n d_H D_s^3} = \frac{d_H^3 (n^4 - 1)}{n D_s^3} \tag{2}$$

As the weight, material and length of both the shafts are same,

$\therefore$ Cross-sectional area of solid shaft = Cross-sectional area of hollow shaft

$$A_s = A_H$$

$$\frac{p}{4} D_s^2 = \frac{p}{4} (D_H^2 - d_H^2)$$

or

$$D_s = \sqrt{D_H^2 - d_H^2}$$

$$D_s^3 = (D_H^2 - d_H^2)\left(\sqrt{D_H^2 - d_H^2}\right)$$

$$D_s^3 = (n^2 d_H^2 - d_H^2)\sqrt{n^2 d_H^2 - d_H^2}$$

$$D_s^3 = d_H^3 (n^2 - 1)\sqrt{n^2 - 1} \tag{3}$$

Substituting the value D_s^3 of in equation (2), we get

$$\frac{T_H}{T_s} = \frac{d_H^3 (n^4 - 1)}{n d_H^3 (n^2 - 1)\sqrt{n^2 - 1}} = \frac{(n^2 + 1)(n^2 - 1)}{n(n^2 - 1)\sqrt{n^2 - 1}}$$

$$\frac{T_H}{T_s} = \frac{n^2 + 1}{n\sqrt{n^2 - 1}}$$

Since $D_H > d_H$ and $D_H > n$, it is obvious that the value of 'n' is greater than unity.

Suppose, $n = 2$

Then $\dfrac{T_H}{T_s} = \dfrac{2^2+1}{2\sqrt{2^2-1}} = 1.44$

This shows that the torque transmitted by the hollow shaft is greater than that the solid shaft, thereby proving that the hollow shaft is stronger than the solid shaft.

(b) Comparison by Weight

In this case it is assumed that both the shafts have the same length and material. Now if the torque applied to both shafts is same, then, the maximum shear stress will also be same in both the cases

$$\frac{\text{Weight of hollow shaft}}{\text{Weight of solid shaft}} = \frac{W_H}{W_s} = \frac{A_H}{A_s}$$

$$= \frac{\pi}{4} \frac{(D_H^2 - d_H^2)}{\frac{\pi}{4} D_s^2} = \frac{d_H^2(n^2-1)}{D_s^2}$$

Now $\dfrac{\text{Weight of hollow shaft}}{\text{Weight of solid shaft}}$

$$= \frac{W_H}{W_S} = \frac{A_H}{A_S} = \frac{\frac{\pi}{4}(D_H^2 - d_H^2)}{\frac{\pi}{4}D_s^2} = \frac{D_H^2 - d_H^2}{D_s^2} \tag{1}$$

Let $\dfrac{D_H}{d_H} = n \therefore D_H = nd_H$ and substituting this value in equation (1)

We get

$$\frac{W_H}{W_S} = \frac{n^2 d_H^2 - d_H^2}{D_s^2} = \frac{d_H^2(n^2-1)}{D_s^2} \tag{2}$$

Torque applied in both the cases is same $T_s = T_H$

$$\tau \cdot \pi / 16 D_s^3 = \tau \cdot \pi / 16 \left[\frac{D_H^4 - d_H^4}{D_H}\right]$$

$$D_s^3 = \frac{D_H^4 - d_H^4}{D_H} = \frac{n^4 d_H^4 - d_H^4}{nd_H} = \frac{d_H^3(n^4-1)}{n}$$

$$D_s = d_H \left[\frac{n^4-1}{n}\right]^{1/3} \quad \text{or } D_s^2 = d_H^2\left[\frac{n^4-1}{n}\right]^{2/3} \tag{3}$$

Substituting the value of D_s^2 in equation (2) we have

$$\frac{W_H}{W_S} = \frac{d_H^2(n^2-1)}{d_H^2\left(\dfrac{n^4-1}{n}\right)^{2/3}} = \frac{(n^2-1)n^{2/3}}{(n^4-1)^{2/3}} \tag{4}$$

If, $n = 2$

then, $\dfrac{W_H}{W_S} = \dfrac{(2^2-1)\times(2)^{2/3}}{(2^4-1)^{2/3}} = 0.7829$

which shows that for same material, length and given torque, weight of hollow shaft will be less. So hollow shafts are economical compared to solid shafts as regards tension.

SHAFTS IN SERIES

In order to form a composite shaft sometimes two shafts are connected in series. In such cases each shaft transmits the same torque. The angle of twist is the sum of the angle of twist of the two shafts connected in series.

Thus, total angle of twist is given by

$$\theta = \theta_1 + \theta_2 = \frac{Tl_1}{G_1 I_{P_1}} + \frac{Tl_2}{G_2 I_{P_2}} = T\left[\frac{l_1}{G_1 I_{P_1}} + \frac{l_2}{G_2 I_{P_2}}\right] \tag{11.6}$$

where

$\quad T \quad = \quad$ Torque transmitted by each shaft

$\quad l_1, l_2 \quad = \quad$ Respective lengths of the two shafts

$\quad G_1, G_2 \quad = \quad$ Respective moduli of rigidity, and

$\quad I_{p1}, I_{p2} \quad = \quad$ Respective polar moment of inertias

When shafts are made of same material,

$G_1 = G_2 = G$

Then

$$\theta = \frac{T}{G}\left[\frac{l_1}{I_{p1}} + \frac{l_2}{I_{p2}}\right] \tag{11.7}$$

Here, the driving torque is applied at one end and the resisting torque at the other.

SHAFTS IN PARALLEL

The shafts are said to be in parallel when the driving torque is applied at the junction of the shafts and the resisting torque is at the other ends of the shafts. Here, the angle of twist is same for each shaft, but the applied torque is divided between the two shafts.

i.e., $\theta_1 = \theta_2$

$$\frac{T_1 l_1}{G_1 I_{p1}} = \frac{T_2 l_2}{G_2 I_{p2}}$$

and $T = T_1 + T_2$

If the shafts are made of same material

$$G_1 = G_2$$

Then $\dfrac{T_1 l_1}{I_{p1}} = \dfrac{T_2 l_2}{I_{p2}}$ or $\dfrac{T_1}{T_2} = \dfrac{I_{p1} l_2}{I_{p2} l_1}$

When torque is shared equally by both the shafts

$$T_1 = T_2, \text{ then } I_{p1} l_2 = I_{p2} l_1$$

TORSIONAL RESILIENCE

Consider a hollow shaft of external diameter D, internal diameter d and length l, subjected to a gradually applied torque T. Let θ be the angle of twist. Energy is stored in the shaft due to this angular distortion. This is called torsional energy or the torsional resilience, denoted by U.

Torsional energy = workdone by the torque = Average torque × Angular twist

i.e.,
$$U = \frac{T}{2} \cdot \theta \tag{1}$$

From torsion equation

$$\frac{T}{I_p} = \frac{\tau}{R} = \frac{2\tau}{D}$$

or
$$T = I_p \frac{2\tau}{D} = \frac{\pi}{32}(D^4 - d^4)\frac{2\tau}{D} = \frac{\pi\tau}{16}\left[\frac{D^4 - d^4}{D}\right]$$

and
$$\frac{\tau}{R} = \frac{G\theta}{l} \text{ or } \frac{2\tau}{D} = \frac{G\theta}{l}$$

$$\therefore \theta = \frac{2\tau}{D} \cdot \frac{l}{G}$$

Substituting the value of T and θ in equation (1) we get

$$U = \frac{1}{2} \times \frac{\pi}{16} \cdot \tau \left(\frac{D^4 - d^4}{D} \right) \times \frac{2\tau}{D} \cdot \frac{l}{G} = \frac{\tau^2}{4G} \left[\frac{D^2 + d^2}{D^2} \right] \left[\frac{D^2 - d^2}{4} \right] \pi l$$

$$= \frac{\tau^2}{4G} \left(\frac{D^2 + d^2}{D^2} \right) \times \text{Volume of the shaft} \qquad (11.8)$$

Average torsional energy/unit volume

$$= \frac{\tau^2}{4G} \cdot \frac{D^2 + d^2}{D^2}$$

For solid shaft, $d = 0$

$$U = \frac{\tau^2}{4G} \times \text{volume of shaft}$$

For a very thin hollow shaft, D is nearly equal to d

In that case, $U = \dfrac{\tau^2}{2G} \times$ volume of shaft

DESIGN OF CIRCULAR MEMBERS IN TORSION FOR STRENGTH

In designing members for strength, allowable shear stresses must be selected. These depend on the information available from experiments and on the intended application. Accurate information on the capacity of materials to resist shear stresses comes from tests on thin-walled tubes. Solid shafting is employed in routine tests. Moreover, as torsion members are so often used in power equipment, many fatigue experiments are done. Typically, the shear strength of ductile materials is only about half as large as their tensile strength. The ASME (American Society of Mechanical Engineers) code of recommended practice for transmission shafting gives an allowable value in shear stress of 8000 psi for unspecified steel and 0.3 of yield, or 0.18 of ultimate, shear strength, whichever is smaller. In practical designs, suddenly applied and shock loads warrant special considerations.

After the torque to be transmitted by a shaft is determined and the maximum allowable shear stress is selected, according to Equation 11.1. the proportions of a member are given as

$$\frac{I_p}{c} = \frac{T}{t_{max}} \qquad (11.9)$$

where I_p/c is the parameter on which the elastic strength of a shaft depends. For an axially loaded rod, such a parameter is the cross-sectional area of a member. For a solid shaft, $I_p/c = \pi c^3/2$, in which c is the outside radius. By using this expression and Equation 11.9, the required radius of a shaft can be determined. Any number of tubular shafts can be chosen to satisfy Equation 11.9. by varying the ratio of the outer radius to the inner radius, c/b to provide the required value of I_p/c.

The reader should carefully note that large local stresses generally develop at changes in cross sections and at splines and keyways, where the torque is actually transmitted. These are, of critical importance in the design of rotating shafts, and appropriate stress concentration factors should be used to multiply the appropriate stress. Table 3.1 can be used for this purpose and can also be used for additional data.

Members subjected to torque are very widely used as rotating shafts for transmitting power. For ready reference, a formula is derived for the conversion of horsepower, the conventional unit used in the industry, into torque acting through the shaft. By definition, 1 hp does the work of 745.7 N.m/s. One N.m/s is conveniently referred to as a watt (W) in the SI units. Thus, 1 hp can be converted into 745.7 W. Likewise, it will be recalled from dynamics that power is equal to torque multiplied by the angle, measured in radians, through which the shaft rotates per unit of time. For a shaft rotating with a frequency of fHz, the angle is $2p\,f$ rad/s. Hence, if a shaft were transmitting a constant torque T measured in N.m, it would do $2p\,fT$ N.m of work per second. Equating this to the horsepower supplied,

$$hp \times 745.7 = 2p\,fT\,[\text{N.m/s}]$$

$$T = \frac{119 \times hp}{f}[\text{N.m}] \qquad (11.10)$$

$$T = \frac{159 \times kW}{f}\left[\text{N.m}\right] \qquad (11.11)$$

where f is the frequency in Hertz of the shaft transmitting the horsepower, hp or kilowatts, kW. These equations, convert the applied power into applied torque.

PROBLEM 11.1

Find the diameter of a solid shaft for a 10-hp motor operating at 30 Hz. The maximum stress is limited to 55 MPa. Hertz : 1 cycle per second (cps)

Solution

We know from equation (11.10)

$$T = \frac{119 \times hp}{f} = \frac{119 \times 10}{30} = 39.7 \text{ N.m}$$

also from equation (11.9)

$$\frac{I_p}{c} = \frac{T}{\tau_{max}} = \frac{39.7 \times 10^3}{55} = 722 \text{ mm}^3$$

$$\frac{I_p}{c} = \frac{\pi c^3}{2} \text{ or } c^3 = \frac{2\,I_p}{\pi\ c} = \frac{2 \times 722}{\pi} = 460 \text{ mm}^3$$

Hence, $c = 7.72\,\text{mm}$ or $d = 2c = 15.4\,\text{mm}$

For practical purposes, a 16-mm shaft would probably be selected.

PROBLEM 11.2

Determine the diameters of solid shafts to transmit 150 kW each without exceeding a shear stress of 70 MPa. One of these shafts operates at a frequency of 0.30 Hz and the other at a frequency of 300 Hz.

Solution

Subscript 1 applied to the low-speed shaft and 2 to the high-speed shaft.

From Equation 11.11

$$T_1 = \frac{159 \times \text{kW}}{f_1} = \frac{159 \times 150}{0.30} = 79{,}500 \text{ N.m}$$

Similarly,

$$T_2 = 79.5\,\text{N.m}$$

From equation 11.9

$$\frac{I_{p1}}{c} = \frac{T_1}{\tau_{max}} = \frac{79{,}500}{70} = 1.14 \times 10^6\,\text{mm}^3$$

$$\frac{I_{p1}}{c} = \frac{p d_1^3}{16} \quad \text{or} \quad d_1^3 = \frac{16}{\pi}(1.14 \times 10^6) = 5.81 \times 10^6\,\text{mm}^3$$

Hence,

$$d_1 = 180\,\text{mm} \quad \text{and} \quad d_2 = 18\,\text{mm}$$

PROBLEM 11.3

Find the relative rotation of section B–B with respect of section A–A of the solid elastic shaft shown in Figure when a constant torque T is being transmitted through it. The polar moment of inertia of the cross-sectional area I_p is constant.

Solution

In this case, $T_x = T$ and I_p is constant: hence, from torsion equation (11.1)

$$\phi = \int_A^B \frac{T_x dx}{I_p G} = \int_O^L \frac{T dx}{I_p G} = \frac{T}{I_p G}\int_O^L dx = \frac{TL}{I_p G} \tag{1}$$

That is,

$$\phi = \frac{TL}{I_p G} \qquad (2)$$

In applying equation (2), note particularly that the angle ϕ is expressed in radians. Also observe the great similarity of this relation to equation $\delta = PL / AE$ for axially loaded bars. Here, analogous to this equation, equation (2) can be recast to express the torsional spring constant, or torsional stiffness, k_t as

$$k_t = \frac{T}{\phi} = \frac{I_p G}{L} \frac{\text{N.m}}{\text{rad}} \qquad (3)$$

This constant represents the torque required to cause a rotation of 1 radian $(\text{i.e.}, \phi = 1)$. It depends only on the material properties and size of the member. As for axially loaded bars, one can visualize torsion members as springs; See Figure below.

The reciprocal of k_t defines the torsional flexibility f_t. Hence, for a circular solid or hollow shaft,

$$f_t = \frac{1}{k_t} = \frac{L}{I_p G} \left[\frac{\text{rad}}{\text{N.m}} \right] \qquad (4)$$

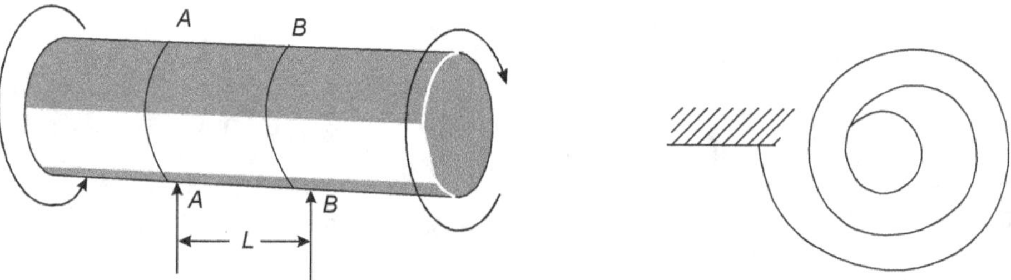

Schematic representation of a torsion spring

This constant defines the rotation resulting from application of a unit torque (i.e., $T = 1$). On multiplying by the torque T, one obtains equation (2).

If, in the analysis, a shaft must be subdivided into a number of regions, appropriate identifying subscripts should be attached to the definitions given by equations (3) and (4). For example, for the ith segment of a bar, one can write $(k_t)_i = I_p G_i / L_i$ and $(f_t)_i = L_i / I_p G_i$.

The previous equations are widely used in mechanical vibration analyses of transmission shafts, including crank shafts. These equations are also useful for solving statically indeterminate problems. These equations are required in the design of members for torsional stiffness when it is essential to limit the amount of twist. For such applications, note that I_p rather than the I_p / c used in strength calculations, is the governing parameter. In axially loaded bar problems, the cross-sectional area A serves both purposes.

Lastly, it should be noted that since in a torsion test ϕ, T, L and can be measured or calculated from the dimensions of a specimen, the shear modulus of elasticity for a specimen can be determined from equation (2) since $G = TL / I_p \phi$.

PROBLEM 11.4

Consider the stepped shaft shown in Figure (a) rigidly attached to a wall E, and determine the angle twist of the end A when the two torques at B and at D are applied. Assume the shear modulus G to be 80 GPa, a typical value for steels.

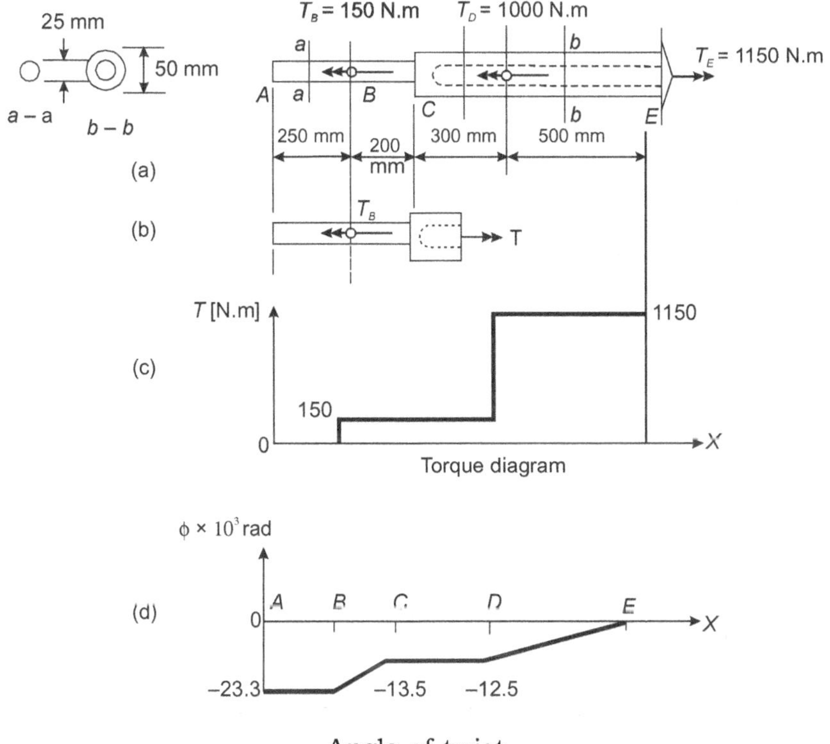

Angle-of-twist

Solution

Except for the difference in parameters, the solution of this problem is very similar to that for an axially loaded bar. First, the torque of E is determined to assure equilibrium. Then internal torques at arbitrary sections, isolating the left segment of a shaft, such as shown in Figure (b), are examined. If the direction of the torque vector T coincides with that of the positive x axis, it is taken as positive, or vice versa. This leads to the conclusion that between A and B there is no torque, whereas between B and D the torque is + 150 Nm. The torque between D and E is + 1150 Nm. The torque diagram is drawn in Figure (c). The internal torques, identified by the subscripts for the various segments are

$$T_{AB} = 0, \ T_{BD} = T_{BC} = T_{CD} - 150\,\text{N.m}, \ \text{and} \ T_{DE} = 1150\,\text{N.m}$$

The polar moments of inertia for the two kinds of cross sections occurring in this problem are found using the following equations.

$$(I_p)_{AB} = (I_p)_{BC} = \frac{pd^4}{32} = \frac{p' \; 25^4}{32} = 38.3' \; 10^3 \, \text{mm}^4$$

$$(I_p)_{CD} = (I_p)_{DE} = \frac{p}{32}(d_0^4 - d_i^4) = \frac{p}{32}(50^4 - 25^4) = 575 \times 10^3 \, \text{mm}^4$$

To find the angle of twist of the end A, Equation (11.1) is applied for each segment and the results summed. The limits of integration for the segments occur at points where the value of T or I_p change abruptly.

$$\phi = \int_A^E \frac{T_x dx}{I_{px}G} = \int_A^B \frac{T_{AB} dx}{(I_p)_{AB}G} + \int_B^C \frac{T_{BC} dx}{(I_p)_{BC}G} + \int_C^D \frac{T_{CD} dx}{(I_p)_{CD}G} + \int_D^E \frac{T_{DE} dx}{(I_p)_{DE}G}$$

In the last group of integrals, T's and I_p's are constant between the limits considered, so each integral reverts to known solution , equation for ϕ. Hence,

$$\phi = \sum_i \frac{T_i L_i}{(I_p)_i G_i} = \frac{T_{AB} L_{AB}}{(I_p)_{AB}G} + \frac{T_{BC} L_{BC}}{(I_p)_{BC}G} + \frac{T_{CD} L_{CD}}{(I_p)_{CD}G} + \frac{T_{DE} L_{DE}}{(I_p)_{DE}G}$$

$$= 0 + \frac{150 \times 10^3 \times 200}{38.3 \times 10^3 \times 80 \times 10^3} + \frac{150 \times 10^3 \times 300}{575 \times 10^3 \times 80 \times 10^3} + \frac{1150 \times 10^3 \times 500}{575 \times 10^3 \times 80 \times 10^3}$$

$$= 0 + 9.8 \times 10^{-3} + 1.0 \times 10^{-3} + 12.5 \times 10^3 = 23.3 \times 10^{-3} \, \text{rad}$$

As can be noted from the preceding equations, the angle of twist for the four shaft segments starting from the left end are 0 rad, 9.8×10^{-3} rad, $1.0' \; 10^{-3}$ rad, and 12.5×10^{-3} rad. Summing these quantities beginning from A, in order to obtain the function for the angle of twist along the shaft, gives the broken line from A to E, shown in Figure (d). Since no shaft twist can occur at the built-in end, this function must be zero at E, as required by the boundary condition. Therefore, according to the adopted sign convention, the angle of twist at A is -23.3×10^{-3} rad occurring in the direction of applied torques.

No doubt local disturbances in stresses and strain occur at the applied concentrated torques and the change in the shaft size, as well as at the bulit-in end. However, these are local effects having limited influence on the overall behaviour of the shaft according to St. Venant principle. If however, should these localized effects be considered, one must resort to a 3-D theory of elasticity solution.

PROBLEM 11.5

Determine the torsional stiffness k_τ for the rubber bushing shown in Figure. Assume that the rubber is bonded to the steel shaft and the outer steel tube, which is attached to a machine housing. The shear modulus for the rubber is G. Neglect deformations in the metal parts of the assembly.

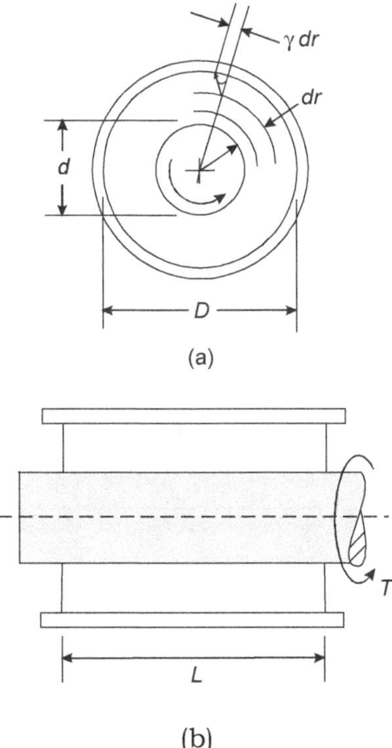

(a)

(b)

Solution

Due to the axial symmetry of the problem, on every imaginary cylindrical surface of rubber of radius r, the applied torque T is resisted by constant shear stresses τ. The area of the imaginary surface is $2\pi rL$. On this basis, the equilibrium equation for the applied torque T and the resisting torque developed by the shear stresses τ acting at a radius r is

$$T = (2\pi rL,)\tau r \; [\text{area} \times \text{stress} \times \text{arm}]$$

From this relation, $T = 2pr^2Lt$. Hence, by using the Hooke's law the shear strain γ (gamma sign) can be determined for an infinitesimal tube of radius r and thickness dr, Figure (a), from the following relations:

$$g = \frac{t}{G} = \frac{T}{2p\,LGr^2}$$

PROBLEM 11.6

Shaft *BC* is hollow with inner and outer diameter of 90 mm and 120 mm respectively. Shafts *AB* and *CD* are solid and of diameter *d*. For the loading shown, determine (a) the maximum and miniimum shearing stress in shaft BC, (b), the required diameter *d* of shafts *AB* and *CD* if the allowable shearing stress in these shafts is 65 MPa.

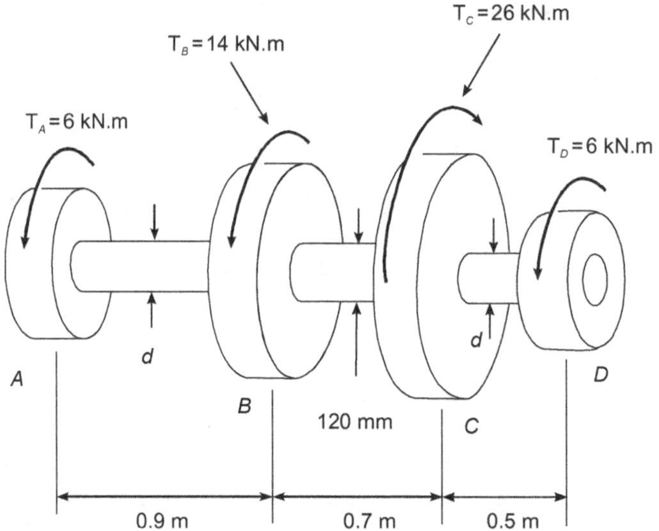

Equations of statics Denoting by T_{AB} the torque in shaft AB, we pass a section through shaft AB, and for the free body shown, we write

$$\Sigma M_x = 0: \qquad (6\,\text{kN.m}) - T_{AB} = 0 \quad T_{AB} = 6\,\text{kN.m}$$

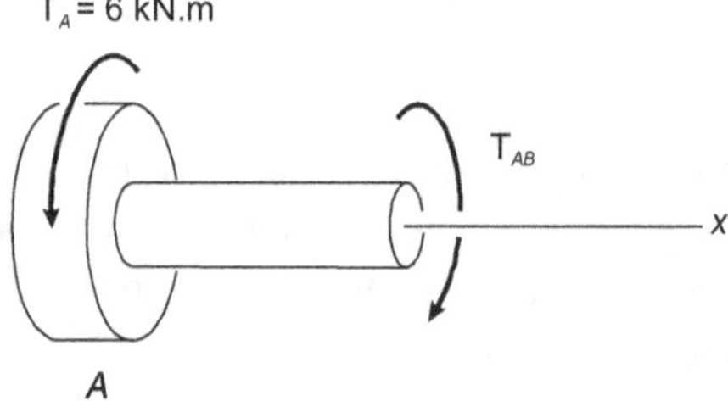

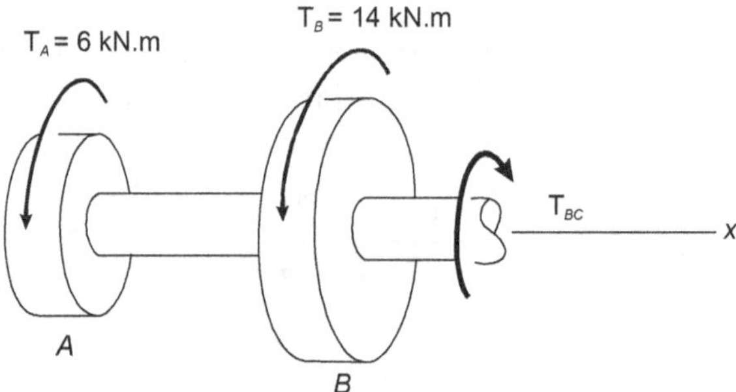

We now pass a section through shaft BC and, for the free body shown, we have

$$\Sigma M_x = 0: \qquad (6\,\text{kN.m}) + (14\,\text{kN.m}) - T_{BC} = 0 \quad T_{BC} = 20\,\text{kN.m}$$

(a) *Shaft BC* For this hollow shaft we have

$$J = \frac{\pi}{2}(c_2^4 - c_1^4) = \frac{\pi}{2}[(0.060)^4 - (0.045)^4] = 13.92 \times 10^{-6}\,\text{m}^4$$

Maximum shearing stress   On the outer surface, we have

$$\tau_{max} = \tau_2 = \frac{T_{BC}c_2}{J} = \frac{(20\,\text{kN.m})(0.060\,\text{m})}{13.92 \times 10^{-6}\,\text{m}^4}$$

$$\tau_{max} = 86.2\,\text{MPa}$$

Minimum shearing stress   We note that the stresses are proportional to the distance from the axis of the shaft

$$\frac{t_{min}}{t_{max}} = \frac{c_1}{c_2} \quad \frac{t_{min}}{86.2\,\text{MPa}} = \frac{45\,\text{mm}}{60\,\text{mm}} \quad t_{min} = 64.7\,\text{MPa}$$

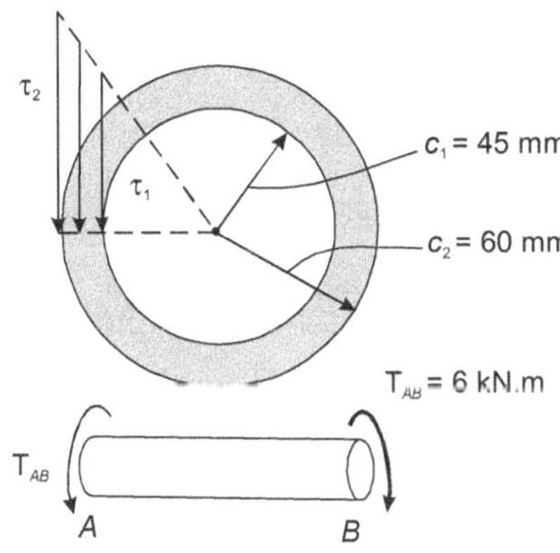

(b) *Shafts AB and CD*   We note that in both of these shafts the magnitude of the torque is $T = 6$ kN. m and $t_{all} = 65$ MPa. Denoting by c the radius of the shafts, we get

$$\tau = \frac{Tc}{J}$$

$$65\,\text{MPa} = \frac{(6\,\text{kN.m})c}{\dfrac{\pi}{2}c^4}$$

From this equation, we obtain

$$c^3 = 58.8 \times 10^{-6}\,\text{m}^3 \quad c = 38.9 \times 10^{-3}\,\text{m}$$

$$d - 2c = 2(38.9\,\text{mm}) \quad d = 77.8\,\text{mm}$$

PROBLEM 11.7

The preliminary design of a large shaft connecting a motor to a generator calls for the use of a hollow shaft with inner and outer diameters of 100 mm and 150 mm respectively. Knowing that the allowable shearing stress is 85 MPa. Determine the maximum torque which may be transmitted (a) by the shaft as designed, (b) by solid shaft of the same weight, (c) by a hollow shaft of the same weight and of 200 mm outside diameter.

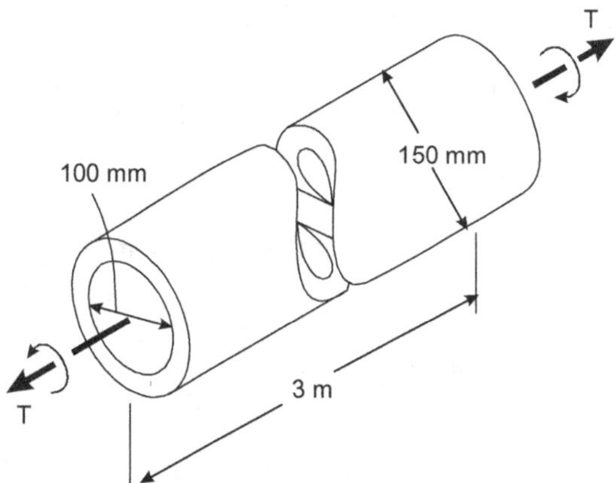

(a) Hollow shaft as designed: For the hollow shaft we have

$$J = \frac{\pi}{2}(c_2^4 - c_1^4) = \frac{\pi}{2}[(0.075\,\text{m})^4 - (0.050\,\text{m})^4] = 39.9 \times 10^{-6}\,\text{m}^4$$

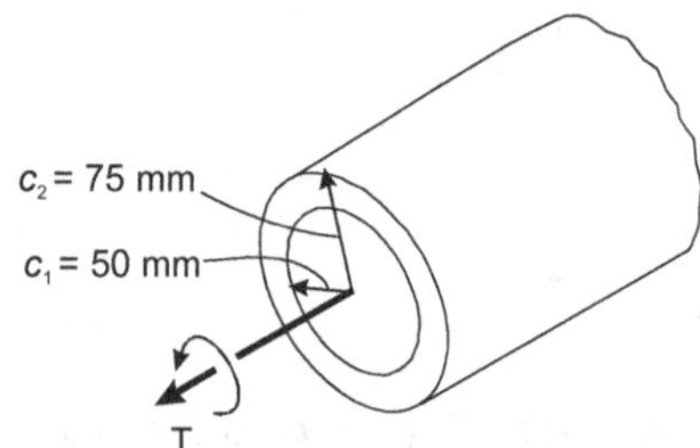

Using torsion equation, we get

$$\tau_{max} = \frac{Tc_2}{J}$$

$$85\,\text{MPa} = \frac{T(0.075\,\text{m})}{39.9 \times 10^{-6}\,\text{m}^4}$$

$$T = 45.2\,\text{kN.m}$$

(b) Solid shaft of equal weight: For the shaft as designed and to have the same weight and length, their cross-sectional areas must be equal.

$$A_{(a)} = A_{(b)}$$

$$\pi[(75\,\text{mm})^2 - (50\,\text{mm})^2] = \pi c_3^2$$

$$c_3 = 55.9\,\text{mm}$$

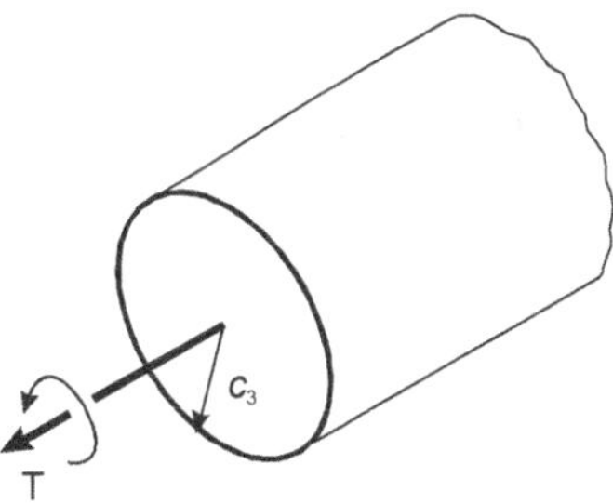

Since $t_{all} = 85\,\text{MPa},$ we write

$$\tau_{max} = \frac{Tc_3}{J}$$

$$85\,\text{MPa} = \frac{T(0.0559\,\text{m})}{\dfrac{\pi}{2}(0.0559\,\text{m})^4}$$

$$T = 23.3\,\text{kN.m}$$

c. Hollow shaft of 200-mm diameter: For equal weight, the cross-sectional areas must be equal. We determine the inside diameter using

$$A_{(a)} = A_{(c)}$$

$$\pi[(75\,\text{mm})^2 - (50\,\text{mm})^2] = \pi[(100\,\text{mm})^2 - c_5^2]$$

$$c_5 = 82.92\,\text{mm}$$

For $c_5 = 82.92$ mm and $c_4 = 100$ mm,

$$J = \frac{p}{2}[(0.100\,\text{m})^4 - (0.08292\,\text{m})^4] = 82.82' \ 10^{-6}\,\text{m}^4$$

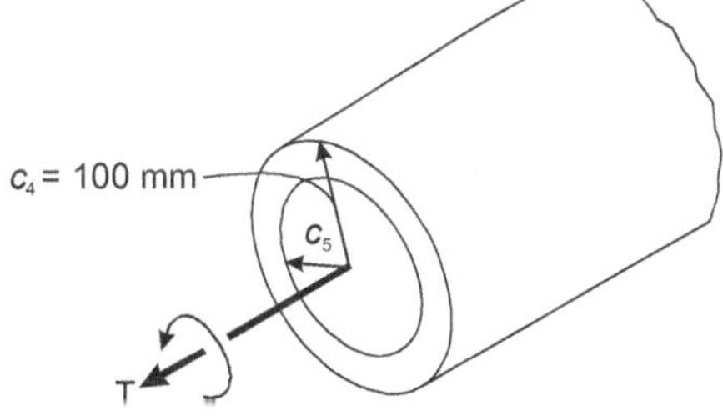

With $t_{all} = 85\,\text{MPa}$ and $c_4 = 100\,\text{mm},$

$$\tau_{max} = \frac{Tc_4}{J}$$

$$85\,\text{MPa} = \frac{T(0.100\,\text{m})}{82.82 \times 10^{-6}\,\text{m}^4}$$

$$T = 70.4\,\text{kN m}$$

PROBLEM 11.8

The vertical shaft AD is attached to fixed base at D and is subjected to the torque shown. A 44 mm-diameter hole has been drilled into portion CD of the shaft. Knowing that the entire shaft is made of steel for which $G = 80$ GPa determine the angle of twist at end A.

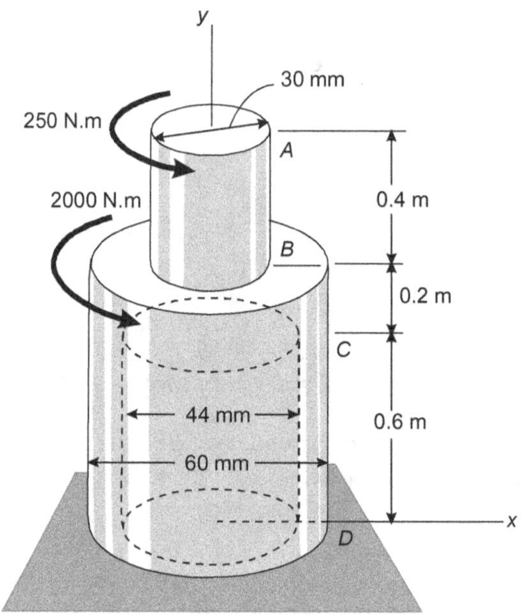

Solution

The shaft consists of three portions AB, BC and CD each uniform cross section and each with a constant internal torque.

Statics   Passing a section through the shaft between A and B and using the free body shown, we get

$$(250\,\text{N.m}) - T_{AB} = 0$$

$$\Sigma M_y = 0: \qquad T_{AB} = 250\,\text{N.m}$$

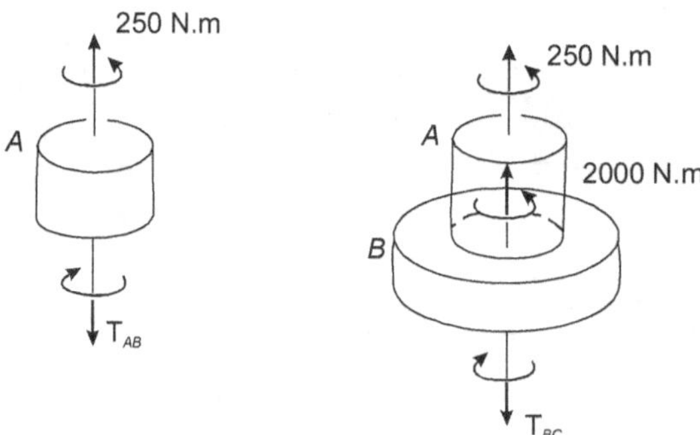

Passing now a section between B and C, we have

$$\sum M_y = 0: \qquad (250\,\text{N.m}) + (2000\,\text{N.m}) - T_{BC} = 0 \qquad T_{BC} = 2250\,\text{N.m}$$

Since no torque is applied at C,

$$T_{CD} = T_{BC} = 2250\,\text{N.m}$$

Polar moments of inertia

$$J_{AB} = \frac{\pi}{2}c^4 = \frac{\pi}{2}(0.015\,\text{m})^4 = 0.0795 \times 10^{-6}\,\text{m}^4$$

$$J_{BC} = \frac{\pi}{2}c^4 = \frac{\pi}{2}(0.030\,\text{m})^4 = 1.272 \times 10^{-6}\,\text{m}^4$$

$$J_{CD} = \frac{\pi}{2}(c_2^4 - c_1^4) = \frac{\pi}{2}[(0.030\,\text{m})^4 - (0.022\,\text{m})^4] = 0.904 \times 10^{-6}\,\text{m}^4$$

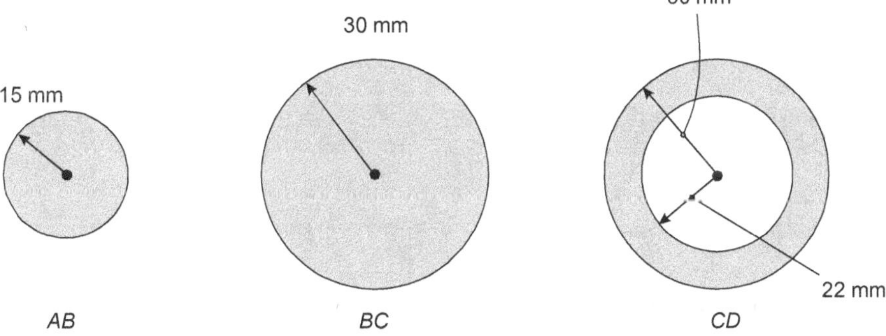

Angle of twist   Using torsion equation and recalling that $G = 80$ GPa for the entire shaft, we have

$$\phi_A = \sum_i \frac{T_i L_i}{J_i G} = \frac{1}{G}\left(\frac{T_{AB} L_{AB}}{J_{AB}} + \frac{T_{BC} L_{BC}}{J_{BC}} + \frac{T_{CD} L_{CD}}{J_{CD}} \right)$$

$$\phi_A = \frac{1}{80\,\text{GPa}}\left[\frac{(250\,\text{N.m})(0.4\,\text{m})}{0.0795 \times 10^{-6}\,\text{m}^4} + \frac{(2250)(0.2)}{1.272 \times 10^{-6}} + \frac{(2250)(0.6)}{0.904 \times 10^{-6}} \right]$$

$$= 0.01572 + 0.00442 + 0.01867 = 0.0388\,\text{rad}$$

$$\phi_A = (0.0388\,\text{rad})\frac{360^\circ}{2\pi\,\text{rad}}$$

$$\phi_A = 2.22^\circ$$

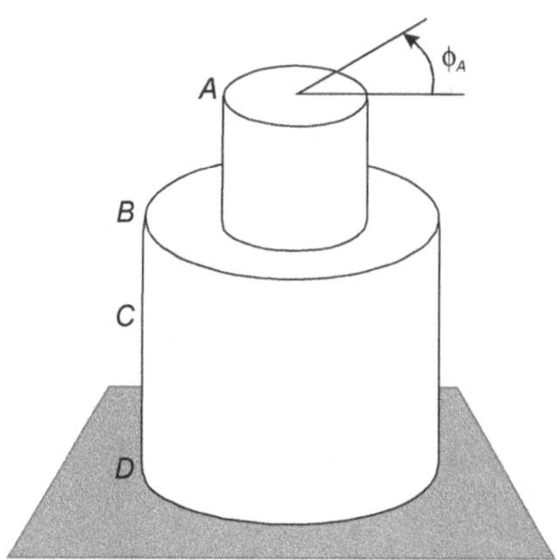

PROBLEM 11.9

Two solid steel shafts are connected by the gears shown. Knowing for each shaft that $G = 80\,\text{GPa}$ and that the allowable shearing stress is 55 MPa, determine (a) the largest torque T_0 which may be applied to end A of shaft AB, (b) the corresponding angle through which end A of shaft AB rotates.

Statics Denoting by F the magnitude of the tangential force between gear teeth, we have

Gear B $\Sigma M_B = 0:$ $F(20\,\text{mm}) - T_0 = 0$

Gear C $\Sigma M_c = 0:$ $F(56\,\text{mm}) - T_{CD} = 0$ (1)

$$T_{CD} = 2.8\,T_0$$

Kinematics Noting that the peripheral motions of the gears are equal, we write

$$r_B \phi_B = r_C \phi_C$$

$$\phi_B = \phi_c \frac{r_C}{r_B}$$

$$= \phi_c \frac{56\,\text{mm}}{20\,\text{mm}} = 2.8\phi_c \qquad (2)$$

(a) Torque T_0

    Shaft AB With $T_{AB} = T_0$ and $c = 0.009$ m, together with a maximum permissible shearing stress of 55 MPa,

$$\tau = \frac{T_{AB}\,c}{J}$$

$$55\,\text{MPa} = \frac{T_0(0.009\,\text{m})}{\dfrac{\pi}{2}(0.009\,\text{m})^4}$$

$$T_0 = 63.0 \text{ N.m}$$

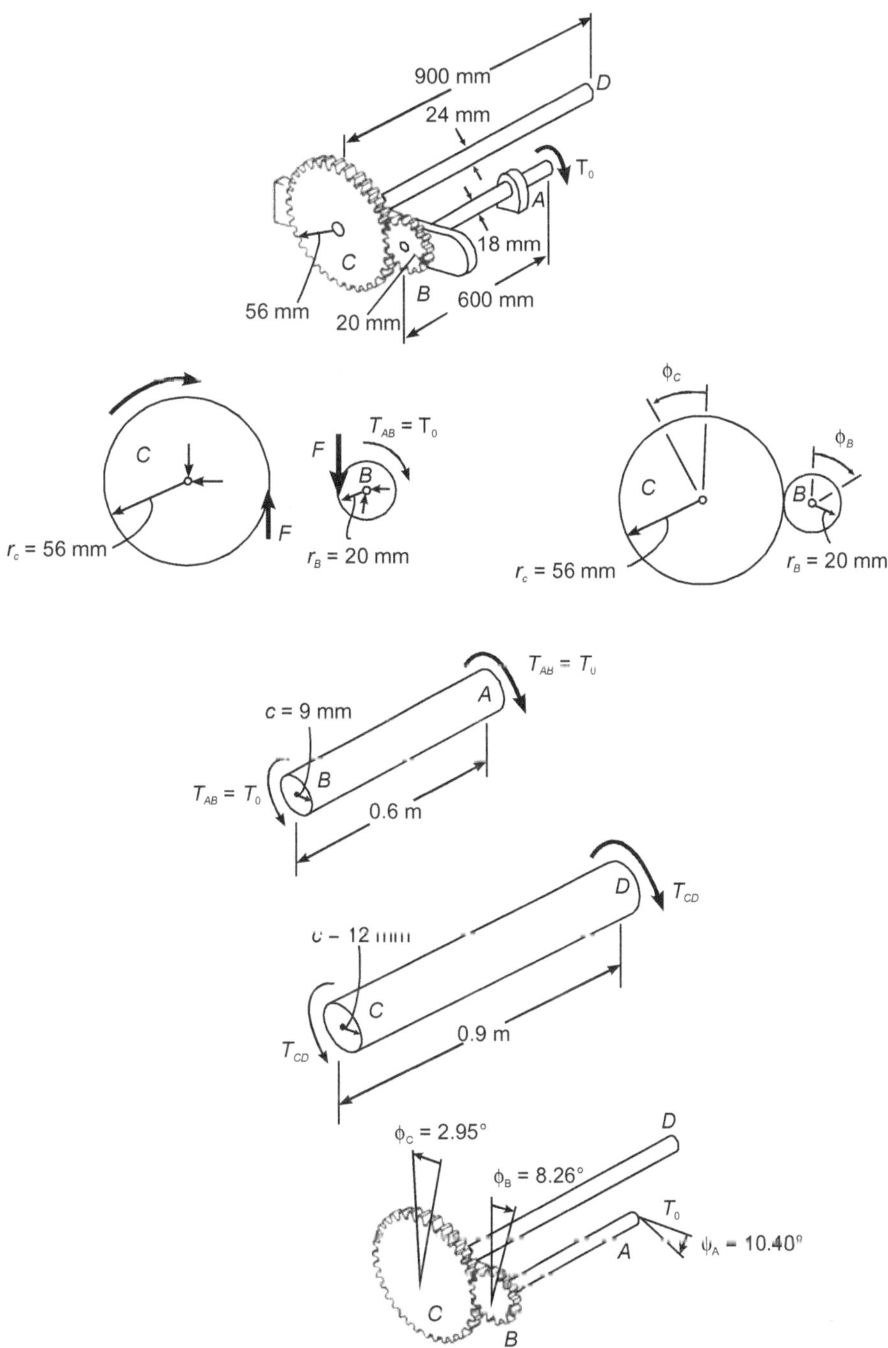

Shaft CD   From (1)

$$T_{CD} = 2.8T_o. \text{ With } c = 0.0112 \text{ m and } \tau_{all} = 55 \text{ MPa}$$

$$\tau = \frac{T_{CD}c}{J}$$

$$55 \text{ MPa} = \frac{2.8T_o(0.012 \text{ m})}{\dfrac{\pi}{2}(0.012 \text{ m}^4)}$$

$$T_o = 53.3. \text{ N.m}$$

Maximum permissible torque   We choose the smaller value obtained for T_o

$$T_o = 53.3 \text{ N.m}$$

b.   *Angle of rotation at end A*   We first compute the angle of twist for each shaft.
Shaft AB   For $T_{AB} = T_o = 53.3$ N.m, we have

$$f_{A/B} = \frac{T_{AB}L}{GJ} = \frac{(53.3 \text{ N.m})(0.06 \text{ m})}{(80 \text{ GPa})\dfrac{p}{2}(0.009 \text{ m})^4} = 0.0388 \text{ rad} = 2.22°$$

Shaft CD $T_{CD} = 2.8T_o = 2.8(53.3 \text{ N.m})$

$$\phi_{C/D} = \frac{T_{CD}L}{GJ} = \frac{2.8(53.3 \text{ N.m})(0.9 \text{ m})}{(80 \text{ GPa})\dfrac{\pi}{2}(0.012 \text{ m})^4} = 0.0515 \text{ rad} = 2.95°$$

Since end D of shaft CD is fixed, we have

$\phi_c = \phi_{C/D} = 2.95°$ Using (2) we find the rotation of gear B is

$$\phi_B = 2.8\phi_C = 2.8(2.95°) = 8.26°$$

For end A of shaft AB, we have

$$\phi_A = \phi_B + \phi_{A/B} = 8.26° + 2.22°$$
$$\phi_A = 10.48°$$

PROBLEM 11.10

A steel shaft and an aluminum tube are connected to a fixed support and to a rigid disc as shown in the cross section. Knowing that the initial stresses are zero, determine the maximum torque T_o which may be applied to the disc if the allowable stresses are 120 MPa in the steel shaft and 70 MPa in the aluminum tube. Use $G = 80$ GPa for steel and $G = 27$ GPa for aluminum.

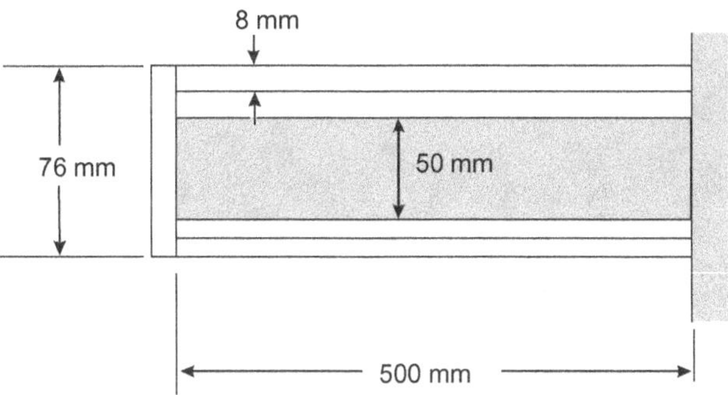

Statics   Free body of discL Denoting by T_1 the torque exerted by the tube on the disc and by T_2 the torque exerted by the shaft, we find

$$T_o = T_1 + T_2 \tag{1}$$

Deformations   Since both the tube and the shaft are connected to the rigid disc, we have

$$\phi_1 = \phi_2$$

$$\frac{T_1 L_1}{J_1 G_1} = \frac{T_2 L_2}{J_2 G_2}$$

$$\frac{T_1(0.5\,\text{m})}{(2.003 \times 10^{-6}\,\text{m}^4)(27\,\text{GPa})} = \frac{T_2(0.5\,\text{m})}{(0.614 \times 10^{-6}\,\text{m}^4)(80\,\text{GPa})} \tag{2}$$

$$T_2 = 0.908\,T_1$$

Shearing stresses   We shall assume that the requirement $\tau_{\text{alum}} \leq 70\,\text{MPa}$ is critical. For the aluminum tube, we have

$$T_1 = \frac{\tau_{\text{alum}} J_1}{c_1} = \frac{(70\,\text{MPa})(2.003 \times 10^{-6}\,\text{m}^4)}{0.038\,\text{m}} = 3690\,\text{N.m}$$

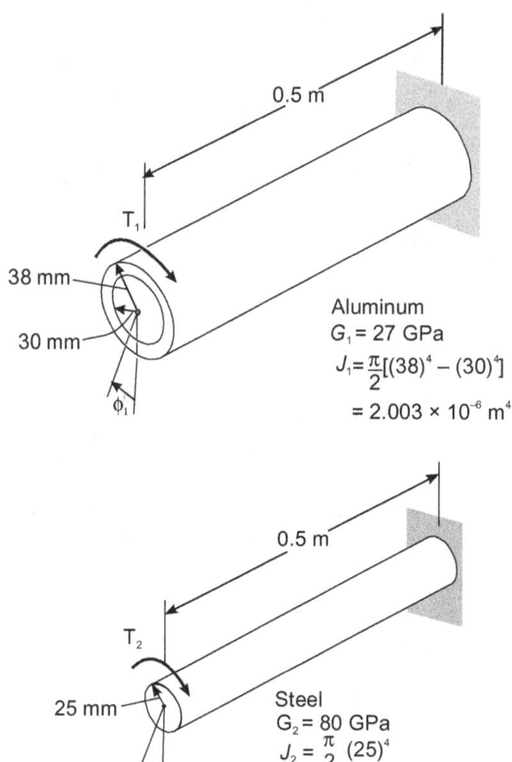

Using equation (2), we compute the corresponding value T_2 and then find the maximum shearing stress in the steel shaft.

$$T_2 = 0.908\, T_1 = 0.908(3690) = 3350\,\text{N.m}$$

$$\tau_{steel} = \frac{T_2 c_2}{J_2} = \frac{(3350\,\text{N.m})(0.025\,\text{m})}{0.614 \times 10^{-6}\,\text{m}^4} = 136.4\,\text{MPa}$$

We note that the allowable steel stress of 120 MPa is exceeded; our assumption was wrong. Thus the maximum torque T_0 will be obtained by making $t_{steel} = 120\text{MPa}$. We first determine the torque

$$T_2 = \frac{t_{steel} J_2}{c_2} = \frac{(120\,\text{MPa})(0.614' \, 10^{-6}\,\text{m}^4)}{0.025\,\text{m}} = 2950\ \text{N.m}$$

From equation (2), we have

$$2950\,\text{N.m} = 0908\, T_1$$

$$T_1 = 3250\,\text{N.m}$$

Using equation (1), we obtain the maximum permissible torque

$$T_0 = T_1 + T_2 = 3250\,\text{N.m} + 2950\,\text{N.m}$$
$$T_0 = 6.20\,\text{kN.m}$$

REVIEW QUESTIONS

SHORT QUESTIONS

1. What do you mean by strength of a shaft?
2. Find the expressions for polar modulus for a solid and hollow shaft?
3. What is Polar Modulus? Give the expression for Polar Modulus for a solid shaft and for a hollow shaft.
4. Define Torsional rigidity of a shaft.
5. Write down the expression for the polar moment of inertia of hollow circular section.
6. A solid shaft is subjected to a torque of 15 kNm. Find the minimum diameter rwquired for the shaft if the permissible shear stress is limited to 60 N/mm^2.
7. What are the two conditions to be satisfied in the design of a circular shaft?
8. List the loads normally acting on a shaft.
9. Write the equation of torsion acting in a circular shaft.
10. Write down the simple torsion formula with the meaning of each symbol for circular corss section.
11. Write the assumption for finding out the shear stress of a circular shaft, subjected to torsion.
12. What is polar modulus? Write the polar modulus value of a rectangle.

LARGE QUESTIONS

1. A steel shaft is required to transmit 75 kW power at 100 r.p.m. and the maximum twisting moment is 30% greater than the mean. Find the diameter of the steel shaft if the maximum stress is 70 N/mm^2. Also determine the angle of twist in a length of 3 m of the shaft. Assume the modulus of rigidity for steel as 90 kN/mm^2.

2. A hollow steel shaft of 100 mm internal diameter and 150 mm external diameter is to be replaced by a solid alloy shaft. If the polar modulus has the same value for both, calculate the diameter of the latter and ratio of their torsional rigidities.

3.   i. Obtain a relation for the torque and power, a solid shaft can transmit.

   ii. A solid steel shaft has to transmit 100 kW at 160 r.p.m. Taking allowable shear stress as 70 MPa, find the suitable diameter of the shaft. The maximum torque transmitted in each revolution exceeds the mean by 20%.

4. A steel shaft of diameter 200 mm runs at 300 rpm. The steel shaft has a 30 mm thick bronze bushing shrunk over its entire length of 8 m. If the maximum shearing stress in the steel shaft is not to exceed 12 MPa. Take G_{steel} = 84 GPa, G_{bronze} = 42GPa. Determine,

    i. Torsional rigidity of the shaft.

    ii. Power of the engine.

5. A solid circular shaft transmits 75 kW power at 200 rpm. Calculate the shaft diameter, if the twist in the shaft is not exceed 1° in 2 meter length of the shaft and the shear stress is limited to 50 N/mm². Take $C = 1 \times 10^5$ n/mm².

6. A bar of magnesium alloy 28 mm in diameter was tested on a gauge length of 25 cm in tension and in torsion. A tensile load of 5 tonnes produced an extension of 0.4 mm and a torque of 1250 kg –cm produced a twist of 1.51 degrees. Determine the (i) Young's modulus (ii) Modulus of rigidity (iii) Bulk modulus (iv) Poisson's ratio for the material under test.

7. A hollow shaft is of 160 mm external diameter and 80 mm internal diameter. It transmits 350 kW at 220 rpm. What bending moment could be carried in addition if the maximum shearing stress must not exceed 35 MPa?

8. A hollow steel shaft 10 cm external diameter and 5 cm internal diameter transmits 800 kW at 5000 r.p.m and is subjected to an end thrust of 40,000 N. Find the bending moment that may be safely applied to the shaft if the greater principal stress is not to exceed 100 N/mm².

12

COMBINED BENDING AND TORSION

Generally we assume shaft is subjected to torsion. But due to weight of pulley, couplings, pull in belts and ropes, weight of heavy dises and blades as in the case of steam and gas turbines, the shaft is subjected to bending also. Thus actually both shear stress due to torsion and direct stress due to bending are induced in addition to shear stresses caused by transvese loads. We start with the torsion equation.

$$\frac{T}{I_P} = \frac{\tau}{R} \qquad T = \tau \frac{\pi}{16} \cdot D^3 \text{ (solid shaft) (Torsion equation)}$$

$$\frac{M}{I} = \frac{\sigma_b}{y} \text{ (Bending equation)}$$

$$M = \frac{\sigma_b I}{y} = \frac{\sigma_b \pi D^4}{64 \times D/2} = \frac{\sigma_b \pi D^3}{32} \text{ (solid shaft)}$$

If a direct stress σ_d and shear stress τ acts at a certain point,

Then, we have seen in chapter on combined stresses

$$\sigma_{max} = \frac{\sigma_d}{2} + \sqrt{\left(\frac{\sigma_d}{2}\right)^2 + \tau^2} \qquad (12.1)$$

$$\tau_{max} = \sqrt{\left(\frac{\sigma_d}{2}\right)^2 + \tau^2} \qquad (12.2)$$

Multiplying both sides by $\dfrac{\pi D^3}{32}$ we get

$$\sigma_{max} \cdot \frac{\pi D^3}{32} = \frac{\sigma_d}{2} \cdot \frac{\pi D^3}{32} + \sqrt{\left[\frac{\sigma_d}{2} \cdot \frac{\pi D^3}{32}\right]^2 + \left(\tau \cdot \frac{\pi D^3}{32}\right)^2}$$

$$M_e = \frac{M}{2} + \sqrt{\left(\frac{M}{2}\right)^2 + \left(\frac{T}{2}\right)^2} \qquad \left(\because T = \tau \times \frac{\pi}{16} \cdot D^3\right)$$

$$M_e = \frac{M + \sqrt{M^2 + T^2}}{2}$$ (12.3)

where M_e is the Equivalent bending moment.

$$\tau_{max} = \sqrt{\left(\frac{\sigma_d}{2}\right)^2 + \tau^2}$$

Multiplying both sides by $\dfrac{\pi D^3}{16}$, we get

$$= \tau_{max} \cdot \frac{\pi D^3}{16} = \sqrt{\left(\frac{\sigma_d}{2} \cdot \frac{\pi D^3}{16}\right)^2 + \left(\tau \frac{\pi D^3}{16}\right)^2}$$

$$T_e = \sqrt{M^2 + T^2}$$ (12.4)

where, T_e is the Equivalent torque

PROBLEM 12.1

Two forces $\mathbf{P}_1$ and $\mathbf{P}_2$, of magnitude $P_1 = 15$ kN and $P_2 = 18$ kN, are applied as shown to the end A of bar AB, which is welded to a cylindrical member BD of radius $c = 20$ mm (Figure (a)). Knowing that the distance from A to the axis of member BD is $a = 50$ mm, determine the normal and shearing stresses at points H and K of the transverse section of member BD located at a distance $b = 60$ mm from end B. Assume all stresses to remain below the proportional limit of the material.

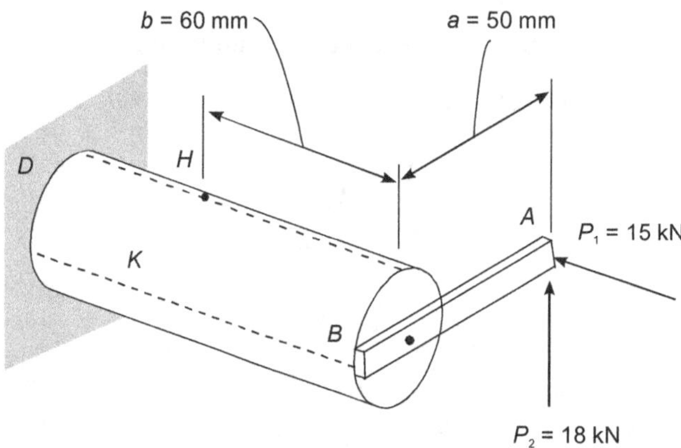

Figure (a)

Internal forces in section HK   We first replace the forces $\mathbf{P}_1$ and $\mathbf{P}_2$ by an equivalent system of forces and couples applied at the centre C of the section containing points H and K (Figure b). This system, which represents the internal forces in the section, consists of the following forces and couples.

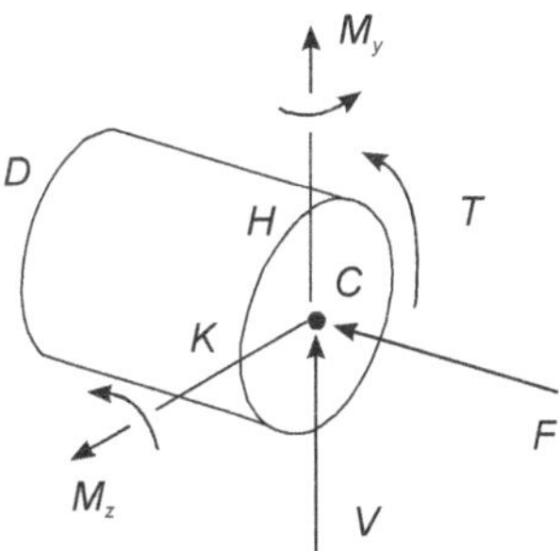

Figure (b)

1. A centric axial force **F** equal to the force $\mathbf{P}_1$, of magnitude
$$F = P_1 = 15\,\text{kN}$$

2. A shearing force **V** equal to the force $\mathbf{P}_2$, of magnitude
$$V = P_2 = 18\,\text{kN}$$

3. A twisting couple **T** of torque T equal to the moment of $\mathbf{P}_2$ about the axis of member BD:
$$T = P_2 a = (18\,\text{kN})(50\,\text{mm}) = 900\,\text{N.m}$$

4. A bending couple $\mathbf{M}_y$ of magnitude M_y equal to the moment of P_1 about a vertical axis through C:
$$M_y = P_1 a = (15\,\text{kN})(50\,\text{mm}) = 750\,\text{N.m}$$

5. A bending couple $\mathbf{M}_z$ of magnitude M_z equal to the moment of P_2 about a transverse, horizontal axis through C:
$$M_z = P_2 b = (18\,\text{kN})(60\,\text{mm}) = 1080\,\text{N.m}$$

Each of these forces and couples may produce a normal or shearing stress at points H and K of the section. Our purpose is to compute separately each of these stresses, and then to add the normal stresses and add the shearing stresses at each of the two points. First, however, we shall determine the geometric properties of the section. We have

$$A = \pi c^2 = \pi(0.020\,\text{m})^2 = 1.257 \times 10^{-3}\,\text{m}^2$$

$$I_y = I_z = \frac{1}{4}\pi c^4 = \frac{1}{4}\pi(0.020\,\text{m})^4 = 125.7 \times 10^{-9}\,\text{m}^4$$

$$I_p = \frac{1}{2}\pi c^4 = \frac{1}{2}\pi(0.020\,\text{m})^4 = 251.3 \times 10^{-9}\,\text{m}^4$$

Stresses at H   We observe that normal stresses σ_x are produced at H by the centric force **F** and the bending couple M_z, and that a horizontal shearing stress τ_{xz} is caused by the twisting couple **T** (Figure c). On the other hand, the bending couple M_y does not produce any normal stress at H, since H is located on the corresponding neutral axis, and the vertical shearing force **V** does not produce any shearing stress at H, since H is located at the top of the section. Determining the sign of each stress from the figure, we write

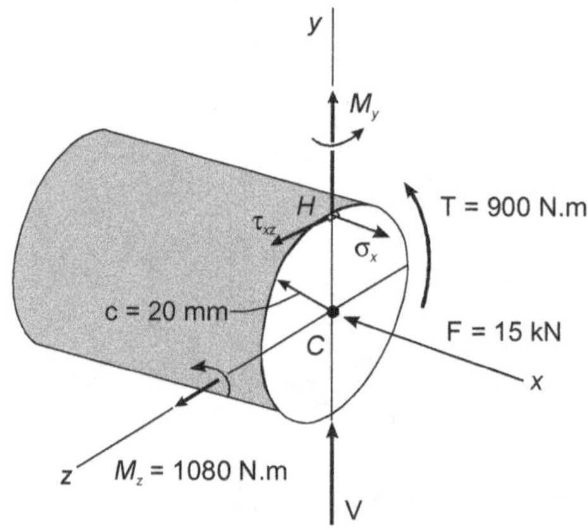

Figure (c)

$$\sigma_x = (\sigma_x)_{centric} + (\sigma_x)_{bending} = -\frac{F}{A} - \frac{M_z c}{I_z}$$

$$= -\frac{15\,kN}{1.257 \times 10^{-3}\,m^2} - \frac{(1080\,N.m)(0.020\,m)}{125.7 \times 10^{-9}\,m^4}$$

$$= -11.9\,MPa - 171.9\,MPa$$

$$\sigma_x = -183.8\,MPa$$

$$\text{and } \tau_{xz} = (\tau_{xz})_{twist} = \frac{Tc}{Jc} = \frac{(900\,N.m)(0.020\,m)}{251.3 \times 10^{-9}\,m^4}$$

$$\tau_{xz} = 71.6\,MPa$$

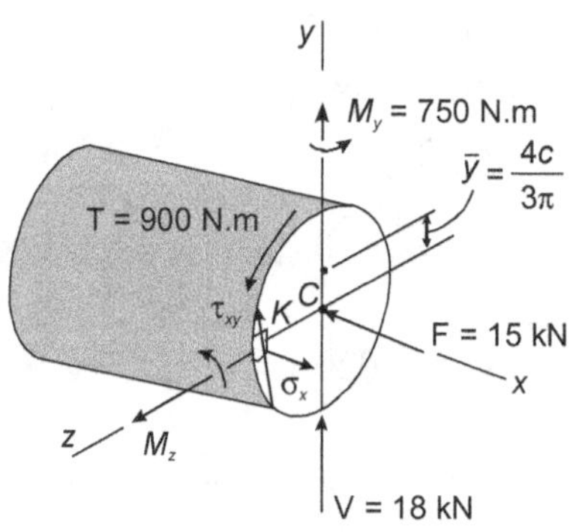

Figure (d)

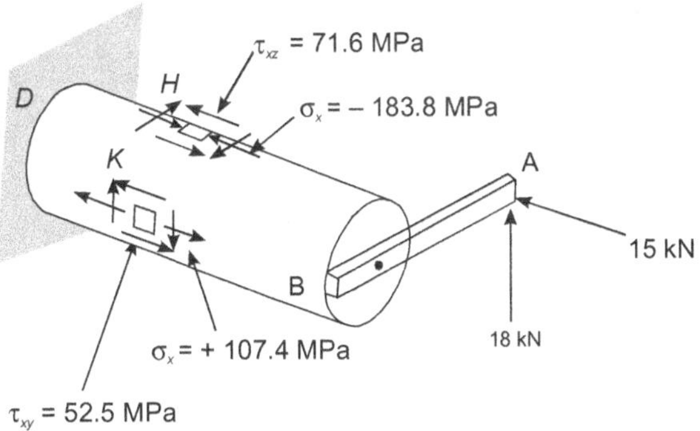

Figure (e)

Stresses at K   We note that normal stresses σ_x are produced at K by the centric force F and the bending couple M_y, and that vertical shearing stresses τ_{xy} are caused by the twisting couple $\mathbf{T}$ and the shearing force $\mathbf{V}$ (Figure d). We obtain

$$\sigma_x = \frac{-F}{A} + \frac{M_y c}{I_y} = -11.9\,\text{MPa} + \frac{(750\,\text{N.m})(0.020\,\text{m})}{125.7\times10^{-9}\,\text{m}^4}$$

$$= -11.9\,\text{MPa} + 119.3\,\text{MPa}$$

$$\sigma_x = +107.4\,\text{MPa}$$

In order to compute the shearing stress due to $\mathbf{V}$, we must determine the first moment Q and width t of the shaded area shown in Figure d. Recalling that $\bar{y} = 4c/3\pi$ for a semicircle of radius c, we have

$$Q = A'y = \left(\frac{1}{2}\pi c^2\right)\left(\frac{4c}{3\pi}\right) = \frac{2}{3}c^3 = \frac{2}{3}(0.020\,\text{m})^3$$

$$= 5.33\times10^{-6}\,\text{m}^3$$

$$\text{and } t = 2c = 2(0.020\,\text{m}) = 0.040\,\text{m}$$

We get,

$$(\tau_{xy})_v = +\frac{VQ}{I_z t} = +\frac{(18\times10^3\,\text{N})(5.33\times10^{-6}\,\text{m}^3)}{(125.7\times10^{-9}\,\text{m}^4)(0.040\,\text{m})}$$

$$= +19.1\,\text{MPa}$$

Noting that $(\tau_{xy})_{twist} - (\tau_{xz})_{twist}$ we have

$$\tau_{xy} = (\tau_{xy})v - (\tau_{xy})_{twist} = +19.1\,\text{MPa} - 71.6\,\text{MPa}$$

$$\tau_{xy} = -52.5\,\text{MPa}$$

Square elements located at *H* and *K* on the surface of the cylindrical member are used in Figure 5.50 to summarize the results obtained. Note that shearing stresses acting on the longitudinal sides of the elements have been included.

PROBLEM 12.2

Three forces are applied as shown at points *A*, *B*, and *D* of a short steel post. Knowing that the horizontal cross section of the post is a 40 × 140-mm rectangle, determine the normal and shearing stress at point *H*.

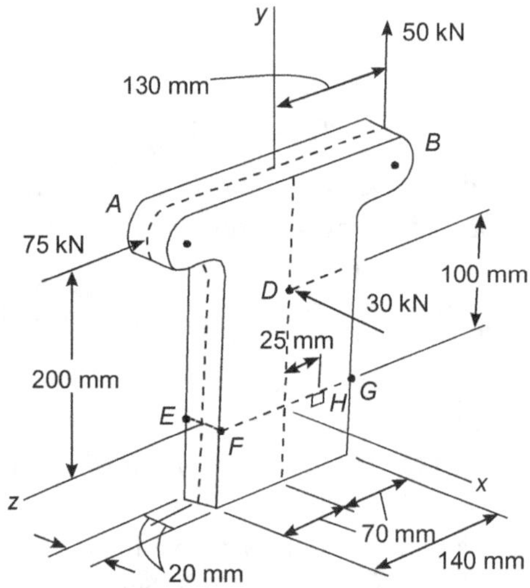

Internal forces in section EFG   We replace the three applied forces by an equivalent force–couple system at the centre *C* of the rectangular section *EFG*. We have

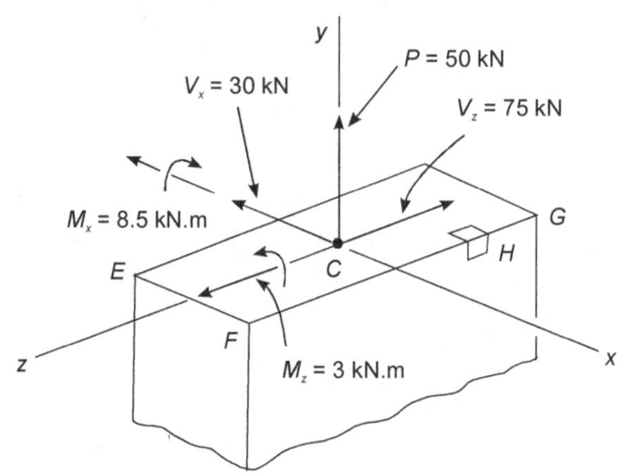

$$V_x = -30\,\text{kN} \quad P = 50\,\text{kN} \quad V_z = -75\,\text{kN}$$

$$M_x = (50\,\text{kN})(0.130\,\text{m}) - (75\,\text{kN})(0.200\,\text{m}) = -8.5\,\text{kN.m}$$

$$M_z = (30\,\text{kN})(0.100\,\text{m}) = 3\,\text{kN.m}$$

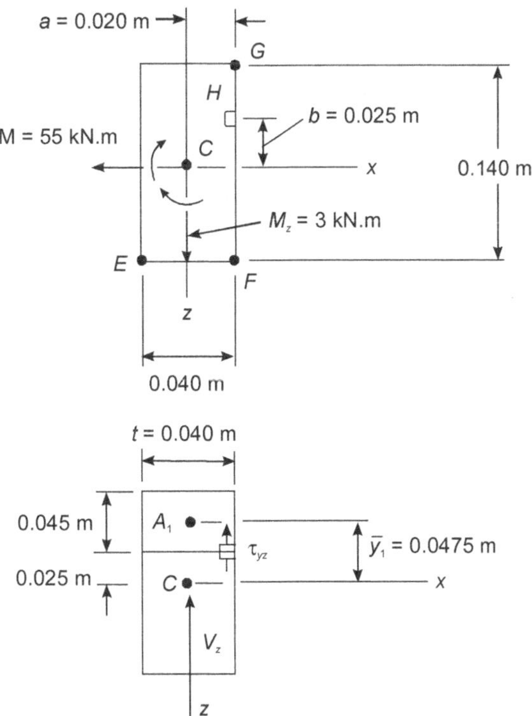

We note that there is no twisting couple about the y-axis. The geometric properties of the rectangular section are

$$A = (0.040\,\text{m})(0.140\,\text{m}) = 5.6 \times 10^{-3}\,\text{m}^2$$

$$I_x = \frac{1}{12}(0.040\,\text{m})(0.140\,\text{m})^3 = 9.15 \times 10^{-6}\,\text{m}^4$$

$$I_z = \frac{1}{12}(0.140\,\text{m})(0.040\,\text{m})^3 = 0.747 \times 10^{-6}\,\text{m}^4$$

Normal stress at H We note that normal stresses σ_y are produced by the centric force **P** and by the bending couple $\mathbf{M_x}$ and $\mathbf{M_z}$. We determine the sign of each stress by carefully examining the sketch of the force–couple system at C.

$$\sigma_y = +\frac{P}{A} + \frac{|M_z|a}{I_z} - \frac{|M_x|b}{I_x}$$

$$= \frac{50\,\text{kN}}{5.6 \times 10^{-3}\,\text{m}^2} + \frac{(3\,\text{kN.m})(0.020\,\text{m})}{0.747 \times 10^{-6}\,\text{m}^4} - \frac{(8.5\,\text{kN.m})(0.025\,\text{m})}{9.15 \times 10^{-6}\,\text{m}^4}$$

$$\sigma_y = 8.93\,\text{MPa} + 80.3\,\text{MPa} - 23.2\,\text{MPa}$$

$$\sigma_y = 66.0\,\text{MPa}$$

Shearing stress at H   Considering first the shearing force $\mathbf{V}_x$, we note that $Q = 0$ with respect to the z-axis, since H is on the edge of the cross section. Thus $\mathbf{V}_x$ produces no shearing stress at H. The shearing force $\mathbf{V}_z$ does produce a shearing stress at H and we write

$$Q = A_1 \overline{Y_1} = \left[(0.040\,\text{m})(0.045\,\text{m})\right](0.0475\,\text{m}) = 85.5 \times 10^{-6}\,\text{m}^3$$

$$\tau_{yz} = \frac{V_z Q}{I_x t} = \frac{(75\,\text{kN})(85.5 \times 10^{-6}\,\text{m}^3)}{\left(9.15 \times 10^{-6}\,\text{m}^4\right)(0.040\,\text{m})} \tau_{yz} = 17.52\,\text{MPa}$$

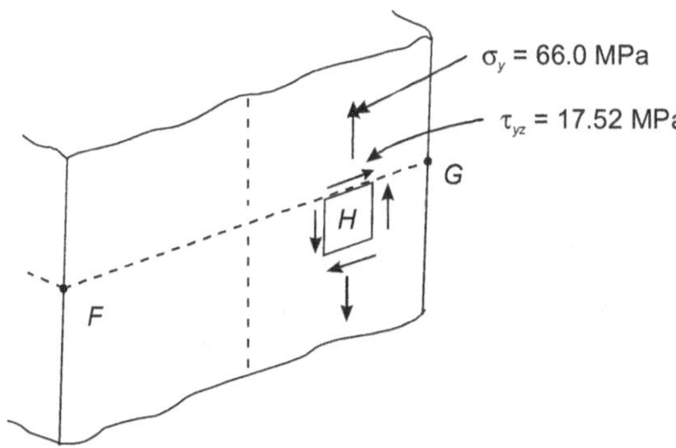

Summary   We now show both σ_y and τ_{yz} on a square element at H. Note that the direction of the shearing stress on the upper face of the element is the same as that of the shearing force V_z exerted on the cross section.

PROBLEM 12.3

A horizontal 2.5-kN force acts at point D of crankshaft AB which is held in static equilibrium by a twisting couple $\mathbf{T}$ and by reactions at A and B. Knowing that the bearings are self-aligning and exert no couples on the shaft, determine the normal and shearing stresses at points H, J, K and L located at the ends of the vertical and horizontal diameters of a transverse section located 48 mm to the left of bearing B.

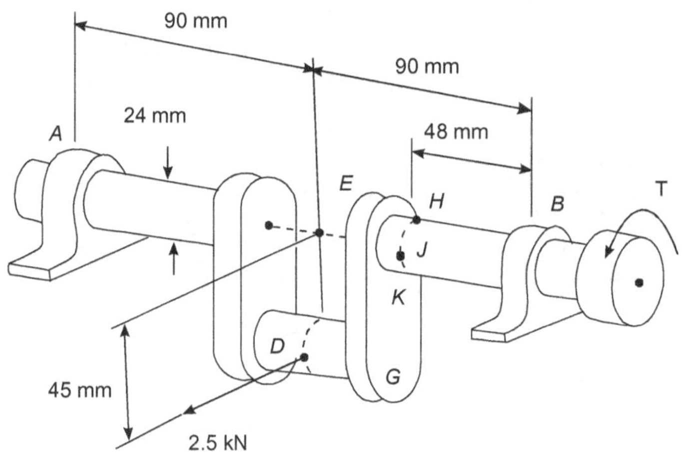

Free body, entire crankshaft.

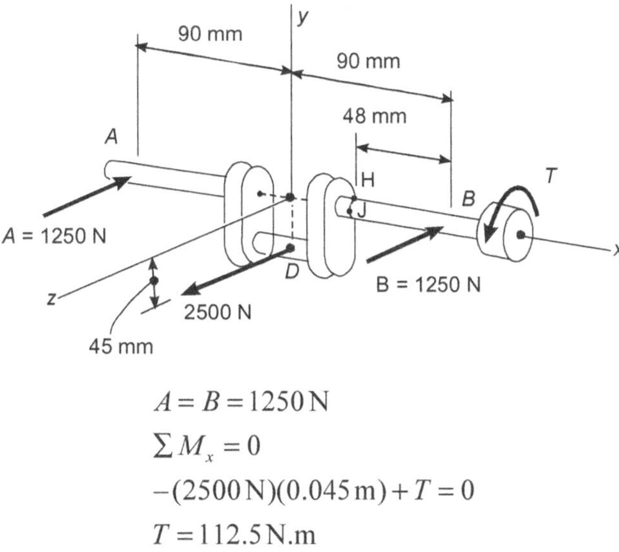

$$A = B = 1250\,\text{N}$$
$$\Sigma M_x = 0$$
$$-(2500\,\text{N})(0.045\,\text{m}) + T = 0$$
$$T = 112.5\,\text{N.m}$$

Internal forces in transverse section   We replace the reaction B and the twisting couple **T** by an equivalent force–couple system at the centre C of the transverse section containing H, J, K and L.

$$T = 112.5\,\text{N.m}$$
$$V = B = 1250\,\text{N} \quad T = 112.5\,\text{N.m}$$
$$M_y = (1250\,\text{N})(0.048\,\text{m}) = 60\,\text{N.m}$$

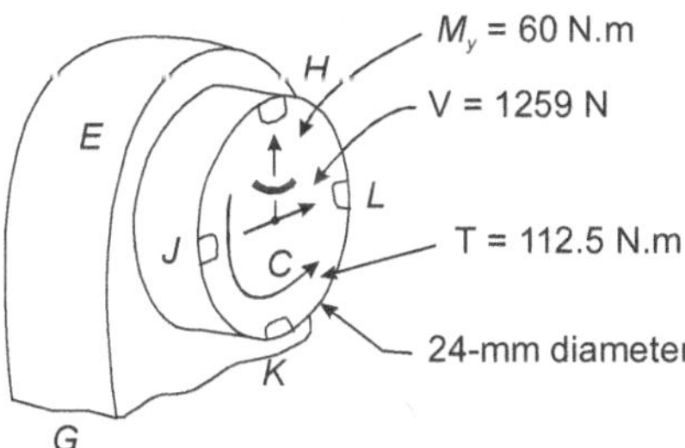

The geometric properties of the 24-mm diameter section are

$$A = \pi(0.012\,\text{m})^2 - 452 \times 10^{-6}\,\text{m}^4 \qquad I = \frac{1}{4}\pi(0.012\,\text{m})^4 = 16.29 \times 10^{-9}\,\text{m}^4$$

$$c = 12\,\text{mm} = 0.012\,\text{m} \qquad J = \frac{1}{2}\pi(0.012\,\text{m})^4 = 32.6 \times 10^{-9}\,\text{m}^4$$

Stresses produced by twisting couple T   Using torsion equations, we determine the shearing stresses at points H, J, K and L and show them in Figure (a).

$$\tau = \frac{Tc}{J} = \frac{(112.5\,\text{N.m})(0.012\,\text{m})}{32.6\times10^{-9}\,\text{m}^4} = 41.4\,\text{MPa}$$

Stresses produced by shearing force V   The shearing force **V** produces no shearing stresses at points J and L. At points H and K we first compute Q for a semicircle about a vertical diameter and then determine the shearing stress produced by the shear force $V = 1250$ N. These stresses are shown in Figure (b).

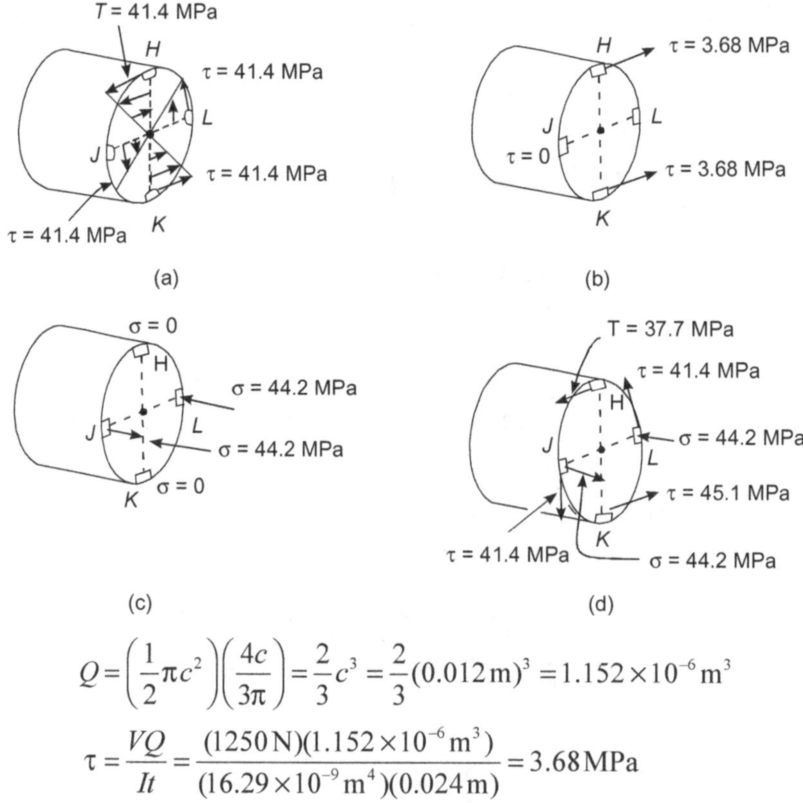

$$Q = \left(\frac{1}{2}\pi c^2\right)\left(\frac{4c}{3\pi}\right) = \frac{2}{3}c^3 = \frac{2}{3}(0.012\,\text{m})^3 = 1.152\times10^{-6}\,\text{m}^3$$

$$\tau = \frac{VQ}{It} = \frac{(1250\,\text{N})(1.152\times10^{-6}\,\text{m}^3)}{(16.29\times10^{-9}\,\text{m}^4)(0.024\,\text{m})} = 3.68\,\text{MPa}$$

Stresses produced by the bending couple M_y   Since the bending couple $\mathbf{M}_y$ acts in a horizontal plane, it produces no stresses at H and K. Using bending equation, we determine the normal stresses at points J and L and show them in Figure (c).

$$\sigma = \frac{|M_y|c}{I} = \frac{(60\,\text{N.m})(0.012\,\text{m})}{16.29\times10^{-9}\,\text{m}^4} = 44.2\,\text{MPa}$$

Summary   We add the stresses shown and obtain the total normal and shearing stresses at points H, J, K, and L.

PROBLEM 12.4

The solid shaft AB in Figure rotates at 480 rpm and transmits 30 kW from the motor M to machine tools connected to gears G and H; 20 kW is taken off a gear G and 10 kW at gear H. Knowing that $\tau_{all} = 50$ MPa determine the smallest permissible diameter for shaft AB.

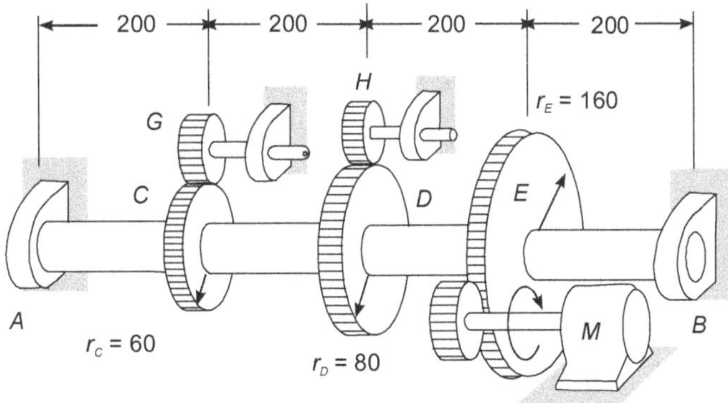

Dimensions in mm

Solution

Noting that f = 480rpm = 8Hz, we determine the torque exerted on gear E:

$$T_E = \frac{P}{2\pi f} = \frac{30\,\text{kW}}{2\pi\,(8\text{Hz})} = 597\,\text{N.m}$$

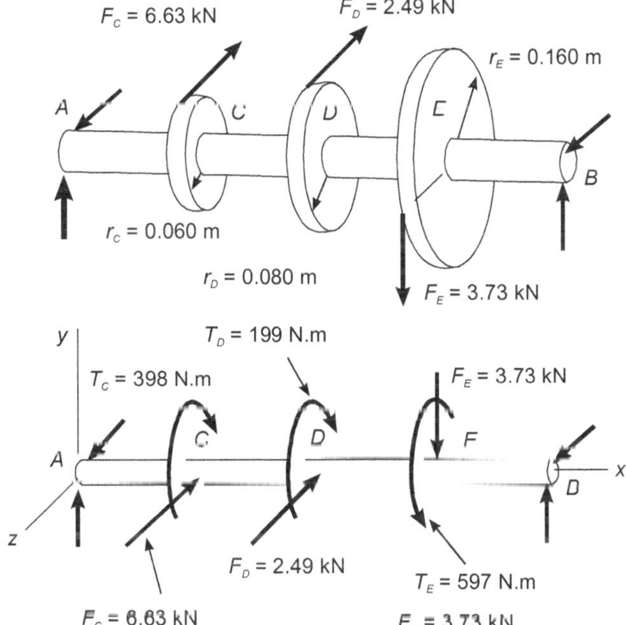

The corresponding tangential force acting on the gear is

$$F_E = \frac{T_E}{r_E} = \frac{597\,\text{N.m}}{0.16\,\text{m}} = 3.73\,\text{kN}$$

A similar analysis of gears C and D yields

$$T_c = \frac{20\,\text{kW}}{2\pi(8\text{Hz})} = 398\,\text{N.m} \qquad F_c = 6.63\,\text{kN}$$

$$T_D = \frac{10\,\text{kW}}{2\pi(8\text{Hz})} = 199\,\text{N.m} \qquad F_D = 2.49\,\text{kN}$$

We now replace the forces on the gears by equivalent force–couple systems.

Bending-Moment and Torque Diagrams

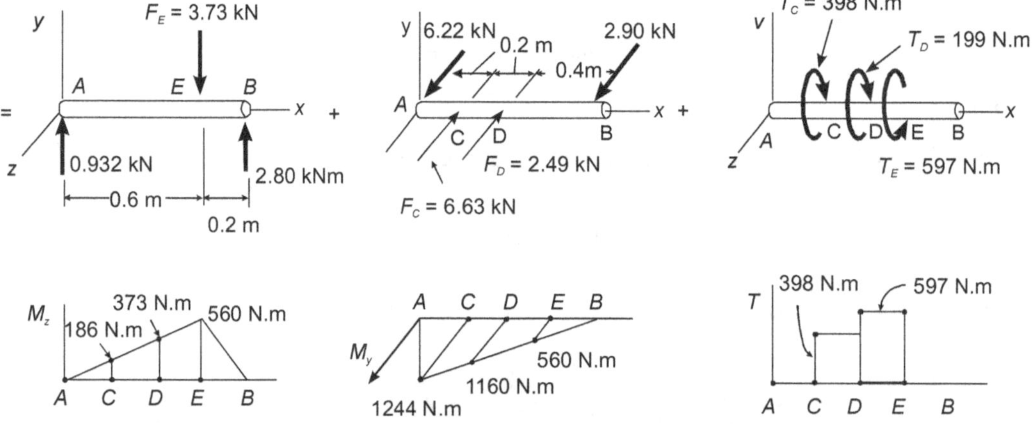

Critical transverse section By computing $\sqrt{M_y^2 + M_z^2 + T^2}$ at all potentially critical sections, we find that its maximum value occurs just to the right of D:

$$\left(\sqrt{M_y^2 + M_z^2 + T^2}\right)_{\text{max}} = \sqrt{(1160)^2 + (373)^2 + (597)^2} = 1357\,\text{N.m}$$ (note that this is the equivalent torque)

Diameter of shaft For $\tau_{\text{all}} = 50$ MPa, torsion equation when substituted $T = T_{\text{max}}$ gives

$$\frac{J}{c} = \frac{\left(\sqrt{M_y^2 + M_z^2 + T^2}\right)_{\text{max}}}{\tau_{\text{all}}} = \frac{1357\,\text{N.m}}{50\,\text{MPa}} = 27.14 \times 10^{-6}\,\text{m}^3$$

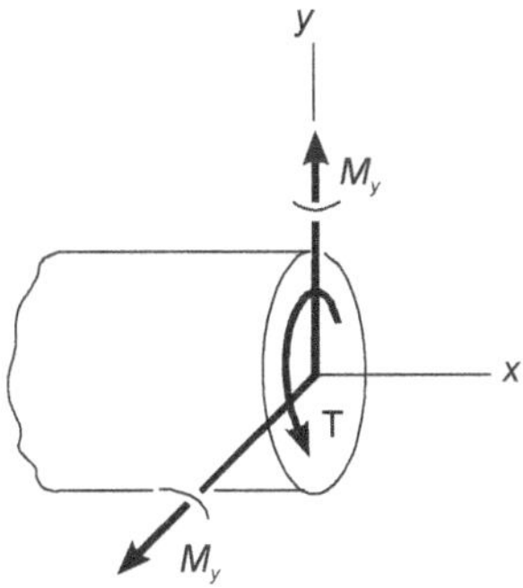

For a solid circular shaft of radius c, we have

$$\frac{J}{c} = \frac{\pi}{2}c^3 = 27.14 \times 10^{-6} \quad c = 0.02585\,\text{m} = 25.85\,\text{mm}$$

Diameter $= 2c = 51.7$ mm

PROBLEM 12.5

Figure shows a crank loaded by a force F = 1.3 kN that causes twisting and bending of a 20-mm-diameter shaft fixed to a support at the origin of the reference system. In reality, the support may be an inertia that we wish to rotate, but for the purposes of a stress analysis we can consider this a statics problem.

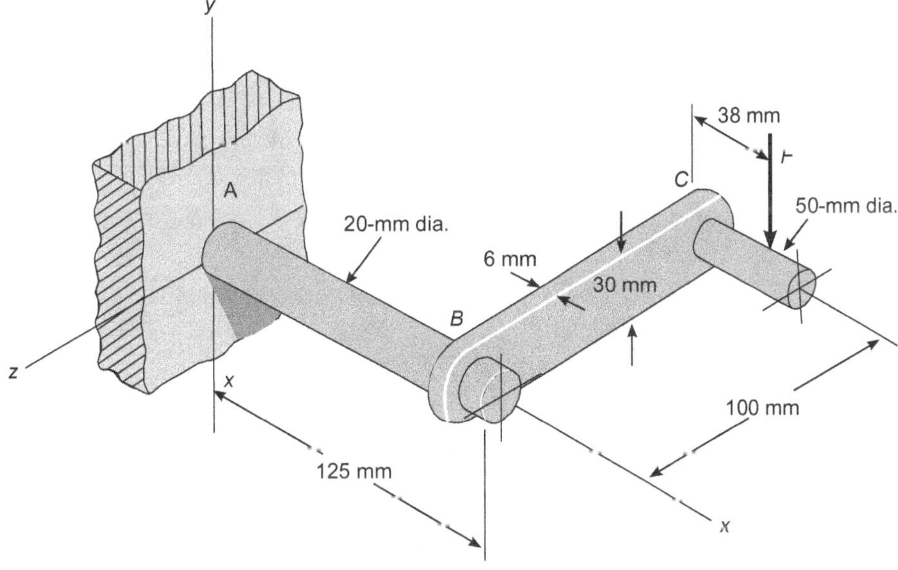

Figure

(a) Draw separate free-body diagrams of the shaft AB and the arm BC, and compute the values of all forces, moments, and torques that act. Label the directions of the coordinate axes on these diagrams.

(b) Compute the maxima of the torsional stress and the bending stress in the arm BC and indicate where these act.

(c) Locate a stress element on the top surface of the shaft at A, and calculate all the stress components that act upon this element.

(d) Deteremine the maximum normal and shear stresses at A.

Solution

(a) The two free-body diagrams are shown in Figure (a). The results are

At end C of arm BC: $F = -1.3j$ kN, $T_c = -0.05$ k kN.m

At end B of arm BC: $F = 1.3j$ kN, $M_1 = 0.13i$ k kN.m, $T_1 = 0.05$k kN.m

At end B of arm AB: $F = -1.3j$ kN, $T_2 = -0.13i$ kN.m, $M_2 = -0.05$k kN.m

At end A of arm AB: $F = 1.3j$ kN, $M_A = 0.66$k kN.m, $T_A = 0.13i$ kN.m

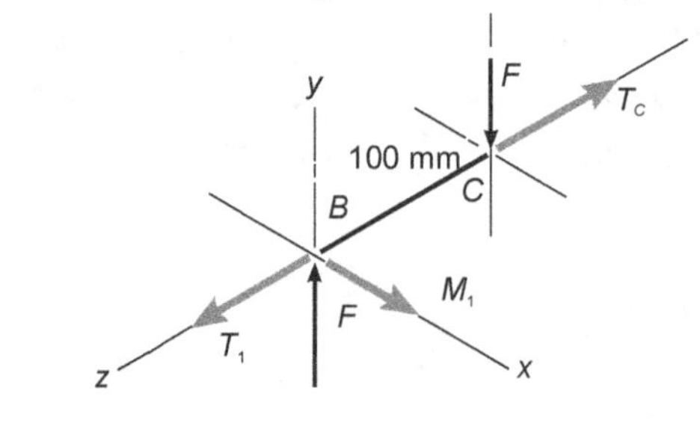

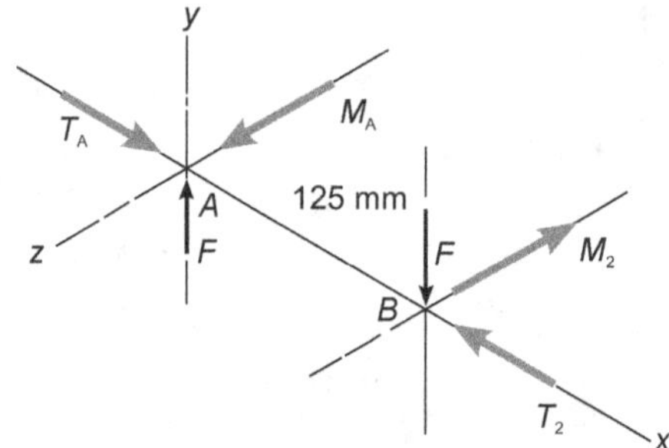

Figure (a)

(b) For arm BC, the bending moment will reach a maximum near the shaft at B. If we assume this is 0.13 kN.m, then the bending stress for a rectangular section will be

Answer

$$\sigma = \frac{M}{I/c} = \frac{6M}{bh^2} = \frac{6(130)}{0.006(0.03)^2} = 144.4\,\text{MPa}$$

Of course, this is not exactly correct, because at B the moment is actually being transferred into the shaft, probably through a weldment. For the torsional stress,

Answer

$$\tau_{max} = \frac{T}{bc^2}\left(3 + \frac{1.8}{b/c}\right) = \frac{50}{0.03(0.006)^2}\left(3 + \frac{1.8}{0.03/0.006}\right) = 155.6\,\text{MPa}$$

This stress occurs at the middle of the 30-mm side.

(c) For a stress element at A, the bending stress is tensile and is

Answer

$$\sigma_x = \frac{M}{I/c} = \frac{32M}{\pi d^3} = \frac{32(660)}{\pi(0.02)^3} = 840.3\,\text{MPa}$$

The torsional stress is

$$\tau_{xz} = \frac{-T}{J/c} = \frac{-16T}{\pi d^3} = \frac{-16(130)}{\pi(0.02)^3} = -82.8\,\text{MPa}$$

Answer

where the reader should verify that the negative sign accounts for the direction of τ_{xz}.

(d) Point A is in a state of plane stress where the stresses are in the xz plane. Thus the principal stresses are given with subscripts corresponding to the x, z axes.

Answer

The maximum normal stress is then given by

$$\sigma_1 = \frac{\sigma_x + \sigma_z}{2} + \sqrt{\left(\frac{\sigma_x - \sigma_z}{2}\right)^2 + \tau_{xz}^2}$$

$$= \frac{840.3 + 0}{2} + \sqrt{\left(\frac{840.3 - 0}{2}\right)^2 + (-82.8)^2} = 848.4\,\text{MPa}$$

Answer

The maximum shear stress at A occurs on surfaces, different than the surfaces containing the principal stresses or the surfaces containing the bending and torsional shear stresses. The maximum shear stress is given again with modified subscripts, by

$$\tau_1 = \sqrt{\left(\frac{\sigma_x - \sigma_z}{2}\right)^2 + \tau_{xz}^2} = \sqrt{\left(\frac{840.3 - 0}{2}\right)^2 + (-82.8)^2} = 428.2\,\text{MPa}$$

PROBLEM 12.6

The 40-mm diameter solid steel shaft shown in Figure (a) is simply supported at the ends. Two pulleys are keyed to the shaft where pulley B is of diameter 100 mm and pulley C is of diameter 200 mm. Considering bending and torsional stresses only, determine the locations and magnitudes of the greatest tensile, compressive, and shear stresses in the shaft.

Solution

Figure 4b shows the net forces, reactions, and torsional moments on the shaft. Although this is a three-dimensional problem and vectors might seem appropriate, we will look at the components of the moment vector by performing a two-plane analysis. Figure c shows the loading in the xy plane, as viewed down the z axis, where bending moments are actually vectors in the z direction. Thus we label the moment diagram as M_z verus x. For the xz plane, we look down the y axis, and the moment diagram is M_y versus x as shown in Figure d.

The net moment on a section is the vector sum of the components. That is,

$$M = \sqrt{M_y^2 + M_z^2}$$

At point B,

$$M_B = \sqrt{225^2 + 900^2} = 928\,\text{N.m}$$

At point C,

$$M_C = \sqrt{450^2 + 450^2} = 636\,\text{N.m}$$

Thus the maximum bending moment is 928 N.m and the maximum bending stress at pulley B is

$$\sigma = \frac{Md/2}{\pi d^4/64} = \frac{32M}{\pi d^3} = \frac{32(928)}{\pi(0.04)} = 147.7\,\text{MPa}$$

The maximum torsional shear stress occurs between B and C and is

$$\tau = \frac{Td/2}{\pi d^4/32} = \frac{16T}{\pi d^3} = \frac{16(180)}{\pi(0.04^3)} = 14.3\,\text{MPa}$$

The maximum bending and torsional shear stresses occur just to the right of pulley B at points E and F as shown in Figure. At point E, the maximum tensile stress will be σ_1 given by

Answer

$$\sigma_1 = \frac{\sigma}{2} + \sqrt{\left(\frac{\sigma}{2}\right)^2 + \tau^2} = \frac{147.7}{2} + \sqrt{\left(\frac{147.7}{2}\right)^2 + 14.3^2} = 149.1\,\text{MPa}$$

At point F, the maximum compressive stress will be σ_2 given by,

Answer

$$\sigma_2 = -\frac{\sigma}{2} - \sqrt{\left(\frac{-\sigma}{2}\right)^2 + \tau^2} = \frac{-147.7}{2} - \sqrt{\left(\frac{-147.7}{2}\right)^2 + 14.3^2} = -1.4\,\text{MPa}$$

The extreme shear stress also occurs at E and F and is

Answer

$$\tau_1 = \sqrt{\left(\frac{\pm\sigma}{2}\right)^2 + \tau^2} = \sqrt{\left(\frac{\pm 147.7}{2}\right)^2 + 14.3^2} = 75.2\,\text{MPa}$$

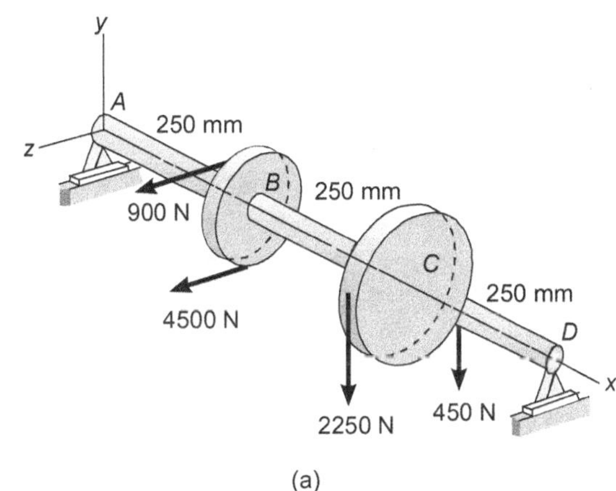

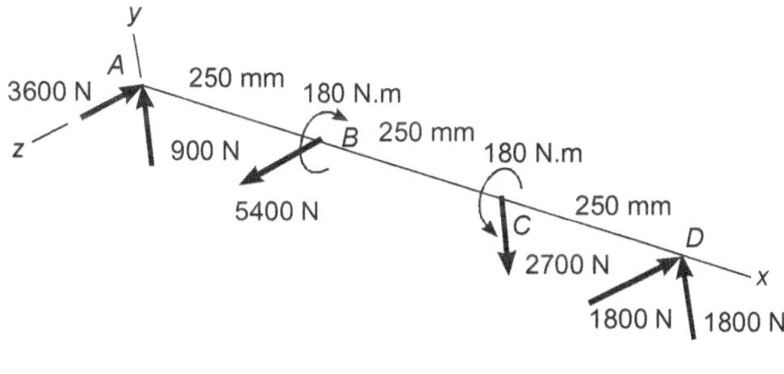

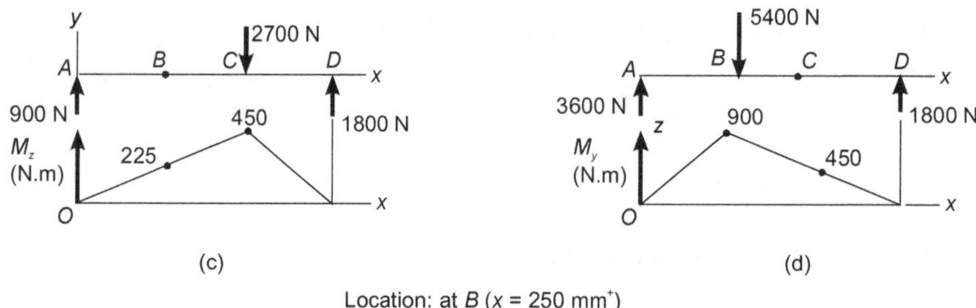

Location: at B ($x = 250$ mm$^+$)

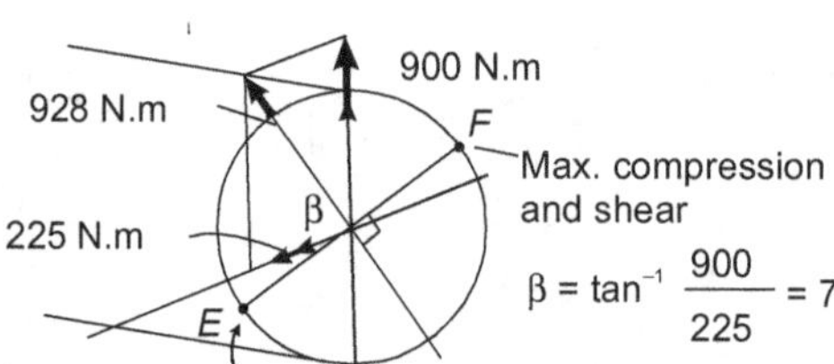

Figure

13

DEFLECTION OF BEAMS UNDER TRANSVERSE LOADS

We had seen earlier the stresses produced in beams due to axial loads, bending moments and shear forces.

Deflection is another important design parameter that needs to be kept under check. For example in a turbogenerator rotor excessive deflection may cause rotating elements to rub against the casing.

This chapter discusses certain analytical methods to determine deflections in laterally loaded beams. The deflected shape is referred to as elastic curve. The sign convention, relation between shear forces and bending moments, the basic beam bending equation are already dealt with and hence not repeated, but they are necessary to understand this chapter.

DIFFERENTIAL EQUATION OF ELASTIC CURVE

To derive the equation for the elastic curve, we take recources to the curvature formula for any curve as derived in mathematics. Consider any curve as shown in Figure 13.1 and take any point P on the curve. Set in the X-Y axis system, P' is appoint at an incremental distance dx along the curve. Also dx and dy are the distances along the X-and Y axes from P to P'. As the distance ds is very small, we can take

$$ds^2 = dx^2 + dy^2 \qquad (13.1)$$

If the radius of the curve at P is R and the angle between the radii to P and P' is $d\alpha$ then $Rd\alpha$ is equal to if $d\alpha$ is in radians

Now

$$ds = \sqrt{(dx^2 + dy^2)} - Rd\alpha \qquad (13.2)$$

or

$$\frac{1}{R} = \frac{d\alpha}{\sqrt{(dx^2 + dy^2)}}$$

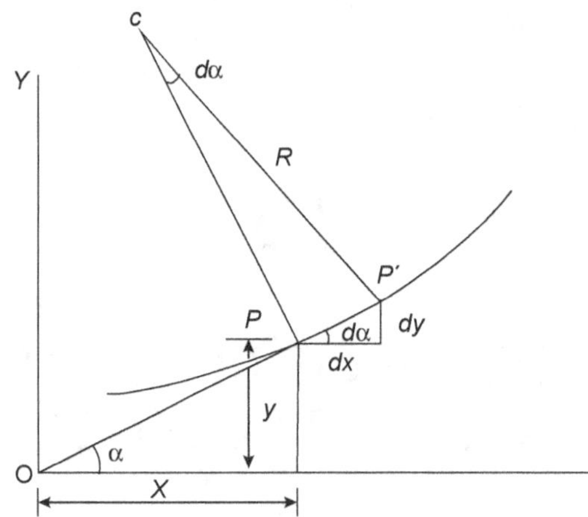

Figure 13.1 Elastic curve

$$dx\sqrt{[1+(dy/dx)^2]} = \frac{R(d^2y/dx^2)dx}{[1+(dy/dx)^2]}$$

or
$$\frac{1}{R} = \frac{d^2y/dx^2}{[1+(dy/dx)^2]^{3/2}} \left(\frac{1}{R} \text{ is called curvature}\right) \qquad (13.3)$$

The application of this equation to the elastic curve yields a relationship between the BM and the deflection y (Figure 13.2). Some simplification of this expression for the curvature is justified in the case of the elastic curve because the deflection as well as the slope of the beams are extremely small. Thus dy/dx, which gives the slope of the elastic curve at a point, is very small and its square is still smaller. Therefore, the term $[1+(dy/dx)^2]^{3/2}$ can be taken equal to 1, reducing the curvature expression to d^2y/d^2x. Thus,

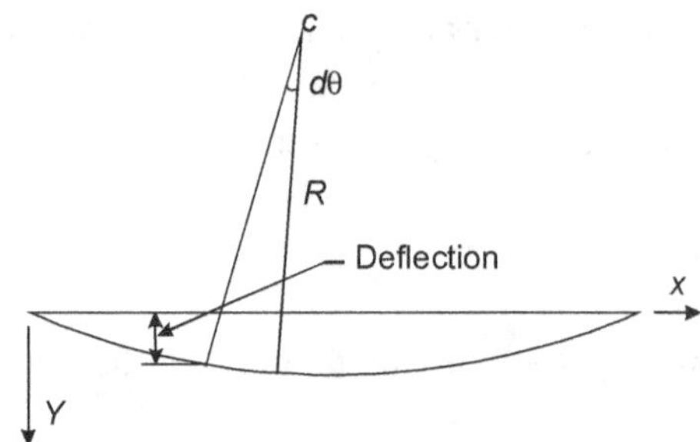

Figure 13.2

$$\frac{1}{R} = \frac{d^2y}{dx^2} \text{ (for small deflections and small slopes)}$$

$$M = \frac{EI}{R} = EI\frac{d^2y}{dx^2} \qquad (13.4)$$

This second-order differential equation, on solution, will yield a relationship between y and x, which is the equation for the elastic curve.

METHODS FOR CALCULATING DEFLECTION

Several methods are available for calculating the deflection and slope in beams. They are all based upon the fundamental expression we have derived for curvature. We will discuss three of them in this chapter. They are:

i. The double integration method based upon the fact d^2y/dx^2 when integrated twice yields y,

ii. The area-moment method, wherein the slope and deflection are calculated from the area of the BM diagram and the moment of the area of the BM diagram, and methods (i) and (ii) are discussed in detail and illustrated through several solved examples.

iii. The conjugate-beam method whereby a conjugate beam derived from the actual beam is loaded with the M/EI diagram, and the slope and deflection in the actual beam are calculated as SF and BM, respectively, of the conjugate beam. The reader may use other references in bibliograhy to obtain more details, as this is not discussed in this text.

Each method has some advantage, depending upon the problem. Let us start with the double integration method, which is a very general method that yields an equation for the elastic curve.

THE DOUBLE INTEGRATION METHOD

The basic principle in the double integration method is to integrate the expression EId^2y/dx^2 twice to obtain an equation for y in terms of x.

$$EI\frac{d^2 y}{dx^2} = -M_x \tag{13.5}$$

where M_x is the expression for BM at x. Integrating,

$$EI\frac{dy}{dx} = \int(-M_x dx) + C_1 \tag{13.6}$$

where C_1 is a constant of integration. In this equation, $dy/dx = \tan\theta$, where θ is the slope of the elastic curve at x. Since $\tan\theta$ is very small, $\tan\theta = \theta$ in radians (slope is, therefore, obtained in radians as mentioned earlier). Integrating once more,

$$EIy = \int\left[\int -M_x dx + C_1\right] + C_2 \tag{13.7}$$

where C_2 is the second constant of integration.

Thus, once the expression for M_x, which will be in terms of x (obtained from bending moment diagram or BMD), is integrated twice, we get an expression for y in terms of x, i.e., the equation for the elastic curve.

The constants of integration, C_1 and C_2 are evaluated from the boundary conditions. More precisely, they are obtained from conditions which apply due to restraints imposed on the bending by the supports. These conditions can be in terms of either slopes or deflections.

BOUNDARY CONDITIONS

i. *Simply supported beam*   The two ends of the beam do not deflect where they are supported. The conditions are, therefore, $x = 0$, $y = 0$, and $x = l$, $y = 0$. (roof beams, bridges, rotor on two bearings).

ii. *Cantilever*   The end of the beam, having a fixed support, is not deflected. (e.g., tall columns, towers) nor does the axis have any slope at that point. The tangent at point A remains horizontal. The conditions are, therefore, $x = 0$, $y = 0$ and $x = 0$, $dy/dx = 0$.

iii. *Overhanging beam*   The overhanging beam has the conditions $x = 0$, $y = 0$ and $x = l$, $y = 0$. The beam is supported at $x = 1$ but extend beyonds 1.

iv. *Doubly overhanging beam*   In this case, the conditions are $x = a$, $y = 0$ and $x = 1 + a, y = 0$. The supports are at $x = a$ and $x = 1 + a$ and beam extends on both sides of the support.

v. *Clamped beam*   Up to this point the boundary conditions relate to geometry and are called kinematic or geometric BC's. Now we have natural BC's which we relate to force and moment. slope = deflection = 0 at $x = 0$ and $x = l$

vi. *Free-Free beam*   Shear force = Bending moment = 0 at $x = 0$ and $x = l$. At the free end no external force or moment exists. (e.g., ships, aircrafts, missiles which have free ends).

vii. Guided end: slope = 0 e.g., overloaded handling cranes. and shear force = 0

Only these combinations are permitted and compatible. For and example it is not possible to prescribe both displacement and shear force at the same point. Similar is the condition for slope and bending moment. If we prescribe one, the other gets fixed by the equations of mechanics. This point is very important and must be remembered while using BC's.

The following example problems illustrate double integration method. Table 13.1 gives deflection and slopes of beam for commonly used loadings.

PROBLEM 13.1

The cantilever beam AB is of uniform cross section and carries a load **P** at its free end A. Determine the equation of the elastic curve and the deflection and slope at A.

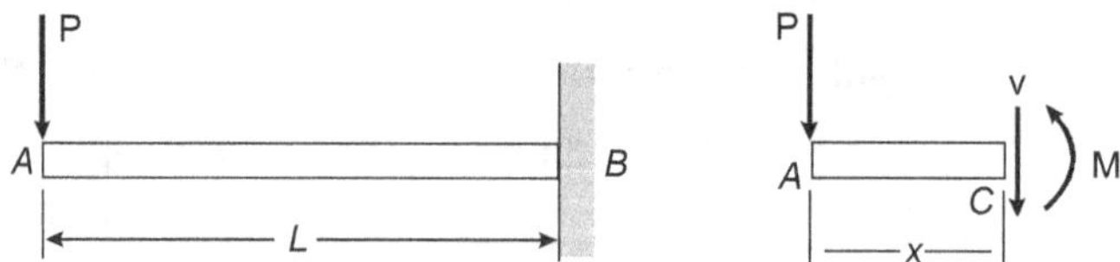

Table 13.1

Beam and loading	Elastic curve	Maximum deflection	Slope at end	Equation of elastic curve
1 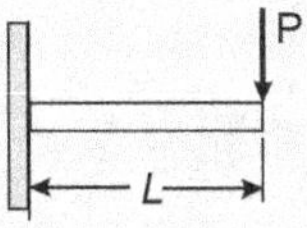	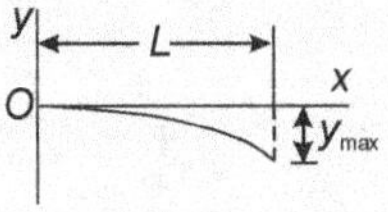	$-\dfrac{PL^3}{3EI}$	$-\dfrac{PL^2}{2EI}$	$y = \dfrac{P}{6EI}(x^3 - 3Lx^2)$
2 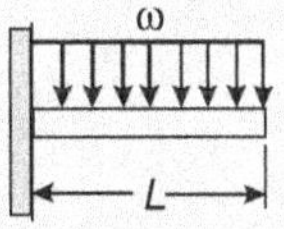	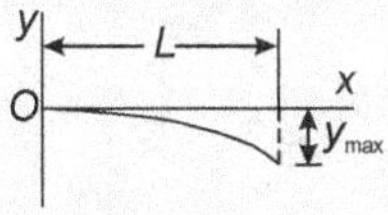	$-\dfrac{\omega L^4}{8EI}$	$-\dfrac{\omega L^3}{6EI}$	
3 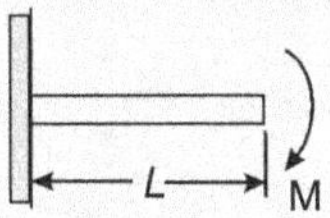	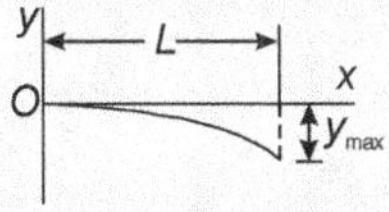	$-\dfrac{ML^2}{2EI}$	$-\dfrac{ML}{EI}$	$y = -\dfrac{M}{2EI}x^2$
4 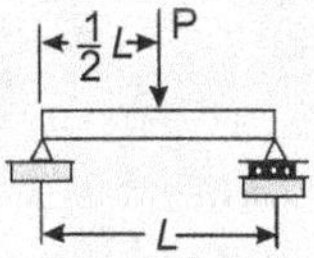	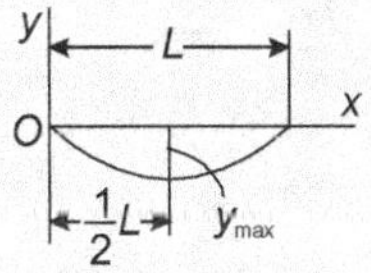	$-\dfrac{PL^3}{48EI}$	$-\dfrac{PL^2}{16EI}$	For $x \le \dfrac{1}{2}L$: $\quad y = \dfrac{P}{48EI}(4x^3 - 3L^2x)$
5 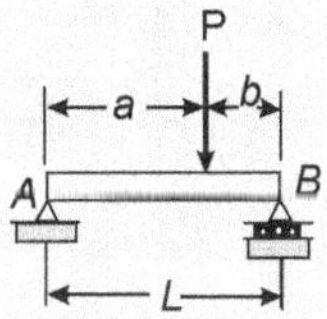	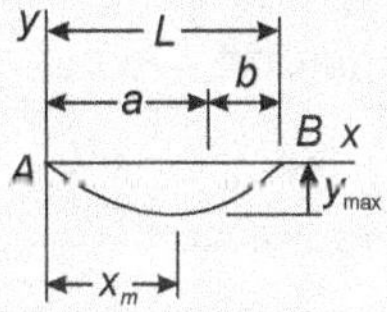 For $a > b$: $\quad -\dfrac{Pb(L^2 - b^2)^{3/2}}{9\sqrt{3}EIL}$ at $x_m = \sqrt{\dfrac{L^2 - b^2}{3}}$		$\theta_A = -\dfrac{Pb(L^3 - b^2)}{6EIL}$ $\theta_B = +\dfrac{Pa(L^2 - a^2)}{6EIL}$	For $x < a$: $\quad y = \dfrac{Pb}{6EIL}[x^3 - (L^2 - b^2)x]$ For $x = a$: $\quad y = -\dfrac{Pa^2b^2}{3EIL}$
6 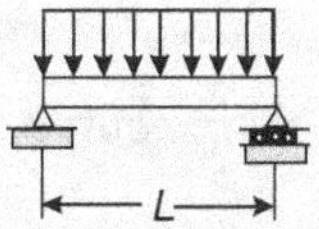	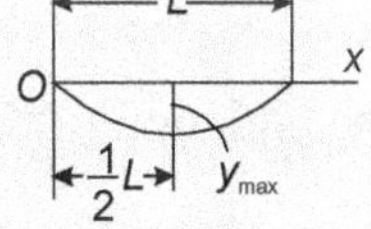	$-\dfrac{5\omega L^3}{384EI}$	$\pm\dfrac{\omega L^3}{24EI}$	$y = -\dfrac{\omega}{24EI}(x^4 - 2Lx^2 + L^2)$
7 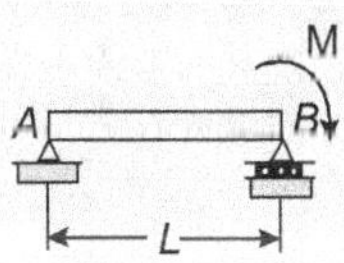	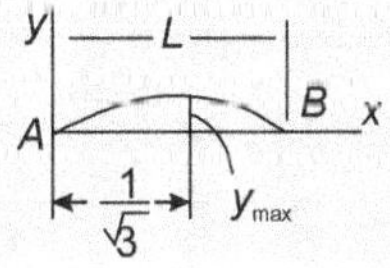	$\dfrac{ML^2}{9\sqrt{3}EI}$	$\theta_A = +\dfrac{ML}{6EI}$ $\theta_B = -\dfrac{ML}{3EI}$	$y = \dfrac{M}{6EI}(x^3 - L^2x)$

Using the free-body diagram of the portion AC of the beam (Figure), where **C** is located at a distance x from end A, we find

$$M = -Px$$

Substituting for M into equation (13.5) and multiplying both members by the constant EI, we write

$$EI\frac{d^2 y}{dx^2} = -Px$$

Integrating in x, we obtain

$$EI\frac{dy}{dx} = -\frac{1}{2}Px^2 + C_1$$

We now observe that at the fixed end B we have $x = L$ and $\theta = dy/dx = 0$ (Figure). Substuting these values into equation (13.6) and solving for C_1, we have

$$C_1 = \frac{1}{2}PL^2$$

which when inserted back into equation (13.6) we get $EI\frac{dy}{dx} = -\frac{1}{2}Px^2 + PL^2$

Integrating both members of equation above, we get

$$EL\ y = \frac{1}{6}Px^3 + \frac{1}{2}PL^2 x + C_2$$

But, at B we have $x = L$, $y = 0$. Substituting into above equation, we have

$$0 = -\frac{1}{6}PL^3 + \frac{1}{2}PL^3 + C_2$$

$$C_2 = -\frac{1}{3}PL^3$$

Inserting the value of C_2 in equation for $EI\ y$, we obtain the equation of the elastic curve

$$EI\ y = \frac{1}{6}Px^3 + \frac{1}{2}PL^2 x - \frac{1}{3}PL^3$$

or

$$y = \frac{P}{6EI}(-x^3 + 3L^2 x - 2L^3)$$

The deflection and slope at A are obtained by letting $x = 0$ in equation and we find

$$y_A = -\frac{PL^3}{3EI} \quad \text{and} \quad \theta_A = \left(\frac{dy}{dx}\right)_A = \frac{PL^2}{2EI}$$

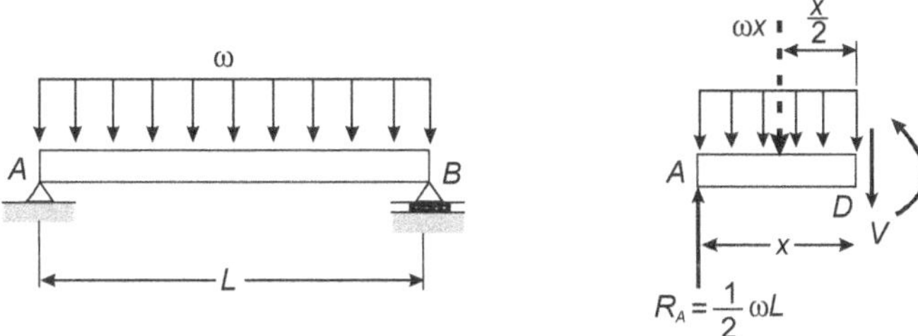

PROBLEM 13.2

A simply supported prismatic beam AB carries a uniformly distributed load w per unit length (Figure). Determine the equation of the elastic curve and the maximum deflection of the beam.

Drawing the free-body diagram of the portion AD of the beam (Figure) and taking moments about D, we find that

$$M = \frac{1}{2}wLx - \frac{1}{2}wx^2 \tag{1}$$

Substituting for M into equation (13.5), we write

$$EI\frac{d^2y}{dx^2} = -\frac{1}{2}wx^2 + \frac{1}{2}wLx \tag{2}$$

Integrating twice in x, we have

$$EI\frac{dy}{dx} = -\frac{1}{6}wx^3 + \frac{1}{4}wLx^2 + C_1 \tag{3}$$

$$EI\,y = -\frac{1}{24}wx^4 + \frac{1}{12}wLx^3 + C_1x + C_2 \tag{4}$$

Observing that $y = 0$ at both ends of the beam (Figure), we first let $x = 0$ and $y = 0$ in equation and obtain $C_2 = 0$. We then make $x = L$ and $y = 0$ in the same equation and write

$$0 = -\frac{1}{24}wL^4 + \frac{1}{12}wL^4 + C_1L$$

$$C_1 = -\frac{1}{24}wL^3$$

Putting the values of C_1 and C_2 back into equation, we obtain the equation of the elastic curve

$$EI\, y = -\frac{1}{24}wx^4 + \frac{1}{12}wLx^3 - \frac{1}{24}wL^3x$$

or

$$y = \frac{w}{24EI}(-x^4 + 2Lx^3 - L^3x) \tag{5}$$

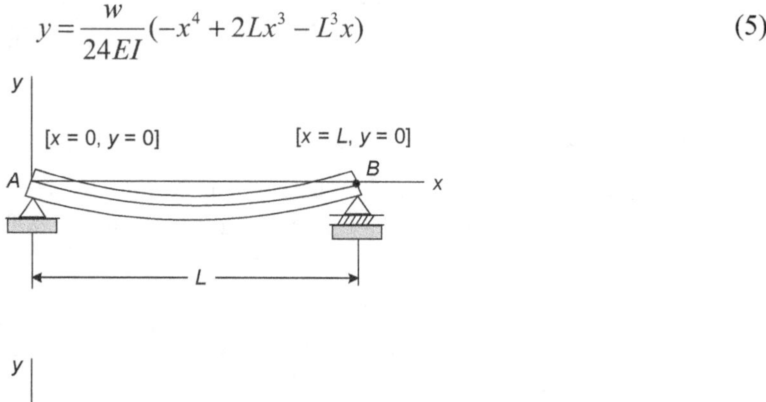

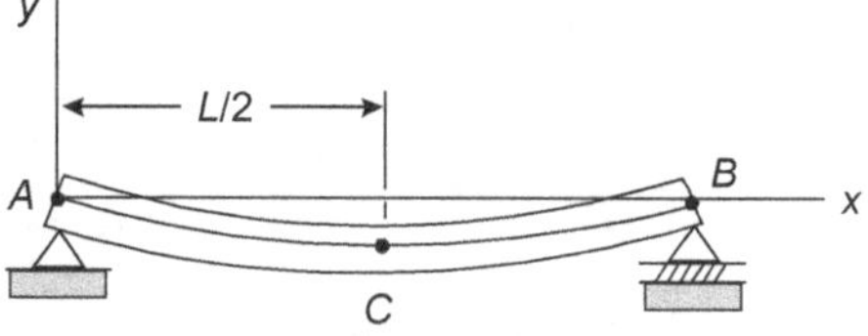

Substituting into equation (3) the value obtained for C_1, we check that the slope of the beam is zero for $x = L/2$ and that the elastic curve has a minimum at the midpoint C of the beam (Figure). Letting $x = L/2$ in equation 4, we have

$$y_c = \frac{w}{24EI}\left(-\frac{L^4}{16} + 2L\frac{L^3}{8} - L^3\frac{L}{2}\right) = -\frac{5wL^4}{384EI}$$

The maximum deflection or, more precisely, the maximum absolute value of the deflection, is thus

$$\left|y_{max}\right| = \frac{5wL^4}{384EI}$$

In each of the two examples considered so far, only one free-body diagram was required to determine the bending moment in the beam. As a result, a single function of x was used to represent M throughout the beam. However, the reactions at supports, or concentrated loads or discontinuities in a distributed load will make it necessary to divide the beam into several portions, and to represent the bending moment by a different function $M(x)$ in each of these portions of beam. Each of the functions $M(x)$ will then lead to a different expression for the slope $\theta(x)$ and for the deflection $y(x)$. Since each of the expression obtained for the deflection must contain two constants of integration, a large number of constants will have to be determined. As we shall see in the next example, the required additional boundary conditions may be obtained by observing that, while the shear and bending moment can be discontinuous at several points in a beam, the deflection and the slope of the beam cannot be discontinuous at any point.

PROBLEM 13.3

For the prismatic beam and the loading shown in Figure, determine the slope and deflection at point *D*.

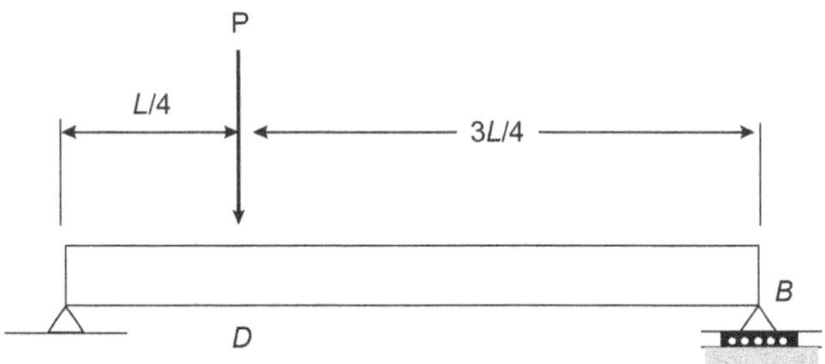

We must divide the beam into portions, *AD* and *DB*, and determine the function $y(x)$ which defines the elastic curve for each of these portions.

1. From A to D ($x < L/4$) We draw the free-body diagram of a portion of beam *AE* of length $x < L/4$ (Figure). Taking moments about *E*, we have

$$M_1 = \frac{3P}{4}x \tag{1}$$

or, recalling equation (13.5),

$$EI\frac{d^2y_1}{dx^2} = \frac{3}{4}Px \tag{2}$$

where $y_1(x)$ is the function which defines the elastic curve for portion AD of the beam. Integrating in *x*, we write

$$EI\theta_1 = EI\frac{dy_1}{dx} = \frac{3}{8}Px^2 + C_1 \tag{3}$$

$$EI\,y_1 = \frac{1}{8}Px^3 + C_1x + C_2 \tag{4}$$

2. From D to B($x>L/4$) Refer to the free-body diagram of the portion of the beam of AE length $x>L/4$

$$M_2 = \frac{3px}{4} - P\left(x - \frac{L}{4}\right) \tag{5}$$

$$EI\frac{d^2y}{dx^2} = -\frac{1}{4}P_x + \frac{1}{4}PL \tag{6}$$

where $y_2(x)$ defines elastic curve of portion DB of the beam. Integrating in *x*, write

$$EI\theta_2 = EI\frac{dy_2}{dx} = -\frac{1}{8}Px^2 + \frac{1}{4}PLx + C_3 \tag{7}$$

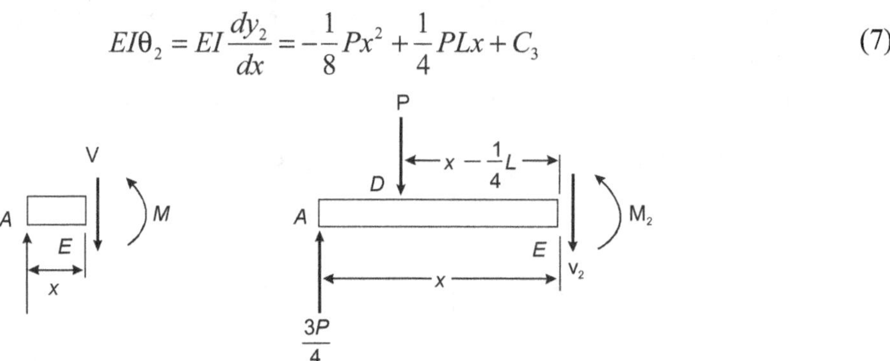

Determination of the constants of integration. The conditions which must be satisfied by the constants of integration have been summarized in Figure. At the support A, where the deflection is defined by equation (4), we must have $x = 0$ and $y_1 = 0$. At the support B, where the deflection is defined by equation (8), we must have $x = L$ and $y_2 = 0$. Also, the fact that there can be no sudden change in deflection or in slope at point D requires that $y_1 = y_2$ and $\theta_1 = \theta_2$ when $x = L/4$. We have therefore:

$$EIy_2 = -\frac{1}{24}px^3 + \frac{1}{8}PLx^2 + C_3x + C_4 \tag{8}$$

$$[x = L,\ y_2 = 0],\ \text{equation (4)}: 0 = C_2 \tag{9}$$

$$[x = L,\ y_2 = 0],\ \text{equation (8)}: 0 = \frac{1}{2}PL^3 + C_3L + C_4 \tag{10}$$

$$[x = L/4,\ \theta_1 = \theta_2],\ \text{equations (3) and (7)}:$$

$$\frac{3}{128}PL^2 + C_1 = \frac{7}{128}PL^2 + C_3 \tag{11}$$

$[x = L/4,\ y_1 = y_2]$, equations (4) and (8):

$$\frac{PL^3}{512}+C_1\frac{L}{4}=\frac{11PL^3}{1536}+C_3\frac{L}{4}+C_4 \tag{12}$$

Solving these equations simultaneously, we find

$$C_1=-\frac{7PL^2}{128},C_2=0,C_3=-\frac{11PL^2}{128},C_4=\frac{PL^3}{384}$$

Substituting for C_1 and C_2 into equations (3) and (4), we write that for $x\le L/4$

$$EI\theta_1=\frac{3}{8}Px^2-\frac{7PL^2}{128}$$

$$EI\,y_1=\frac{1}{8}Px^3-\frac{7PL^2}{128}x$$

Letting $x=L/4$ in each of these equations, we find that the slope and deflection at point D are, respectively,

$$\theta_D=-\frac{PL^2}{32EI}\qquad\text{and}\qquad y_D=-\frac{3PL^3}{256EI}$$

We note that, since $\theta_D\ne 0$, the deflection at D is not the maximum deflection of the beam, although the load P is applied at D.

PROBLEM 13.4

The overhanging steel beam ABC carries a concentrated load P at end C. For portion AB of the beam,(a)derive the equation of the elastic curve, (b) determine the maximum deflection, (c) evaluate y_{max} for the following data:

$$I=301\times10^{-6}\,\text{m}^4\qquad E=200\ \text{GPa}$$

$$P=250\,\text{kN}\quad L=5\text{m}\quad a=1.2\text{m}$$

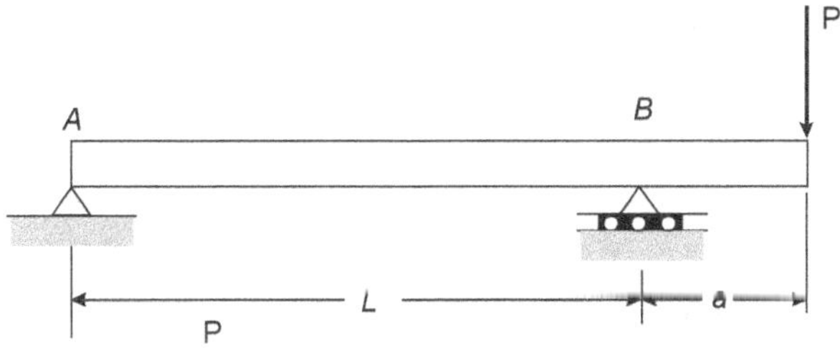

Free-Body Diagrams Reactions $R_A=Pa/L\downarrow$ $R_B=P(1+a/L)\uparrow$

Using the free-body diagram of the portion of beam AD of length x, we find

$$M=P\frac{a}{L}x\qquad\qquad(0<x<L)$$

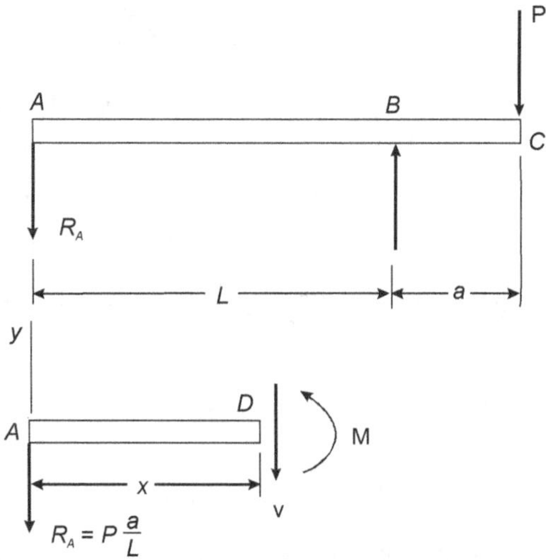

Differential equation of the elastic curve. We use equation for M and write

$$EI\frac{d^2y}{dx^2} = -P\frac{a}{L}x$$

Noting that the flexural rigidity EI is constant, we integrate twice and find

$$EI\frac{dy}{dx} = -\frac{1}{2}P\frac{a}{L}x^2 + C_1 \tag{1}$$

$$EI\,y = -\frac{1}{6}P\frac{a}{L}x^3 + C_1x + C_2 \tag{2}$$

Determination of constants. For the boundary conditions shown, we have

$[x = 0, y = 0]$: From equation (2), we find $C_2 = 0$

$[x = L, y = 0]$: Again using equation (2), we write

$$EI(0) = -\frac{1}{6}P\frac{a}{L}L^3 + C_1L \qquad C_1 = +\frac{1}{6}PaL$$

(a) *Equation of the elastic curve*   Substituting for C1 and C2 into equations (1) and (2), we have

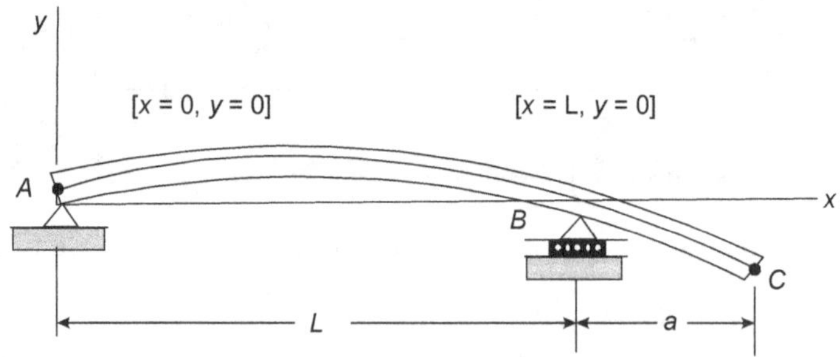

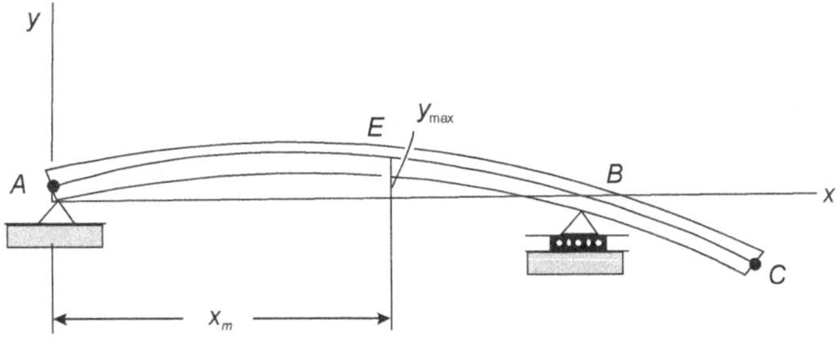

$$EI\frac{dy}{dx} = -\frac{1}{2}P\frac{a}{L}x^2 + \frac{1}{6}PaL \qquad \frac{dy}{dx} = \frac{PaL}{6EI}\left[1 - 3\left(\frac{x}{L}\right)^2\right] \tag{3}$$

$$EI\,y = -\frac{1}{6}P\frac{a}{L}x^3 + \frac{1}{6}PaLx \qquad y = \frac{PaL^2}{6EI}\left[\frac{x}{L} - \left(\frac{x}{L}\right)^3\right] \tag{4}$$

(b) *Maximum deflection in portion AB*　The maximum deflection y_{max} occurs at point E where the slope of the elastic curve is zero. Setting $dy/dx = 0$ in equation (3), we obtain

$$0 = \frac{PaL}{6EI}\left[1 - 3\left(\frac{x_1}{L}\right)^2\right] \qquad x_1 = L/\sqrt{3} = 0.577L$$

We substitute $x_1/L = 0.577$ into equation (4) and have

$$y_{max} = \frac{PaL^2}{6EI}[(0.577) - (0.577)^3] = 0.0642\frac{PaL^2}{EI}$$

(c) *Evaluation of* y_{max}　For the data given, the value of y_{max} is

$$y_{max} = 0.0642\frac{(250\,\text{kN})(1.2\,\text{m})(5\,\text{m})^2}{(200\,\text{GPa})(301\times10^{-6}\,\text{m}^4)} = 8.00\,\text{mm}$$

PROBLEM 13.5

For the beam and loading shown, determine (a) the equation of the elastic curve, (b) the slope at end A,(c) the maximum deflection.

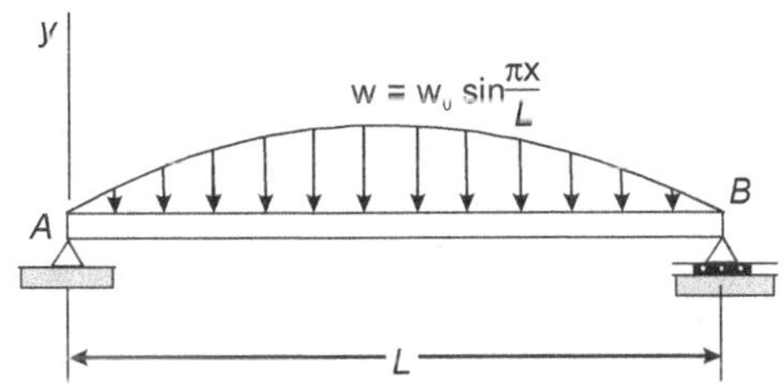

Note

This loading pattern is considered important as any arbitrary continuous loading in the interval 0 to *l* can be expressed in terms of Fourier series consisting of addition of several harmonic functions with decreasing span lengths like $L/2$, $L/3$, $L/4$ and so on which will ultimately converge to the given loading. Hence this problem is solved here.

Differential equation of the elastic curve. From equation,

$$EI\frac{d^4 y}{dx^4} = -w(x) = -w_0 \sin\frac{\pi x}{L} \tag{1}$$

Integrating equation (1) twice:

$$EI\frac{d^3 y}{dx^3} = V = +w_0 \frac{L}{\pi}\cos\frac{\pi x}{L} + C_1 \tag{2}$$

$$EI\frac{d^2 y}{dx^2} = M = +w_0 \frac{L^2}{\pi^2}\sin\frac{\pi x}{L} + C_1 x + C_2 \tag{3}$$

Boundary conditions

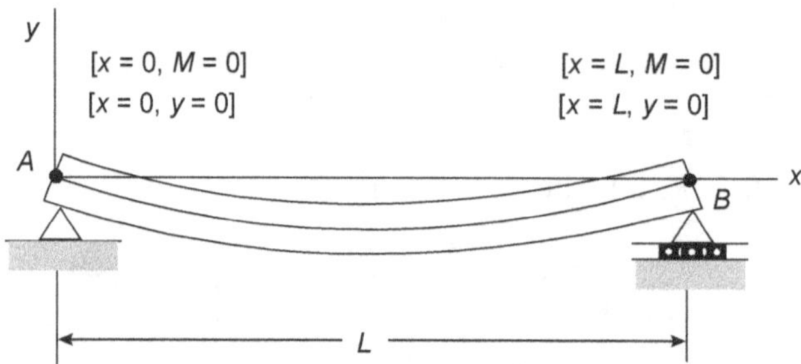

$[x = 0, M = 0]$: From equation (3),we find $C_2 = 0$. Note here that we are using the natural BC at x = 0, M = 0.

$[x = L, M = O]$: Again using equation (3), we write

$$0 = w_0 \frac{L^2}{\pi^2}\sin\pi + C_1 L \qquad C_1 = 0$$

Thus

$$EI\frac{d^2 y}{dx^2} = +w_0 \frac{L^3}{\pi^2}\sin\frac{\pi x}{L} \tag{4}$$

Integrating equation (4) twice

$$EI\frac{dy}{dx} = EI\theta = -w_0 \frac{L^3}{\pi^3}\cos\frac{\pi x}{L} + C_3 \tag{5}$$

$$EIy = -w_0 \frac{L^4}{\pi^4} \sin\frac{\pi x}{L} + C_3 x + C_4 \tag{6}$$

$[x = 0,\ y = 0]$: Using equation (6), we find $C_4 = 0$

$[x = L,\ y = 0]$: Again using equation (6), we find $C_3 = 0$

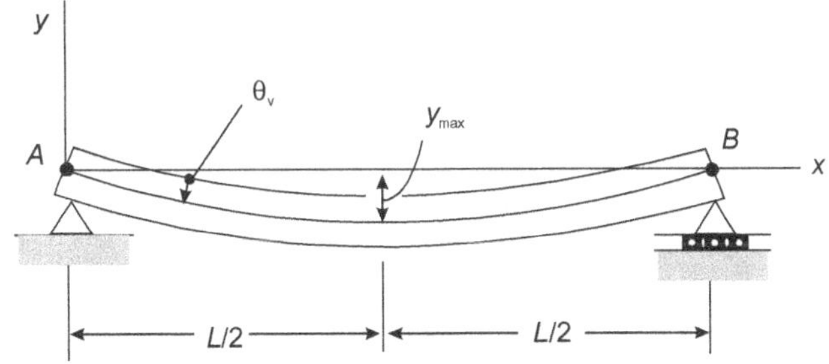

(a) Equation of elastic curve

$$Ely = -w_o \frac{L^4}{\pi^4} \sin\frac{\pi x}{L}$$

(b) Slope at end A. For $x = 0$, we have

$$EI\theta_A = \frac{w_o L^3}{\pi^3} \cos\theta$$

$$\theta_A = \frac{\omega_0 L^3}{\pi^3 EI}$$

(c) Maximum Deflection. For $x = \frac{1}{2}L$,

$$Ely_{max} = -w_o \frac{L^4}{\pi^4} \sin\frac{\pi}{2}$$

$$y_{max} = \frac{w_o L^4}{\pi^4 EI}$$

PROBLEM 13.6

For the uniform beam AB, (a) determine the reaction at A, (b) derive the equation of the elastic curve, (c) determine the slope at A. (Note that the beam is statically indeterminate to the first degree.)

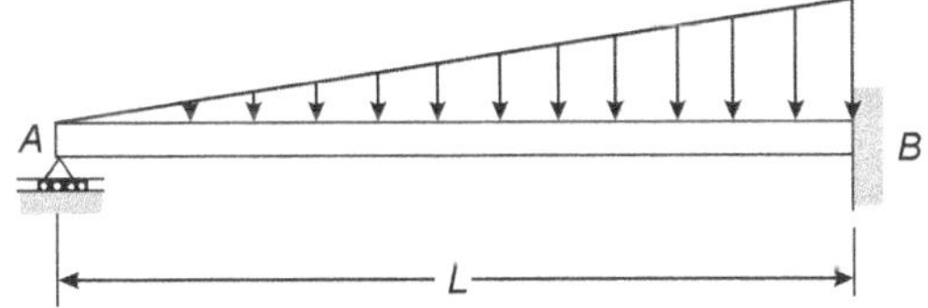

Bending moment. Using the free-body shown, we write

$$+\Sigma M_D = 0; \quad R_A x - \frac{1}{2}\left(\frac{w_0 x^2}{L}\right)\frac{x}{3} - M = 0 \qquad M = R_A x - \frac{w_0 x^3}{6L}$$

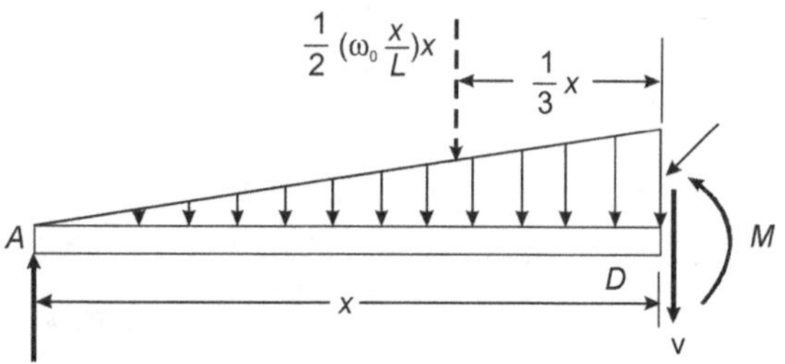

Differential equation of the elastic curve. We use equation (13.4) and write

$$EI = \frac{d^2 y}{dx^2} = R_A x - \frac{w_0 x^3}{6L}$$

Noting that the flexural rigidity EI is constant, we integrate twice and find

$$EI\frac{dy}{dx} = EI\theta = \frac{1}{2}R_A x^2 - \frac{w_0 x^4}{24L} + C_1 \tag{1}$$

$$EIy = \frac{1}{6}R_A x^3 - \frac{w_0 x^5}{120L} + C_1 x + C_2 \tag{2}$$

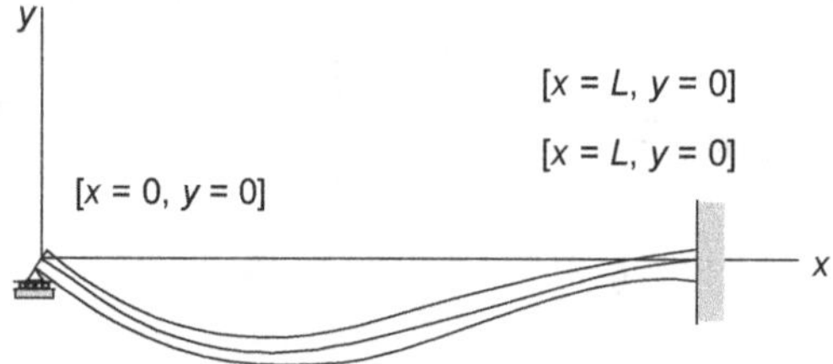

Boundary conditions   The three boundary conditions which must be satisfied are shown on the sketch

$$[x=0, y=0]: \qquad C_2 = 0 \tag{3}$$

$$[x=L, \theta=0]: \qquad \frac{1}{2}R_A L^2 - \frac{w_0 L^3}{24} + C_1 = 0 \tag{4}$$

$$[x=L, y=0]: \qquad \frac{1}{6}R_A L^3 - \frac{w_0 L^4}{120} + C_1 L + C_2 = 0 \tag{5}$$

(a) Reaction at A. Multiplying equation (4) by L, subtracting equation (5) member by member from the equation obtained, and noting that $C_2 = 0$ we have

$$\frac{1}{3}R_A L^3 - \frac{1}{30}w_o L^4 = 0$$

$$R_A = \frac{1}{10}w_o L$$

We note that the reaction is independent of E and I. Substituting $R_A = \frac{1}{10}w_o L$ into equation (4), we have

$$\frac{1}{2}\left(\frac{1}{10}w_0 L\right)L^2 - \frac{1}{24}w_0 L^3 + C_1 = 0$$

$$C_1 = -\frac{1}{120}w_0 L^3$$

(b) Equation of the elastic curve. Substituting for R_A, C_1 and C_2 into equation (2), we have

$$EIy = \frac{1}{6}\left(\frac{1}{10}w_0 L\right)x^3 - \frac{w_0 x^5}{120L} - \left(\frac{1}{120}w_0 L^3\right)x$$

$$y = \frac{w_0}{120EIL}(-x^5 + 2L^2 x^3 - L^4 x)$$

(c) Slope at A differentiating the above equation with respect to x, we write

$$\theta = \frac{dy}{dx} = \frac{w_0}{120EIL}(-5x^4 + 6L^2 x^2 - L^4)$$

Making $x = 0$, we have $\theta_A = -\dfrac{w_0 L^3}{120EI}$

PROBLEM 13.7

For the beam and loading shown Figure and using singularity functions, (a) express the slope and deflection as functions of the distance x from the support at A, (b) determine the deflection at the midpoint D. Use $E = 200$ GPa and $I = 1.024 \times 10^{-6}$ m^4

Readers should familiarize with singularity function representation of point load and point moments. The given distributed loading is replaced by the two equivalent open-ended loadings shown in Figure and that the following expressions are obtained for the shear and bending moment:

The concentrated loads and moments are taken care by the singular functions which have finite value at the print of application and vanish at other locations. $V(x)$ and $M(x)$ are as under.

$$V(x) = -1.5(x-0.6)^1 + 1.5(x-1.8)^1 + 2.6 - 1.29\,(x-0.6)^0 \tag{1}$$

$$M(x) = -0.75(x-0.6)^2 + 0.75(x-1.8)^2 + 2.6x - 1.2(x-0.6)^1 - 1.44(x-2.6)^0 \tag{2}$$

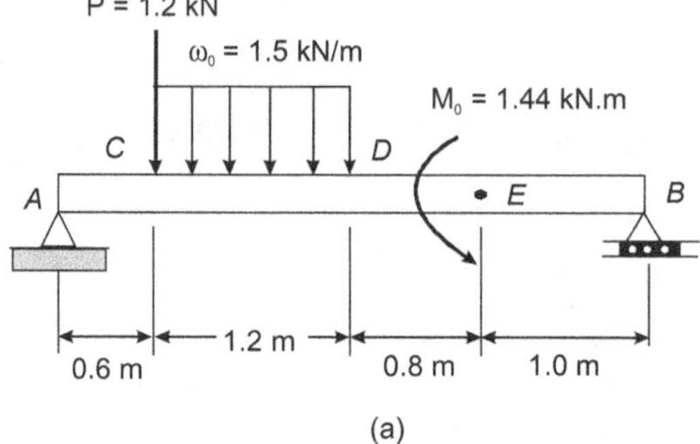

(a)

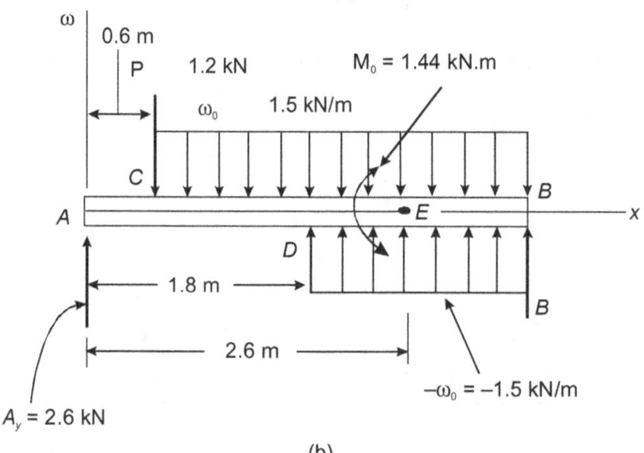

(b)

Integrating the last expression twice, we obtain

$$EI\theta = -0.25(x-0.6)^3 + 0.25(x-1.8)^3 + 1.3x^2 - 0.6(x-0.6)^2 - 1.44(x-2.6)^1 + C_1 \tag{3}$$

$$EIy = -0.0625(x-0.6)^4 + 0.0625(x-1.8)^4 + 0.4333x^3$$
$$- 0.2(x-0.6)^3 - 0.72(x-2.6)^2 + C_1 x + C_2 \tag{4}$$

The constants C_1 and C_2 may be determined from the boundary conditions shown in Figure. Letting $x = 0$, $y = 0$ in equation (8.49) and noting that all the brackets contain negative quantities and, therefore, are equal to zero, we conclude that $C_2 = 0$. Letting now $x = 3.6$, $y = 0$, and $C_2 = 0$ in equation (4), we write.

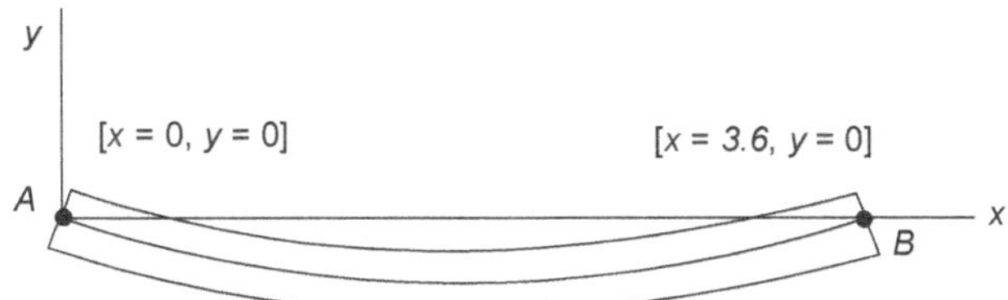

$$0 = -0.0625(3.0)^4 + 0.0625(1.8)^4 + 0.4333(3.6)^3 - 0.2(3.0)^3 - 0.72(1.0)^2 + C_1(3.6) + 0$$

Solving for C_1, we find $C_1 = -2.692$.

(d) Substituting for C_1 and C_2 into equation (4) and making $x = x_D = 1.8$ m, we find that the deflection at point D is defined by the relation

$$EIy_D = -0.0625(1.2)^4 + 0.0625(0)^4 + 0.4333(1.8)^3 - 0.2(1.2)^3 - 0.72(-0.8)^2 - 2.6929 \quad (1.8)$$

We have

$$EIy_D = -0.0625(1.2)^4 + 0.0625(0)^4 + 0.4333(1.8)^3 - 0.2(1.2)^3 - 0 - 2.6929(1.8) = -2.794$$

Recalling the given numerical values of E and I, we write

$$(200 \text{ GPa})(1.024 \times 10^{-6} \text{ m}^4)y_D = -2.794 \text{ kN.m}^3$$

$$y_D = -13.64 \times 10^{-3} \text{ m} = -13.64 \text{ mm}$$

PROBLEM 13.8

For the prismatic beam and loading shown, determine (a) the equation of the elastic curve, (b) the slope at A, (c) the maximum deflection.

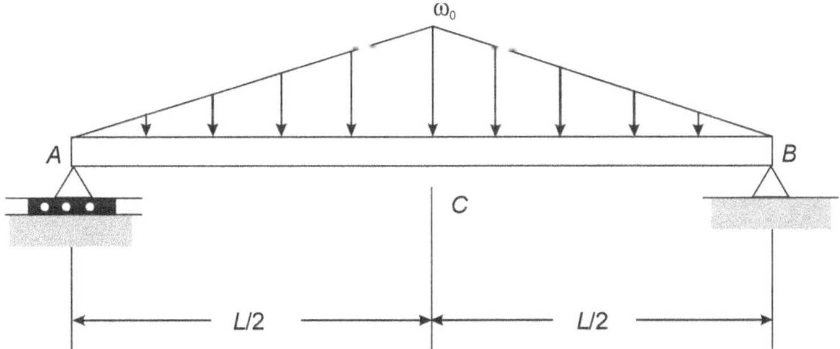

Bending moment   The equation defining the bending moment of the beam was obtained in sample problem. Using the modified loading diagram shown, we have

$$M(x) = -\frac{w_0}{3IL}x^3 + \frac{2w_0}{3L}\left(x - \frac{1}{2}L\right)^3 + \frac{1}{4}w_0 Lx$$

(a) *Equation of the elastic curve*   Using equation (13.4), we write

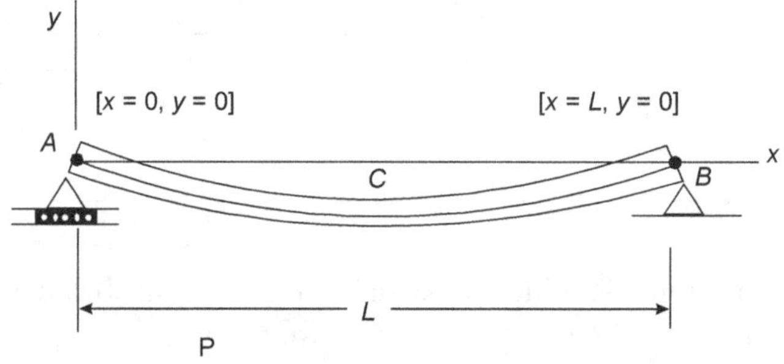

$$EI\frac{d^2y}{dx^2} = \frac{w_0}{3L}x^3 + \frac{2w_0}{3L}\left(x - \frac{1}{2}L\right)^3 + \frac{1}{4}w_0Lx \tag{1}$$

And, integrating twice in x

$$EI\theta = -\frac{w_0}{12L}x^4 + \frac{w_0}{6L}\left(x - \frac{L}{2}\right)^4 + \frac{w_0L}{8}x^2 + C_1 \tag{2}$$

$$EIy = -\frac{w_0}{60L}x^5 + \frac{w_0}{30L}\left(x - \frac{L}{2}\right)^5 + \frac{w_0L}{24}x^3 + C_1x + C_2 \tag{3}$$

Boundary conditions

$[x = 0, y = 0]$: Again using equation (3), we find $C_2 = 0$

$[x = L, y = 0]$: Again using equation (3), we write

$$0 = -\frac{w_0L^4}{60}x^4 + \frac{w_0}{30L}\left(\frac{L}{2}\right)^5 + \frac{w_0L^4}{24} + C_1L$$

$$C_1 = -\frac{5}{192}w_0L^3$$

Substituting C_1 and C_2 into equation (2) and (3), we have

$$EI\theta = -\frac{w_0}{12L}x^4 + \frac{w_0}{6L}\left(x - \frac{1}{2}L\right)^4 + \frac{w_0L}{8}x^2 - \frac{5}{192}w_0L^3 \tag{4}$$

$$EI\,y = -\frac{w_0}{60L}x^5 + \frac{w_0}{30L}\left(x - \frac{1}{2}L\right)^5 + \frac{w_0 L}{24}x^3 - \frac{5}{192}w_0 L^3 x \tag{5}$$

(b) Slope at A. Substituting $x = 0$ into equation (4), we find

$$EI\theta_A = -\frac{5}{192}w_0 L^3$$

$$\theta_A = \frac{5w_0 L^3}{192\,EI}$$

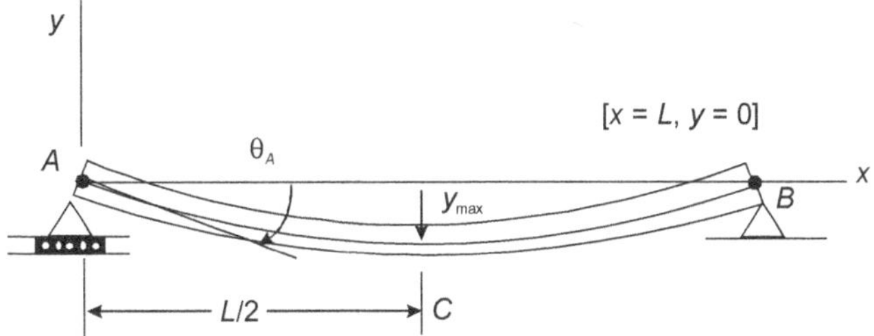

(c) *Maximum deflection*   Because of the symmetry of the supports and loading, the maximum deflection occurs at point C, where $x = L/2$. Substituting into equation (5) we obtain

$$EI\,y_{max} = w_0 L^4\left[-\frac{1}{60(32)} + 0 + \frac{1}{24(8)} - \frac{5}{192(2)}\right] = -\frac{w_0 L^4}{120\,EI} \qquad y_{max} = \frac{w_0 L^4}{120\,EI}$$

PROBLEM 13.9

A beam fixed at both ends supports a uniformly distributed downward load (*see* Figure). The EI for the beam is constant. (a) Find the expression for the elastic curve using the fourth-order governing differential equation. (b) Verify the results found using the second-order differential equation.

SOLUTION

This beam is statically indeterminate to the second degree since horizontal reactions are assumed to be zero. The solution is obtained by four successive integrations. Then the constants of integrations are found from the boundary conditions.

$$EI\frac{d^4 y}{dx^4} = a(x) = -w_0 \tag{1}$$

$$EI\frac{d^3 y}{dx^3} = -w_0 x + C_1 \tag{2}$$

$$EI\frac{d^2 y}{dx^2} = \frac{w_0 x^2}{2} + C_1 x + C_2 \tag{3}$$

$$EI\frac{dy}{dx} = \frac{w_0 x^3}{6} + C_1\frac{x^2}{2} + C_2 x + C_3 \qquad (4)$$

$$EIy = -\frac{w_0 x^4}{24} + C_1\frac{x^3}{6} + C_2\frac{x^2}{2} + C_3 x + C_4 \qquad (5)$$

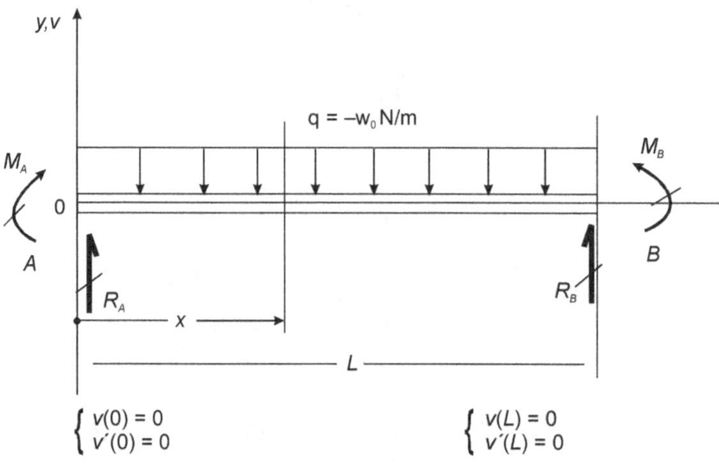

Four kinematic boundary conditions are available for determining the constants of integration

$$EIy(0) = EIy_A = 0 = C_4$$

$$EIy'(0) = EIy'_A = 0 = C_3$$

$$EIy(L) = EIy_B = 0 = -\frac{w_0 L^4}{24} + C_1\frac{L^3}{6} + C_2\frac{L^2}{2}$$

$$EIy'(L) = EIy'_B = 0 = -\frac{w_0 L^3}{6} + C_1\frac{L^2}{2} + C_2 L$$

Constants C_3 and C_4 do not enter the last two equations since they are zero. By solving the last two equations simultaneously

$$C_1 = \frac{w_0 L}{2} \quad \text{and} \quad C_2 = -\frac{w_0 L^2}{12}$$

By substituting these constants into the equation for the elastic curve, after algebraic simplifications

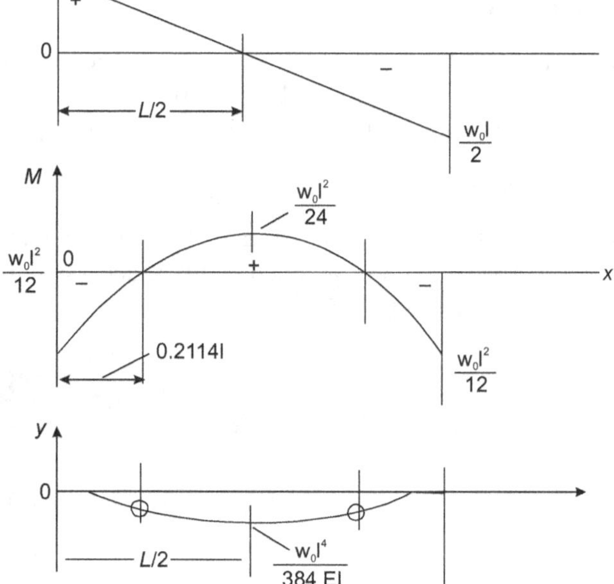

$$y = -\frac{w_0 x^2}{24EI}(L - x)^2 \tag{6}$$

According to equations the basic, EI times the second and third derivatives of the deflection $y(x)$ gives, respectively, $M(x)$ and $V(x)$. At $x = 0$, these relations define the reactions at A. Hence, C_1 is the vertical reaction and C_2 is the moment at this support. In this case, because of symmetry, the vertical reactions can be found directly from statics. However, this is not necessary in this typical solution of a boundary-value problem The moment and shear at B can be found from the same expressions at $x=L$.

Shear, moment, and deflection diagrams for this beam are shown in Figure. The absolute maximum deflection occurring in the middle of the span is

$$y_{max} = \frac{w_0 L^4}{384EI} \tag{7}$$

(b) This solution is found using second-order differential equation (13.4), and although the vertical reaction at A can be determined directly from statics, it will be treated as an unknown. On this basis,

$$EI\frac{d^2 y}{dx^2} = M(x) = M_A + R_A x - \frac{w_0 x^2}{2}$$

Integrating twice,

$$EI\frac{dy}{dx} = M_A x + R_A \frac{x^2}{2} - \frac{w_0 x^3}{6} + C_3$$

$$EIy = M_A \frac{x^2}{2} + R_A \frac{x^3}{6} - \frac{w_0 x^4}{24} + C_3 x + C_4$$

Constants C_3 and C_4 as well as R_A and M_A are found from the four kinematic boundary conditions
$$EIy(0) = EIy_A = 0 = C_4$$
$$EIy'(0) = EIy'_A = 0 = C_3$$
$$EIy(L) = EIy_B = 0 = -M_A \frac{L^2}{2} + R_A \frac{L^3}{6} - \frac{w_0 L^4}{24}$$
$$EIy'(L) = EIy'_B = 0 = -M_A L + R_A \frac{L^2}{2} - \frac{w_0 L^4}{6}$$

Solving the last two equations simultaneously,

$$R_A = \frac{w_0 L}{2} \quad \text{and} \quad M_A = -\frac{w_0 L^2}{12}$$

Substituting these expressions into the equation for deflection with $C_3 = C_4 = 0$, equation (6) is obtained as required. Substituting for R_A and M_A in equation for $M(x)$ and setting $M(x) = 0$, we get the location at which $M_x = 0$ i.e., at $x = 0.2114l$ and $0.7886l$.

PROBLEM 13.10

Determine the equation of the elastic curve for the uniformly loaded continuous beam shown in Figure (a). Use the second-order differential equation. *EI* is constant.

Solution

Because of symmetry, the solution can be confined to determining the deflection for either span. Also, because of symmetry, it can be concluded that at the middle support, not only is the deflection zero, but since the elastic curve cannot rotate in either direction, its slope is also zero. In this manner, the problem can be reduced to the one-degree statically indeterminate problem shown in Figure (b) with known boundary conditions.

We proceed in the same manner as before. Formulate expression for $M(x)$. Use this in the differential equation and perform two successive integrations. Boundary conditions provide the necessary information for determining the constants of integration and unknown reaction R_A.

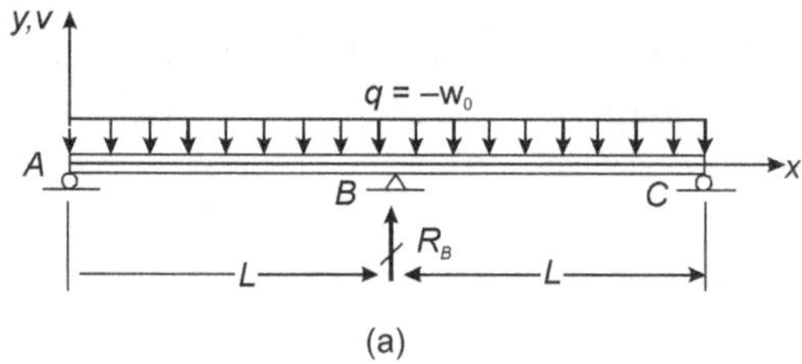

(a)

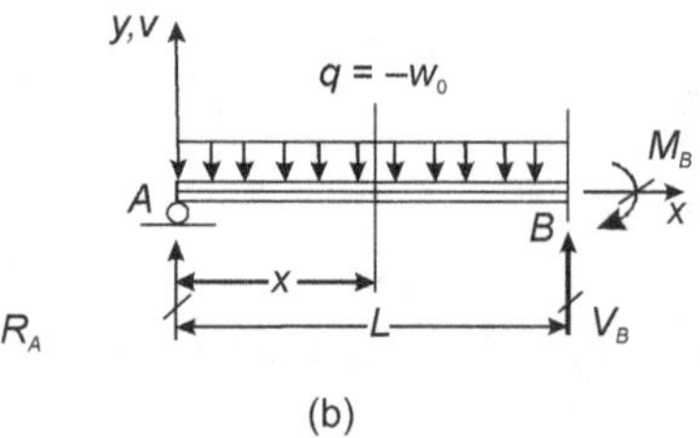

(b)

Second-order differential-equation solution

$$EI\frac{d^2 y}{dx^2} = M(x) = M_A + R_A x - \frac{w_0 x^2}{2} \tag{1}$$

$$EI\frac{dy}{dx} = R_A\frac{x^2}{2} - \frac{w_0 x^3}{6} + C_3 \tag{2}$$

$$EIy = R_A\frac{x^3}{6} - \frac{w_0 x^4}{24} + C_3 x + C_4 \tag{3}$$

Boundary conditions

$$EIy(0) = EIy_A = 0 = C_4$$

$$EIy(L)=EIy'_B = 0 = R_A \frac{L^2}{2} - \frac{w_0 L^3}{6} + C_3 \qquad (4)$$

$$EIy'(L) = EIy'_B = 0 = R_A \frac{L^3}{6} - \frac{w_0 L^4}{24} + C_3 L$$

By solving the last two equations simultaneously,

$$R_A = \frac{3w_0 L}{8} \quad \text{and} \quad C_3 = -\frac{w_0 L^3}{48} \qquad (5)$$

which upon substitution into the equation for the elastic curve, leads to

$$y = -\frac{w_0 x}{48 EI}(L^3 - 3Lx^2 + 2x^3) \qquad (6)$$

From symmetry, the reactions at A and C are equal, and by using statics the reactions at B is

$$R_B = \frac{5w_0 L}{4} \qquad (7)$$

This reaction is also numerically equal to $2V_B$. (V_B refers to reaction at B.(vertical reaction))

THE AREA–MOMENT METHOD

The area-moment method involves the use of a simple technique of finding slopes and deflections due to bending, from the BM diagram. Starting from the basic equation $M/I = E/R$, it relates the area and the moment of area of the BM diagram to the deformations in beams. While the method does not directly give the slope and deflection, it is very useful, particularly when one is interested in finding slopes and deflections at specified points only. It is also useful when the cross-sectional dimensions of the beam vary, giving different values of moment of inertia at different lengths or in a continuously varying pattern.

THE FIRST AREA–MOMENT THEOREM

This theorem states that the angle between the tangents to the elastic curve between any two points is equal to the area of the M/EI diagram between the two points.

As a second aspect of the same beam, consider the two points 1 and 2 of the elastic curve [Figure 13.3]. Draw the tangent at point 1 to the elastic curve and a vertical line through point 2 are drawn. Let the two lines intersect at point 3 as in Figure 13.3(a). Also consider an elemental length subtending an angle $d\theta$ at the centre. The tangents at the two ends of the elemental length intersect the vertical line through 2 at two points, giving a vertical intercept dv. Since the angles involved are very small , the length dv can be considered equal to $x_1 d\theta$ where x_1 is the distance of point 2 from the elemental length ds, along the beam. Therefore, from Figure 13.3(b),

$$dv = x_1 d\theta, \quad dv = x_1 \frac{M}{EI} dx$$

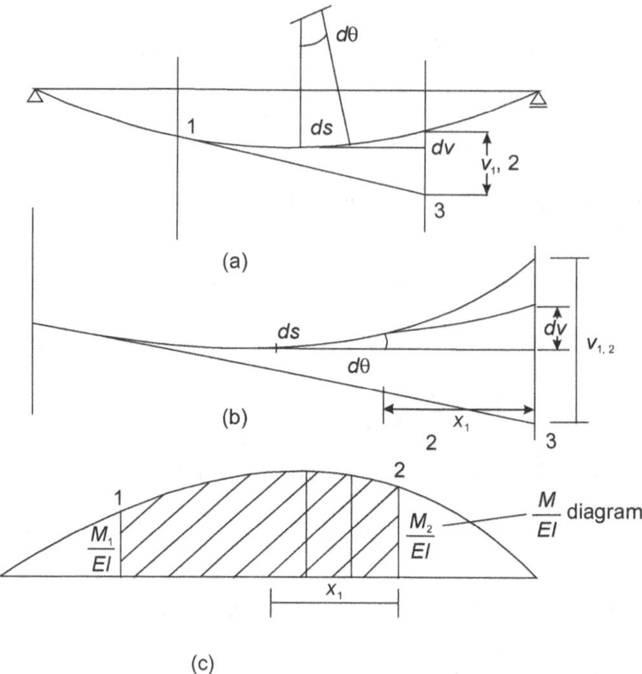

Figure 13.3

(*M/EI*)*dx* being the area of the *M/EI* diagram over the length *dx,* the quantity on the right side of this equation is the moment of the area of the *M/EI* diagram about point 2. If we integrate both these expressions over the length 1–2,

$$\int_1^2 dv = \int_1^2 \frac{M}{EI}\,dx\,x_1$$

The left side integrated over 1–2 gives us the intercept 3–2 while the right side is the moment of the area of the *M/EI* diagram between 1 and 2 about 2. Figures 13.3(b) and (c) show the elemental length and integrated *M/EI* diagrams. This leads us to the second area-moment theorem.

THE SECOND AREA–MOMENT THEOREM

This theorem states that the moment of the area of the M/EI diagram between two points of a beam about one of these points is equal to the vertical intercept made by the tangent drawn at one point on a vertical line through the second point (about which moment is taken). Note that the intercept will be made at the point about which the moment is taken. If we conside $v_{1,2}$ as the intercept made by the tangent at 1 on the vertical through 2, $v_{2,1}$ is the intercept made by the tangent at 2 on the vertical through 1 (Figure). These two intercepts need not be equal.

To summarize, the first area-moment theorem states that the area of the *M/EI* diagram between any two points 1 and 2 (as shown in Figure) is equal to the angle between the tangents drawn to the elastic curve at points 1 and 2. The second area-moment theorem states that the vertical intercept between the tangent drawn at one point and the vertical line drawn through the second point is equal to the moment of the area of the *M/EI* diagram between points 1 and 2 about the second point.

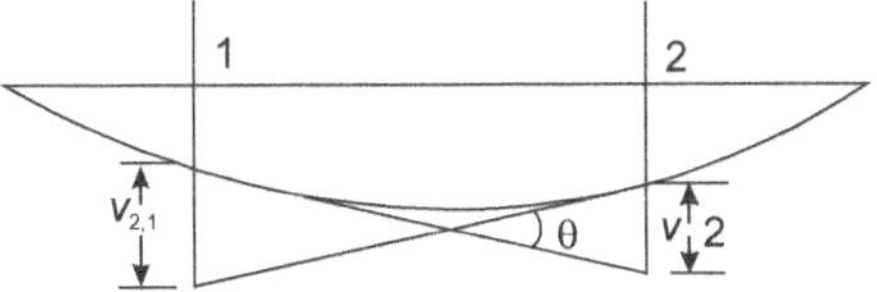

Figure 13.4

The two area-moment theorems help us find the slopes and deflections in beams. As mentioned earlier, the area-moment theorems do not directly give slopes and deflections. But they can be used to find slopes and deflections at specific locations.

PROBLEM 13.11

Find the deflection due to the concentrated force *P* applied as shown in Figure (a) at the centre of a simply supported beam. The flexural rigidity *EI* is constant.

Solution

The bending-moment diagram is in Figure(b). Since *EI* is constant, the *M/EI* diagram need not be made, as the areas of the bending-moment diagram divided by *EI* give the necessary quantities for use in the moment-area theorems. The elastic curve is shown in Figure(c). It is concave upward throughout its length as the bending moments are positive. This curve must pass through the points of the support at *A* and *B*.

It is apparent from the diagram of the elastic curve that the desired quantity is represented by distance CC'. Moreover, from purely geometrical or kinematic considerations $CC' = C'C'' - C'C$, where distance $C''C$ is measured from a tangent to the elastic curve passing through the point of support *B*.

However, since the deviation of a support point from a tangent to the elastic curve at the other support may always be computed by the second moment-area theorem, a distance such as $C'C''$ may be found by proportion from the geometry of the Figure.

In this case, $t_{A/B}$ follows by taking the whole *M/EI* area between *A* and *B* and multiplying it by its x measured from a vertical through *A*; hence $C'C'' = \dfrac{1}{2} t_{A/B}$. By another application of the second theorem, $t_{C/B}$ which is equal to $C''C$, is determined. For this case, the *M/EI* area is hatched in Figure(b), and , for it, x is measured from *C*. Since the right reaction is *P/4* and the distance $CB = 2a$ the maximum ordinate for the shaded triangles is $+ Pa/2$.

$$vc = C'C'' - C''C = t_{A/B}/2 - t_{C/B}$$

$$t_{A/B} = \phi_1 x_1 = \frac{1}{EI}\left(\frac{4a}{2}\frac{3Pa}{4}\right)\frac{a+4a}{3} = +\frac{5Pa^3}{2EI}$$

$$t_{C/B} = \phi_2 x_2 = \frac{1}{EI}\left(\frac{2a}{2}\frac{Pa}{2}\right)\frac{2a}{3} = +\frac{Pa^3}{3EI}$$

$$v_C = \frac{t_{A/B}}{2} - t_{C/B} = \frac{5Pa^3}{4EI} - \frac{Pa^3}{3EI} = \frac{11Pa^3}{12EI}$$

The positive signs of $t_{A/B}$ and $t_{C/B}$ indicate the points A and C lie above the tangent through B. As may be seen from Figure C, the deflection at the centre of the beam is in a downward direction.

The slope of the elastic curve at C can be found from the slope of one of the ends. For point B on the right,

$$\theta_B = \theta_c + \Delta\theta_{B/C} \qquad \text{or} \qquad \theta_C = \theta_B - \Delta\theta_{B/C}$$

$$\theta_C = \frac{t_{A/B}}{L} - \phi_2 = \frac{5Pa^2}{8EI} - \frac{Pa^2}{2EI} = \frac{Pa^2}{8EI} \qquad \text{(counterclockwise)}$$

The previous procedure for finding the deflection of a point on the elastic curve is generally applicable. For example, if the deflection of point E, Figure (d), at a distance e from B is desired, the solution may be formulated as

$$v_E = E'E'' - E'E = (e/L)t_{A/B} - t_{E/B}$$

By locating point E, at a variable distance x from one of the supports, the equation of the elastic curve can be obtained.

To simplify the arithmetical work, some care in selecting the tangent at a support must be exercised. Thus, although $v_C = t_{B/A}/2 - t_{C/A}$ (not shown in the diagram), this solution would involve the use of the unshaded portion of the bending-moment diagram to obtain $t_{C/A}$ which is more tedious.

PROBELM 13.12

For a prismatic beam loaded as in the preceding example, find the maximum deflection caused by applied force P; see Figure (a).

Solution

The bending-moment diagram and the elastic curve are shown in Figure (b) and (c), respectively. The elastic curve is concave up throughout its length and the maximum deflection occurs where the tangent to the elastic curve is horizontal. This point of tangency is designated in the figure by D and is located by the unknown horizontal distance d measured from the right support B. Then, by drawing a tangent to the elastic curve through the point B at the support, one see that $\Delta\theta_{B/D} = \theta_B$ since the line passing through the supports is horizontal. However, the slope θ_B of the elastic curve at B may be determined by $t_{A/B}$ and dividing it by the length of the span. On the other hand, by using the first moment-area theorem, $\Delta\theta_{B/D}$ may be expressed in terms of the shaded area in Figure (b). Equating $\Delta\theta_{B/D}$ to θ_B and solving for d locates the horizontal tangent at D. Then, again from geometrical considerations, it is seen that the maximum deflection represented by DD' is equal to the tangential deviation of B from a horizontal tangent through D.(i.e., $t_{B/D}$)

$$t_{A/B} = \phi_1 x_1 = +\frac{5Pa^3}{2EI} \qquad\qquad (see\ \text{problem}\ 13.4)$$

$$\theta_B = \frac{t_{A/B}}{L} = \frac{t_{A/B}}{4a} = \frac{5Pa^2}{8EI}$$

$$\Delta\theta_{B/D} = \frac{1}{EI}\left(\frac{d}{2}\frac{Pd}{4}\right) = \frac{Pd^2}{8EI} \qquad\qquad (\text{area between } D \text{ and } B)$$

since

$$\theta_B = \theta_D + \Delta\theta_{B/D} \text{ and it is required that } \theta_D = 0$$

$$\Delta\theta_{B/D} = \theta_B \qquad \frac{Pd^2}{8EI} = \frac{5Pa^2}{8EI} \text{ hence,} d = \sqrt{5}a$$

$$v_{\max} = v_D = DD' = t_{B/D} = \phi_3 x_3$$

$$= \frac{1}{EI}\left(\frac{d}{2}\frac{Pd}{4}\right)\frac{2d}{3} = \frac{5\sqrt{5}Pa^3}{12EI} = \frac{11.2Pa^3}{12EI}$$

After distance d is found, the maximum deflection may also be obtained as $v_{\max} = v_{A/D}$ or $v_{\max} = v = (d/L)t_{A/D} - t_{D/B}$ (not shown). Also note that using the condition $t_{A/D} = t_{B/D}$, Figure (d), an equation may be set up for d.

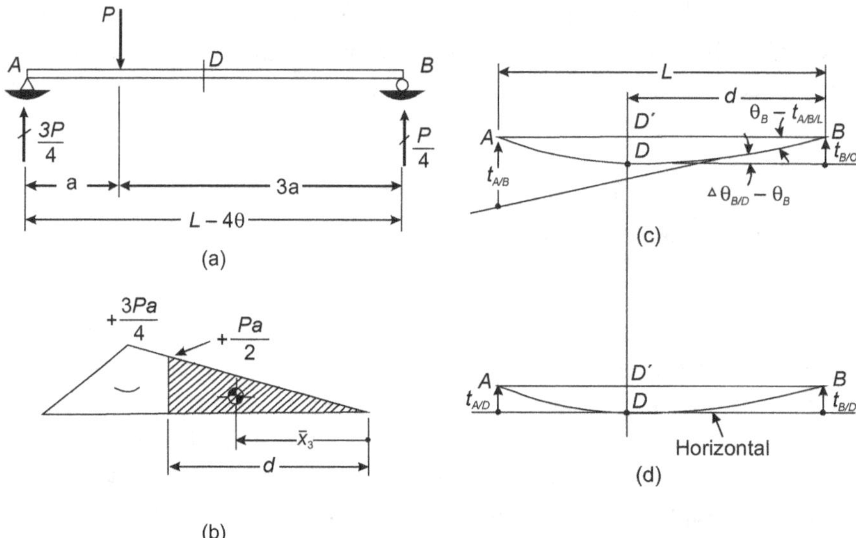

It should be apparent from this solution that it is easier to calculate the deflection at the centre of the beam, which was illustrated in problem, than to determine the maximum deflections. Yet, by examining the end results, one sees that, numerically, the two deflections differ little $v_{centre} = \dfrac{11Pa^3}{12EI}$ as

opposed to $v_{max} = \dfrac{11.2Pa^3}{12EI}$. For this reason, in many practical problems of simply supported beams,

where all the applied forces act in the same direction, it is often sufficiently accurate to calculate deflection at the centre instead of attempting to obtain the true maximum.

PROBLEM 13.13

The prismatic rods AD and DB are welded together to form the cantilever beam ADB. Knowing that the flexural rigidity is EI in portion AD of the beam and $2EI$ in portion DB, determine, for the loading shown, the slope and deflection at end A.

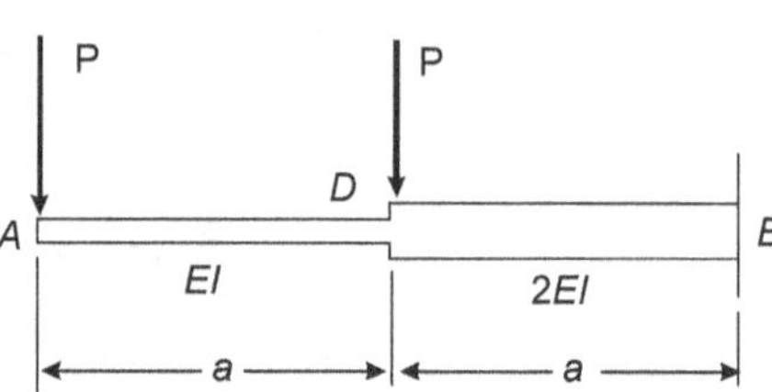

(M/EI) diagram. We first draw the bending-moment diagram for the beam and then obtain the *(M/EI)* diagram by dividing the value of M at each point of the beam by the corresponding value of the flexural rigidity.

Reference tangent   We choose the horizontal tangent at the fixed end B as the reference tangent. Since $\theta_B = 0$ and $y_B = 0$

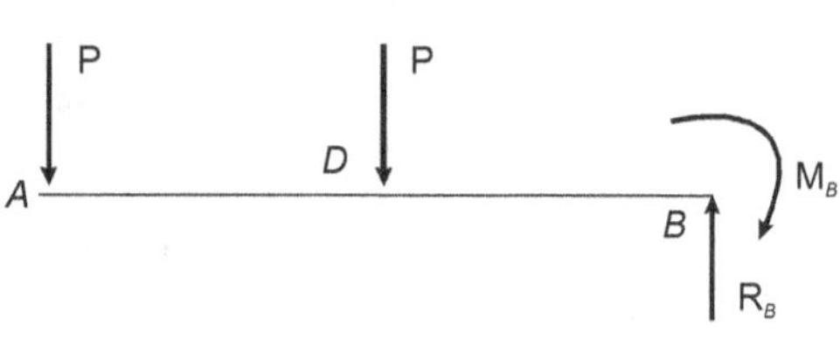

$$\theta_A = -\theta_{B/A}$$

$$y_A = t_{A/B}$$

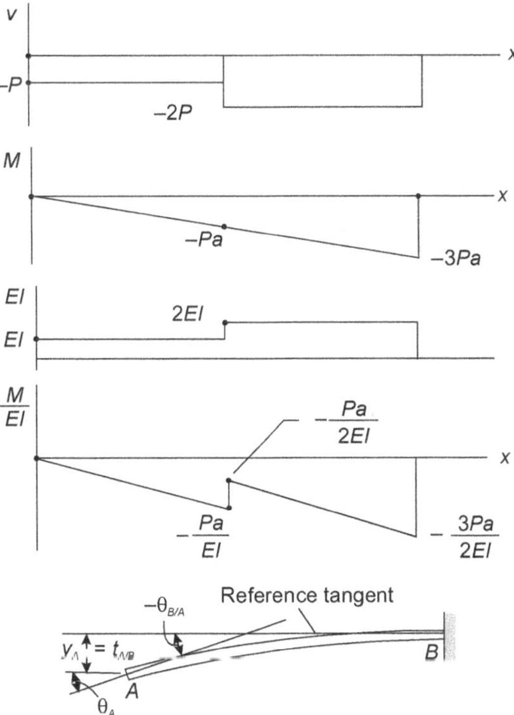

Slope at A   Dividing the (M/EI) diagram into the three triangular portions shown, we write

$$A_1 = -\frac{1}{2}\frac{Pa}{EI}a = -\frac{Pa^2}{2EI}$$

$$A_2 = -\frac{1}{2}\frac{Pa}{2EI}a = -\frac{Pa^2}{4EI}$$

$$A_3 = -\frac{1}{2}\frac{3Pa}{2EI}a = -\frac{3Pa^2}{4EI}$$

Using the first moment-area theorem, we have

$$\theta_{B/A} = A_1 + A_2 + A_3 = -\frac{Pa^2}{2EI} - \frac{Pa^2}{4EI} - \frac{3Pa^2}{4EI} = -\frac{3Pa^2}{2EI}$$

$$\theta_A = -\theta_{B/A} = +\frac{3Pa^2}{2EI}$$

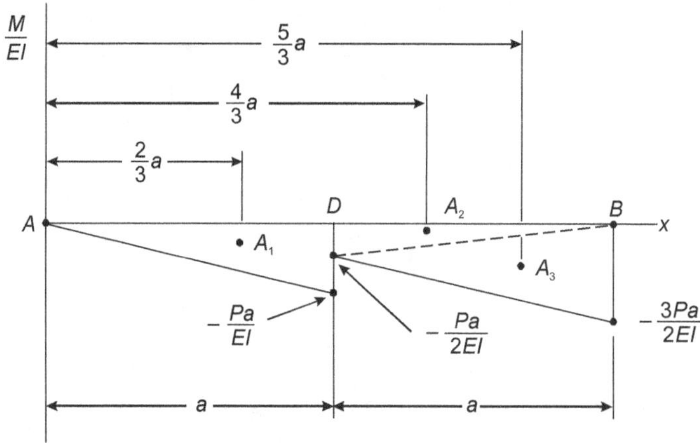

Deflection at A   Using the second moment–area theorem, we get

$$y_A = t_{A/B} = A_1\left(\frac{2}{3}a\right) + A_2\left(\frac{4}{3}a\right) + A_3\left(\frac{5}{3}a\right)$$

$$= \left(-\frac{Pa^2}{2EI}\right)\frac{2a}{3} + \left(-\frac{Pa^2}{4EI}\right)\frac{4a}{3} + \left(\frac{3Pa^2}{4EI}\right)\frac{5a}{3}$$

$$y_A = -\frac{23Pa^3}{12EI}$$

PROBLEM 13.14

For the prismatic overhung beam and loading in the overhung portion as shown, determine the slope and deflection at end *E*. Take *EI* = constant.

(M/EI) diagram   From a free-body diagram of the beam, we determine the reactions and then draw the shear and bending-moment diagrams. Since the flexural rigidity of the beam is constant, we divide each value of *M* by *EI* and obtain the (*M/EI*) diagram shown.

Reference tangent   Since the beam and its loading are symmetric with respect to the midpoint *C*, the tangent at *C* is horizontal and is used as the reference tangent. Referring to the sketch, we observe that, since $\theta_c = 0$,

$$\theta_E = \theta_C + \theta_{E/C} = \theta_{E/C} \tag{1}$$

$$y_E = t_{E/c} - t_{D/C} \tag{2}$$

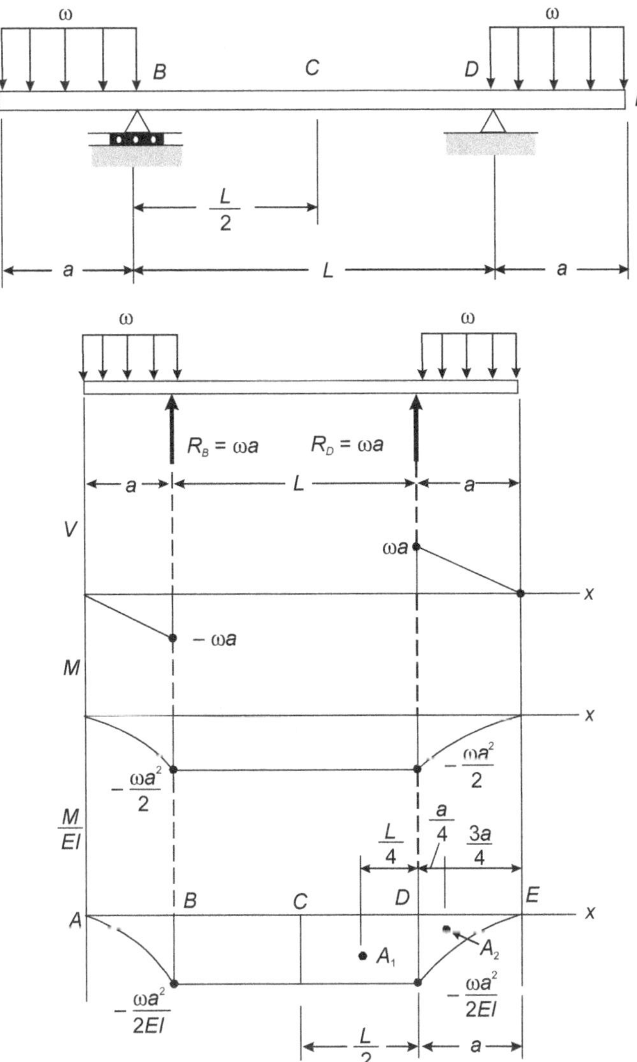

Slope at E Referring to the (M/EI) diagram and using the first moment–area theorem, we write

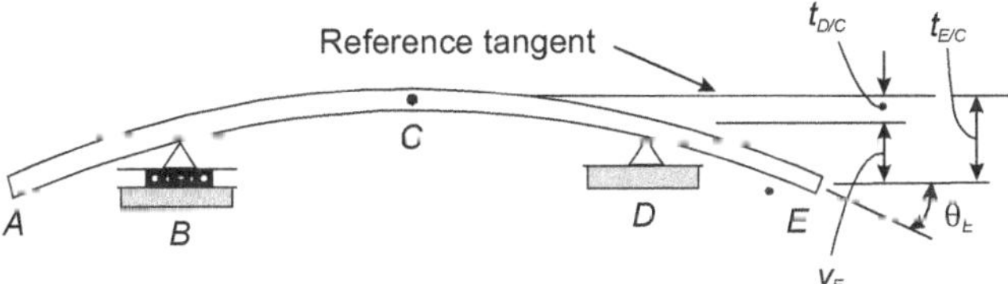

$$A_1 = -\frac{\omega a^2}{2EI}\left(\frac{L}{2}\right) = -\frac{\omega a^2 L}{4EI}$$

$$A_2 = -\frac{1}{3}\left(\frac{\omega a^2}{2EI}\right)(a) = -\frac{\omega a^3}{6EI}$$

Using equation (1), we have

$$\theta_E = \theta_{E/C} = A_1 + A_2 = -\frac{\omega a^2 L}{4EI} - \frac{\omega a^3}{6EI}$$

$$\theta_E = -\frac{\omega a^2}{12EI}(3L + 2a)$$

Deflection at E   Using the second moment–area theorem, we write

$$t_{D/C} = A_1\frac{L}{4} = \left(-\frac{\omega a^2 L}{4EI}\right)\frac{L}{4} = -\frac{\omega a^2 L^2}{16EI}$$

$$t_{E/C} = A_1\left(a + \frac{L}{4}\right) + A_2\left(\frac{3a}{4}\right)$$

$$= \left(-\frac{\omega a^2 L}{4EI}\right)\left(a + \frac{L}{4}\right) + \left(\frac{\omega a^3}{6EI}\right)\left(\frac{3a}{4}\right)$$

$$= -\frac{\omega a^3 L}{4EI} - \frac{\omega a^2 L^2}{16EI} - \frac{\omega a^4}{8EI}$$

Using equation (2), we have

$$y_E = t_{E/C} - t_{D/C} = -\frac{\omega a^3 L}{4EI} - \frac{\omega a^4}{8EI}$$

$$y_E = -\frac{\omega a^3}{8EI}(2L + a)$$

$$y_E = \frac{\omega a^3}{8EI}(2L + a)\downarrow$$

REVIEW QUESTIONS

SHORT QUESTIONS

1. Mention any two methods of finding the slope and deflection of beams.

2. Write down the equations for maximum deflection of a simply supported beam loaded with a central point load.

3. Write the relationship between intensity of load, shear force, bending moment, slope and deflection in a beam.

4. Give the expression for strain energy due to bending of a cantilever beam carrying a point load at free end.

5. State Area Moment theorem.

6. Give the relations between curvature, bending moment, shear force, slope, deflection etc., at a section.

7. Write the expression for the deflection at the free end of a cantilever of length 'L' carries an UDL of w kN/m. Assume uniform cross section throughout.

8. A rectangular R.C simply supported beam of span 3 m and cross section 200 mm × 350 mm carries a point load of 100 KN at its mid span. Find the maximum slope and deflection of the beam if $E = 0.2 \times 10^5$ N/mm^2.

9. State Castigliano's theorem for the deflection of beams.

10. List any four methods of determining slope and deflection of loaded beam.

11. A simply supported beam of length 4 meters carries a uniformly distributed load of 15 kN/m throughout its length. Determine the maximum deflection and slope in the beam. Take flexural rigidity EI = 25000 kNm^2.

12. Write down Maxwells reciprocal Theorem.

13. Give the expressions for strain energy in bending and torsion.

14. What is the use of conjugate beam method over other methods?

LARGE QUESTIONS

1. A cantilever of length 3 m carries two point loads of 2 kN at free end and 4 kN at a distance of 1 m from the free end. Find the deflection at free end using area moment method. Take $E = 2 \times 10^5$ N/mm^2 and $I = 10^8$ mm^4.

2. A beam of length 6 m is simply supported at the ends and carries two point loads of 48 kN at a distance of 1 m and 3 m respectively from the left support. Compute the slope and deflection under each load. Assume EI = 17000 kN–m^2.

3. A steel girder of uniform section, 14 meters long is simply supported at its ends. It carries concentrated loads of 90 kN and 60 kN at two points, 3 metre and 4.5 metre from the two ends respectively. Calculate the deflection of the girder at the points under the two loads and find the maximum deflection. Take $I = 64 \times 10^{-4}$ m^4 and $E = 210 \times 10^6$ kN / m^2.

4. A simply supported beam AB of span 4 m, carrying a load of 100 kN at its mid span C has cross sectional moment of inertia 24×10^6 mm^4 over the left half of the span and 48×10^6 mm^4 over the right half. Find the slope at two supports and the deflection under the load. Take E = 200 GPa.

5. A cantilever AB, 2 m long, is carrying a load of 20 kN at free end and 30 kN at a distance 1 m from the free end. Find the slope and deflection at the free end. Take E = 200 GPa and I = 150×10^6 mm^4.

6. i.   Derive the expression for slope and deflection of a simply supported beam of length '*l*' subjected to a concentrated load of 'W; at distances od ;a' and 'b' from the two ends.

    ii. A beam simply supported at its ends over an 8 m span is loaded with 40 kN, 80 kN and 120 kN at 2, 4 and 6 m respectively from one end. The maximum stress is 90 N/mm^2 and the beam is 300 mm deep. If E = 200 GPa, find the maximum deflection and state where it occurs.

7. A simply supported beam of length 4m carries point loads of KN each at a distance of 1m from each end. If $E = 2 \times 10^5$ N/mm^2 and $I = 10^8$ mm^4 for a beam, then using conjugate beam method, determine:

    i.   Slope at each end and under each load.

    ii. Deflection under each load and at the centre.

14

BENDING OF CURVED BARS

During the discussion on pure bending of beams, beams which originally are straight, i.e., without any initial curvature were considered. The formula derived can still be applied where the radius of curvature is very large as in the case of rail tracks.

However, when we consider components like chain links, crane hooks or rings bend in pipes, the radius of curvature of the members are very small (radius of curvature small and comparable with other dimensions of the component such as depth or width of cross section). Bars considered are curved in the vertical plane carrying the loads (loading plane). However if the bars are curved in plan view, they are subjected to turishing moments in addition to bending moments. This chapter deals primarily with curved bars bent in its own plane.

BENDING IN THE PLANE OF THE CURVE

A straight beam having either a constant cross section or a cross section which changes gradually along the length of the beam, the neutral surface is defined as the longitudinal surface of zero fibre stress when the member is subjected to pure bending. It contains the neutral axis of every section, and these neutral axes pass through the centroids of the respective sections. In this article on bending in the plane of the curve, the use of the many formulas is restricted to those members for which that axis passing through the centroid of a given section and selected normal to the plane of bending of the member is a principal axis. The one exception to this requirement is for a condition equivalent to the beam being constrained to remain in its original plane of curvature such as by frictionless external guides.

To determine the stresses and deformations in curved beams satisfying the restrictions given above, one first identifies several cross sections and then locates the centroids of each. From these centroidal the curved centroidal surface can be defined. For bending in the plane of the curve there will be at each section (1) a force N normal of the cross section and taken to act through the centroid, (2) a shear force V parallel to the cross section in a radial direction, and (3) a sending couple M in the plane of the curve. In addition there will be radial stresses in the curved beam to establish equilibrium. These internal loadings are shown in Figure 14.1a, and the stresses and deformations due to each will be evaluated.

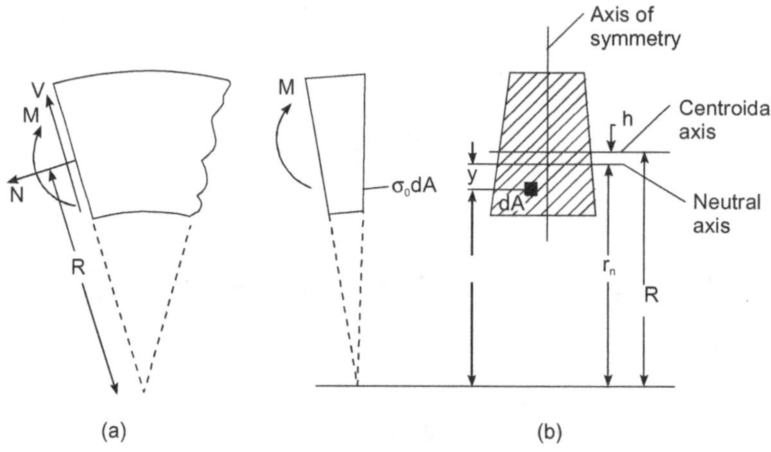

Figure 14.1

Circumferential normal stresses due to pure bending   When a curved beam bent in the plane of initial curvature, plane sections remain plane, because of the different lengths of fibres on the inner and outer portions of the beam, the distribution of unit strain, and therefore stress, is not linear. The neutral axis does not pass through the centroid of the section. The error involved in their use is slight as long as the radius of curvature is more than about eight times the depth of the beam. At that curvature the errors in the maximum stresses are in the range of 4 to 5%. The errors created by using the straight-beam formulae become large for sharp curvatures as shown in Table 14.1 which gives formulas and selected numerical data for curved beams of several cross sections and for varying degrees of curvature.

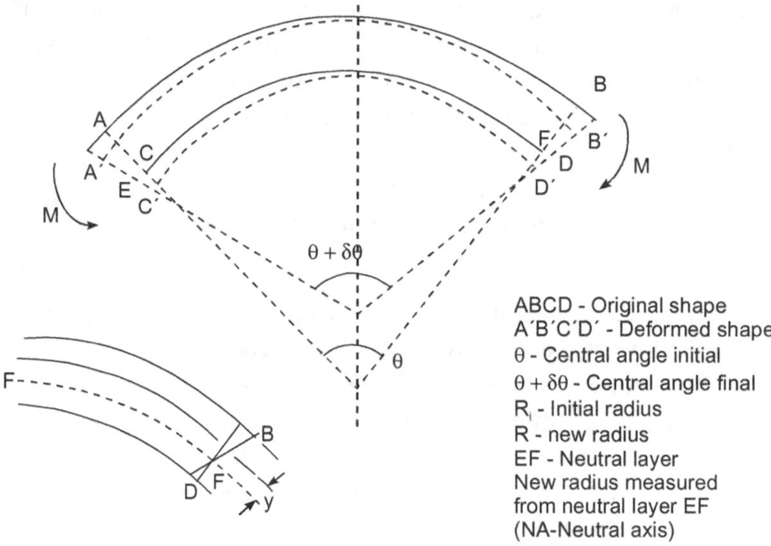

Figure 14.2

We consider a fibre at a distance y from the NA. Original length of this fibre $= (R_i + y)\theta$

Increased length of the fibre $= (R + y)(\theta + \delta\theta)$

$$\varepsilon = \frac{(R + y)(\theta + \delta\theta) - (R_i + y)\theta}{(R_i + y)\theta}$$

$$\varepsilon = \frac{R(\theta + \delta\theta) + y\delta\theta - R_i\theta}{(R_i + y)\theta}$$

From $R_i\theta = R(\theta + \delta\theta)$, we get

$$\frac{R_i - R}{R} = \frac{\delta\theta}{\theta}$$

From these, we have

$$\varepsilon = \frac{y}{R_i + y}\frac{(R_i - R)}{R}$$

In case we consider y to be small compared to R_i we can say that

$$\varepsilon = \frac{y}{R_i}\left(\frac{R_i - R}{R}\right) = y\left(\frac{1}{R} - \frac{1}{R_i}\right)$$

Knowing the value of strain, the stress

$$\sigma = E\varepsilon = Ey\left(\frac{1}{R} - \frac{1}{R_i}\right)$$

Then, from the simple theory of bending from which $\sigma/y = M/I$ we can state that

$$\frac{\sigma}{y} = \frac{M}{I} = E\left(\frac{1}{R} - \frac{1}{R_i}\right) \tag{14.1}$$

In this derivation we have assumed that y is small compared to R_i and the theory of simple bending is considered applicable.

In case R_i is small, as in the case of hooks and rings, we cannot make these assumptions.

BARS WITH LARGE CURVATURE

Bars with large curvature are analysed using the theory developed by Winkler and Bach. The following assumptions are made.

    i. Transverse sections remain plane before and after bending.

ii. Longitudinal fibres bend independently.

iii. Stresses in the beam are within the proportional limit.

iv. The beam section has an axis of symmetry and the loads lie in a plane containing this axis.

v. E is same in tension and compression

Winkler–Bach Formula

We consider the curved beam shown in Figure 14.3 It is subjected to pure bending by couples applied at the ends. The transverse section is symmetric with respect to the Y-axis. O is the centre of curvature initially before bending. AD and BC are transverse sections subtending angle θ at O.

The applied couple is considered positive if it decreases the radius of curvature of the bar and is the radius of the neutral layer. Neutral layer is the layer with no elongation or shortening. Also R is the curvature of the neutral layer after bending and R_y is the radius of a layer at y from the NA.

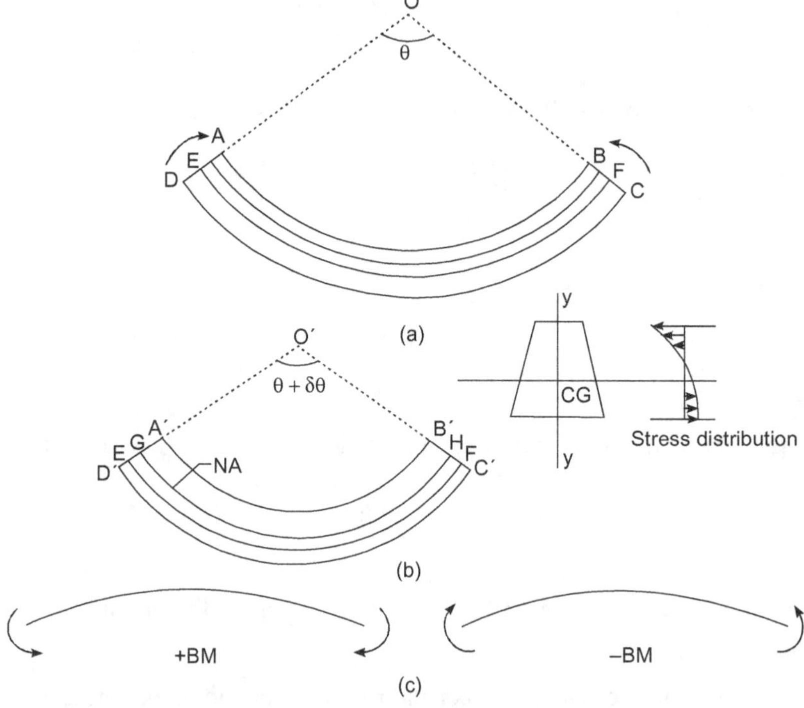

Figure 14.3

On the application of a couple M, the radius of curvature decreases and the new centre of curvature is O' [Figure 14.1(b)]. All the longitudinal fibres rotate in circular arcs. From the nature of rotation of the fibres, it is evident that the outer fibres have increased in length and the inner fibres have decreased in length. There must exist a fibre which has the same length as before. This fibre is represented as EF and is the neutral fibre, and the corresponding line in the cross section is the NA. The initial radius R_i and the new radius of curvature R are referred to this line. We have, therefore, since the length of *EF* remains the same, $R_i\theta = R(\theta + \delta\theta)$.

We now consider a fibre *GH* located y above the neutral surface. Length *GH* decreases to G′H′ after bending. We have, for the deformation of *GH*,

$$G'H - GH = (R - y)(\theta - \delta\theta) - (R_i - y)\theta = -y\delta\theta \tag{14.2}$$

This relation is an indication that the strains do not vary linearly with y. From Hooke's law,

$$\sigma = E\varepsilon = E\frac{\delta\theta}{\theta}\frac{y}{(R_i - y)} \tag{14.3}$$

The stresses like σ, ε, do not vary linearly with y. The stress variation can be seen to be hyperbolic by plotting σ against y. We have now two conditions in the case of pure bending.

i.   The sum of the forces given by $\Sigma\sigma \times \delta A = 0$, and

ii. The sum of the moment of the forces about the NA (equal to the external couple applied), $\Sigma\sigma \times \delta Ay$, must be equal to the moment M, That is

$$-\int_A \sigma\, dA = 0$$

$$\text{and } \int_A (-y\sigma\, dA) = M \tag{14.4}$$

Substituting for σ, the first of these equations gives

$$-\int E\frac{\delta\theta}{\theta}\frac{y}{(R_i - y)}\, dA = 0$$

$$\text{or } \int \frac{y}{(R_i - y)}\, dA = 0$$

If R_y is the radius of the fibre *GH*, then

$$R_y = R - y$$

and

$$\int \frac{R - R_y}{R_y}\, dA = 0$$

or

$$R\int \frac{dA}{R_y} \int -dA = 0 \tag{14.5}$$

This yields

$$R = \frac{\int dA}{\int dA / R_y}$$

This relation, being different from $R = (1/A)\int R_y dA$ which would have been the distance to the centroid, shows that the neutral surface does not contain the centroid of the section. Expressions for R for the neutral surface depend upon the cross-sectional shape, and can be derived from the above relationship.

Substituting for σ in the second condition,

$$\int \sigma \, dAy = M$$

$$\int E \frac{\delta\theta}{\theta} \frac{y}{(R_i - y)} y \, dA = M$$

and in terms of the radius R_y of the fibre *GH,*

$$\int E \frac{\delta\theta}{\theta} \frac{(R - R_y)}{(R_y - R_y)} (R - R_y) dA = M$$

$$\frac{E\delta\theta}{\theta}\left[\int\left(\frac{R^2 + R^2_{\,y} - 2RR_y}{R_y}\right) dA\right] = M$$

$$\frac{E\delta\theta}{\theta}\left(R^2 \int \frac{dA}{R_y} + \int R_y dA - 2R dA\right) = M \tag{14.6}$$

We have seen earlier that

$$R = \frac{A}{\int dA / R_y}$$

$$\int \frac{dA}{R_y} = \frac{A}{R} \tag{14.7}$$

Therefore,

$$\int R^2 \frac{dA}{R_y} = RA$$

and

$$\int R_y dA = \bar{R}A$$

where R is the radius of the centroid of the section. We thus have

$$E\frac{\delta\theta}{\theta}(RA - 2RA - \overline{R}A) = M$$

and

$$E\frac{\delta\theta}{\theta} = \frac{M}{A(\overline{R} - R)}$$

For $M > 0, \delta\theta > 0$ and $\overline{R} - R > 0$ or $R < \overline{R}$. This shows that the neutral surface lies above the centroidal line and towards the centre of curvature. Setting $\overline{R} - R = S$, we get

$$E\frac{\delta\theta}{\theta} = \frac{M}{AS}$$

where S is the shift of the neutral surface from the centroidal plane.

The bending stress at y can now be expressed as

$$\sigma = \frac{My}{AS(R_i - y)} = \frac{M(R - R_y)}{ASR_y} \tag{14.8}$$

This is the Winkler–Bach formula for stress distribution in a curved bar.

In this formula, σ is the stress at a distance y from the neutral surface, A is the area of cross section, S is the shift of neutral surface from the centroidal layer, R_i is the initial radius, and R_y is the radius of the layer at which stress is being found. Also R is the radius of the neutral surface. The term y is measured from the neutral layer.

SIGN CONVENTION

Bending moment is considered positive when it tends to reduce the radius of curvature. Positive and negative bending moments will be as shown in Figure 14.3(c).

Plotting σ against y, we obtain an arc of a hyperbola, which is the stress distribution (Figure 14.3). In case of straight beam, the distribution is linear.

Let us find the curvature R of the bent beam

$$\frac{1}{R} = \frac{1}{R_i}\left(1 + \frac{\delta\theta}{\theta}\right)$$

$$= \frac{1}{R_i}\left(1 + \frac{M}{EAS}\right)$$

$$\frac{1}{R} - \frac{1}{R_i} = \frac{M}{EASR_i} \tag{14.9}$$

is the change in curvature.

LOCATION OF THE NEUTRAL SURFACE

R is the radial distance of the neutral surface from the centre of curvature, and can be derived for different cross-sectional shapes.

$$R = \frac{A}{\int dA / r}$$

For different cross-sectional shapes, the expressions for value of R are given in Table 14.1. For example, in the case of a rectangular section of width b and depth h,

$$\text{Area } A = bh$$

From Figure 14.4(a)

$$\int_{r_1}^{r_2} \frac{dA}{r} = \int_{r_1}^{r_2} \frac{bdr}{r} = b[\log^r{}_e]_{r_1}^{r_2} = b\log e_e\left(\frac{r_2}{r_1}\right)$$

$$R = \frac{A}{\int dA / r} = \frac{bh}{b\log_e(r_2/r_1)} = \frac{h}{\log_e(r_2/r_1)} \qquad (14.10)$$

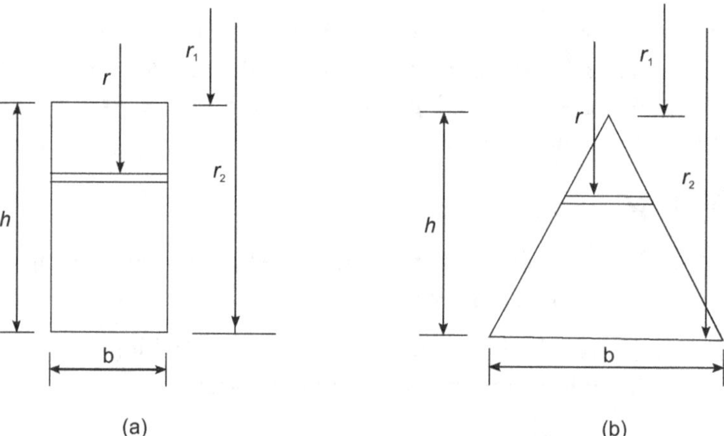

Figure 14.4

At this juncture, it may be metioned that theory of elasticity solution for a curved bar of narrow rectangular section bent in the plane of curvature by couple M applied at the end using stress function approach gives

$$\sigma_r = -\frac{4M}{N}\left(\frac{a^2 b^2}{r^2}\log\frac{b}{a} + b^2\log\frac{r}{b} + a^2\log\frac{a}{r}\right)$$

$$\sigma_\theta = -\frac{4M}{N}\left(\frac{-a^2 b^2}{r^2}\log\frac{b}{a} + b^2\log\frac{r}{b} + b^2 - a^2\right)$$

$$\tau_{r\theta} = 0 \qquad \text{where } N = \left(b^2 - a^2\right)^2 - 4a^2 b^2\left(\log\frac{b}{a}\right)^2$$

which are exact solution to the problem.

By comparing the above exact solution with the cases of linear stress distribution, Hyperbolic stress distribution (pressure treatment of curved bars), the letter formula is quite accurate (error $< 1\%$). This shows the closeness of the present approach to the theory of elasticity approach.

Similarly, in the case of a triangle of base b and depth h, from Figure 14.4(b),

$$\text{Area} = \frac{bh}{2}$$

$$\int_{r_1}^{r_2}\frac{dA}{r} - \int_{r_1}^{r_2}\frac{(b/h)(r_2 - r)dr}{r} = \frac{b}{h}\int_{r_1}^{r_2}\left(\frac{r_2}{r} - 1\right)dr$$

$$= \frac{b}{h}[r_2\log_e(r_2/r_1) - (r_2 - r_1)]$$

$$= \frac{b}{h}[r_2\log_e(r_2/r_1) - h]$$

$$R = \frac{A}{\int dA/r} = \frac{bh}{2b\left[(r_2/h)\log_e\left(r_2/r_1\right) - 1\right]} = \frac{h}{2\left[(r_2/h)\log_e\left(r_2/r_1\right) - 1\right]} \qquad (14.11)$$

TABLE 14.1 Formulas for curved beams subjected to bending in the plane of the curve Ref[1]

Form and dimensions of cross section, reference no.	Formulas	Values of $\dfrac{h}{c}$, k_i, k_0 and for various values of $\dfrac{R}{c}$
1. Solid rectangular section	$\dfrac{h}{c} = \dfrac{R}{c} - \dfrac{2}{ln\dfrac{R/c+1}{R/c-1}}$ (Note : $h/c, k_i,$ and k_o are independent of the width b) $k_i = \dfrac{1}{3h/c}\dfrac{1-h/c}{R/c-1}$ $\displaystyle\int_{area} \dfrac{dA}{r} b = \ln\dfrac{R/c+1}{R/c-1}$ $k_o = \dfrac{1}{3h/c}\dfrac{1+h/c}{R/c+1}$	$\dfrac{R}{c} = 1.20 \quad 1.40 \quad 1.60 \quad 1.80 \quad 2.00 \quad 3.00 \quad 4.00 \quad 6.00 \quad 8.00 \quad 10.00$ $\dfrac{h}{c} = 0.366 \; 0.284 \; 0.236 \; 0.204 \; 0.180 \; 0.115 \; 0.085 \; 0.056 \; 0.042 \; 0.033$ $k_i = 2.888 \; 2.103 \; 1.798 \; 1.631 \; 1.523 \; 1.288 \; 1.200 \; 1.124 \; 0.090 \; 1.071$ $k_o = 0.566 \; 0.628 \; 0.671 \; 0.704 \; 0.730 \; 0.810 \; 0.853 \; 0.898 \; 0.922 \; 0.937$
2. Trapezoidal section	$\dfrac{d}{c} = \dfrac{3(1+b_1/c)}{1+2b_1/b}$ $\qquad \dfrac{c_1}{c} = \dfrac{d}{c} - 1$ $\dfrac{h}{c} = \dfrac{R}{c} - \dfrac{\frac{1}{2}(1+b_1/b)(d/c)^2}{\left[\dfrac{R}{c}+\dfrac{c_1}{c}-\dfrac{b_1}{b}\left(\dfrac{R}{c}-1\right)\right]ln\left(\dfrac{R/c+c_1/c}{R/c-1}\right)-\left(1-\dfrac{b_1}{b}\right)\dfrac{d}{c}}$ $k_i = \dfrac{1}{2h/c}\dfrac{1-h/c}{R/c-1}\dfrac{\left(1+4b_1/b+(b_1/b)^2\right)}{(1+2b_1/b)^2}$ $k_o = \dfrac{c_1/c}{2h/c}\dfrac{c_1/c+h/c}{R/c+c_1/c}\dfrac{\left(1+4b_1/b+(b_1/b)^2\right)}{(2+b_1/b)^2}$ Note: While $h/c, k_i,$ and k_o depend upon the width ratio b_1/b, they are independent of the width b	$\left(\text{When } b_1\big/b = \dfrac{1}{2}\right)$ $\dfrac{R}{c} = 1.20 \quad 1.40 \quad 1.60 \quad 1.80 \quad 2.00 \quad 3.00 \quad 4.00 \quad 6.00 \quad 8.00 \quad 10.00$ $\dfrac{h}{c} = 0.403 \; 0.318 \; 0.267 \; 0.232 \; 0.206 \; 0.134 \; 0.100 \; 0.067 \; 0.050 \; 0.040$ $k_i = 3.011 \; 2.183 \; 1.859 \; 1.681 \; 1.567 \; 1.314 \; 1.219 \; 1.137 \; 1.100 \; 1.078$ $k_0 = 0.544 \; 0.605 \; 0.648 \; 0.681 \; 0.707 \; 0.790 \; 0.836 \; 0.885 \; 0.911 \; 0.927$

(Contd.,)

TABLE 14.1 (*Continues*)

Form and dimensions of cross section, reference no.	Formulas	Values of $\dfrac{h}{c}$, k_i, k_0 and for various values of $\dfrac{R}{c}$
3. Triangular section, base inward	$\dfrac{h}{c} = \dfrac{R}{c} - \dfrac{4.5}{\left(\dfrac{R}{c}+2\right) ln\left(\dfrac{R/c+2}{R/c-1}\right)-3}$ $c = \dfrac{d}{3}$ $k_i = \dfrac{1}{2h/c}\dfrac{1-h/c}{R/c-1}$ $\displaystyle\int_{area}\dfrac{dA}{r} = b\left[\left(\dfrac{R}{3c}+\dfrac{2}{3}\right) ln\left(\dfrac{R/c+2}{R/c-1}\right)-1\right]$ $k_0 = \dfrac{1}{2h/c}\dfrac{2+h/c}{R/c+2}$ (Note : $h/c, k_i,$ and k_o are independent of the width b)	$\dfrac{R}{c} = 1.20 \quad 1.40 \quad 1.60 \quad 1.80 \quad 2.00 \quad 3.00 \quad 4.00 \quad 6.00 \quad 8.00 \quad 10.00$ $\dfrac{h}{c} = 0.434 \quad 0.348 \quad 0.296 \quad 0.259 \quad 0.232 \quad 0.155 \quad 0.117 \quad 0.079 \quad 0.060 \quad 0.048$ $k_i = 3.265 \quad 2.345 \quad 1.984 \quad 1.784 \quad 1.656 \quad 1.368 \quad 1.258 \quad 1.163 \quad 1.120 \quad 1.095$ $k_0 = 0.438 \quad 0.497 \quad 0.539 \quad 0.573 \quad 0.601 \quad 0.697 \quad 0.754 \quad 0.821 \quad 0.859 \quad 0.883$
4. Triangular section, base outward	$\dfrac{h}{c} = \dfrac{R}{c} - \dfrac{1.125}{1.5-\left(\dfrac{R}{c}-1\right) ln\left(\dfrac{R/c+0.5}{R/c-1}\right)}$ $c = \dfrac{2d}{3}$ $k_i = \dfrac{1}{8h/c}\dfrac{1-h/c}{R/c-1}$ $\displaystyle\int_{area}\dfrac{dA}{r} = b_1\left[\left(1-\dfrac{2}{3}\right)\left(\dfrac{R}{c}-1\right) ln\left(\dfrac{R/c+0.5}{R/c-1}\right)-1\right]$ $k_o = \dfrac{1}{4h/c}\dfrac{2h/c+1}{2R/c+2}$ (Note : $h/c, k_i,$ and k_o are independent of the width b_1)	$\dfrac{R}{c} = 1.20 \quad 1.40 \quad 1.60 \quad 1.80 \quad 2.00 \quad 3.00 \quad 4.00 \quad 6.00 \quad 8.00 \quad 10.00$ $\dfrac{h}{c} = 0.151 \quad 0.117 \quad 0.097 \quad 0.083 \quad 0.073 \quad 0.045 \quad 0.033 \quad 0.022 \quad 0.016 \quad 0.013$ $k_i = 3.527 \quad 2.362 \quad 1.947 \quad 1.730 \quad 1.595 \quad 1.313 \quad 1.213 \quad 1.130 \quad 1.094 \quad 1.074$ $k_0 = 0.636 \quad 0.695 \quad 0.735 \quad 0.765 \quad 0.788 \quad 0.857 \quad 0.892 \quad 0.927 \quad 0.945 \quad 0.956$

(Contd.,)

TABLE 14.1 (*Continues*)

Form and dimensions of cross section, reference no.	Formulas	Values of $\dfrac{h}{c}$, k_i, k_0 and for various values of $\dfrac{R}{c}$
5. Diamond	$$\frac{h}{c}=\frac{R}{c}-\frac{1}{\dfrac{R}{c}\,In\left[1-\left(\dfrac{c}{R}\right)^2\right]+ln\dfrac{R/c+1}{R/c-1}}$$ $$k_i=\frac{1}{6h/c}\frac{1-h/c}{R/c-1}$$ $$k_o=\frac{1}{6h/c}\frac{1+h/c}{R/c+1}$$ $$\int_{area}\frac{dA}{r}=b\left[\frac{R}{c}ln\left[1-\left(\frac{c}{R}\right)^2\right]+ln\frac{R/c+1}{R/c-1}\right]$$ (Note: $h/c, k_i,$ and k_0 are independent of the width b)	$\dfrac{R}{c}=1.200$ 1.400 1.600 1.800 2.000 3.000 4.000 6.000 8.000 10.000 $\dfrac{h}{c}=0.175$ 0.138 0.116 0.100 0.089 0.057 0.042 0.028 0.021 0.017 $k_i=3.942$ 2.599 2.118 1.866 1.709 1.377 1.258 1.159 1.115 1.090 $k_0=0.510$ 0.572 0.617 0.652 0.681 0.772 0.822 0.875 0.904 0.922
6. Solid circular or elliptical section	$$\frac{h}{c}=\frac{1}{2}\left[\frac{R}{c}-\sqrt{\left(\frac{R}{c}\right)^2-1}\right]$$ $$k_i=\frac{1}{4h/c}\frac{1-h/c}{R/c-1}\quad(\text{Note}: h/c, k_i, \text{ and } k_0 \text{ are independent of the width } b)$$ $$k_0=\frac{1}{4h/c}\frac{1+h/c}{R/c+1}\quad\int_{area}\frac{dA}{r}=\pi b\left[\frac{R}{c}-\sqrt{\left(\frac{R}{c}\right)^2-1}\right]$$	$\dfrac{R}{c}=1.20$ 1.40 1.60 1.80 2.00 3.00 4.00 6.00 8.000 10.00 $\dfrac{h}{c}=0.268$ 0.210 0.176 0.152 0.134 0.086 0.064 0.042 0.031 0.025 $k_i=3.408$ 2.350 1.957 1.748 1.616 1.332 1.229 1.142 1.103 1.080 $k_0=0.537$ 0.600 0.644 0.678 0.705 0.791 0.837 0.887 0.913 0.929

(*Contd.,*)

7. Solid semicircle or semiellipse, base inward	$R = R_x + c \quad \dfrac{d}{c} = \dfrac{3}{4}$ (Note : $h/c, k_i,$ and k_0 are independent of the width b)	

$$k_i = \frac{0.3879}{h/c} \frac{1 - h/c}{R/c - 1}$$

$$k_0 = \frac{0.2860}{h/c} \frac{h/c + 1.3562}{R/c + 1.3562}$$

For $R_x \geq d : R/c \geq 3.356$ and

$$\int_{area} \frac{dA}{r} = \frac{\pi R_x b}{2d} - b - \frac{b}{d} \sqrt{R_x^2 - d^2} \left(\frac{\pi}{2} - sin^{-1} \frac{d}{R_x} \right)$$

$$\frac{h}{c} = \frac{R}{c} - \frac{(d/c)^2 \Big/ 2}{\dfrac{R}{c} - 2.5 - \sqrt{\left(\dfrac{R}{c} - 1\right)^2 - \left(\dfrac{d}{c}\right)^2 \left(1 - \dfrac{2}{\pi} sin^{-1} \dfrac{d/c}{R/c - 1}\right)}}$$

For $R_x < d : R/c < 3.356$ and

$$\int_{area} \frac{dA}{r} = \frac{\pi R_x b}{2d} - b + \frac{b}{d} \sqrt{d^2 - R_x^2} \, ln \frac{d + \sqrt{d^2 - R_x^2}}{R_x}$$

$$\frac{h}{c} = \frac{R}{c} - \frac{(d/c)^2 \Big/ 2}{\dfrac{R}{c} - 2.5 + \dfrac{2}{\pi} \sqrt{\left(\dfrac{d}{c}\right)^2 - \left(\dfrac{R}{c} - 1\right)^2} \, ln \dfrac{d}{R} \dfrac{c + \sqrt{(d/c)^2 - (R/c - 1)^2}}{c \Big/ - 1}}$$

$\dfrac{R}{c} =$	1.200	1.400	1.600	1.800	2.000	3.000	4.000	6.000	8.000	10.000
$\dfrac{h}{c} =$	0.388	0.305	0.256	0.222	0.197	0.128	0.096	0.064	0.048	0.038
$k_i =$	3.056	2.209	1.878	1.696	1.579	1.321	1.224	1.140	1.102	1.080
$k_C =$	0.503	0.565	0.609	0.643	0.671	0.761	0.811	0.867	0.897	0.916

8. Solid semicircle or semiellipse, base outward	$R = R_x - c_1 \quad \dfrac{d}{c} = \dfrac{3}{3\pi - 4} \quad \dfrac{c_1}{c} = \dfrac{4}{3\pi - 4}$	

(Note : $h/c, k_i,$ and k_0 are independent of the width b)

$$\int_{area} \frac{dA}{r} = \frac{\pi R_x b}{2d} - b - \frac{b}{d} \sqrt{R_x^2 - d^2} \left(\frac{\pi}{2} + sin^{-1} \frac{d}{R_x} \right)$$

$$\frac{h}{c} = \frac{R}{c} - \frac{(d/c)^2 \Big/ 2}{\dfrac{R}{c} + \dfrac{10}{3\pi - 4} \dfrac{c}{R} - \sqrt{\left(\dfrac{R}{c} + \dfrac{c_1}{c}\right)^2 - \left(\dfrac{d}{c}\right)^2 \left(1 + \dfrac{2}{\pi} sin^{-1} \dfrac{d/c}{R/c + c_1}\right)}}$$

$$k_i = \frac{0.2109}{h/c} \frac{1 - h/c}{R/c - 1}$$

$$k_0 = \frac{0.2860}{h/c} \frac{h/c + 0.7374}{R/c + 0.7374}$$

Note: For a semicircle $b/2 = d$)

$\dfrac{R}{c} =$	1.200	1.400	1.600	1.800	2.000	3.000	4.000	6.000	8.000	10.000
$\dfrac{h}{c} =$	0.244	0.189	0.157	0.135	0.118	0.075	0.055	0.036	0.027	0.021
$k_i =$	3.264	2.262	1.892	1.695	1.571	1.306	1.210	1.130	1.094	1.073
$k_0 =$	0.593	0.656	0.698	0.730	0.755	0.832	0.871	0.912	0.933	0.946

NOTATION R = Radius of curvature measured to centroid of section; c = Distance from centroidal axis to extreme fibre on concave side of beam; A = Area of section; h = Distance from centroidal axis to neutral axis measured toward centre of curvature; I = Moment of inertia of cross section about centroidal axis perpendicular to plane of curvature; and $k_i = \sigma_i/\sigma$ and $k_0 = \sigma_0/\sigma$ where σ_i = Actual stress in extreme fiber on concave side, σ_0 = Actual stress in extreme fibre on convex side, and σ = Fictitious unit stress in corresponding fibre as computed by ordinary flexure formula for a straight beam

(Contd.,)

STRESS DISTRIBUTION

The stress distribution as given by the Winkler–Bach formula is hyperbolic as seen by a plot across the depth of the beam. For the positive bending moment, which we have discussed and which decreases the radius of curvature, the fibres on the convex surface are in tension and the fibres on the concave surface are in compression. The following examples illustrate the procedure adopted for solving problems involving curved bars.

PROBLEM 14.1

Curved bar of a rectangular section: maximum stress

A curved bar of rectangular section 60 mm × 30 mm is subjected to a BM of 600 Nm. Find the maximum stresses in the section. The mean radius of curvature, = 100 mm.

Solution   We locate the neutral surface first

$$R = \frac{h}{\log_l(r_2 / r_1)} = \frac{30}{\log_i(100+15)/(100-15)} = \frac{30}{0.3023} = 99.245 \text{ mm}$$

The difference between the centrodial line and the neutral surface is only 0.755 mm. To find the maximum stress, we use the equation

$$\sigma_y = \frac{My}{Ae(R - y)}$$

$$\text{or } \sigma = \frac{M(r - R)}{Aer}$$

where $e = 0.755$ mm.

$r = 100 + 15 = 115$ mm corresponds to the maximum tensile stress. $r = 100 - 15 = 85$ mm corresponds to the maximum compressive stress.

Thus,

$$\sigma_{max} = \frac{600(115 - 99.245) \times 10^3}{60 \times 30 \times 0.755 \times 115} = 60.5 \text{ N/mm}^2 \text{ (tensile)}$$

$$\sigma_{max} = \frac{600(85 - 99.245) \times 10^3}{60 \times 30 \times 0.755 \times 85} = 74 \text{ N/mm}^2 \text{ (compressive)}$$

Let us find the change from the straight bar equations.

$$\sigma_{max} = \frac{M}{I}\,y = \frac{600 \times 10^3}{60 \times 30^3 / 12} \times 15 = 66.66 \text{ N/mm}^2$$

PROBLEM 14.2

Curved beam of a circular section

A curved beam of circular section has an initial radius of 80 mm. The diameter of the bar is 30 mm. Find the maximum tensile and compressive stresses due to a bending moment of 400 Nm.

Solution   We first find the neutral surface by finding radius R. From Table 14.1, the value of R for the section is given by

$$R = (1/2)\left[80 + \sqrt{(80^2 - 20^2)} \right] = 78.72 \text{ mm}$$

Shift = 80 – 78.72 = 1.28 mm, which is the distance between the centrodial axis and the neutral axis (NA). The neutral axis is nearer to the centre of curvature.

Bending moment = 400 Nm = 400 × 10³ Nmm

The stress will be compressive on the concave side and tensile on the convex side.

Maximum stress is given by

$$\sigma = My\left(ASR_y \right)$$

where y is the distance to the extreme fibre from the NA, A is the area of section, S is the shift, and R_y is the radius to the fibre.

$$\text{Area of section } = \pi \times 40^2 / 4 = 1256.64 \text{ mm}^2, \ s = 1.28 \text{ mm}$$

$y = 20 - 1.28 = 18.72$ mm for the compression fibre and $20 + 1.28 = 21.28$ mm to the tension fibre

$R_y = 80 - 20 = 60$ mm to the compression fibre and $80 + 20 = 100$ mm for the tension fibre

$$\text{Maximum compressive stress} = 400 \times 10^3 \times 18.72 / (1256.64 \times 1.28 \times 60)$$
$$= 77.6 \text{ N/mm}^2$$
$$\text{Maximum tensile stress} = 400 \times 10^3 \times 21.28 / (1256.64 \times 1.28 \times 100) = 52.9 \text{ N/mm}^2$$

PROBLEM 14.3

Curved beam: T-section

A section of a curved beam is a T-section shown in Figure. If the initial radius of curvature is 200 mm, find the maximum stresses due to a bending moment of 600 Nm.

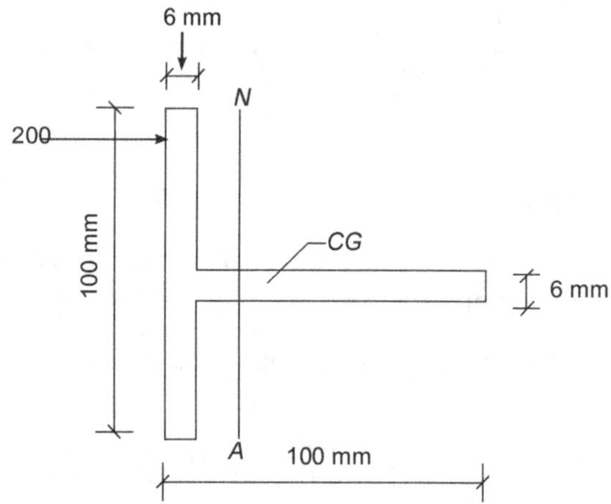

Solution   Area of the section $= 100 \times 6 + 94 \times 6 = 1164$ mm²

We locate the centroid by

$$y = (100 \times 6 \times 3 + 94 \times 6 \times 53) / 1164 = 27.22 \text{ mm from the inner surface}$$

Radius of the neutral surface is given by

$R = A / \int (dA / R_y),$ where R_y is the radius to the elemental fibre dA.

So,

$$\int_A (dA / R_y) = \int_6^{100} \frac{6dy}{(200 + y)} + \int_0^6 \frac{100dy}{(200 + y)}$$
$$= 6[\log_e (200 + y)]_6^{100} + 100[\log_e (200 + y)]_0^6$$
$$= 2.255 + 2.956 = 5.211$$
$$R = 1164 / 5.211 = 223.4 \text{ mm}$$

$$\text{Shift} = 227.22 - 223.4 = 3.8 \text{ mm}$$

The edge of the flange will have maximum compressive stress and the edge of the web will have maximum tensile stress given by

$$\sigma = M \times y / (\text{area} \times \text{shift} \times \text{radius})$$

For maximum compressive stress,

$M = 600 \times 10^3$ Nmm, $y = 23.4$ mm, $A = 1164$ mm², shift $= 3.8$ mm, and radius of the edge $= 200$ mm

$$\sigma_{max} \text{ (compressive)} = 600 \times 10^3 \times 23.4 / [1164 \times 3.8 \times 200] = 15.9 \text{ N/mm}^2$$

Maximum tensile stress is at $y = 76.6$ mm

Also radius = 300 mm

σ_{max} (tensile) = $600 \times 10^3 \times 76.6 / [1164 \times 3.8 \times 300]$ = 34.6 N/mm²

PROBLEM 14.4

Curved beam : I-section

The I-section shown is the section of the hook shown subjected to a load passing through its centre. If the permissible stress is 100 N/mm² in tension/compression, find the maximum load P that can be put on the hook.

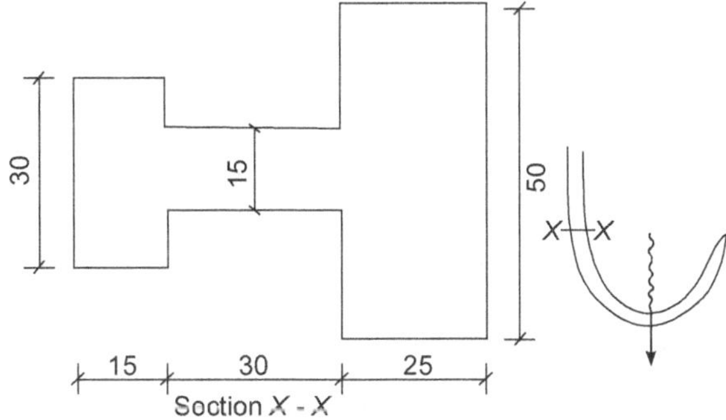

Solution   Area of the section = $30 \times 15 + 30 \times 15 + 50 \times 25 = 2150$ mm²

The centroid of the section is at

$[(30 \times 15 \times 7.5) + (30 \times 15 \times 30) + (50 \times 25 \times 72.5)] / 2150 = 50$ mm from the edge of the smaller flange

The radius of NA,

$$R = A / \int (dA / R_y), \int (dA / R_y)$$

has to be calculated separately for the three rectangles forming the I-section.

$$\int_{0}^{15} \left(\frac{30\,dy}{(105+y)} \right) + \int_{0}^{30} \frac{15\,dy}{(75+y)} + \int_{0}^{25} \frac{50\,dy}{(50+y)}$$

or $\int (dA / R_y) = 30[\log_e (120 / 105)] + 15[\log_e (105 / 75)] + 50[\log_e (75 / 50)]$

$$= 29.32$$

$R = 2150 / 29.32 = 73.32$ mm

$S = 85 - 73.32 = 11.68$ mm

The stress in the section will be a direct stress (tensile) with the bending stress.

Direct stress = $20,000/2150 = 9.3$ N/mm²

Bending stress will be tensile at the inner edge and compressive at the outer edge.

Bending moment $= P \times 73.32$ Nmm

$$\text{Bending stress (tensile)} = \frac{P \times 73.32 \times 28.7}{2150 \times 5.38 \times 50}$$

Therefore,

$$\frac{P}{2150} + \frac{P \times 73.32 \times 28.7}{2150 \times 5.38 \times 50} = 100; \ P = 24 \text{ kN}$$

$$\text{Bending stress (compressive)} = \frac{P \times 73.32 \times 41.3}{2150 \times 5.38 \times 120}$$

Therefore,

$$\frac{P}{2150} + \frac{P \times 73.32 \times 41.3}{2150 \times 5.38 \times 120} = 100 \, P = 37 \text{ kN}$$

The permissible load is the lower of the two values, i.e., P=24kN.

PROBLEM 14.5   Crane hook

A crane hook has a trapezoidal section, as shown in Figure. For the loading shown, find the maximum stresses at section *AA*.

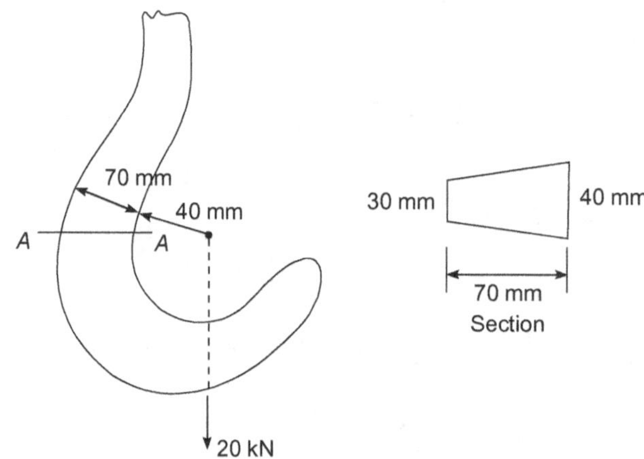

Solution   We first locate the centroid.

$$\bar{y} = \frac{(2 \times 30 + 40)}{30 + 40} \times \left(\frac{70}{3}\right) = 33.33 \text{ mm (from the 40 mm edge)}$$

The radius of the centroidal axis,

$$R = 40 + 33.33 = 73.33 \text{ mm}$$

$$\text{Bending moment at section} = 20 \times 10^3 \times 73.33$$

$$= 1466.6 \times 10^3 \text{ Nmm}$$

To locate R, the radius of the neutral surface, we use the formula for the trapezoidal section

$$R = \frac{h^2(a_1 + a_2)/2}{(a_1 r_1 - a_2 r_1)\log(r_2/r_1) - h(a_1 - a_2)}$$

In our case, $h = 70$ mm, $a_1 = 40$ mm, $a_2 = 30$ mm, $r_2 = 110$ mm, and $r_1 = 40$ mm. Substituting the values,

$$R = \frac{70^2(40+30)/2}{(40 \times 110 - 30 \times 40)\log(110/40) - 70(40-30)} = 67.594$$

Therefore, the shift of the NA from the centroid = 73.333 – 67.594 = 5.739 mm.

$$\sigma_{max}(\text{at } R_y = 110) = \frac{M(R_y - R)}{ASR_y} = \frac{1466 \times 10^3(110 - 67.594)}{2450 \times 5.739 \times 110} = 40.2 \text{ N/mm}^2$$

$$\sigma_{max}(\text{at } R_y = 40) = \frac{1466 \times 10^3(40 - 67.594)}{2450 \times 5.739 \times 40} = 71.9 \text{ N/mm}^2 \text{ (compressive)}$$

PROBLEM 14.6

Plot the distribution of stresses across section AA of the crane hook shown in Figure. The cross section is rectangular, with $b = 18$ mm and and $h = 100$ mm, and the load $F = 22$ kN.

Solution   Since $A = bh$, we have $dA = b\,dr$ and, from

$$r_n = \frac{A}{\int \dfrac{dA}{r}} = \frac{bh}{\int_{r_i}^{r_0} \dfrac{b}{r}\,dr} = \frac{h}{\ln \dfrac{r_0}{r_i}} \tag{1}$$

From Figure, we see that $r_i = 50$ mm, $r_0 = 150$ mm, $r_c = 100$ mm, and $A = 1800$ mm². Thus, from Equation (1),

$$r_n = \frac{h}{\ln(r_0/r_i)} = \frac{100}{\ln \dfrac{150}{20}} = 91 \text{ mm}$$

and so the eccentricity is $e = r_c - r_n = 100 - 91 = 9$ mm. The moment M is positive and is $M = Fr_c = 22\,(0.1) = 2.2$ kNm. Adding the axial component of stress gives

$$\sigma = \frac{F}{A} + \frac{My}{Ae(r_n - y)} = \frac{22 \times 10^3}{1800 \times 10^{-6}} + \frac{(22 \times 10^3)(0.091 - r)}{1800 \times 10^{-6}(0.009)r} \tag{2}$$

Substituting values of r form 50 to 150 mm results in the stress distribution shown in Figure. The stresses at the inner and outer radii are found to be 123.6 and 41.2 MPa, respectively, as shown

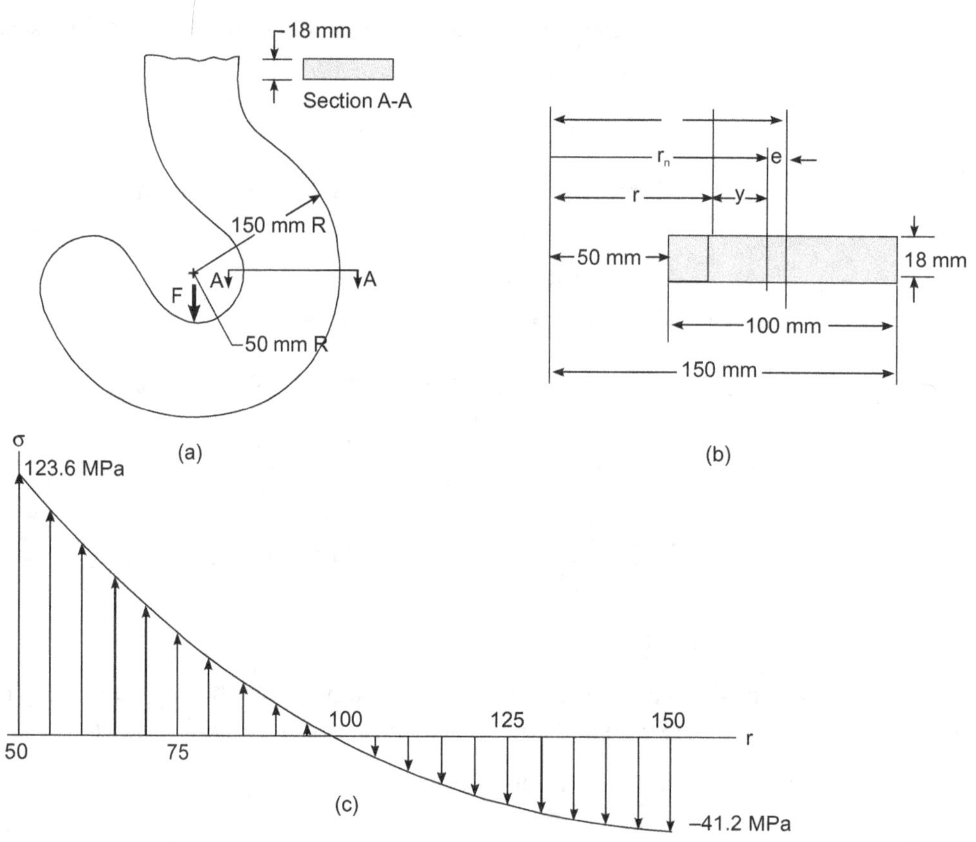

Note in the hook example, the symmetrical rectangular cross section causes the maximum tensile stress to be 3 times greater than the maximum compressives stress. If we wanted to design the hook to use material more effectively, we would use more material at the inner radius and less material at the outer radius. For this reason, trapezoidal, T, or unsymmetric I, cross sections are commonly used. Sections most frequently encountered in the stress analysis of curved beams are shown in Table 14.1

PROBLEM 14.7

A crane hook having an approximate trapezoidal cross-section is shown in Figure. It is made of plain carbon steel 45C8 ($\sigma_{yp} = 380$ N/mm^2) and the factor of safety is 3.5. Determine the load carrying capacity of the hook.

Solution   For the cross-section XX and

$$R_N = \frac{\left(\dfrac{b_i + b_0}{2}\right)h}{\left(\dfrac{b_i R_0 - b_0 R_i}{h}\right)\log_e\left(\dfrac{R_0}{R_i}\right) - (b_i - b_0)}$$

$$R_N = \frac{\left(\dfrac{90+30}{2}\right)(120)}{\left(\dfrac{90\times170-30\times50}{120}\right)\log_e\left(\dfrac{170}{50}\right)-(90-30)}$$

$$= 89.1816 \text{ mm}$$

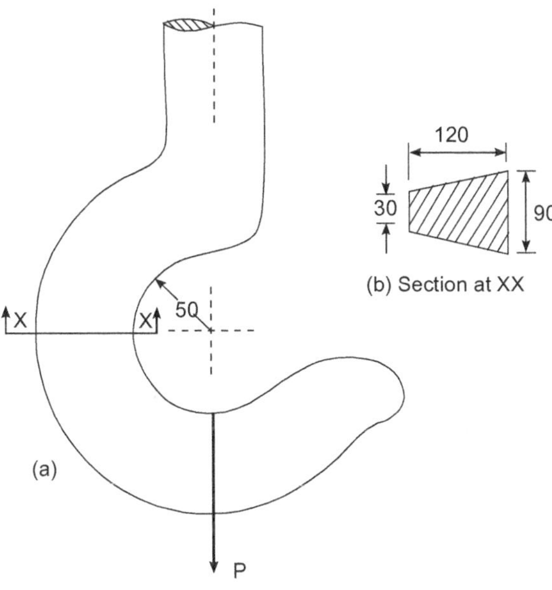

$$R = R_i + \frac{h(b_i + 2b_0)}{3(b_i + b_0)}$$

$$= 50 + \frac{120(90 + 2\times30)}{3(90+30)} = 100 \text{ mm}$$

$$e = R - R_N = 100 - 89.1816 = 10.8184 \text{ mm}$$

$$h_i = R_N - R_i = 89.1816 - 50 = 39.1816 \text{ mm}$$

$$A = \frac{1}{2}\Big[h(b_i + b_0)\Big]$$

$$= \frac{1}{2}\Big[(120)(90+30)\Big] - 7200 \text{ mm}^2$$

$$M_b = PR = (100P)\text{ N - mm}$$

The bending stress at the inner fibre is given by,

$$\sigma_{bi} - \frac{M_b h_i}{AeR_i} - \frac{(100P)(39.1816)}{(7200)(10.8184)(50)}$$

$$= \frac{(7.2435)P}{(7200)} \text{N} / \text{mm}^2 \tag{i}$$

In addition to bending stress, there is direct tensile stress at section *XX*. It is given by,

$$\sigma_i = \frac{P}{A} = \frac{P}{(7200)} \text{N} / \text{mm}^2 \tag{ii}$$

Superimposing the two stress and equation the resultant to perimissible stress, we have

$$\sigma_{bi} + \sigma_i = \frac{\sigma_{yp}}{(fs)}$$

$$\frac{(7.2435)P}{7200} + \frac{P}{7200} = \frac{380}{3.5}$$

$$P = 94827.95 \text{ N}$$

PROBLEM 14.8

A curved link of the mechanism made from a rounded steel bar, is shown in Figure. The material of the link is plain carbon steel 30C8 (σ_{yp} = 400 N/mm²) and the factor of safety is 3.5. Determine the dimension of the link.

Solution

At section *XX*.

$$R = 4D$$

$$R_i = 4D - 0.5D = 3.5D$$

$$R_0 = 4D + 0.5D = 4.5D$$

From Eq. (4.6o),

$$R_N = \frac{\left(\sqrt{R_0} + \sqrt{R_i}\right)^2}{4}$$

$$= \frac{\left(\sqrt{4.5D} + \sqrt{3.5D}\right)^2}{4} = 3.9843D$$

$$e = R - R_N = 4D - 3.9843D = 0.0457D$$

$$h_i = R_N - R_i = 3.9843D - 3.5D = 0.4843D$$

$$A = \frac{\pi}{4}D^2 = \left(0.7854D^2\right)\text{mm}^2$$

$$M_b = 1000 \times 4D = \left(4000D\right) \text{N} - \text{mm}$$

The bending stress at the inner fibre is given by,

$$\sigma_{bi} = \frac{M_b h_i}{AeR_i} = \frac{(4000D)(0.4843D)}{(0.7854D^2)(0.0157D)(3.5D)}$$

$$= \left(\frac{44\,886.51}{D^2}\right) \text{N}/\text{mm}^2$$

In addition to bending stress, there is direct tensile stress at section XX. It is given by,

$$\sigma_t = \frac{P}{A} = \frac{1000}{(0.7854D^2)} = \left(\frac{12273}{D^2}\right) \text{N}/\text{mm}^2$$

Supcrimposing the bending and direct tensile stresses and equating the resultant stress to permissible stress, we have

$$\sigma_{bi} + \sigma_t = \frac{\sigma_{yp}}{(fs)}$$

$$\left(\frac{44886.51}{D^2}\right) + \left(\frac{1273.24}{D^2}\right) = \frac{400}{3.5}$$

$$D = 20.10 \text{ mm}$$

PROBLEM 14.9

The C-frame of a 100 kN capacity press is shown in Figure. The material of the frames is grey cast iron FG 200 and the factor of safety is 3. Determine the dimension of the frame.

Solution

The section at XX is subjected to direct tensile stress and bending stress. Using notations for equation for R_N with the following dimensions.

$$b_i = 3t \qquad\qquad h = 3t \qquad\qquad R_i = 2t$$

$$R_0 = 5t \qquad\qquad t_i = t \qquad\qquad t = 0.75\,t$$

From Equation for R_N, we get

$$R_N = \cfrac{t_i\left(b_i - t\right) + th}{\left(b_i - t\right)\log_e\left(\cfrac{R_i + t_i}{R_i}\right) + t\log_e\left(\cfrac{5t}{2t}\right)}$$

$$= \cfrac{t\left(3t - 0.75t\right) + 0.75t\left(3t\right)}{\left(3t - 0.75t\right)\log_e\left(\cfrac{2t + t}{t}\right) + 0.75t\log_e\left(\cfrac{5t}{2t}\right)}$$

$$= 2.8134t$$

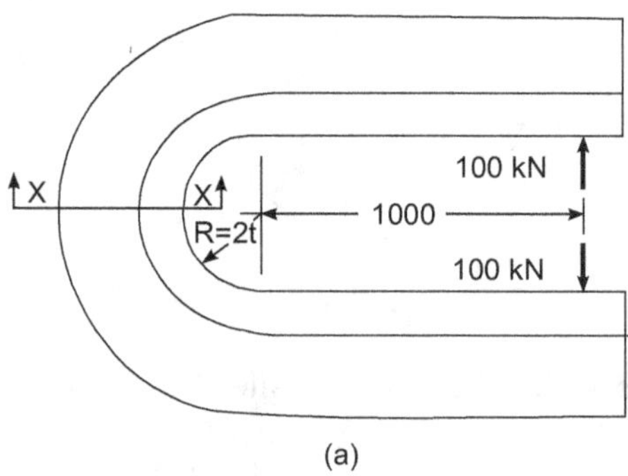

(a)

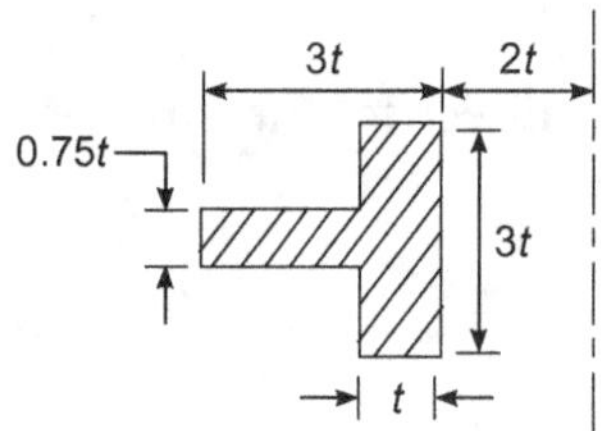

(b) Section at *XX*

From Equation for R, we get

$$R = R_i + \cfrac{\dfrac{1}{2}th^2 + \dfrac{1}{2}t_i^2\left(b_i - t\right)}{th + t_i\left(b_i - t\right)}$$

$$= 2t + \cfrac{\dfrac{1}{2}\left(0.75t\right)\left(3t\right)^2 + \dfrac{1}{2}t^2\left(3t - 0.75t\right)}{\left(0.75t\right)\left(3t\right) + t\left(3t - 0.75t\right)}$$

$$e = R - R_N - R_i = 3t - 2.8134t - 2t = 0.1866t$$

$$h_i = R_N - R_i = 2.8134t - 2t = 0.1834t$$

$$A = (3t)(t) + (0.75t)(2t)\left(4.5t^2\right) \text{ mm}^2$$

$$M_b = 100 \times 10^3 (1000 + R)$$

$$= 100 \times 10^3 (100 + 3t) \text{ N - mm}$$

The bending stress at the inner fibre is given by,

$$\sigma_i = \frac{M_b h_i}{AeR_i} = \frac{100 \times 10^3 (1000 + 3t)(0.8134t)}{\left(4.5t^2\right)(0.1866t)(2t)}$$

$$= \frac{100 \times 10^3 (100 + 3t)(2.1795)}{\left(4.5t^2\right)} \text{N} / \text{mm}^2$$

The direct tensile stress is given by,

$$\sigma_i = \frac{P}{A} = \frac{100 \times 10^3}{\left(4.5t^2\right)} \text{N} / \text{mm}^2$$

Adding the two stresses and equating the resultant stress to permissible stress,

$$\sigma_{bi} + \sigma_t = \frac{S_{ut}}{(fs)} \frac{100 \times 10^3 (1000 + 3t)(2.1795)}{\left(4.5t^2\right)} + \frac{100 + 10^3}{\left(4.5t^2\right)} = \frac{200}{3}$$

$$t^3 - 2512.83t - 726500 = 0$$

Solving the above cubic equation by trial and error method,

$$t = 99.2 \text{ mm or } t = 100 \text{ mm}$$

Shear stress due to the radial shear force V   Although equation $\tau = \dfrac{VAz}{Ib}$ does not apply to curved beams, equation $\tau_{max} = \alpha \dfrac{V}{A}$, used as for a straight beam, gives the maximum shear stress

with sufficient accuracy in most instances. Again an analysis for a rectangular cross section carried out using the theory of elasticity shows that the peak shear stress in a curved beam occurs not at the centroidal axis as it does for a straight beam but toward the inside surface of the beam. For a very sharply curved beam, $R/d = 0.7$, the peak shear stress was 2.04 V/A at a position one-third of the way from the inner surface to the centroid. For a sharply curved beam, $R/d = 1.5$, the peak shear stress was 1.56 V/A at a position 80% of the way from the inner surface to the centroid. These values can be compared to a peak shear stress of 1.5V/A at the centroid for a straight beam of rectangular cross section.

If a strength-of-materials solution for the shear stress in a curved beam is desired, the element in Figure 14.5(b) can be used and moments taken about the centre of curvature. Using the normal stress distribution one can find the shear stress expression to be

$$\sigma_\theta = (N\,/\,A) + (My\,/\,AhR),\tag{14.12}$$

where t_r is the thickness of the section normal to the plane of curvature at the radial position r and

$$A_r = \int_b^r dA_1 \qquad \text{and} \qquad Q_r = \int_b^r r_1 d A_1 \tag{14.13}$$

Equation (14.12) gives conservative answers for the peak values of shear stress in rectangular sections when compared to elasticity solutions. The locations of peak shear stress are the same in both analyses,and the error in magnitude is about 1%.

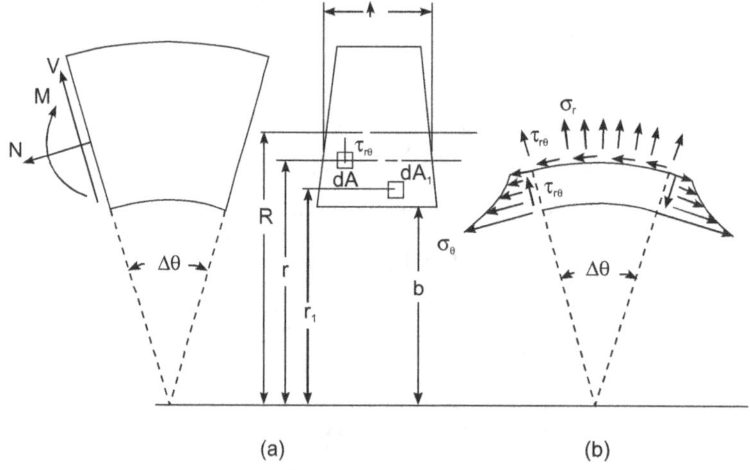

(a)   (b)

Figure 14.5

RADIAL STRESSES DUE TO MOMENT M AND NORMAL FORCE N

Owing to the radial componenets of the fibre stresses, radial stresses are present in a curved beam; these are tensile when the bending moment tends to straighten the beam and compressive under the reverse condition. A strength-of-materials solution may be developed by summing radial forces and summing forces perpendicular to the radius using the element in Figure 14.5.

$$\sigma_r = \frac{R-h}{t_r Ahr}\left[(M - NR)\left(\int_b^r \frac{dA_1}{r_1} - \frac{A_r}{R-h}\right) + \frac{N}{r}(RA_r - Q_r)\right]\tag{14.14}$$

Equation (14.14) is as accurate for radial stress as is equation (14.12) for shear stress when used for a rectangular cross section and compared to an elasticity solution. However, the complexity of equation (14.14) coupled with the fact that the stresses due to N are generally smaller than those due to M leads to the usual practice of omitting the terms involving N. This leads to the equation for radial stress found in many texts.

$$\theta_r = \frac{R-h}{t_r A h r} M \left(\int_b^r \frac{dA_1}{r_1} - \frac{A_r}{R-h} \right) \tag{14.15}$$

Again care must be taken when using equations. (14.12), (14.14), and (14.15) to use an accurate value for h.

Radial stress is usually not a major consideration in compact sections for it is smaller than the circumferential stress and is low where the circumferential stresses are large. However, in flanged sections with thin webs the radial stress may be large at the junction of the flange and web, and the circumferential stress is also large at this position. This can lead to excessive shear stress and possible yielding if the radial and circumferential stresses are of opposite sign. A large compressive radial stress in a thin web may also lead to a buckling of the web. Corrections for curved-beam formulas for sections having thin flanges are needed and also if a section has a thin web and very thick flanges. Under these conditions the individual flanges tend to rotate about their own neutral axes and larger radial and shear stresses are developed.

CURVED BEAMS WITH WIDE FLANGES

In reinforcing rings for large pipes, airplane fuselages, and ship hulls, the combination of a curved sheet and attached web or stiffener forms a curved beam with wide flanges. Formulas for the effective width of flange in such a curved beam are as follows.

When the flange is indefinitely wide (e.g., the inner flange of a pipe-stiffener ring), the effective width is

$$b' = 1.56 \sqrt{Rt}$$

where b′ is the total width assumed effective, R is the mean radius of curvature of the flange, and t is the thickness of the flange.

When the flange has a definite unsupported width b (gross width less web thickness), the ratio of effective to actual width b'/b is a function of qb, where

$$q = 4\sqrt{\frac{3(1-v^2)}{R^2 t^2}}$$

U-shaped members   A U-shaped member having a semicircular inner boundary and a rectangular outer boundary is sometimes used as a punch or riveter frame. Such a member can usually be analysed as a curved beam having a concentric outer boundary, but when the back thickness is large, a more accurate analysis may be necessary.

Deflections   If a sharply curved beam is only a small portion of a larger structure, the contribution to deflection made by the curved portion can best be calculated by using the stresses at the inner and outer surfaces to calculate strains and the strains then used to determine the rotations of the plane sections.

DEFLECTION OF CURVED BEAMS

Deflections of curved beams can generally be found most easily by applying an energy method such as Castigliano's second theorem. One such expression is given by $y = \dfrac{\partial U}{\partial P}$. equation The proper expression to use for the complementary energy depends upon the degree of curvature in the beam.

Deflection of curved beams of large radius   Machine frames, spring, clips, fasteners are quite commonly made in curved shapes. The manner of finding stresses in such members has been discussed earlier with sufficient illustrative example problem. Apart from stresses, in many applications like machine frames, one needs to restrict the deflection also by incorporating sufficient rigidity in the curved beams. Hence deflection also need to be determined. Castigliano's theorem is quite useful for the analysis of deflections in curved members also. If for a curved beam the radius of curvature is large enough such that equations of Chapter (8) and (9) $\left(\text{viz} \, \sigma = \dfrac{MZ}{I} \text{ and } \tau = \dfrac{VA'Z'}{Ib} \right)$ are acceptable, i.e.,

the radius of curvature is greater than 10 times the depth, then the stress distribution across the depth of the beam is very nearly linear and the complementary energy of flexure is given with sufficient accuracy by $U_f = \int \dfrac{M^2}{2EI} dx$. If in addition, the angular span is great enough such that deformations due to axial stress from the normal force N and the shear stresses due to transverse shear V can be neglected, deflections can be obtained by applying Eqs. $U_f = \int \dfrac{M^2}{2EI} dx$ and $y = \dfrac{\partial U}{\partial P}$ and rotations by Eq. $\Delta\theta = \int_a^b \dfrac{M}{EI} dx$. The following example shows how this is done.

$$\Delta\theta = \int_a^b \frac{M}{EI}\, dx$$

Figure 14.6 represents a slender uniform bar curved to form the quadrant of a circle; it is fixed at the lower end and at the upper end is loaded by a vertical force V, a horizontal force $H,$ and a couple M_0. It is desired to find the vertical deflection $D_y,$ the horizontal deflection D_x and the rotation of the upper end denoted here by D_y, D_x and θ respectively.

Solution

According to Castigliano's second theorem, $D_y = \partial U / \partial V, D_x = \partial U / \partial H,$ and $\theta = \partial U / \partial M_0$. Denoting the angular position of any section by x, it is evident that the moment there is $M = VR \sin x + HR (1 - \cos x) + M_0$. Disregarding shear and axial stress, and replacing ds by $R\, dx$, we have

$$U = U_f = \int_0^{\frac{\pi}{2}} \frac{[VR\sin x + HR(1-\cos x) + M_o]^2 \, Rdx}{2EI} \tag{14.16}$$

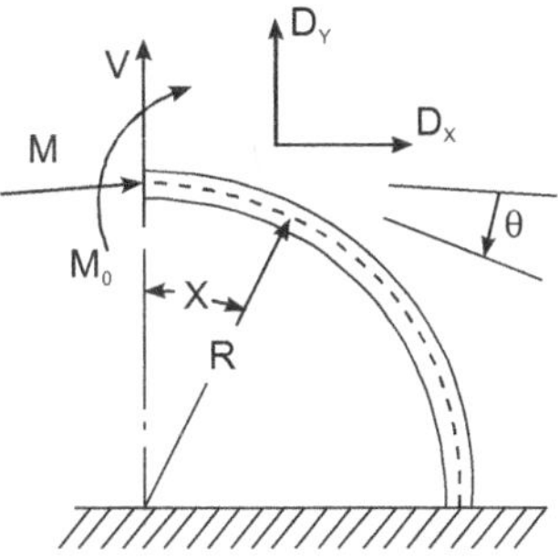

Figure 14.6

Instead of integrating this and then carrying out the partial differentiations, we will differentiate first and then integrate, and for convenience suppress the constant term *EI* until all computations are completed. Thus

$$D_y = \frac{\partial U}{\partial V} = \int_0^{\frac{\pi}{2}} [VR\sin x + HR(1-\cos x) + M_o]^2 \, (R\sin x)\,Rdx$$

$$= VR^3 \left(\tfrac{1}{2}x - \tfrac{1}{2}\sin x\cos x\right) - HR^3 (\cos x + \tfrac{1}{2}\sin^2 x) - M_oR^2 \cos x \Big|_0^{\pi/2} \tag{14.17}$$

$$= \frac{\left((\pi/4)VR^3 + \tfrac{1}{2}HR^3 + M_oR^2\right)}{EI}$$

$$D_x = \frac{\partial U}{\partial H} = \int_0^{\frac{\pi}{2}} [VR\sin x + HR(1-\cos x) + M_0]R(1-\cos x)\,Rdx$$

$$= VR^3(-\cos x - \tfrac{1}{2}\sin^2 x) + HR^3(\tfrac{3}{2}x - 2\sin x + \tfrac{1}{2}\sin x\cos x) + M_oR^2(x - \sin x)\Big|_0^{\pi/2} \tag{14.18}$$

$$= \frac{\tfrac{1}{2}VR^3 + (\tfrac{3}{4}\pi - 2)HR^3 + (\pi/2 - 1)M_oR^2}{EI}$$

$$\theta = \frac{\partial U}{\partial M_o}$$

$$= \int_0^{\frac{\pi}{2}} [VR\sin x + HR(1-\cos x) + M_o]Rdx$$

$$= -VR^2\cos x + HR^2(x - \sin x) + M_oRx\Big|_0^{\pi/2}$$

$$= \frac{VR^2 + (\pi/2 - 1)HR^2 + (\pi/2)M_oR}{EI} \tag{14.19}$$

The deflection produced by any one load or any combination of two loads is found by setting the other load or loads equal to zero; thus, V alone would produce $D_x = \frac{1}{2}VR^3/EI$, and M alone would produce $D_y = M_oR^2/EI$. In this example all results are positive, indicating that D_x is in the direction of H, D_y in the direction of V, and θ in the direction of M_0.

Deflection of a Curved Member Loaded by Forces at Right Angles to the Plane

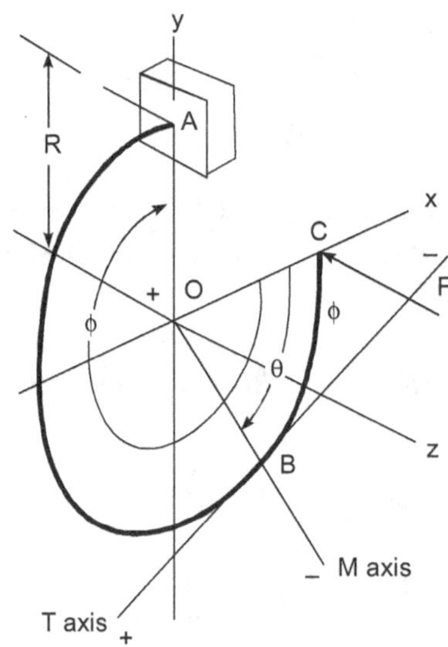

Figure 14.7

Ring ABC in xy plane subject to force F parallel to the z axis, Figure 14.7. Corresponding to a ring segment CB at angle θ from the point of application of F, the moment axis is a line BO and the torque axis is a line in the xy plane tangent to the ring at B. Note the positive directions of the T and M axes.

Determination of the defection of a curved member loaded by forces at right angles to the plane of the member is more difficult, but the method is the same. We shall include here only one of the more useful solutions to such a problem, though the methods for all are similar. Figure 14.7 shows a cantilevered ring segment having a span angle ϕ. Assuming $R/h > 10$, the strain energy neglecting direct shear, is obtained from the equation

$$U = \int_0^\phi \frac{M^2 R\,d\theta}{2EI} + \int_0^\phi \frac{T^2 R\,d\theta}{2GJ} \tag{14.20}$$

The moments and torques acting on a section at B, due to the force F, are

$$M = F R \sin \theta \qquad T = F R (1 - \cos \theta)$$

The defection δ of the ring segment at C and in the direction of F is then found to be

$$\delta = \frac{\partial U}{\partial F} = \frac{FR^3}{2}\left(\frac{\alpha}{EI} + \frac{\beta}{GJ}\right) \tag{14.21}$$

where the coefficient α and β are dependent on the span angle ϕ and are defined as follow:

$$\alpha = \phi - \sin\phi\cos\phi \tag{14.22}$$

$$\beta = 3\phi - 4\sin\phi + \sin\phi\cos\phi \tag{14.23}$$

where ϕ is in radians.

$$\sigma = \frac{MZ}{I}, R = \frac{EI}{M} \text{ inapplicable}$$

DISTORTION OF TUBULAR SECTIONS

In curved beams of thin tubular section, e.g., bends in pipes, the distortion of the cross section produced by the radial components of the fibre stresses reduces both the strength and stiffness. If the beam curvature is not so sharp as to make equation $\sigma = \dfrac{MZ}{I}, R = \dfrac{EI}{M}$ inapplicable the effect of this distortion of the section can be taken into account as follows.

In calculating deflection of curved beams of hollow circular section, replace I by KI, where

$$K = 1 - \frac{9}{10 + 12(tR/a^2)^2}$$

(Here R = Radius of curvature of the beam axis, a = Outer radius of tube section, and t = Thickness of tube wall). In calculating the maximum bending stress in curved beams of hollow circular section, use the formulas.

$$\sigma_{max} = \frac{Ma}{I}\frac{2}{3K\sqrt{3\beta}} \quad \text{at} \quad y = \frac{a}{\sqrt{3\beta}} \quad if \frac{tR}{a^2} < 1.472$$

or

$$\sigma_{max} = \frac{Ma}{I}\frac{1-\beta}{K} \quad \text{at} \quad y = a \quad if \ \frac{tR}{a^2} < 1.472$$

$$\text{where} \qquad \beta = \frac{6}{5 + 6(tR/a^2)^2}$$

and y is measured from the neutral axis. Torsional stresses and deflections are unchanged.

In calculating deflection or stress in curved beams of hollow square section and uniform wall thickness, replace I by

$$\frac{1 + 0.0270\, n}{1 + 0.0656\, n} I$$

where $n = b^4 / R^2 t^2$ (Here R = Radius of curvature of the beam axis, b = Length of the side of the square section, and t = Thickness of the section wall).

TORSION OF CURVED BARS; HELICAL SPRINGS

As an extension we shall discuss tension of curved bars which are applicable to helical springs under direct tension or compression. Readers can later compare the results given here with those given in Chapter 15 on springs. Account must be taken of the influence of curvature and slope for sharply curved bars such as helical springs. Among others, Wahl has discussed this problem, and the former presents charts which greatly facilitate the calculation of stress and deflection for springs of non-circular section.

Let R = radius of coil measured form spring axis to centre of section (Figure 14.8), d = Diameter of circular section , $2b$ = Thickness of square section, P = Load (either tensile or compressive), n = Number of active turns in spring, α = Pitch angle of spring, f = Total stretch or shortening of spring, and t = Maximum shear stress produced. Then for a spring of circular wire,

$$f = \frac{64PR^3n}{Gd^4}\left[1 - \frac{3}{64}\left(\frac{d}{R}\right)^2 + \frac{3+v}{2(1+v)}(\tan\alpha)^2\right] \tag{14.24}$$

$$\tau = \frac{16PR}{\pi d^3}\left[1 + \frac{5}{8}\frac{d}{R} + \frac{7}{32}\left(\frac{d}{R}\right)^2\right] \tag{14.25}$$

For a spring of square wire,

$$f = \frac{2.789PR^3n}{Gb^4} \qquad \text{for} \quad c > 3 \tag{14.26}$$

$$\tau = \frac{4.8PR}{8b^3}\left(1 + \frac{1.2}{c} + \frac{0.56}{c^2} + \frac{0.5}{c^3}\right) \tag{14.27}$$

where $C = R / b$.

For a spring of rectangular wire, section $2a \times 2b$ where $a > b$,

$$f = \frac{3\pi PR^3n}{8Gb^4}\,\frac{1}{a/b - 0.627\left[t\,an\,h(\pi b/2a) + 0.004\right]} \tag{14.28}$$

For $c > 3$ if the long dimension $2a$ is parallel to the spring axis or for $c > 5$ if the long dimension $2a$ is perpendicular to the spring axis,

$$\tau = \frac{PR(3b + 1.8a)}{8b^2a^2}\left(1 + \frac{1.2}{c} + \frac{0.56}{c^2} + \frac{0.5}{c^3}\right) \tag{14.29}$$

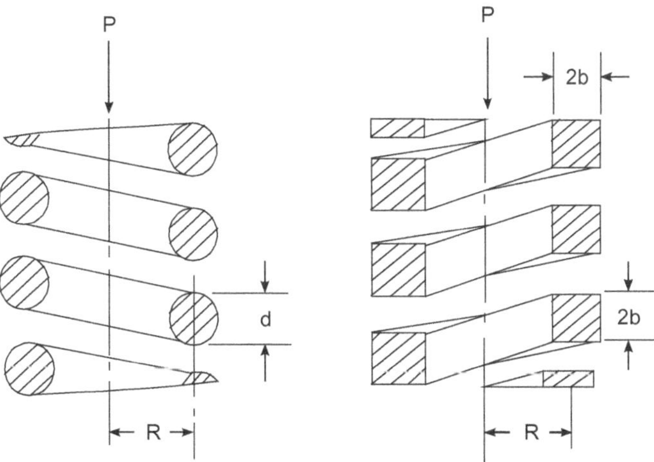

Figure 14.8

It should be noted that in each of these cases the maximum stress is given by the ordinary formula for the section in question (from Table 14.1) multiplied by a corrective factor that takes account of curvature,and these corrective factors can be used for any curved bar of the corresponding cross section. Also,for compression springs with the end turns ground down for even bearing, n should be taken as the actual number of turns (including the tapered end turns) less 2. For tension springs n should be taken as the actual number of turns or slightly more.

Unless laterally supported, compression springs that are relatively long will buckle when compressed beyond a certain critical deflection. This critical deflection depends on the ratio of L,the free length, to D, the mean diameter. Consideration of coil closing before reaching the critical deflection is necessary for design check.

15

SPRINGS

INTRODUCTION

Springs are elastic members which deflect or distort under load and regain their original shape upon removal of load. They are used in suspensions of road and rail vehicles, governors, massive foundations, and mechanisms.

The functions of the springs are

i. to absorb shock or impact loadings as in carriage springs.

ii. to store energy as in clock springs.

iii. to apply forces to and control motions as in brakes and clutches.

iv. to change the characteristics of a member as in flexible mounting of motors.

The springs are made of high carbon steel (0.7 % – 1%) or medium carbon alloy steels. Phosphor bronze, brass, 18/8 stainless steel, monel and other metal alloys are used for corrosion-resistant springs.

Various types of springs are employed for different purposes, some of them are as follows:

1. Helical springs
    i. Close-coiled
    ii. Open-coiled
    iii. Tension helical springs
    iv. Compression helical springs
2. Leaf springs
    i. Full-elliptic
    ii. Semi-elliptic
    iii. Cantilever
3. Torsion springs

4. Circular springs

5. Belleville springs

6. Flat springs

HELICAL SPRINGS—CLOSE-COILED HELICAL SPRINGS

A helical spring is a length of wire or bar wound into a helix. There are mainly two types of helical springs, i. closed-coiled ii. open-coiled.

Close-coiled helical spring with 'Axial Load'

Circular section of wire springs

Let us consider a close-coiled helical spring loaded with an axial load W which is shown in Figure 15.1

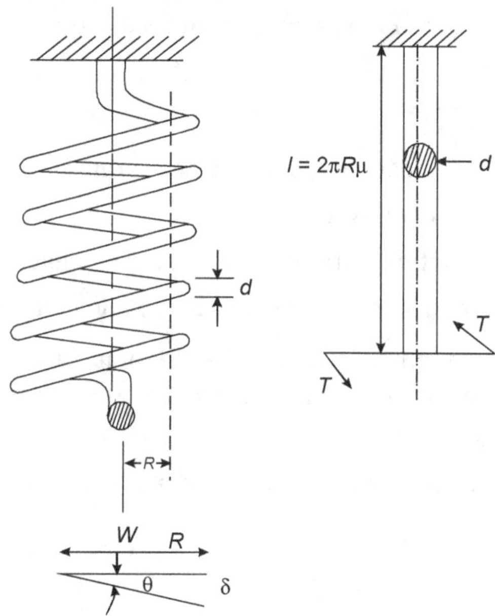

Figure 15.1   Close-coiled helical spring

where,

R—Radius of coil

d—Diameter of the wire of the coil

δ—Deflection of coil under the load W

G—Modulus of rigidity

n—No. of coils of turns

θ—Angle of twist

l—Length of wire $2\pi Rn$

τ—Shear stress and

I_p—Polar moment of inertia of wire $= \dfrac{\pi}{32} d^4$

It may be noted that under axial load, each section of the coil is subjected to torsion, but there are bending and shearing stresses which being small are usually neglected.

Shear stress τ

$$\frac{T}{I_P} = \frac{G\theta}{l} = \frac{\tau}{r} \quad \text{(From torsion equation)}$$

$$\frac{T}{I_P} = \frac{\tau}{r}$$

$$T = \frac{\tau I_P}{r}$$

$$= \frac{\tau \times \pi d^4}{32} \times \frac{2}{d}$$

$$= \tau \cdot \frac{\pi}{16} d^3$$

$$\tau = \frac{16T}{\pi d^3}$$

$$\text{or } \tau = \frac{16WR}{\pi d^3} \qquad (\because T = WR)$$

Deflection,

Again, $\dfrac{T}{I_P} = \dfrac{G\theta}{l}$

$$\theta = \frac{Tl}{GI_P} = \frac{WR \times 2\pi Rn \times 32}{G \times \pi d^4} = \frac{64WR^2 n}{Gd^4}$$

But $\delta = R \times \theta$

$$\delta = \frac{64WR^3 n}{Gd^4} \tag{15.1}$$

Wahl's correction factor

In the above derivation, effect of curvature of spring and direct shear is neglected. The above equation is modified to the following by introducing a factor K called Wahl's correction factor.

$$\tau = \frac{16WR}{\pi d^3}.K$$

$$K = \frac{4S-1}{4S-4} + \frac{0.615}{S}$$

Where $S = \dfrac{D}{d} = $ Spring index

where D = Mean diameter of coil

$$\text{Stiffness of spring } k = \frac{W}{\delta} = \frac{W}{\left(\dfrac{64WR^3 n}{Gd^4}\right)} = \frac{Gd^4}{64R^3 n} \tag{15.2}$$

Energy stored (U)

$$\text{Energy stored, } U = \frac{1}{2}T \times \theta = \frac{1}{2}W.R \times \frac{64WR^2 n}{Gd^4}$$

$$= \frac{1}{2}\cdot\frac{1}{2}\frac{16WR}{Gd^3}\cdot\frac{8WR^2 n}{d} = \frac{1}{4G}\frac{16WR}{\pi d^3}\cdot\frac{16WR}{\pi d^3}\left[2\pi Rnd^2 \times \frac{\pi}{4}\right] \tag{15.3}$$

$$= \frac{1}{4G}\cdot\tau^2 \;\; \text{Volume of wire}$$

$$\text{i.e., } \frac{\tau^2}{4G} \times \text{Volume of wire}$$

$$\text{Again, energy stored, } U = \frac{1}{2}T.\theta$$

$$= \frac{1}{2}\cdot W.R.\frac{\delta}{R}$$

$$= \frac{1}{2}\cdot W.\delta \;(\because \delta = R\theta)$$

PROBLEM 15.1

1. For a close-coiled helical spring subjected to an axial load of 300 N, having 12 coils of wire diameter of 16 mm, and made with coil diameter of 250 mm, find

   i. Axial deflection

   ii. Strain energy stored

   iii. Maximum torsional shear stress in the wire

   iv. Maximum shear stress using Wahl's correction factor.

   Take $G = 80$ GN/m²

Solution

Modulus of rigidity G = 80 GN/m², number of coils $n = 12$, wire diameter $d = 16$ mm = 0.016m, coil diameter $D = 250$ mm = .25m, axial load, W = 300 N.

Axial deflection, δ

i. $\delta = \dfrac{64WR^3 n}{Gd^4} = \dfrac{64 \times 300 \times (0.25/2)^3 \cdot 12}{80 \times 10^9 \times (0 \cdot 016)^4}$

$$= 0.0858\,\text{m} = 85.8\,\text{mm}$$

ii. Strain energy stored, $U = \dfrac{1}{2}W\delta = \dfrac{1}{2}300 \times 0.0858 = 12.87$ Nm

iii. τ = Max. torsional shear stress

$\tau = \dfrac{16WR}{\pi d^3} = \dfrac{16 \times 300 \times (0.25/2)}{\pi \times (0.016)^3} \times 10^{-6}$ MN/m² = 46.63 MN/m²

$\tau = 46.63$ MN/m²

iv. Maximum shear stress using Wahl's correction factor, τ

$$\tau = \frac{16WR}{\pi d^3} \times K$$

where,

$$K = \frac{4S-1}{4S-4} + \frac{0.615}{S}$$

But S (Spring index) $= \dfrac{D}{d} = \dfrac{250}{16} - 15.625$

$K = \dfrac{4 \times 15.625 - 1}{4 \times 15.625 - 4} + \dfrac{0.615}{15.625} = 1.0513 + 0.0394$

$$= 1.0907$$

$\tau = \dfrac{16 \times 300 \times (0.25/2)}{\pi (0.016)^3} \times 1.0907 \times 10^{-6}$ MN/m²

$$= 50.85 \text{ MN/m}^2$$

i.e., $\tau - 50.85$ MN/m²

PROBLEM 15.2

1. It is required to design a helical compression spring subjected to a maximum force of 1250 N. The deflection of the spring corresponding to the maximum force should be approximately 30 mm. The spring index can be taken as 6. The spring is made of patented and cold-drawn steel wire. The ultimate tensile strength and modulus of rigidity of the spring material are 1090 and

81370 N/mm² respectively. The permissible shear stress for the spring wire should be taken as 50% of the ultimate tensile strength. Design the spring and calculate:

  i. Wire diameter;

 ii. Mean coil diameter

iii. Number of active coils;

 iv. Total number of coils;

  v. Free length of the spring; and

 vi. Pitch of the coil.

Solution

The permissible shear stress is given by,

$$\tau = 0.5\,S_{ut} = 0.5(1090) = 545\,\text{N/mm}^2$$

$$K = \frac{4S-1}{4S-4} + \frac{0.615}{S} = \frac{4(6)-1}{4(6)-4} + \frac{0.615}{6} = 1.2525$$

$$\tau = K\left(\frac{8PC}{\pi d^2}\right) \text{ or } 545 = (1.2525)\left\{\frac{8(1250)(6)}{\pi d^2}\right\}$$

$$\therefore d = 6.63 \text{ or } 7 \text{ mm} \tag{i}$$

$$D = S = 6(7) = 42 \text{ mm} \tag{ii}$$

$$\delta = \frac{8PD^3N}{Gd^4} \quad \text{or} \quad 30 = \frac{8(1250)(42)^3\,N}{(81370)(7)^4}$$

$$N = 7.91 \text{ or } 8 \text{ coils} \tag{iii}$$

It is assumed that spring has square and ground ends. The number of active coils is 2. Therefore,

$$N_1 = N + 2 = 8 + 2 = 10 \text{ coils} \tag{iv}$$

The actual deflection of the spring is given by,

$$\delta = \frac{8\pi D^3N}{Gd^4} = \frac{8(1250)(42)^3(8)}{(81370)(7)^4} = 30.34\,\text{mm}$$

Solid length of spring $= N_t d = 10(7) = 70$ mm

It is assumed that there will be a gap of 1 mm between consecutive coils when the spring is subjected to the maximum force. The total number of coils is 10.

The total axial gap between the coils will be $(10 - 1) \times 1 = 9$ mm.

$$\text{Free length} = \text{Solid length} + \text{Total axial gap} + \delta$$
$$= 70 + 9 + 30.34$$
$$= 109.34 \text{ or } 110 \text{ mm} \tag{v}$$

$$\text{Pitch of coil} = \frac{\text{Free length}}{(N_t - 1)} = \frac{109.34}{(10 - 1)}$$
$$= 12.15 \text{ mm}$$

PROBLEM 15.3

A railway wagon weighing 40 kN and moving with a speed of 8 km/hr is stopped by a buffer of 4 springs whose allowable maximum compression is 150 mm. Find out the number of turns in each spring, if the diameter of the spring wire is 14 mm and the diameter of the coil is 80 mm. Assume $G = 84 \text{ GN/m}^2$.

Solution

Weight of railway wagon $W = 4o$ kN;

$$\text{Speed of the wagon } V = \frac{8 \text{ Km}}{\text{h}} = \frac{8 \times 1000}{60 \times 60} = 2.22 \text{ m/s}$$

$$\text{Kinetic energy of wagon} = \frac{1}{2} \frac{W}{g} V^2 = \frac{1}{2} \times \frac{40 \times 1000}{9.81} \times (2.22)^2 = 10048 \text{ MN}$$

$$\text{Energy absorbed by each buffer spring} = \frac{10048}{4} = 2512 \text{ MN}$$

Also, energy absorbed = Work done

$$2512 = \frac{1}{2} W\delta = \frac{1}{2} W \times \frac{150}{1000}$$

$$W = \frac{2512 \times 2 \times 1000}{150} = 33493 \text{ N}$$

No. of coils n is found using the relation,

$$\delta = \frac{64WR^3 n}{Gd^4}$$

$$0.15 = \frac{64 \times 33493 \times (0.04)^3 \times n}{84 \times 10^9 \times (0.014)^4}$$

$$n = \frac{0.15 \times 84 \times 10^9 \times (0.014)^4}{64 \times 33493 \times (0.04)^3}$$

$$n = 3.53$$

Closed-coiled Spring Subjected to Axial Twist

When a twisting couple is applied to the spring parallel to the axis of the spring wire, it produces a bending effect on it. Depending on the direction of application of the twisting couple or turning moment, the spring coils will close or open out. In both cases the radius of coils will close or open out. In both cases the radius of the coils changes and bending stresses will be induced.

Let

n_1 = Number of coils before application of twisting moment

n_2 = Number of coils after application of twisting moment

ϕ = Angle of rotation

I = Moment of inertia of coil section

R_1 = Mean radius of coil

R_2 = Changed radius of coil

σ_b = Bending stress, and

E = Young's modulus of elasticity.

$$\text{Initial curvature} = \frac{1}{R_1}$$

$$\text{Final curvature} = \frac{1}{R_2}$$

$$\text{Change in curvature} = \frac{1}{R_2} - \frac{1}{R_1}$$

As per bending equation,

$$\frac{M}{I} = \frac{E}{R} \quad \text{or} \quad \frac{1}{R} = \frac{M}{EI}$$

$$\frac{M}{EI} = \frac{1}{R_2} - \frac{1}{R_1}$$

Since the length of wire remains unchanged before and after applying twisting couple,

$$l = 2\pi R_1 n_1 = 2\pi R_2 n_2$$

but ϕ = Final helix angle – Initial helix angle = $2\pi n_2 - 2\pi n_1$

$$\frac{M}{EI} = \frac{1}{R_2} - \frac{1}{R_1} = \frac{1}{\left(\dfrac{l}{2\pi n_2}\right)} - \frac{1}{\left(\dfrac{l}{2\pi n_1}\right)} = \frac{2\pi}{l}(n_2 - n_1) = \frac{\phi}{l}$$

$$\therefore \phi = \frac{Ml}{EI}$$

$$= \frac{M \times 2\pi R n}{E\left(\dfrac{\pi}{64} d^4\right)} = \frac{128 MRn}{Ed^4} \tag{15.4}$$

Also,

$$\sigma_b = \frac{M}{2} = \frac{My}{I} = \frac{Md/2}{\dfrac{\pi d^4}{64}} = \frac{32M}{\pi d^3} \tag{15.5}$$

Now, energy stored $U = \dfrac{1}{2} M\phi$

$$= \frac{1}{2} M \frac{Ml}{EI} = \frac{1}{2} \frac{M^2 l}{EI}$$

$$= \frac{\pi^2 d^6 \sigma_b^2 l \times 64}{2 \times 32 \times 32 \times 6 \times E \times \pi d^4}$$

$$= \frac{\sigma_b^2 \times \pi d^2 l}{4 \times 8E} = \frac{\sigma_b^2}{8E} \times \frac{\pi d^2 l}{4}$$

$$U = \frac{\sigma_b^2}{8E} \times \text{Volume of spring wire} \tag{15.6}$$

Hence, energy stored,

$$U = \frac{\sigma_b^2}{8E} \times \text{Volume of spring wire}$$

OPEN-COILED HELICAL SPRINGS

Open-Coiled with Axial Load

Applying the same symbols as used in previous types the slope of coils α is introduced additionally.

Now the wire length $= l = 2\pi R \sec\alpha \times n$ where R is the radius of the coil.

The couple applied to the material under the applied load W will be WR and at each point along centre line of wire this couple may be resolved into two components, one of torsion and one of bending.

Couple producing torsion $= T = WR\cos\alpha$

Couple producing bending $= M = WR\sin\alpha$

The couple WR will act in a plane passing through the axes OY and OX, the centre line of the wire being at an angle α to OX. The bending moment will tend to wind the coils of the spring more tightly and to a smaller radius of curvature.

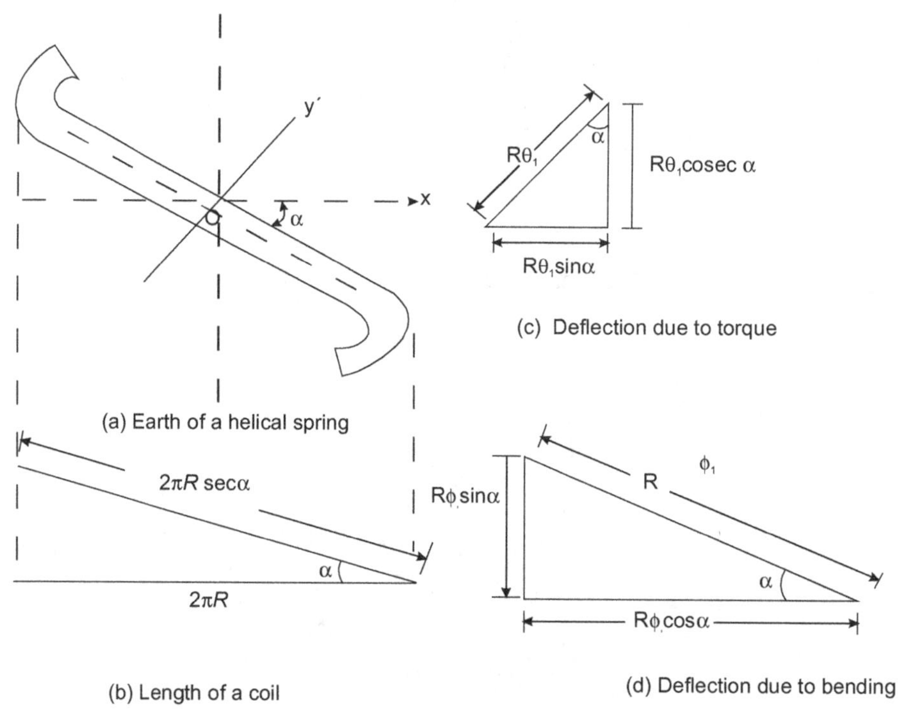

Figure 15.2

The axial extension of the spring may be most easily calculated by equating the workdone by the load to the internal strain energy of the material.

The strain energy due to bending $= \dfrac{M^2 l}{2EI}$ (15.7)

$\therefore$ Strain energy for twisting $= \dfrac{T^2 l}{2CI_P}$

where the deflection is δ

$$\frac{1}{2} W.\delta = \frac{T^2 l}{2GI_P} + \frac{M^2 l}{2EI}$$

$$= \frac{W^2 R^2 \cos^2 \alpha.l}{2GI_P} + \frac{W^2 R^2 \sin^2 \alpha.l}{2EI}$$

$$= \frac{W^2 R^2 l}{2}\left[\frac{\cos^2 \alpha}{GI_P} + \frac{\sin^2}{EI}\right]$$

$$= \frac{W^2 R^2\, 2\pi\, Rn \sec \alpha}{2}\left[\frac{\cos^2 \alpha}{GI_P} + \frac{\sin^2 \alpha}{EI}\right]$$

$$\text{Deflection } \delta = 2WR^3 n\pi \sec\alpha \left[\frac{\cos^2\alpha}{GI_p} + \frac{\sin^2\alpha}{EI}\right] \qquad (15.8)$$

$$\text{If } \alpha = 0, \quad \delta = \frac{2WR^3 n\pi}{GI_p}$$

$$= \frac{64WR^3 n\pi}{Gd^4}$$

Alternative Method

Let θ_1 = Angle of twist caused by torque, $(T = WR \cos\alpha)$

Then, $R\,\theta_1$ = Total twisting effect along OY′

and ϕ_1 = Angular change due to bending moment $M\,(= WR \sin\alpha)$

Then $R\,\phi_1$ = Total bending effect along the axis of the wire.

Now, total downward deflection [Figure 15.2(c) and (d)].

$$\delta = \theta_1 \cos\alpha + R\phi_1 \sin\alpha$$

But, $\theta = \dfrac{Tl}{GI_p}$ and $\phi = \dfrac{Ml}{EI}$

$$\delta = \frac{RTl}{GI_p}\cos\alpha + \frac{RMl}{EI}\sin\alpha$$

$$= \frac{R \times WR\cos\alpha \cdot l \times \cos\alpha}{GI_p} + \frac{R \times WR\sin\alpha \cdot l \times \sin\alpha}{EI}$$

$$= WR^2 l \left[\frac{\cos^2\alpha}{GI_p} + \frac{\sin^2\alpha}{EI}\right]$$

$$= WR^2 \times 2\pi R\sec\alpha \times n \left[\frac{\cos^2\alpha}{GI_p} + \frac{\sin^2\alpha}{EI}\right] \qquad (\because l = 2\pi R\sec\alpha \times n)$$

$$\text{or } \delta = 2WR^3 n\pi \sec\alpha \left[\frac{\cos^2\alpha}{GI_p} + \frac{\sin^2\alpha}{EI}\right]$$

The result is the same as in equation (15.8)

Let ψ = resultant of rotations θ_1 and α_1

$$\psi = \theta_1 \sin\alpha - \phi_1 \cos\alpha$$

$$= \frac{Tl}{GI_p} \sin\alpha - \frac{Ml}{EI} \cos\alpha$$

$$= \frac{WR\cos\alpha \times 2\pi R\sec\alpha \times n\sin\alpha}{GI_p} - \frac{WR\sin\alpha \times 2\pi R\sec\alpha \times n\cos\alpha}{EI}$$

$$= 2WR^2 n\pi \sin\alpha \left(\frac{1}{GI_p} - \frac{1}{EI} \right) \qquad (15.9)$$

Open-Coiled spring with Axial Thrust

Let the torque T be applied about axis of spring OY' be resolved about OX' and OY' (Figure 15.2)

Component about OX', $T' = \alpha \sin$ (causes torsion of spring)

Component about OY', $M' = \alpha \cos$ (causes bending of the coil)

If ϕ' is the angular twist due to T' and θ' due to M', and ϕ due to T, then by principle of conservation of energy we have,

$$\frac{1}{2}T\phi = \frac{1}{2}T'\phi' + \frac{1}{2}M'\theta' = \frac{1}{2}T' \times \frac{T'l}{GI_p} + \frac{1}{2}M' \times \frac{M'l}{EI}$$

$$= \frac{1}{2}T'^2\frac{l}{GI_p} + \frac{1}{2}M'^2\frac{l}{EI} = \frac{1}{2}\frac{T^2\sin^2\alpha.l}{GI_p} + \frac{1}{2}\frac{T^2\cos^2\alpha \times l}{EI}$$

$$\therefore \phi = \frac{T\sin^2\alpha \cdot l}{GI_p} + \frac{T\cos^2\alpha \cdot l}{EI} \qquad (15.10)$$

But $l = 2\pi R\sec\alpha \times n$

$$\phi = 2TRn\pi \sec\alpha \left[\frac{\sin^2\alpha}{GI_p} + \frac{\cos^2\alpha}{EI} \right]$$

For axial deflection/extension resolve rotations as before

$$\delta = TRl\sin\alpha\cos\alpha \left(\frac{1}{GI_p} - \frac{1}{EI} \right)$$

$$= TR \times 2\pi Rn\sec\alpha \times \sin\alpha\cos\alpha \left[\frac{1}{GI_p} - \frac{1}{EI} \right]$$

$$\delta = 2TR^2 n\pi \sin\alpha \left[\frac{1}{GI_p} - \frac{1}{EI} \right] \qquad (15.11)$$

Stresses in Circular Wire of Open-coil Spring

Consider an open coil spring of mean coil radius R, helix angle α and wire diameter d. The stresses in the circular wire are calculated as follows.

Case I *Subjected to an axial load W*

On any section of the spring wire,

Twisting moment, $T = WR\cos\alpha$

Bending moment, $M = WR\sin\alpha$

Maximum torsional stress on any section

$$\tau_1 = \frac{16T}{\pi d^3} = \frac{16WR\cos\alpha}{\pi d^3} \tag{15.12}$$

Direct shear stress due to axial load

$$\tau_2 = \frac{W}{\pi/4\, d^2} = \frac{4W}{\pi d^2}$$

Maximum shear stress at inner coil radius

$$\tau = \tau_1 + \tau_2 = \frac{16WR\cos\alpha}{\pi d^3} + \frac{4W}{\pi d^2} \tag{15.13}$$

Minimum shear stress at outer coil radius

$$\tau = \tau_1 - \tau_2 = \frac{16WR\cos\alpha}{\pi d^3} - \frac{4W}{\pi d^2} \tag{15.14}$$

Maximum stress due to bending

$$\sigma_b = \frac{M}{Z} = \frac{WR\sin\alpha}{\left(\dfrac{\pi d^3}{32}\right)} = 32\frac{WR\sin\alpha}{\pi d^3} \tag{15.15}$$

Maximum principal stress occurs at the inner coil radius.

$$\text{Principal stresses: } \sigma_{max}, \sigma_{min} = \frac{\sigma_b}{2} \pm \sqrt{\left(\frac{\sigma_b}{2}\right)^2 + \tau^2} \tag{15.16}$$

Case II *Subjected to an axial torque I*

Twisting moment $T' = T\sin\alpha$

Bending moment $M' = T\cos\alpha$

Maximum torsional shear stress due to T'

$$\tau = \frac{16T'}{\pi d^3} = \frac{16T\sin\alpha}{\pi d^3} \qquad (15.17)$$

Maximum stress due to bending

$$\sigma_b = \frac{32M'}{\pi d^3} = \frac{32T\cos\alpha}{\pi d^3} \qquad (15.18)$$

Principal stresses at the extreme radii (inner and outer radii of coil)

$$\sigma_{max}, \sigma_{min}, = \frac{\sigma_b}{2} \pm \sqrt{\left(\frac{\sigma}{2}\right)^2 + \tau^2} \qquad (15.19)$$

SPRINGS IN SERIES

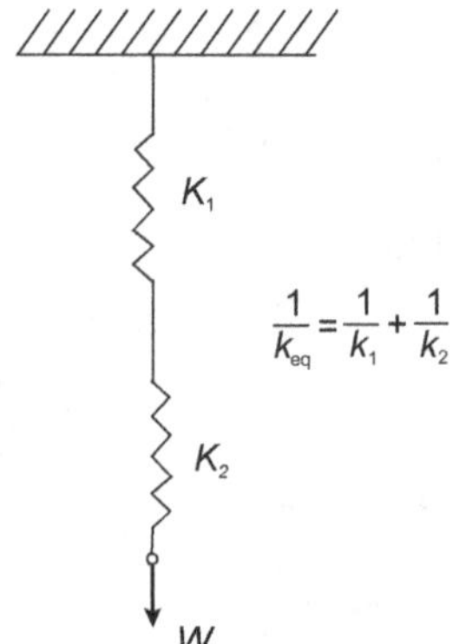

Figure 15.3 Spring in series

Figure 15.3 shows two springs connected in series

Let

W = Load applied

k_2 = stiffness of spring 2

δ_2 = Extension of spring 2

k_1 = stiffness of spring 1

δ_2 = Extension of spring 1

k = stiffness of composite spring

It can be easily imagined that each spring will be subjected to load W and the total extension produced will be the sum of extensions of two springs.

$$\therefore \text{Total extension}, \delta = \delta_1 + \delta_2$$

$$\frac{W}{k_{eq}} = \frac{w}{k_1} + \frac{w}{k_2} \quad (\because \delta = \frac{W}{k})$$

$$\text{or } \frac{1}{k_{eq}} = \frac{1}{k_1} + \frac{1}{k_2}$$

SPRINGS IN PARALLEL

Figure 15.4 shows two springs, connected in parallel. When subjected to load W, they will extend equally say by an amount δ. The load will be shared such that $W = W_1 + W_2$.

$$\delta \cdot k = \delta \cdot k_1 + \delta \cdot k_2 \text{ or } k = k_1 + k_2$$

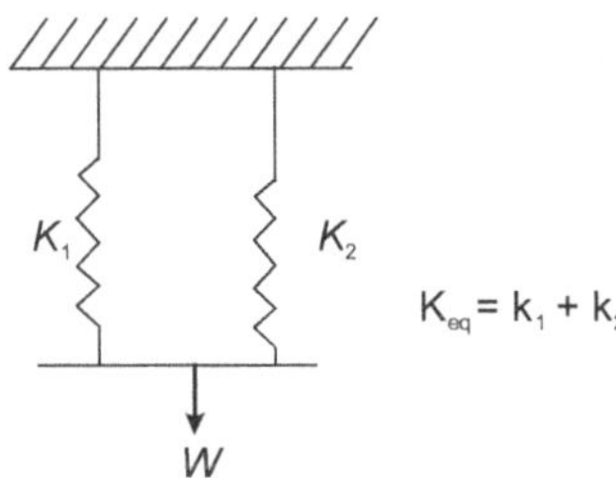

Figure 15.4   Springs in parallel

FLAT SPIRAL SPRING

Figure 15.5 shows a flat spiral spring which consists of a uniform thin rectangular metallic strip (of thickness t and width b) wound into a spiral in one plane, the outer end being anchored to a pin D and the inner end being attached to the winding spindle C for winding the spring. Let H and V be respectively the reactions at D along $\perp r$ and to the line joining the axis of the spindle to the centre of the pin D. Let AB be a small length dl with coordinates x,y.

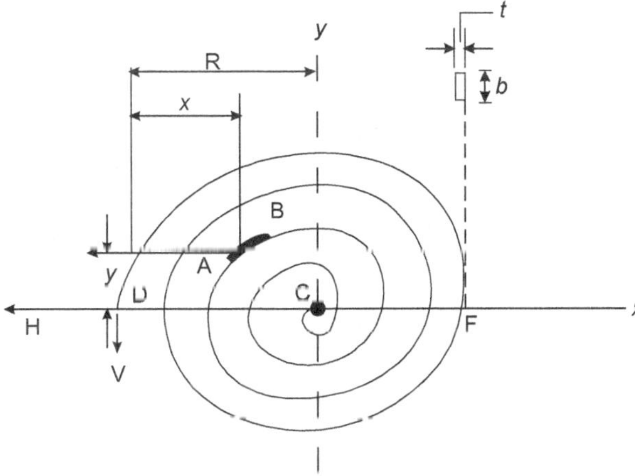

Figure 15.5

Bending moment at the element (taken positive if the number of turns increase) is given by

$$M = V \cdot x - H \cdot y$$

$$(15.20)$$

Let $d\phi$ = Rotation of the end B with respect to end A of the element dl due to the bending moment M

Then $d\phi = \dfrac{Mdl}{EI}$ or $d\phi = \dfrac{V \cdot x - H \cdot y}{EI} \cdot dl$

Integrating both sides, we get

$$\phi = \frac{V}{EI}\int xdl - \frac{H}{EI}\int ydl$$

$$(15.21)$$

If the centroid of the spiral is assumed to be at the centre of spindle C, the first moment of length l about the line CD will be zero.

$$\text{i.e.,} \int ydl = 0$$

$$(15.22a)$$

and the moment of the length about YY lines gives

$$\int ydl = lR$$

$$(15.22b)$$

where

l = length of the spring strip.

R = Distance between points C and D

Also, the spring strip will be in equilibrium, when

$$T = V\!.R$$

$$(15.23)$$

where T = winding torque

Substituting equations 3 and 4 in 2 we have

$$\phi = \frac{(T/R)}{EI} \cdot (l.R) = \frac{Tl}{EI}$$

$$(15.24)$$

From equation 15. we find that the maximum bending moment will be obtained on such a section of strip where x is maximum and y is minimum. Obviously the point F is such a point where y is zero and x is maximum.

Let $DF = 2R$

Then from equation (4) we get

$$M_{max} = M_F = \frac{T}{R} \cdot (2R) = 2T$$

$$(15.25)$$

The corresponding maximum bending stress on the cross-section of the strip

$$\sigma_{max} = \frac{2T}{I} \times \frac{t}{2} = \frac{12Tt}{bt^3} = \frac{12T}{bt^2} \qquad (15.26a)$$

$$T = \frac{bt^2}{12}.\sigma_{max} \qquad (15.26b)$$

(If σ_{max} is known, value of T can be calculated)

Energy stored (U)

The energy stored (corresponding to the torque T) can be calculated as follows.

$$U = \frac{1}{2}T\phi$$

$$= \frac{1}{2}T.\frac{Tl}{EI} = \frac{T^2l}{2EI}$$

$$= \frac{T^2l}{2E\left(\dfrac{bt^3}{12}\right)} = \frac{6T^2l}{Ebt^3}$$

$$\text{i.e.,} U = \frac{6T^2l}{Ebt^3} \qquad (15.27)$$

To find maximum energy which can be stored (when the stress reaches max. value) will be obtained by substituting the value of T from equation 15.26(b) in equation 15.27.

$$U_{max} = \frac{6l}{Ebt^3}\left[\frac{bt^2\sigma_{max}}{12}\right]^2$$

$$= \frac{\sigma_{max}^2}{24E}(btl) = \frac{\sigma_{max}^2}{24E}\,\text{Volume of spring} \qquad (15.28)$$

Resilience of spring = Energy per unit volume

$$= \frac{\sigma_{max}^2}{24E} \qquad (15.29)$$

LAMINATED SPRINGS

Leaf or semi-elliptical leaf or carriage springs used commonly in suspensions of trucks, trains, trolleys, etc. They consist of a number of leaves of spring steel held together at the centre with clamps. The plates are provided with curvature initially and the ends of the top plate are pin jointed to the chassis of the vehicle. The load at which the plates become straight in called 'proof load'.

Semi-Elliptical Springs

Let

b = Width of each plate

t = Thickness of each plate

a = Overlap at each end

N = Number of plates in spring

l = Spring span length

σ_b = Bending stress

E = Young's modulus of elasticity

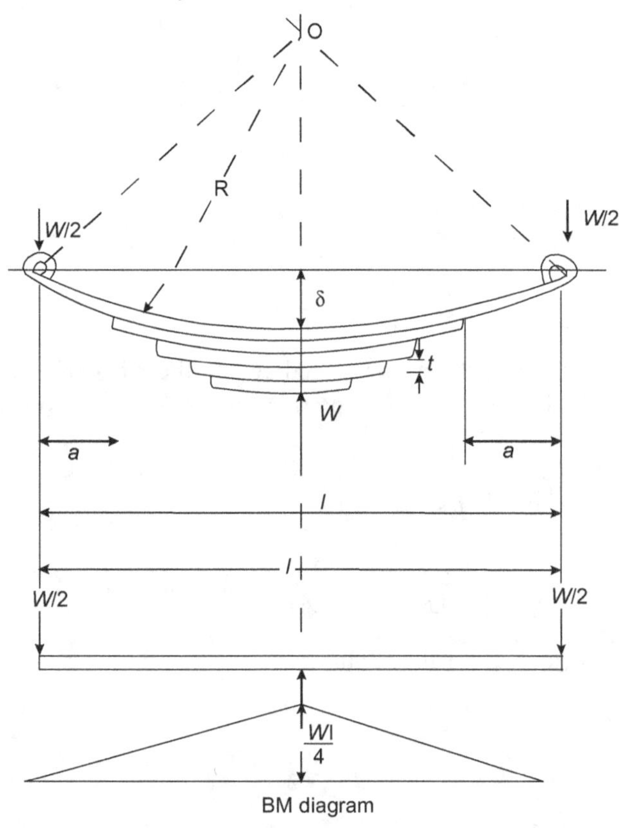

Figure 15.6

The maximum bending moment $\dfrac{Wl}{4}$ occurs at the centre and is resisted by all N plates equally.

Resisting moment of each plate $M = \dfrac{Wl}{4N}$

Now $\sigma_b = \dfrac{M_y}{I}$

$$\sigma_b = \frac{Wl}{4N} \cdot \frac{t}{2} \cdot \frac{1}{\left(\dfrac{bt^3}{12}\right)}$$

$$\text{or}\quad \sigma_b = \frac{3Wl}{2Nbt^2} \tag{1}$$

Let

δ = Initial deflection of plates

R_c = Radius of curvature of the plates

From Figure 15.6 we have

$$R_c^2 = \left(\frac{l}{2}\right)^2 + (R_c - \delta)^2$$

$$R_c^2 = \left(\frac{l}{2}\right)^2 + R_c^2 + \delta^2 - 2R_c\delta$$

Neglecting δ^2, we get

$$2R_c \cdot \delta = \left(\frac{l}{2}\right)^2, \text{ or } \delta = \frac{l}{8R_c}$$

$$\text{Also,} \frac{M}{I} = \frac{F}{R_c}$$

$$\text{or } R_c = \frac{EI}{M} = \frac{Ebt^3}{12 \times \dfrac{Wl}{4N}} = \frac{ENbt^3}{3Wl}$$

$$\delta = \frac{l^2}{8 \times \dfrac{EN}{3}\dfrac{bt^3}{Wl}}$$

$$\delta = \frac{3Wl^3}{8ENbt^3} \tag{2}$$

Strain enregy $U = \dfrac{M^2}{2EI} \times$ total length of leaves

$$= \frac{\left[\dfrac{Wl}{4N}\right]^2 \times 12}{2Ebt^3} \times \text{total length of leaves}$$

$$= \frac{3W^2l^2}{8En^2bt^3} \times \text{total length of leaves}$$

$$= \left(\frac{3Wl}{2Nbt^2}\right)^2 \times \frac{bt}{2\times 3E} \times \text{total length of leaves}$$

$$= \frac{\sigma_b^{\,2}}{6E} \times bt \times \text{total length of leaves}$$

$$= \frac{\sigma_b^{\,2}}{6E} \times \text{Volume of spring}$$

$$\text{i.e., } U = \frac{\sigma_b^{\,2}}{6E} \times \text{Volume of spring} \tag{3}$$

Quarter Elliptical Spring

These springs are called cantilever laminated springs. In figure 15.7 is shown a spring of this type.

It has an effective span length l and carries a load W. If observed carefully it will be noted that the spring is equivalent to a laminated semi-elliptical type spring of length $2l$ carrying a load $2W$ at the centre. Hence by replacing l by $2l$ and W by $2W$ the deflection and stress can be obtained as follows:

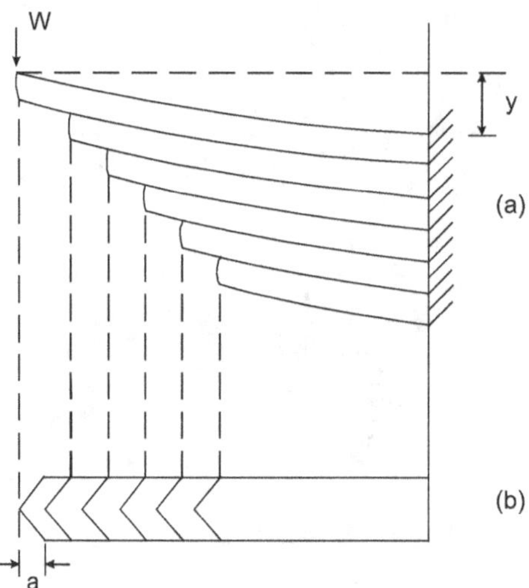

Figure 15.7

$$\delta = \frac{3(2W)\cdot(2l)^3}{8ENbt^3}$$

$$\text{or } \delta = \frac{6Wl^3}{ENbt^3} \tag{1}$$

Similarly, $\sigma_b = \dfrac{3(2W)2l}{2Nbt^2}$

$$\sigma_b = \frac{6Wl}{Nbt^2} \tag{2}$$

PROBLEM 15.4

A carriage spring is made of 600 mm long and 9.5 mm thickness steel plates and 50 mm broad. How many plates are required to carry a load of 4.5 kN without the stress exceeding 230 MN/m²? What would be the central deflection and the initial radius of curvature, if the plates straighten under the load for plate?

Solution

$$E = 200\,\text{GN/m}^2$$

Span length $l = 600$ mm $= 0.6$ m

Thickness of each plate $= t = 9.5$ mm $= 0.0095$ m

Width of each plate $b = 50$ mm $= 00.05$ m

Load $= W = 4.5$ KN

$E = 200$ GN/m²

$\sigma_b = 230$ MN/m²

Number of plates, N

We know,

$$M = \frac{Wl}{4N} \text{ where } M = \text{Bending moment per plate}$$

$$\text{But } M = \sigma_b \times Z \; \left(Z = \frac{1}{6}bt^2 \text{section modulus of one plate}\right)$$

$$\therefore \frac{Wl}{4N} = \sigma_b \times \frac{bt^2}{6}$$

$$\frac{4.5\times10^3 \times 0.6}{4N} = 230\times10^6 \times \frac{0.05\times(0.0095)^2}{6}$$

$$N = \frac{4.5 \times 10^3 \times 0.6 \times 6}{4 \times 230 \times 10^6 \times 0.05 \times (0.0095)^2}$$

$$N = 3.9 \text{ rounded off to } 4$$

Initial radius of Curvature (R_c)

Using the relation

$$\frac{\sigma_b}{y} = \frac{E}{R_c}, \text{ we get}$$

$$\frac{230 \times 10^6}{t/2} = \frac{200 \times 10^9}{R_c}$$

$$R_c = \frac{200 \times 10^9 \times 0.0095}{230 \times 10^6 \times 2}$$

$$R_c = 4.13 \text{ m}$$

Central deflecion, δ :

$$\delta = \frac{3Wl^3}{8ENbt^3} = \frac{3 \times 4.5 \times 10^3 \times (0.6)^3}{8 \times 200 \times 10^9 \times 4 \times 0.05 \times (0.0095)^3}$$

$$\delta = 10.6 \text{ mm}$$

REVIEW QUESTIONS

SHORT QUESTIONS

1. Define proof load of springs.

2. State the expression for maximum shear stress and deflection of close-coiled helical spring when subjected to axial load W.

3. What is meant by stiffness of spring?

4. What is a laminated spring?

5. Give any two practical applications of helical springs.

6. Write down the equation for Wahl factor.

LARGE QUESTIONS

1. A helical spring, in which the mean diameter of the coils is 12 times the wire diameter, is to be designed to absorb 300 J energy with an extension of 150 mm. The maximum shear stress is not to exceed 140 N/mm². Determine the mean diameter of the spring, diameter of the wire which forms the spring and the number of turns. Assume the moduls of rigidity of the material of the spring as 80 kN/mm².

2.   A close coiled helical spring of 100 mm mean diameter is made of 10 mm diameter rod and has 20 turns. The spring carries an axial load of 200 N. Determine the shearing stress. Taking the value of modulus of rigidity as 84 x 10^9 N /m. Determine the deflection when carrying this load. Also calculate the stiffness of the spring and frequency of free vibrations for a mass hanging from it.

3.   i.   Derive a relation for deflection of a closely coiled helical spring subjected to an axial downward load W.

     ii.   A quarter elliptic leaf spring 60 cm long is made of steel plates of width 10 times the thickness. The spring is to carry a load of 3 KN and the end deflection is limited to 5 cm. The bending stress of the plates must not exceed 3000 N/mm². Find suitable values of the size and number of plates to be used. Take E = 2 ×10^5 N/mm².

4.   A close-coiled helical spring is made of a round steel wirc. It carries an axial load of 150 N and is to just get over a rod of 36 mm. The deflection in the spring is not to exceed 25 mm. The maximum allowable shearing stress developed in the spring wire is 200 N/mm² and G = 80,000 n/mm². Find the mean coil diameter, wire diameter and number of turns.

5.   A closely coiled helical spring made of 10 mm diameter steel wire has 15 coils of 100 mm mean diameter. The spring is subjected to an axial load of 100 N. Calculate:

     i.   the maximum shear stress

     ii.   the deflection and

     iii.   stiffness of the spring.

     Take modulus of rigidity, $C = 8.16 \times 10^4$ N/mm².

6.   A closely coiled helical spring is made of 6 mm wire. The maximum shear stress and the deflection under a load of 200 N is not to exceed 90 N/mm² and 1.1 cm respectively. Determine the number of coils and their mean radius. Take N or C = 0.84 ×10^5 N/mm².

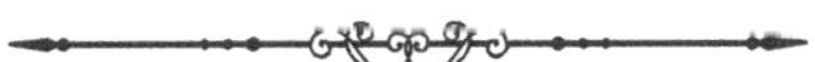

16

COLUMNS AND STRUTS

INTRODUCTION

In Chapter 3 and 6, we dealt with compression members. We assumed that these members were short. This assumption was necessary, because we could not have applied the equations derived in these chapters if the members were long. Most columns, however, are not short, and their behaviour is governed more by stability than strength. In this chapter, we will discuss the behaviour of columns, and the mathematical and empirical method for their design. Most design codes for columns are based upon these principles for the behaviour of long columns, to be derived later in this chapter, and are modified according to material properties, practical end condition and experimental results.

Axially Loaded Compression Members

Axially loaded compression members can thus be classified into long, medium and short columns. Short columns generally fail by compression. The compressive stress can be calculated as load/ area. Thus in Chapter 3, when we calculate the load-carrying capacity of a concrete column, we specify it as a short compression member. Also the load shared by steel and concrete is in the modular ratio (or the ratio of modulii of elasticity of the materials). This does not hold good if the compression member is long. As the load is increased, the compressive stress reaches a value that causes crushing of materials and the column fails.

Medium and long columns are governed by compression as well as bending. The combined stress causes failure of the column by a phenomenon known as buckling. Such columns are governed by a different method of computing allowable or failure load.

Buckling

Buckling is the phenomenon by which a medium or long column fails by a combination of bending and compression. As the axial load or increased, the column tend to buckle or bend. For small values of the load the column remains in stable equilibrium returning to its straight position when the load is removed. As the load is increased, the column gets a permanent deflection even after the lateral load removed. The column is in deflected, neutral equilibrium position. Further increase in the load causes more deflection and the column fails. Such a failure is known as buckling.

Critical or Buckling Load

The maximum value of load, p_{cr}, up to which the column remains in a position of neutral equilibrium, is known as the critical load for the column. At the critical load, the column can remain in equilibrium either in a straight or a deflected position.

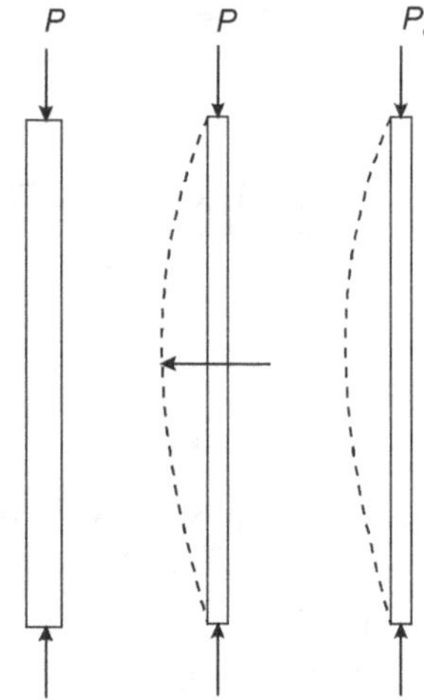

Figure 16.1

It must be clear by now that if a column is geometrically perfect and the load is applied perfectly along the axis, with ideal end conditions, then upon the action of an axial load, the column will fail by pure compression and not buckling. Buckling requires eccentricity or shape imperfections, there a presence of a slight lateral load causes a lateral deflection which in turn causes the axial load to create a blinding moment

EULER'S THEORY ON COLUMNS

Of the many theories developed for columns, theory developed by Leonard Euler (1707–1783) is the one most commonly quoted and many design codes have been based on this basic analysis of column behaviour. There is limitation of applicability of Euler's formulae but has been used to derive many empirical formulae based on this basic analysis.

Assumptions in Euler's Theory

Euler's theory is based on the following assumptions.

    i. the column is initially perfectly straight.

    ii. The load on the column is truly axial.

    iii. The column is prismatic, has the same cross section throughout.

    iv. The material of the column is perfectly elastic, homogeneous and isotropic.

    v. The column is long, its length being much larger compared to sectional dimensions.

    vi. The column fails by buckling.

INTERESTING POINTS TO PONDER

Tall cylindrical pressure vessels are employed in Refineries, Petrochemical and Fertilizer plants. The design for internal pressure is done using either thin or thick shell theory. However such tall vessels are subjected to wind and/or earthquake loads. These loads cause bending. Bending stresses are induced by such loads. These are determined by treating the tall vessel as a line element along the vessel centre line idealised as a beam. The loads can be lumped at different points along the height. Using beam formulation, the bending and shear stresses are determined at various sections. The direct net downward compressive stresses at each section due to the weight of the portion of vessel above the section is also to be considered and added to the longitudinal bending stress, longitudinal pressure stress, taking into account the net tensile, compressive stresses.

The stresses due to bending are more in the lower part of the vessel and designs may involve variable thickness shells with higher thickness provided at the bottom section. The welding of such vessels are done at site with cylindrical portions welded on top of one another.

Likewise combining the stresses in the three directions, the net stresses in x, y, z directions are determined. Then either maximum shear stresses or Vonmises stresses are determined. This is to be checked against codal values as per ASME sec VIII.

Critical Load for columns with Hinged Ends

Consider a column with hinged ends subjected to an axial load P, and remaining in a deflected position shown by dotted lines (Figure 16.2). If y is the deflection at a distance x from the hinged end, assuming the deflections to be small, we can state the equations for the elastic curve as given in Chapter 13 as

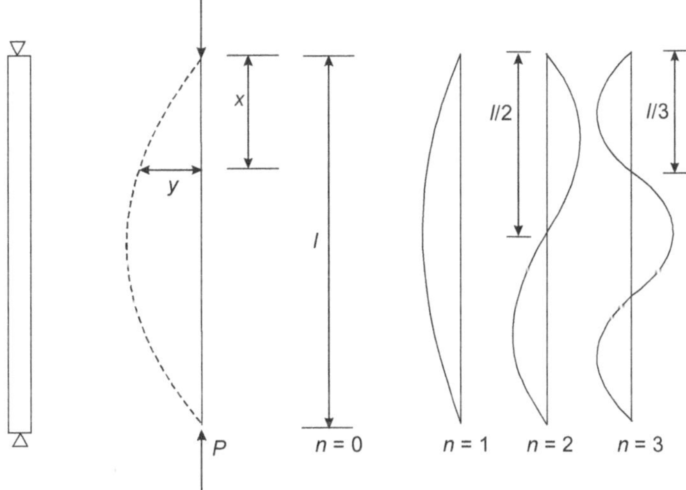

Figure 16.2

$$EI\frac{d^2y}{dx^2} = -Py$$

$$\frac{d^2y}{dx^2} + \frac{P}{EI}y = 0 \tag{16.1}$$

This is a second-order ordinary differential equation.

The solution for this equation is obtained as

$$y = C_1\cos\sqrt{\frac{P}{EI}}x + C_2\sin\sqrt{\frac{P}{EI}}x \tag{16.2}$$

The constants C_1 and C_2 can be evaluated from the end conditions of the column as (i) $y = 0$ at $x = 0$ and (ii) $y = 0$ at $x = 1$, $y = 0$ gives $C_1 = 0$ $x = 1$, $y = 0$ gives $C_2\sin(P/EI)^{1/2}l = 0$. C_2 cannot be zero because if C_1 and C_2 are both zero, the column is not in the bent configuration, and $y = 0$ everywhere. Therefore,

$$\sin\sqrt{\frac{P}{EI}}l = 0 \tag{16.3}$$

gives $(P/EI)^{1/2}l = n\pi$ where $n = 0,1,2,3,\dots.$

$$p = \frac{n^2\pi^2EI}{l^2}$$

Thus

$$y = C_2(\sin n\pi/l)x \tag{16.4}$$

which is the equation for the elastic curve of the column.

For different values of n, the deflected shapes are shown in Figure 16.2. $n = 0$ means no deflection and no axial load. $n = 1$ gives the deflected shape with the elastic curve lying entirely to one side of the straight line. $n = 2$ gives the elastic shape in two halves on either side, and so on. The configurations $n = 2$, $n = 3$ and $n = 4$, etc. are possible only when the column is braced. Further, they give larger values of the load P compared to the one obtained with $n = 1$. Thus the critical load, P_{cr}, for such a case is given by

$$P_{cr} = \frac{\pi^2EI}{l^2} \tag{16.5}$$

This critical load is for a column of length l, with both ends hinged, which we will consider to be the standard case. For the other end condition shown in Fig. 16.3. we will adopt a similar approach.

We may have a doubt at this point. We saw the column's buckled configuration shown on one side in Figure 16.2. It could as well buckle on the right side or any other direction as well. In reality due to imperfections either in the coaxiality of loading implying eccentric loading or imperfection in geometric dimension like out of roundness, the columns will buckle in the plane of least moment of inertia of the section.

We shall revert to the differential equation (16.1).

$$EI\frac{d^2y}{dx^2} = -py \quad y = f(x) \tag{16.6}$$

Recognize this as a second-order ordinary differential equation very similar to the one governing the free vibrations of a single degree of freedom system executing oscillations in the vertical direction, i.e.,

$$m\frac{d^2y}{dt^2} = -ky; \quad y = f(t)$$

We solved for natural frequency $w = \sqrt{\dfrac{k}{m}}$. In the case of columns under axial load, we cannot determine the exact shape but solve for critical load just as in free vibration, we cannot decide the amplitude but can only solve for natural frequencies.

Such problems come under the category of eigen value problems. The word 'eigen' means 'unique'. Complex structural problems formulated using matrix method or finite element methods either for stability or vibration analysis finally results in solution to eigen value problems for which ready made computer-based mathematical procedures are available.

Columns with One End Fixed and the Other End Hinged

Consider figure 16.3. The best form is modified by the fixity at one end, where the slope is zero. The fixing moment M_A at A will give rise to horizontal reactions H at each end such that $H = M_A/l$. From the deflected shape shown in Figure 16.3, we can say that

$$EI\frac{d^2y}{dx^2} = -Py + H(l - x)$$

$$\frac{d^2y}{dx^2} + \frac{P}{EI}y = \frac{P}{EI}\frac{H}{P}(l - x) \tag{16.7}$$

or

Setting $(P/EI)^{\frac{1}{2}} = m,$ of this differential equation is

$$y = C_1 \cos mx + C_2 \sin mx + \frac{H}{P}(l - x) \tag{16.8}$$

$$\frac{dy}{dx} = -C_1 m \sin mx + C_2 m \cos mx - \frac{H}{P}$$

The end conditions in this case are $y = 0$ at $x = 0$ and $dy/dx = 0$ at $x = 1$, $y = 0$ at $x = 0$ gives $C_1 = Hl/P$, $dy/dx = 0$ at $x = 1$ gives $C_2 = H/Pm$. Also at $x = 1$, $y = 0$, This yields

$$C_1 \cos ml + C_2 \sin ml = 0$$

Substituting for C_1 and C_2,

$$-\frac{Hl}{P} \cos ml + \frac{H}{Pm} \sin ml = 0, \tan ml = ml$$

This equation can be solved by trial and error to give $ml = 4.4934$ rad as the smallest value.

$(P/EI)^{\frac{1}{2}}l = 4.4934$ and $P = 2\pi^2 EI/l^2$ because $(4.4934)^2 << 2\pi^2$. The critical load is thus

$$P_{cr} = \frac{2\pi^2 EI}{l^2} \tag{16.9}$$

Columns with One End Fixed and the Other End Free—Cantilever column

Tall process towers, chimneys and stacks can be modelled as cantilever columns with the base grounded on to the foundation and top left free.

Figure 16.3 shows a column with the bottom end fixed and the top end free. The maximum deflection is at the free end and the BM throughout is negative. Thus,

$$EI\frac{d^2 y}{dx^2} = -M = P(\delta - y)$$

$$P = \frac{n^2\pi^2 EI}{4l^2}$$

The smallest value of $n = 1$ gives the smallest value of buckling load P_{cr},

$$P_{cr} = \frac{\pi^2 EI}{4l^2} \tag{16.10}$$

Columns with Both Ends Fixed

For the fixed ended column shown in Figure 16.3, the deflected shape is such that the slopes at both ends are zero. There will be fixed end moments at both ends as shown, which are equal and hence do not give rise to any horizontal reaction. As above,

$$EI\frac{d^2 y}{dx^2} = -Py + M$$

$$\frac{d^2 y}{dx^2} + \frac{P}{EI} y = \frac{P}{EI} \frac{M}{P} \tag{16.11}$$

Setting $m^2 = P/EI$ we get

$$\frac{d^2 y}{dx^2} + m^2 y = \frac{M}{P} m^2$$

This differential equation when solved with boundary conditions (slope = deflection = 0 at both ends) yields P_{cr}'

$$P_{cr} = \frac{4\pi^2 EI}{l} \tag{16.12}$$

Equivalent Lengths for Columns with Different End Conditions

The Euler buckling loads for columns have basically the same form for all end conditions. We can express this as

$$P_{cr} = \frac{\pi^2 EI}{(l_e)^2}$$

where l_e is called the equivalent length of the column.

$l_e = l$ if both ends are hinged

$l_e = \dfrac{l}{\sqrt{2}} = 0.7071l$ if one end is fixed and the other end hinged

$l_e = 2l$, for a column with one end fixed and the other end free,

$l_e = \dfrac{1}{2} l$ for a column with both ends fixed.

where l is the actual length of the column. The equivalent length can be substituted in the general formula for critical load, and obtain value for critical loads for different end conditions.

The concept of equivalent length can also be understood from the deflected shapes shown in Figure 16.3, where the effective length is the length between the points of inflection on the deflected shape.

Slenderness ratio   The critical Euler load is given by the general formula

$$P_{cr} = \frac{\pi^2 EI}{l_e^2}$$

This can be stated as, by substituting $I = Ar^2$

$$P_{cr} = \frac{\pi^2 EAr^2}{l^2}$$

$$\frac{P_{cr}}{A} = \frac{\pi^2 E}{(l/r)^2}$$

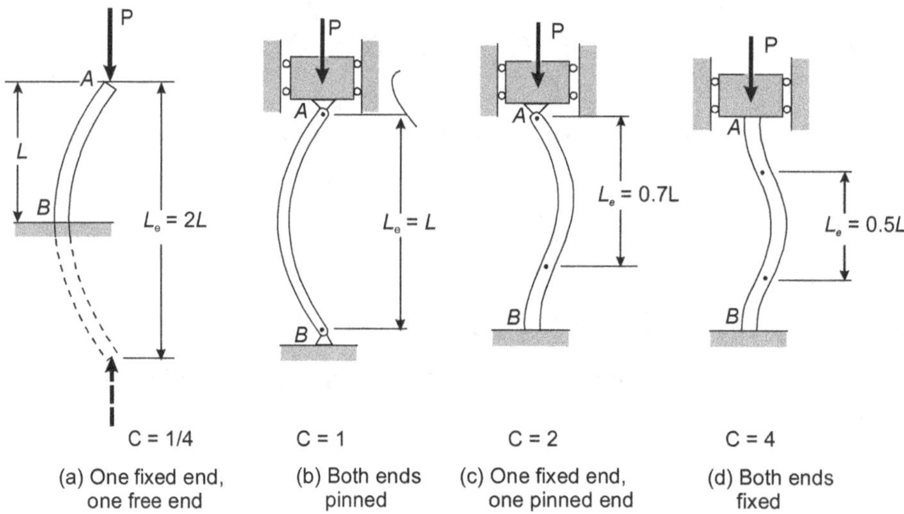

Figure 16.3

where r is the radius of gyrati0n and A is the area of cross section of the member. P_{cr}/A is the critical average stress σ_{cr} in the member.

$$\sigma_{cr} = \frac{\pi^2 E}{(l/r)^2}$$

where l/r is known as the slenderness ratio

$$\text{Slenderness ratio} = \frac{\text{Length}}{\text{Radius of gyration}}$$

The critical compressive stress is thus inversely proportional to the square of the slenderness ratio. A plot of σ_{cr} against l/r will give appear is shown in Figure 16.4. The curve is asymptotic to both the σ_{cr} and (l/r) axes, σ_{cr} tending to infinity as l/r tends to zero and σ_{cr} tending to become zero as l/r tends to infinity.

If design check indicates possibility of buckling, the following alternatives can be considered to improve the column design. Improvements can be implemented based on practical and site constraints.

1.   Increase section area (I is increased)
2.   Provide more fixity (L_e is changed)
3.   Increase the number of columns (load is reduced)
4.   Additional mid way restraints (L_e is reduced)

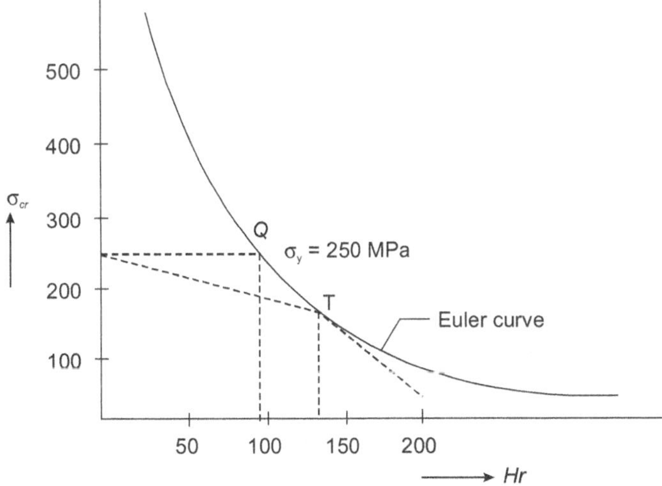

Figure 16.4

PROBLEM 16.1

An aluminum column of length L and rectangular cross section has a fixed end B and supports a centric load at A. Two smooth and rounded fixed plates restrain end A from moving in one of the vertical planes of symmetry of the column, but allow it to move in the other plane. (a) Determine the ratio a/b of the two sides of the cross section corresponding to the most efficient design against buckling. (b) Design the most efficient cross section for the column, knowing that $L = 500$mm, $E = 70$ Gpa, $P = 20$ kN, and that a factor of safety of 2.5. is required.

Buckling in xy plane Referring to Figure, we note that we effective length of the column with respect to buckling in this plane is $L_e = 0.7L$. The radius of gyration r_z of the cross section is obtained by writing

$$I_z = \frac{1}{12}ba^3 \quad A = ab$$

and, since $I_z = Ar_z^2, r_z^2 = \dfrac{I_z}{A} = \dfrac{\frac{1}{12}ba^3}{ab} = \dfrac{a^2}{12} = a/\sqrt{12}$

The effective slenderness ration of the column with respect to bucking in the xy plane is

$$\frac{L_e}{r_z} = \frac{0.7L}{a/\sqrt{12}}$$

Buckling in xz plane   The effective length of the column with respect to buckling in this plane is $L_e = 2L$, and the corresponding radius of gyration is $r_y = b/\sqrt{12}$. Thus

$$\frac{L_e}{r_y} = \frac{2L}{b / \sqrt{12}} \tag{2}$$

(a) *Most efficient design*   The most efficient design is that for which the critical stresses corresponding to the two possible modes of buckling are equal. We note that this will be the case if the two values obtained above for the effective slenderness ratio are equal. We write

$$\frac{0.7L}{a / \sqrt{12}} = \frac{2L}{b / \sqrt{12}}$$

and, solving for the ratio a/b, $\dfrac{a}{b} = \dfrac{0.7}{2} = 0.35$

(b) *Design for Given Data*   Since F.S = 2.5 is required,

$$P_{cr} = (F.S.)P = (2.5)(20kN) = 50kN$$

Using a=0.35 *b*, we have $A = ab = 0.35b^2$ and

$$\sigma_{cr} = \frac{P_{cr}}{A} = 50 \times 10^3 \, N$$

Making $L = 0.500$ m in eq.(2), we have substituting for $L_e / r_y = 3.464 / b$ and substituting for L_e/r and σ_{cr} we get

$$\sigma_{cr} = \frac{\pi^2 E}{(L_e / r)^2} \quad \frac{50 \times 10^3 \, N}{0.35b^2} = \frac{\pi^2 (70 \times 10^9 \, Pa)}{(3.464 / b)^2}$$

$$b = 39.7 \, \text{mm} \quad a = 0.35b = 13.9 \, \text{mm}$$

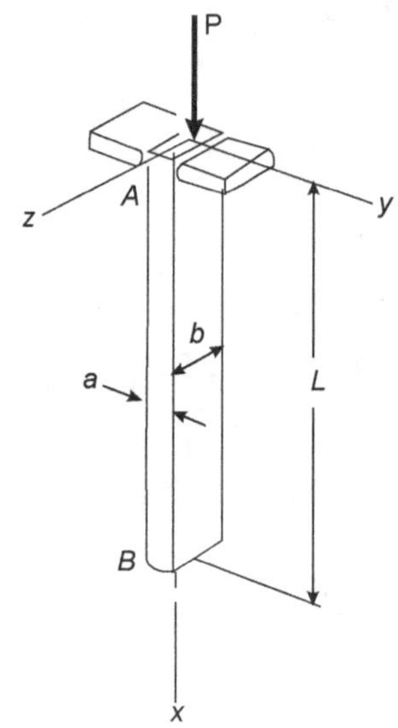

PROBLEM 16.2

Develop specific Euler equations for the sizes of columns having

(a) Round cross sections

(b) Rectangular cross sections

Solution

(a) Using $A = \pi d^2 / 4$ and $k = \sqrt{I / A} = [(\pi d^4 / 64) / (\pi d^2 / 4)]^{1/2} = d / 4$ equation for P_{er}

$$\frac{P_{cr}}{A} = \frac{C\pi^2 E}{(l / r)^2}; d = \left(\frac{64 P_{cr} l^2}{\pi^3 CE}\right)$$

(b) For the rectangular column, we specify a cross section h × b with the restriction that $h \leq b$. If the end conditions are the same for buckling in both directions, then buckling will occur in the direction of the least thickness. Therefore

$$I = \frac{bh^3}{12} \quad A = bh \quad k^2 = I / A = \frac{h^2}{12}$$

Substituting these in equation for p_{er} gives

$$b = \frac{12 P_{cr} l^2}{\pi^2 CEh^3}$$

Note, however, that rectangular columns do not generally have the same end conditions in both directions.

INTERMEDIATE—LENGTH COLUMNS WITH CENTRAL LOADING

Over the years there have been a number of columb formulas proposed and used for the range of l/k values for which the Euler' formula is not suitable. Many of these are based on the use of a single material; others, on a so-called safe unit load rather than the critical value. Most of these formulas are based on the use of a single material; others, on a so-called safe unit load rather than the critical value. Most of these formulas are bsed on the use of a linear relationship between the slenderness ratio and the unit load. The *parabolic* or *J.B. Johnson* formula now seems to be the preferred one among designers in the machine, automotive, aircraft, and structural-steel constructon fields.

The general form of the parabolic formula is

$$\frac{P_{cr}}{A} - a - b\left(\frac{l}{R^2}\right)^2 \tag{16.13}$$

where a and b are constants that are evaluated by fitting a parabola to the Euler curve of Figure 16.4 shown by the dashed line ending at T. If the parabola is begun at σ_y, then $a = \sigma_y$. If the point T is selected as previously noted, then equation (16.13) gives the value of $(l/k)_1$ and the constant b is found to be

$$b = \left(\frac{\sigma_y}{2\pi}\right)^2 \frac{1}{CE}$$

where constant C depends on the end condition as shown Figure 16.3.

Upon substituting the known values of a and b into equation 16.13 we obtain, for the parabolic equation,

$$\frac{P_{cr}}{A} = \sigma_y - \left(\frac{\sigma_y}{2\pi}\frac{l}{k}\right)^2 \frac{1}{CE}\frac{l}{k} \leq \left(\frac{l}{k}\right) \tag{16.14}$$

COLUMNS WITH ECCENTRIC LOADING

We have noted before that deviations from an ideal column, such as load eccentricities or crookedness, are likely to occur during manufacture and assembly. Though these deviations are quite small, it is still convenient to have a method of dealing with them. Frequently, too, problems occur in which load eccentricities are unavoidable.

Figure 16.5 shows a column in which the linie of action of the column force is separated from the centroidal axis of the column by the eccentricity e. This problem is developed by using equation $\dfrac{M}{EI} = \dfrac{d^2 y}{dx^2}$ and the free-body diagram of Figure 16.5.

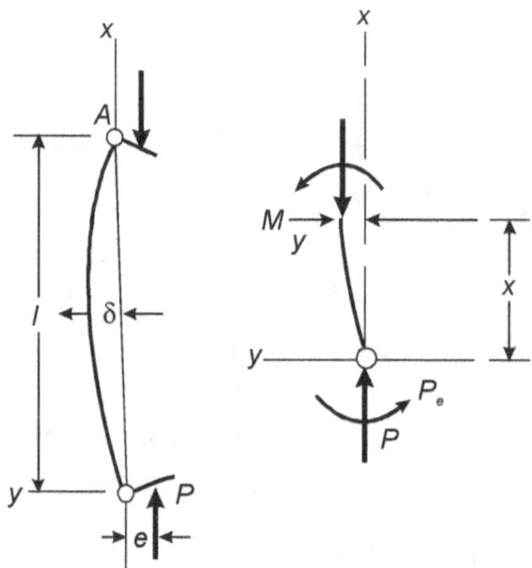

Figure 16.5

This results in the differential equation

$$\frac{d^2 y}{dx^2} + \frac{P}{EI} y = -\frac{Pe}{EI} \tag{16.15}$$

The solution of equation (16.15), for the boundary conditions that $y = 0$ at $x = 0, l$ is

$$y = e\left[\tan\left(\frac{l}{2}\sqrt{\frac{P}{EI}}\right)\sin\left(\sqrt{\frac{P}{EI}}x\right) + \cos\left(\sqrt{\frac{P}{EI}}x\right) - 1 \right] \tag{16.16}$$

By substituting $x = l/2$ in equation (16.16) and using a trigonometric identity, we obtain

$$\delta = e\left[\sec\left(\sqrt{\frac{P}{EI}}\frac{l}{2}\right) - 1 \right] \tag{16.17}$$

The maximum bending moment also occurs at midspan and is

$$M_{\text{max}} = -P(e + \delta) = -P_e \sec\left(\frac{l}{2}\sqrt{\frac{P}{EI}}\right) \tag{16.18}$$

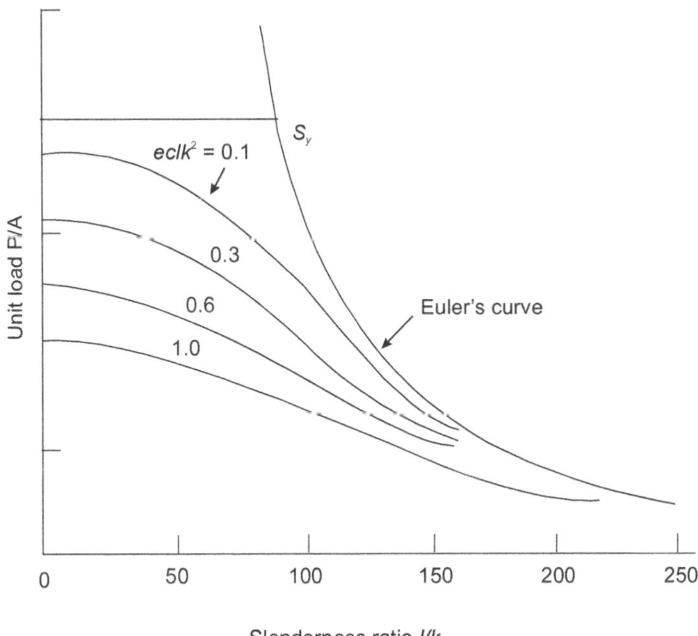

Figure 16.6

The magnitude of the maximum *compressive* stress at the midspan is found by superposing the axial component and the bending component. This gives

$$\sigma_c = \frac{P}{A} - \frac{Mc}{I} = \frac{P}{A} - \frac{Mc}{Ak^2} \tag{16.19}$$

Substituting M_{max} from equation (16.18) yields

$$\sigma_c = \frac{P}{A}\left[1 + \frac{ec}{k^2}\sec\left(\frac{1}{2k}\sqrt{\frac{P}{EA}}\right)\right]$$

(16.20)

By imposing the compressive yield strength σ_{yc} as the maximum value of σ_c, we can write equation (16.20) in the form

$$\frac{P}{A} = \frac{\sigma_{yc}}{1 + (ec/k^2)\sec[(l/2k)\sqrt{P/AE}]}$$

(16.21)

This is called the *secant column formula*. The term ec/k^2 is called the eccentricity ratio. Figure 16.6 is a plot of equation (16.21) for a steel having a compressive (and tensile) yield strength of 280 MPa. Note how the P/A contours asymptotically approach the Euler curve as l/k increases.

Equation 16.21 cannot be solved explicitly for the load P. Design charts, on the fashion of Figure 16.6 can be prepared for a single material if much column design is to be done. Otherwise, a root-finding technique using numerical methods must be used.

STRUTS OR SHORT COMPRESSION MEMBERS

A short bar loaded in pure compression by a force P acting along the centroidal axis will shorten in accordance with Hooke's law, until the stress reaches the elastic limit of the material. At this point, permanent set is introduced and usefulness as a machine member may be at an end. If the foce P is increased still more, the material either becomes "barrel-like" or fractures. When there is eccentricity in the loading, the elastic limit is encountered at smaller loads.

A strut is a short compression member such as the one shown in Figure 16.7. The magnitude of the maximum compressive stress in the x direction at point B in an intermediate section is the sum of a simple component P/A and a flexural component Mc/I; that is,

$$\sigma_c = \frac{P}{A} + \frac{Mc}{I} = \frac{P}{A} + \frac{PecA}{IA} = \frac{P}{A}\left(1 + \frac{ec}{k^2}\right)$$

(16.22)

where $k = (I/A)^{1/2}$ and is the radius of gyration, c is the coordinate of point B, and e is the eccentricity of loading.

Note that the length of the strut does not appear in equation (16.22). In order to use the equation for design or analysis, we ought, therefore, to know the range of lengths for which the equation is valide. In other words, how long is a shoft member?

The different between the secant formula equation (16.21) and equation (16.22) is that the secant equation, unlike equation (16.22), accounts for an increased bending moment due to bending deflection. Thus the secant equation shown the eccentricity to be magnified by the bending deflection. This difference between the two formulas suggests that one way of differentiating between a "secant column" and strut, or short compression member, is to say that in a strut, the effect of bending deflection must be limited to a certain small percentage of the eccentricity. If we decide that the limiting percentage is to be 1 per cent of e, then, from equation (16.17) the limiting slenderness ratio turns out to be

$$\left(\frac{1}{k}\right)_2 = 0.282\left(\frac{AE}{P}\right)^{1/2} \tag{16.23}$$

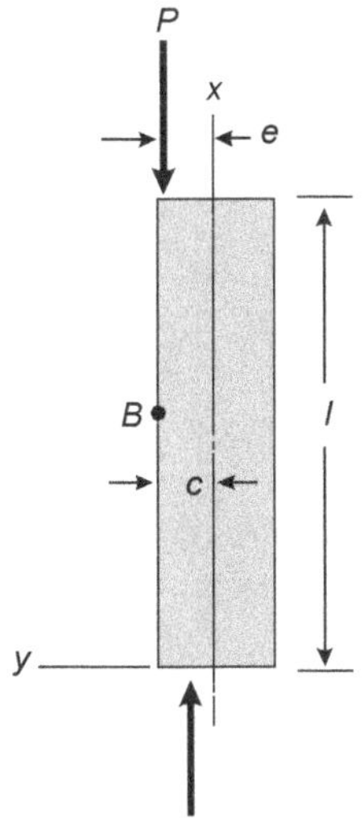

Figure 16.7 Eccentrically loaded strut

This equation then gives the limiting slenderness ratio equation (16.22). If the actual slenderness ratio is greater than $(l/k)_2$, then use the secant formula; otherwise, use equation (16.22).

PROBLEM 16.3

Figure a shows a workpiece clamped to a milling machine table by a bolt tightened to a tension of 8.9 kN. The clamp contact is offset from the centroidal axis of the strut by a distance $e = 2.5$ mm, as shown in part b of the figure. The strut, or block, is steel, 25 mm square and 0.1 m long, as shown. Determine the maximum compressive stress in the block.

Solution

First we find $A = bh = 0.025(0.025) = 625 \times 10^{-6}\,\text{m}^2$, $I = bh^3/12 = 0.025(0.025)^3/12 = 32.55 \times 10^{-9}\,\text{m}^4$, $k^2 = I/A = 32.55 \times 10^{-9}/625 \times 106^{-6} = 52.1 \times 10^{-6}\,\text{m}^2$, and $l/k = 0.1/(52.1 \times 10^{-6})1/2 = 13.9$. Equation 16.23 gives the limiting slenderness ratio as

$$\left(\frac{1}{k}\right)_2 - 0.282\left(\frac{AE}{P}\right)^{1/2} = 0.282\left[\frac{625 \times 10 - 6(210 \times 10^9)}{8900}\right]^{1/2} = 34.2$$

Thus the block could be as long as

$$l = 34.2k = 34.2(52.1 \times 10 - 6)^{1/2} = 0.24\,\text{m}$$

before it needed to be treated by using the secant formula. So equation (16.22) applies and the maximum compressive stress is

$$\sigma_c = \frac{P}{A}\left(1 + \frac{ec}{k^2}\right) = \frac{8900}{625 \times 10^{-6}}\left[1 + \frac{0.0025(0.0125)}{52.1 \times 10^{-6}}\right] = 22.8\,\text{MPa}$$

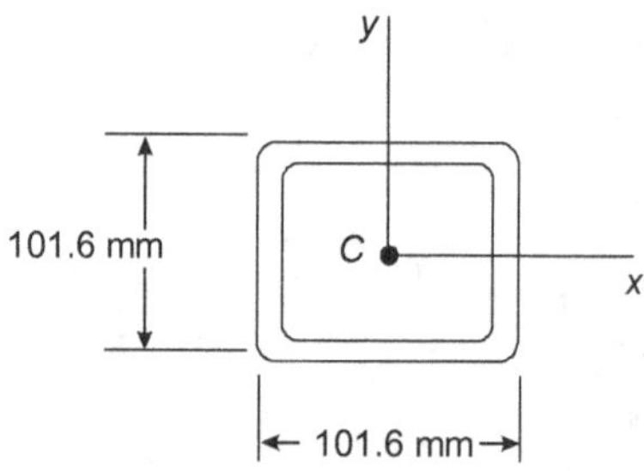

PROBLEM 16.4

The uniform column *AB* consists of a 2.4 m section of structural tubing having the cross section shown. (a) Using Euler's formula and a factor of safety of two, determine the allowable centric load for the column and the corresponding normal stress. (b) Assuming that the allowable load, found in part a, is applied as shown at a point 18 mm from the geometric axis of the column, determine the horizontal deflection of the top of the column and the maximum normal stress in the column. Use $E = 200$ GPa.

Given data

$A = 2320$ mm^2

$I = 3.42 \times 10^6$ mm^4

$r = 38.4$ mm, $c = 50.8$ mm

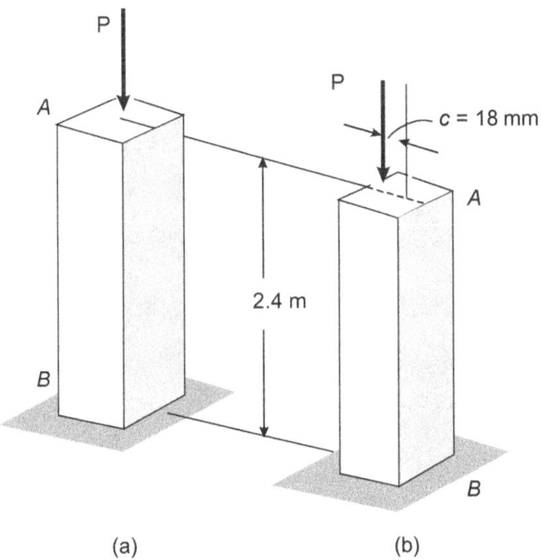

Solution

Effective length Since the column has one end fixed and one end free, its effective length is

$$L_e = 2(2.4m) = 4.8\,\text{m}$$

Critical load Using Euler's formula, we write

$$P_{cr} = \frac{\pi^2 EI}{L_e^2} = \frac{p^2(200\,\text{GPa})(3.42\times10^{-6}\,\text{m}^4)}{(4.8)^2} = 293\,\text{kN}$$

(a) *Allowable load and stress* For a factor of safety of 2, we find

$$P_{all} = \frac{Pcr}{F.S} = \frac{293\,kN}{2} = 146.5\ \text{kN}$$

and

$$\sigma = \frac{P_{all}}{A} = \frac{146.5\,\text{kN}}{2320\times10^{-6}\,m^2} = 63.1\,\text{MPa}$$

(b) *Eccentric load* We observe that column AB and its loading are identical to the upper half of the column of Figure 16.5 (b) which was used in the derivation of the secant formulas; We use secant formula here. Recalling that $P_{all}/P_{cr} = \dfrac{1}{2}$ and using equation (16.17), we compute the horizontal deflection of point A.

$$y_m = \left[\sec\left(\frac{\pi}{2}\sqrt{\frac{P}{Pcr}} \right) - 1 \right] = \left(18\,\text{mm}\left[\sec\left(\frac{\pi}{2\sqrt{2}} \right) - 1 \right] \right)$$

$$(18\,\text{mm})[2.252-1]\quad y_m = 22.5\,\text{mm}$$

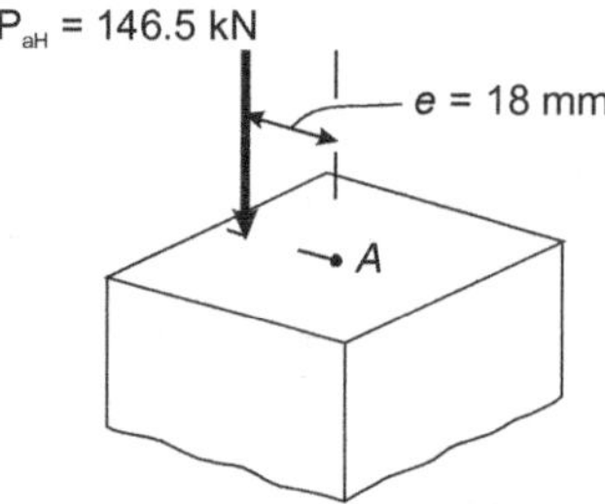

The maximum normal stress is obtained as:

$$\sigma_m = \frac{P}{A}\left[1 + \frac{ec}{r^2}\sec\left(\frac{\pi}{2}\sqrt{\frac{P}{P_{cr}}}\right)\right]$$

$$= \frac{146.5\,\text{kN}}{2320\times10^{-6}\,\text{m}^2}\left[1 + \frac{(18\,\text{mm})(50.8\,\text{mm})}{(38.4\,\text{mm}^2)}\sec\left(\frac{\pi}{2\sqrt{2}}\right)\right]$$

$$= (63.15\,\text{MPa})[1 + 0.620(2.252)]\quad \sigma_m = 151.3\,\text{MPa}$$

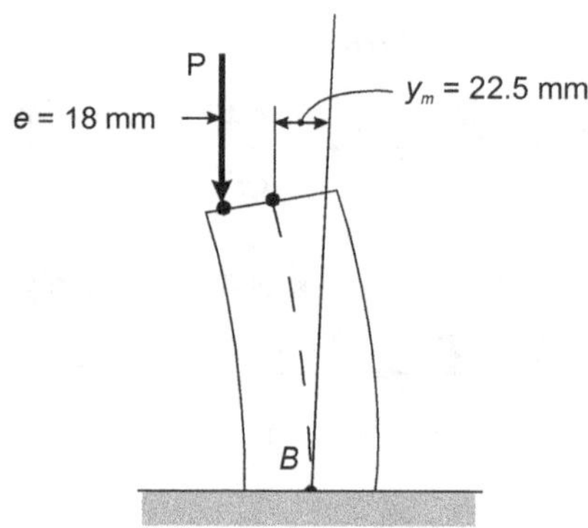

REVIEW QUESTIONS

SHORT QUESTIONS

1. What is slenderness ratio?

2. What are the assumptions made in Euler's column theory?

3. Define 'effective length' of a column.

4. State the limitations of Euler's equation.

5.   Write down the Rankine formula for columns.

6.   The actual length of a column is 10 m. Determine its effective length if both the ends of the column are rigidity fixed.

LARGE QUESTIONS

1.   A hollow cast iron strut 150 mm outer and 100 mm in diameter and 8 m long, one end pin jointed and other end is fixed, is subjected to an axial compressive load. Taking factor of safety as 5 and Rankine's constants 550 N/mm2 and 1/1600, calculate the safe load.

2.   Compare the crippling loads given by Rankine's and Euler's formula for tubular strut 2.25 m long having outer and inner diameters of 37.5 mm and 32.5 mm loaded through pin-joint at both ends. Take yield stress at 315 x 106 N/m2, a 1/7500 and E = 20 x 1010 N/m2.

3.   i.   Derive the Euler's critical load for column of length 'l' with uniform cross section for support condition one end fixed and other end free, subjected to a axial load of 'P'.

    ii.   For what length of a mild steel bar 50 mm diameter used as column the Euler's theory is applicable if the ultimate compressive strength is 350 N/mm2 and E = 210 GPa, when both ends are hinged.

17

THIN AND THICK SHELLS

INTRODUCTION

Thin and thick shells are used extensively in the design of pressure vessels, steam generators (boilers) heavy piping in chemical and thermal power plants. The ASME standards and design codes use the formulae derived in this chapter. The theory is based on simplified treatment. Aircraft fuselage, submarines, turbine casings, heavy pipes with nozzle or having bends, automotive tyres, are also examples of shells which are of complex geometries. For such configurations one may have to use other techniques like F.E.M. It may be noted that, in cases where shells are formed by axisymmetric rotation of thin curved areas, equations can be simplified for such cases, provided loading, material, geometry, constraints are also axially symmetric.

Readers may consult literature on general theory of shells (Refer Bibliography) for more information. Basic principle underlying theory of thin shells is that they resist the external loading (in this case internal pressure) by stretching and no bending. However a plate resists transverse loads by bending. In cases where there are deviations from axial symmetry, at such localized zones like supports, joints of two different pressure vessels, nozzles, etc., in addition to resistance by stretching or membrane action, local bending resistance is also present. We shall now concentrate on membrane action only and not on bending action.

Thin Cylindrical shells

A Cylindrical vessel or shell may be thin or thick depending upon the thickness of the plate in relation to the internal diameter of the cylinder. The ratio $\dfrac{t}{d} = \dfrac{1}{20}$ can be considered suitable line of

demarcation between thin and thick cylinders.

In thin cylinders, the stress may be assumed uniformly distributed over the wall thickness. Boiles, tanks, steam pipes, water pipes, etc., are usually considered as thin cylinders. Thin cylinders are frequently required to operate under pressures up to 30 MN/m^2 or more, for high pressures such as 250 MN/m^2 or more, thick cylinders are used.

When thin cylinders are subjected to internal fluid pressure the following two types of stresses are developed.

1. *Hoop or circumferential stresses* These act in a tangential direction to the circumference of the shell.

2. *Longitudinal stresses* These act parallel to the longitudinal axis of the shell.

3. *Radial stresses* These act radially and are too small and can be neglected. These three stresses are mutually, perpendicular and are principal stresses.

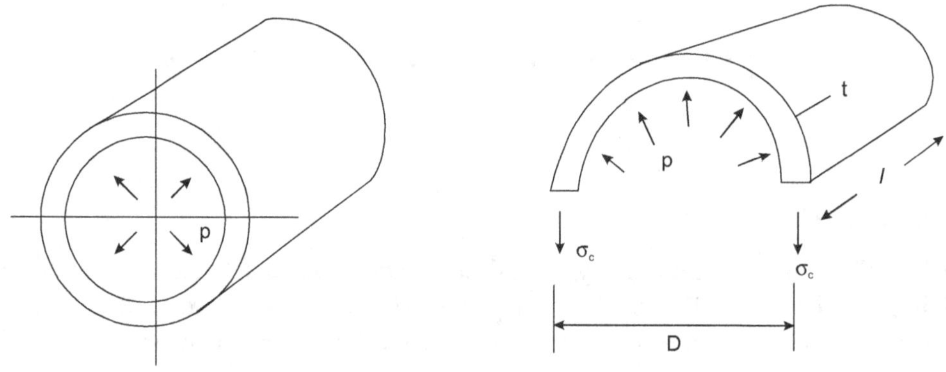

Figure 17.1

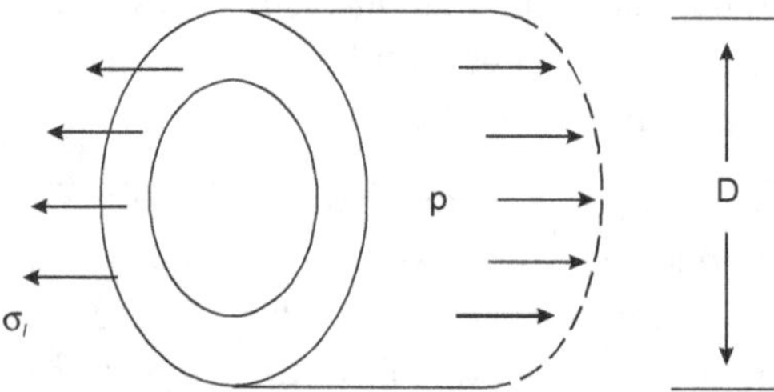

Figure 17.2

Circumferential or Hoop Stresses

Let d = Internal diameter of the cylinder

 t = thickness of the cylinder

 p = Internal pressure (gauge) in the cylinder, and

 σ_c = circumferential or hoop stress. (also symbol σ_h is used for hoop stress)

 Bursting force (pressure) = Resisting strength

$$Pdl = 2\,lt\,\sigma_c \text{ or } \sigma_c = \frac{Pd}{2t} \tag{17.1}$$

Longitudinal Stresses

Let σ_l = Longitudinal stress produced in the shell.

Now, pressure on the ends = Resisting force

$$P \times \pi/4d^2 = \pi dt \cdot \sigma_l \text{ from which } \sigma_l = \frac{Pd}{4t} \tag{17.2}$$

Therefore, the circumferential stress or hoop stress (σ_c) is twice as great as the longitudinal stress (σ_l) and in no case should the hoop stress be greater than the permissible stress in the material of the cylinder.

Maximum Shear Stress

In a cylindrical shell, at any point on its circumference, there is a set of two mutually perpendicular stresses (σ_c) and (σ_l) which are principal stresses and as such the planes in which these act are the principal planes.

The maximum shear stress is found as follows:

Maximum shear stress,

$$\tau_{max} = \frac{\sigma_c - \sigma_l}{2}$$

$$= \frac{Pd}{2t} - \frac{Pd}{4t} = \frac{Pd}{4t}$$

$$i.e., \tau_{max} = \frac{Pd}{4t} \tag{17.3}$$

Thin Cylinder

PROBLEM 17.1

A cylindrical pressure vessel of diameter 1 m and length 2 m, is subjected to an internal pressure of 2 MPa. If the hoop stress is limited to 42 MPa and the longitudinal stress to 28 MPa, find the minimum thickness required. What will be the change in volume of the cylinder under this pressure? E = 200 GPa and v = 0.3.

Solution

The hoop stress is given by

$$\sigma_h = \frac{Pd}{4t}$$

Therefore,

$$\frac{2 \times 1000}{4t} = 42, \ t = 23.81 \, \text{mm}$$

$$\sigma_l = \frac{pd}{4t}, \ \frac{2 \times 1000}{4t} = 28 \qquad t = 17.86 \, \text{mm}$$

Minimum thickness required = 23.81 mm

Volumetric strain,

$$\varepsilon_V = 2\varepsilon_h + \varepsilon_l$$

$$= \frac{pd}{4tE}(5 - 4v)$$

$$\text{Volume of cylinder} = \pi \frac{d^2}{4} L = \frac{\pi \times (1000)^2}{4} \times 2000$$

$$V = 15.708 \times 10^8 \, \text{mm}^3$$

$$\varepsilon_v = \frac{2 \times 1000(5 - 4 \times 0.3)}{4 \times 23.81 \times 2 \times 105} = 0.39899 \times 10^{-3}$$

$$\delta v = \varepsilon_v V = 0.39899 \times 10^{-3} \times 15.708 \times 10^8$$

$$= 6.267 \times 10^5 \, \text{mm}^3$$

PROBLEM 17.2

A cylinder has an internal diameter of 1.2 m and a length of 2.5 m. The internal pressure in the cylinder is 1.5 MPa. The longitudinal joint in the cylinder has an efficiency of 80% and the circumferential joint one of 50 %. Find the minimum thickness required if the stresses are not to exceed 48 MPa in the circumferential direction and 32 MPa in the longitudinal direction.

Solution

Longitudinal stress,

$$\sigma_l = \frac{Pd}{4t\eta}$$

$$32 = \frac{1.5 \times 1200}{4 \times t \times 0.5}$$

$$t = 28.12 \, \text{mm}$$

Circumferential stress,

$$\sigma_c = \frac{Pd}{2t\eta}$$

$$48 = \frac{1.5 \times 1200}{4 \times t \times 0.5}$$

$$t = 23.43$$

The minimum thickness required is 28.12 mm.

PROBLEM 17.3

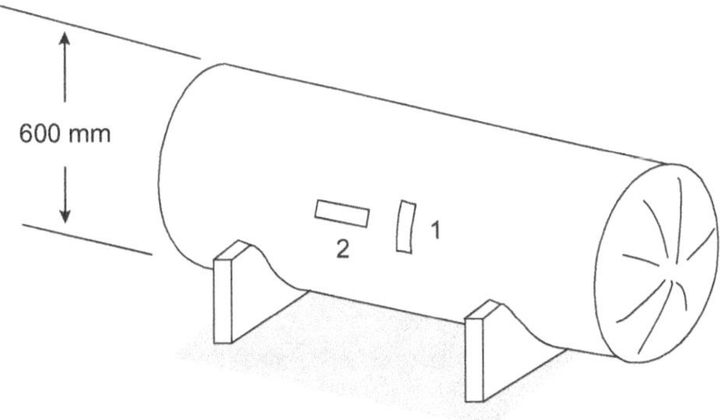

A cylindrical storage tank used to transport gas under pressure has an inside diameter of 600 mm and a wall thickness of 20 mm. Strain gages attached to the surface of the tank in transverse and longitudinal directions indicate strains 255 μ and 60μ of respectively. Knowing that a torsion test has shown that the modulus of rigidity of the material used in the tank is $G = 80$ GPa, determine (a) the gage pressure inside the tank, (b) the principal stresses and the maximum shearing stress in the wall of the tank.

a.   *Gage pressure inside tank*   We note that the given strains are the principal strains at the surface of the tank. Plotting the corresponding points A and B, we draw Mohr's circle for strain.

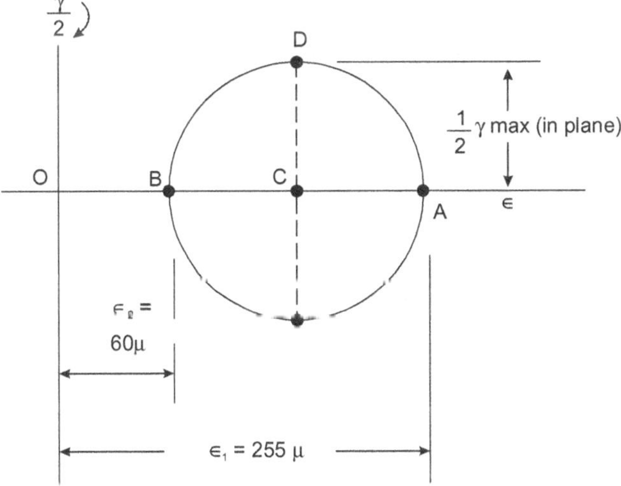

The maximum in-plane shearing strain is equal to the diameter of the circle.

$$\gamma_{max\,(in\,plane)} = \epsilon_1 - \epsilon_2 = 255\mu - 60\mu = 195\mu$$

From Hooke's law for shearing stress and strain, we have

$$\tau_{max(in\,plane)} = G\gamma_{max(in\,plane)}$$

$$= (80 \times 10^9\,Pa)(195 \times 10^{-6}) = 15.60\,MPa$$

Substituting this value and the given data for τ_{max}, we write

$$\tau_{max\,(in\,plane)} = \frac{pr}{4t}$$

$$15.60\,MPa = \frac{p(0.3\,m)}{4(0.02\,m)}$$

Solving for the gage pressure p, we have

$$p = 4.16\ MPa$$

b. Prinicipal stresses and maximum shearing stress: Recalling that, for a thin-walled cylindrical pressure vessel, $\sigma_1 = 2\sigma_2$, we draw Mohr's circle for stress and obtain

$$\sigma_2 = 2\tau_{max\,(in\,plane)} = 2(15.60\,MPa) \qquad \sigma_2 = 31.2\ MPa$$

$$\sigma_1 = 2\sigma_2 = 2(31.2\ MPa) \qquad \sigma_1 = 62.4\ MPa$$

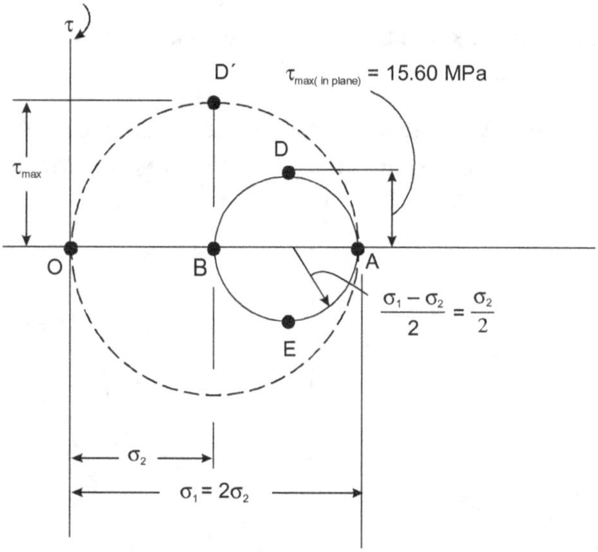

The maximum shearing stress is equal to the radius of the circle of diameter OA and corresponds to a rotation of 45° about a longitudinal axis.

$$\tau_{max} = \frac{1}{2}\sigma_1 = \sigma_2 = 31.2\,MPa \qquad \tau_{max} = 31.2\,MPa$$

Change in Volume of Thin Cylindrical Vessel

PROBLEM 17.4

A thin cylindrical pressure vessel has an internal diameter of 150 mm and a wall thickness of 5 mm. It is subjected to an internal pressure of 7 N/mm². If the cylinder is 900 mm long and $E = 200$ GPa, find the value of Poisson's ratio for the material if the change in volume under this pressure is 15,000 mm³.

Solution

Change in Volume $\delta V \times V$, where δV is the volumetric strain and V is the volume of the cylinder.

$$V = \frac{\pi D^2}{4} l; \quad \delta V \frac{\pi}{4}(L \times 2D\delta D + D^2 \delta L)$$

$$\varepsilon_v = \frac{\delta V}{v} = \frac{\pi}{4}(2 \times 2D\delta D + D^2 \delta L) / \left(\frac{\pi}{4}D^2 L\right) = 2\frac{\delta D}{D} + \frac{\delta L}{L} = 2\varepsilon_h + \varepsilon_l$$

$$\varepsilon_h = \frac{\sigma_h}{E} - v\frac{\sigma_l}{E}; \quad \varepsilon_l = \frac{\sigma_l}{E} - v\frac{\sigma_h}{E}$$

$$\sigma_h = \frac{PD}{4t}; \quad \sigma_l = \frac{PD}{2t}$$

$$\therefore \varepsilon_v = 2\varepsilon_h + \varepsilon_l = \frac{1}{E}(2\sigma h - 2v\sigma_l + \sigma_l - v\sigma_h)$$

$$= \frac{2\sigma_h(2-v) + \sigma_l(1-2v)}{E}$$

Substituting for σ_h and σ_l and simplifying,

$$\delta V = \frac{PD}{4tE}(4 - 2v + 1 - 2v) = \frac{PD}{4tE}(5 - 4v) = \frac{PD}{4tE}(5 - 4v)$$

Therefore,

$$\delta V = \frac{pd}{4tE}(5 - 4v)$$

$$\text{Change in volume} = \frac{pd}{4tE}(5 - 4v) = \frac{\pi d^2}{4}l$$

$$\frac{\pi pd^3}{16tE}(5 - 4v)l = 15,500$$

$$900 \times \frac{\pi \times 7 \times (150)^3}{16 \times 5 \times 200,000}(5 - 4v) = 15,500$$

$$v - 0.322$$

PROBLEM 17.5

Calculate the increase in volume of a boiler shell 1m long and 1m in diameter, when subjected to an internal pressure of 2.5 N/mm². The wall thickness is such that the maximun tensile stress is not to exceed 32 N/mm². $E = 0.21' \ 10^6$ N/mm², $v = 0.3$.

Solution

Given data

$$l = 1\,\text{m} = 1000\,\text{mm} \quad \sigma_1 = \sigma_{max} = 32 \ \text{N}/\text{mm}^2$$

$$d = 1\,\text{m} = 1000\,\text{mm} \quad E = 0.21 \times 106 \ \text{N}/\text{mm}^2$$

$$p = 2.5\,\text{N}/\text{mm}^2 \qquad v = 0.3$$

Circumferential stress = Max. stress $\sigma_1 = \dfrac{pd}{2t} = 32\,\text{N}/\text{mm}^2$

Longitudinal stress $\sigma_2 = \dfrac{\sigma_1}{2} = 16\,\text{N}/\text{mm}^2$

Circumferential strain $\varepsilon_1 = \dfrac{1}{E}\left[\sigma_1 - v\sigma_2\right]$

$$\varepsilon_1 = \frac{32 - 0.3 \times 16}{0.21 \times 10^6} = 1.295 \times 10^{-4}$$

Longitudinal strain $\varepsilon_2 = \dfrac{1}{E}\left[\sigma_2 - v\sigma_1\right] = \dfrac{16 - 0.3 \times 32}{0.21 \times 10^6} = 3.048 \times 10^{-5}$

Volumetric starin $\varepsilon_v = \varepsilon_2 + 2\varepsilon_1$

$$= 3.048 \times 10^{-5} + 2 \times 1.295 \times 10^{-4}$$

$$= 2.8948 \times 10^{-4}$$

Volume of cylinder $V = \dfrac{\pi}{4}d^2 l$

$$= \frac{\pi}{4} \times 1000^2 \times 1000$$

$$= 785.4 \times 10^6 \ \text{mm}^3$$

$$\delta V = \varepsilon_v . V$$

$$\delta V = 2.8948 \times 10^{-4} \times 785.4 \times 10^6$$

$$= 227357\,\text{mm}^3$$

Increase in volume of the boiler = $227357\,\text{mm}^3$

Thin-Walled Spherical Vessels

This type of spherical vessels also known as Horton spheres are used in liquid or gas storage vessels and found in LPG storage yards and chemical plants. In the case of the spherical vessel shown in Figure 17.3 due to symmetry along all diametrical planes, the stress developed is the same in two mutually perpendicular directions. The bursting force across any diametrical plane is

$$F = P_i \frac{\pi d^2}{4} \tag{17.4}$$

This is resisted by the stress s acting along the circumference of thickness t.

$$\sigma \pi d t = P_i \frac{\pi d^2}{4}$$

$$\text{and } \sigma = \frac{p_i d}{4t}$$

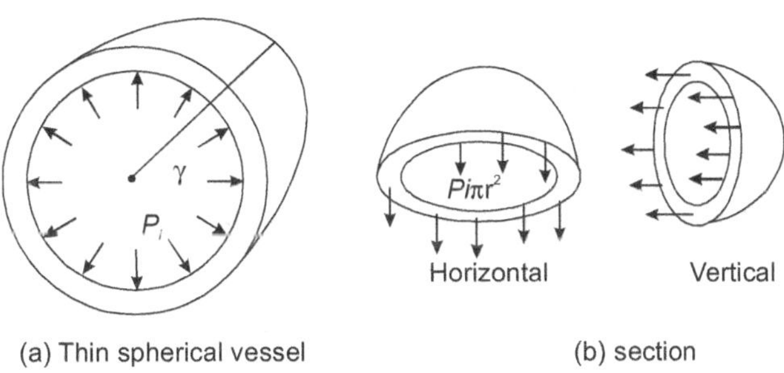

Figure 17.3

Note that the maximum stress in a spherical vessel is half that in a cylindrical vessel. The stresses in two mutually perpendicular directions are the same and so are the strains.

$$\varepsilon = \frac{\sigma}{E}(1 - v) = \frac{p_i d}{4tE}(1 - v)$$

Volumetric strain Volume of the spherical vessel is given by,

$$V = \frac{4}{3}\pi r^3 = \frac{\pi d^3}{6} \tag{17.5}$$

New diameter $d' = d(1 + \varepsilon)$

New volume $V' = \dfrac{\pi (d')^3}{6}$

Volume strain $\varepsilon_v = \dfrac{V' - V}{V}$

Alternately,

$$V = \frac{\pi d^3}{6}$$

$$\delta V = \frac{\pi \times 3d^2 \delta d}{6}$$

$$\delta V = \frac{\pi \times 3d^2 \delta d}{6\pi d^3/6} = 3\frac{\delta d}{d} = 3\varepsilon$$

$$\varepsilon_v = 3\varepsilon \tag{17.6}$$

Thus the volumetric strain is three times the diametrical strain.

Thin Spherical Shell

PROBLEM 17.6

A thin spherical shell has a diameter of 1.2 m and is subjected to an internal pressure of 2.5 MPa. Determine the minimum thickness required if the stress is not exceed 40 MPa. Find the increase in the diameter of the sphere, and the change in volume. $E = 200$ GPa and $v = 0.3$.

Solution

Due to symmetry about its centre, the stress in the sphere in all directions is the same. The stress is equal to

$$\sigma_h = \frac{pd}{4t} = \frac{2.5 \times 1200}{4 \times t}$$

$$\sigma h \leq 40\,\text{N}/\text{mm}^2$$

Therefore,

$$t = \frac{2.5 \times 1200}{4 \times 40} = 18.75\text{mm}$$

Change in diameter = Diameter × Circumferential strain

$$\text{Circumferential strain} = \frac{\sigma_h}{E} - \frac{\sigma_h v}{E} = \frac{\sigma_h}{E}(1-v)$$

$$= \frac{40}{200,000}(1-0.3) = 1.4 \times 10^{-4}$$

Change in diameter = $1200 \times 1.4 \times 10^{-4} = 0.168$mm

Change in volume = V × Volumetric strain

$$\text{Volumetric strain} = \frac{\text{Final volume} - \text{Original volume}}{\text{Original volume}}$$

$$= \frac{(\pi/6)(d + \delta d)^3 - \pi d^3/16}{\pi d^3/6}$$

$$= \frac{d^3 + 3d^2 \delta d + 3d(\delta d)^2 + \delta d^3 - d^3}{d^3}$$

$$= \frac{3\delta d}{d}$$

$$= 3 \times diametrical\ strain$$

(We have neglected the terms of the second order and above)

$$\text{Change in volume} = \frac{\pi \times (1200)^3}{6} \times 3 \times 1.4 \times 10^{-4}$$

$$= 380{,}000\ \text{mm}^3$$

Cylindrical Shell With Hemispherical Ends

This type of vessels are used either in horizontal configuration as in tanker vehicles, liquid storage tanks or in vertical configurations as in chemical or power plants.

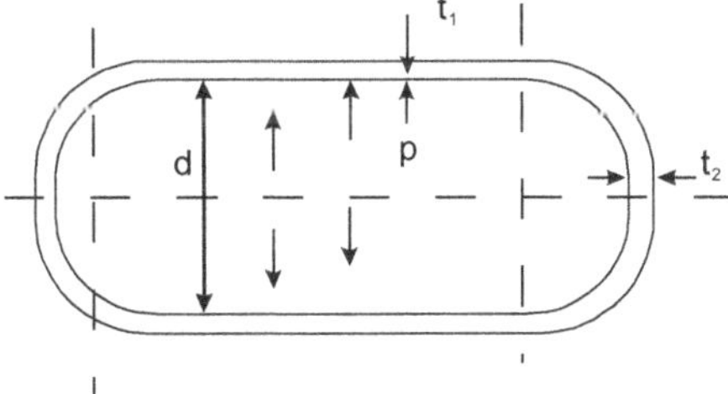

Figure 17.4

Let d be the internal diameter

t_1 = Thickness of cylindrical shell

t_2 = Thickness of hemispherical shell.

Circumferential stress in cylindrical portion

$$\sigma_{c_1} = \frac{pd}{2t_1}$$

Longitudinal stress in cylindrical portion

$$\sigma_{l_1} = \frac{pd}{4t_1}$$

Circumferential strain in cylindrical portion

$$e_{c_1} = \frac{\sigma_{c_1}}{E} - v\frac{\sigma_{l_1}}{E}$$

$$= \frac{2\sigma_{l_1}}{E} - v\frac{\sigma_{l_1}}{E}$$

$$= \frac{pd}{4t_1 E}(2-v)$$

Hemispherical Shell Portion

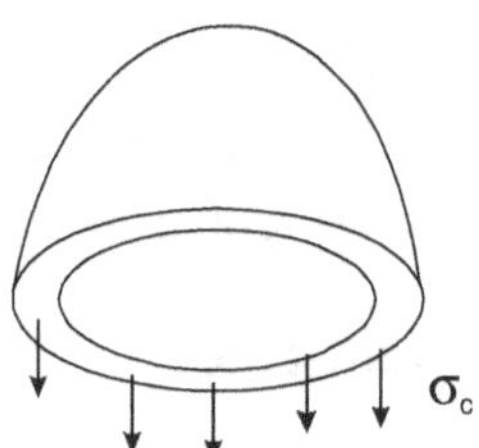

Figure 17.5

Circumferential stress developed in hemispherical portion

$$\sigma_{c_2} = \frac{pd}{4t_2}$$

Circumferential strain in hemispherical portion

$$e_{c_2} = \frac{\sigma_{c2}}{E} - v\frac{\sigma_{c2}}{E}$$

$$e_{c_2} = \frac{pd}{4t_2 E}(1-v)$$

In order that there is no distortion at the junction of cylindrical and hemispherical shells, the circumferential strains have to be equal

$$e_{c_1} = e_{c_2}$$

$$\frac{pd}{4t_1 E}(2-v) = \frac{pd}{4t_2 E}(1-v) \tag{17.7}$$

$$\frac{t_2}{t_1} = \frac{(1-v)}{(2-v)} \tag{17.8}$$

Since $\dfrac{(1-v)}{(2-v)}$ is less than 1, $t_2 < t_1$ for a given pressure, the hemispherical shell will be thinner than the cylindrical shell.

$$\text{For the same stress} = \dfrac{\sigma_{c1} = \sigma_{c2}}{\dfrac{t_2}{t_1}} = 0.5$$

A point to be noted here is that in reality, due to geometry change, there will be local bending stresses introduced in the junction between cylindrical and hemispherical portions. Additional reinforcements may be called for at the junction to contain bending stresses which are localised. The membrane solution outlined in this example does not deal with local bending stresses. Bending theory of shells is to be applied to estimate such local bending stresses.

PROBLEM 17.7

A thin cylindrical vessel has the following geometric and mechanical properties.

Diameter = 1.8m

Length of cylindrical portion = 2.4 m

One end is closed with a flat plate and other end is closed with a hemispherical dome.

Thickness for all components = 6 mm

Young modulus = 200/kN/mm² $\quad$ Poisson's ratio = 0.32

If the cylinder is subjected to an internal fluid pressure of 15000 N/m², determine the change in volume of the container.

Solution

Diameter of the cylinder = 1.8 m = 1800 mm

Length of cylindrical portion = 2.4 m = 2400 mm

thickness, t = 6 mm, $\gamma = 0.32$

Fluid pressure, p = 15000 N/m² = 0.015 N/mm²

$$E = 200\,\text{kN/mm}^2 = 200 \times 10^3 \,\text{N/mm}^2$$

Change in volume of cylinder

$$\frac{\delta V_1}{V} = \frac{pd}{2tE}\left[\frac{5}{2} - 2v\right]$$

$$= \frac{0.015 \times 1800}{2 \times 6 \times 200 \times 10^3}\left[\frac{5}{2} - 2 \times 0.32\right]$$

$$\delta V_1 = \frac{0.0209}{10^3} \times v = 0.0209 \times 10^{-3} \times \pi \times \frac{1800^2}{4} \times 2400$$

$$= 127,642 \, \text{mm}^3$$

Change in volume of the hemisphere $\delta v_2 = \varepsilon_v . v$

$$\frac{\delta V_2}{V} = \frac{pd}{4tE}(1-v) \times \frac{2}{3}\frac{\pi d^3}{8}$$

$$= \frac{0.015 \times 1800}{4 \times 6 \times 200 \times 10^3}(1-0.32) \times \frac{2}{3}\pi\frac{1800^3}{8} = 5840 \, \text{mm}^3$$

$$\delta V = \delta V_1 + \delta V_2 = 133,482 \, \text{mm}^3$$

This ignores the possible bending of the flat diaphragm. This also assumes that the cylinder and the hemisphere expand independantly.

Built up Cylindrical Shells

The formulas derived for (σ_c) and (σ_l) have been obtained with the presumption that the shell is seamless(solid drawn). But in actual practice , cylindrical shells of large diameters, such as steam generator, drums, headers are not seamless, but instead, are built up by longitudinal and circumferential joints. The longitudinal joints reduce the resisting strength of the shell plate against bursting and circumferential joints reduce strength against pressure on the end plates.

Let η_1 = Efficiency of the longitudinal joint

η_c = Efficiency of the circumferential joint

Bursting force = Resisting strength

$$pdl = 21 \, l_c \eta_1$$

$$\sigma_c = \frac{pd}{2t\eta_1}$$

Similarly $\sigma_l = \dfrac{pd}{2t\eta_1}$

Change in dimensions of a thin cylindrical shell due to an internal pressure

Direct strain (due to σ_c) $= \dfrac{\sigma_c}{E}$

Direct strain (due to σ_l) $= \dfrac{\sigma_l}{E}$

Net circumferential strain $= e_c = \dfrac{\sigma_c}{E} - v\dfrac{\sigma_l}{E}$

$$e_c = \dfrac{\sigma_c}{E} - v\dfrac{\sigma_l}{2E} \qquad \sigma_l = \dfrac{\sigma_c}{2}$$

$$= \dfrac{pd}{2tE}(1 - 0.5v)$$

Net longitudinal strain $= e_1 = \dfrac{\delta l}{l} = \dfrac{\sigma_1}{E} - v\dfrac{\sigma_c}{E}$

$$= \dfrac{pd}{4tE}(1 - 2v)$$

Volumetric strain $= e_v$

$\qquad =$ Algebraic sum of all strains in all axes

$\qquad =$ Net longitudinal strain + 2 * net circumferential strain

The net circumferential strain has been taken twice because it has the effect of changing the diameter of the shell necessarily tooth in xx and yy axis

$$e_v = \dfrac{\delta_v}{V} = \dfrac{\text{Change in volume}}{\text{Original volume}}$$

$$\delta V = e_v \times V = (e_l + 2e_c)V$$

$$= \left[\dfrac{pd}{4tE}(1 - 2v) + 2\dfrac{pd}{2tE}(1 - v)\right]V$$

$$= \dfrac{pd}{2tE}\left(\dfrac{5}{2} - 2v\right) \qquad\qquad \left(\text{since } V = \dfrac{\pi d^2 l}{4}\right)$$

Change in length $\delta_l = e_l \times d = \dfrac{pd}{4tE}(1 - 2v) = \dfrac{pdl}{4tE}(1 - 2v)$

$$\delta_d = e \times d = \dfrac{pd}{2tE}(1 - 0.5v)d = \dfrac{pd^2}{2tE}(1 - 0.5v)$$

Wire Wound Shells

In order to increase the resistance of shells against bursting due to internal pressure, the shells are strengthened by wires wound in layers and wires are under tension. This also increase, the pressure carrying capacity of the cylinder.

The resultant circumferential stress in cylinder

= Sum of initial compressive stress due to wire winding + tensile stress due to internal pressure.

The resultant circumferential stress in wire

= Sum of initial tensile stress due to winding tension + hoop stress in shell due to internal pressure.

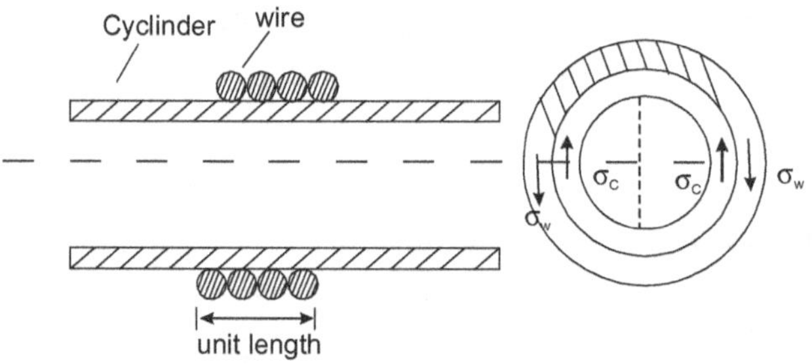

Figure 17.6

d—diameter of cylinder

t—Wall thickness of cylinder

d_w—Diameter of wire

n—Number of turns of wire per unit length $\dfrac{l}{d_w}$

σ_w —Initial tension with which wire is wound

σ_c — Compressive circumferential stress developed in the cylinder

$\acute{\sigma}_c$ — Circumferential stress developed in cylinder

$\acute{\sigma}_w$ — Stress developed in wire

$\acute{\sigma}_l$ — longitudinal stress developed in cylinder

Before admitting fluid into the cylinder

Tensile force exerted by wire per unit length

$$= 2 \times \frac{\pi}{4} d^2{}_w \sigma_w \times n \tag{17.9}$$

Compressive force developed in the cylinder $= 2 \times t \times 1 \times \sigma_c = 2\sigma_c t$

For equilibrium,

$$= 2 \times \frac{\pi}{4} d^2_{\ w} \sigma_w \times n = 2t\sigma_c \qquad (17.10)$$

But

$$n = \frac{1}{d_w}$$

$$\sigma_c = \frac{\pi d_w}{4t} \sigma_w$$

After admitting fluid into the cylinder

When the wire wound cylinder is subjected to internal pressure,

Longitudinal bursting force

$$= p \times \frac{\pi d_w^{\ 2}}{4} = \sigma'_l \pi dt$$

Diametrical bursting force

$$= p.d.l$$

$$= \sigma'_c 2tl + \sigma'_w \frac{\pi d_w}{2}$$

$$= pd = \sigma'_c 2t + \sigma'_w \frac{\pi d_w}{2}$$

Also for compatibility, circumferential strain in wire should be equal to the circumferential strain in cylinder.

$$\frac{\sigma'_c}{E_c} - v\frac{\sigma'_l}{E_c} = \frac{\sigma'_w}{E_w}$$

$$\frac{\sigma'_c}{E_c} - v\frac{pd}{4tE_c} = \frac{\sigma'_w}{E_w} \qquad (17.11)$$

Resultant stress in wire $= \sigma_w + \sigma'_w$
Resultant circumferential stress in cylinder $= \sigma_c - \sigma'_c$

Stress in Wire-wound Thin Cylinder

PROBLEM 17.8

A steel cylinder has a diameter of 320 mm and a thickness of 14 mm. It is initially wound with steel wire of diameter 5 mm and tensioned to a stress of 40 MPa. It is then subjected to an internal pressure of 3 MPa. Determine the stresses in the cylinder wall and wire when it is subjected to this pressure. E = 200 GPa for the cylinder and wire.

Solution

Area of wire,

$$A_w = \frac{\pi \times 5^2}{4} = \frac{25}{4}\pi = 19.63 \, \text{mm}^2$$

There are 1000/5 = 200 turns/metre length. If σ_{c_1} is the initial compressive stress in the cylinder then for equilibrium, the compressive force in the cylinder and tensile force in wire for a given length must be equal. Taking 1 m length of the cylinder,

$$2 \times 14 \times 1000 \times \sigma_{c_1} = 2 \times 200 \times 19.63 \times 40$$

$$\sigma_{c_1} = 11.22 \, N/mm^2$$

The bursting force on the cylinder due to internal pressure must be equal to the tensile force in the cylinder wall and the wire.

Bursting force = pdl = 3 × 320 × 100/m

Hoop tension in the cylinder = $\sigma_{c_2} = 2lt$

$$= 2 \times 1000 \times 14 \times \sigma_{c_2}$$

Hoop tension in the wire = $2n\sigma_{w_2} A_{w_2}$

$$= 2 \times 200 \times 19.63 \times \sigma_{w_2}$$

where σ_{c_2} and σ_{w_2} are tensile stresses in the cylinder walls and the wire. Therefore,

$$28000\sigma_{c_2} + 7852\sigma_{w_2} = 960000$$

Longitudinal stress,

$$\sigma_l = \frac{pd}{4t} = \frac{3 \times 320}{4 \times 14} = 17.14 \, N/mm^2$$

Hoop strain in the cylinder = $\dfrac{\sigma_{c_2}}{E} - v\dfrac{\sigma_l}{E}$

$$= \frac{1}{E}(\sigma_{c_2} - 0.3 \times 17.14)$$

Hoop strain in the wire $= \dfrac{\sigma_{w_2}}{E}$

These two strains must be equal

$$\frac{1}{E}(\sigma_{c_2} - 5.14) = \frac{\sigma_{w_2}}{E}, \quad \sigma_{c_2} = \sigma_{w_2} + 5.14$$

Substituting this value σ_{c_2} in the first equation,

$$28000(\sigma_{w_2} + 5.14) + 7852\,\sigma_{w_2} = 960000$$

$$\sigma_{w_2} = 30.8\,\text{N}/\text{mm}^2$$

$$\sigma_{c_2} = 35.94\,\text{N}/\text{mm}^2$$

Net stress in cylinder wall = 35.94 – 11.22 = 24.72 N/mm²

Net stress in the wire = 30.8 + 40 = 70.8 N/mm²

Both these stresses are tensile.

PROBLEM 17.9

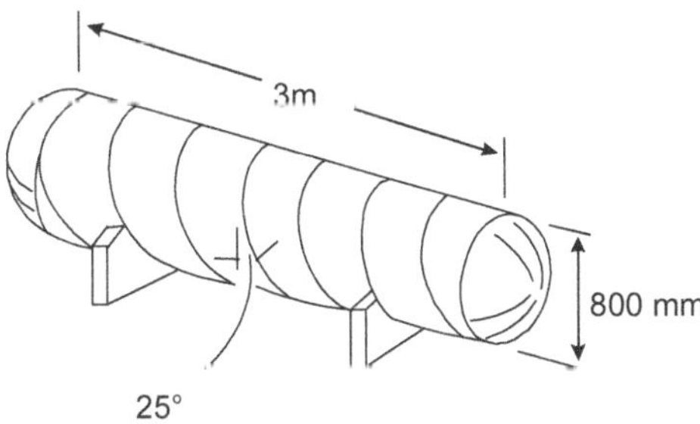

A compressed-air tank is supported by two cradles as shown: one of the cradles is designed so that it does not exert any longitudinal force on the tank. The cylindrical body of the tank is fabricated from a 10 mm steel plate by butt welding along a helix which forms an angle of 25° with a transverse plane. The end caps are spherical and have a uniform wall thickness of 8 mm. For an internal gauge pressure of 1.2 MPa, determine (a) the normal stress and the maximum shearing stress in the spherical caps, (b) the stresses in the directions perpendicular and parallel to the helical weld.

a.   Spherical cap. Using equation.(17.1), we write

$$p = 1.2\,\text{MPa} \quad r = 400\,\text{mm} \quad t = 8\,\text{mm}$$

$$\sigma_1 = \sigma_2 = \frac{pr}{2t} = \frac{(1.2\,\text{MPa})(0.400\,\text{m})}{2(0.008\,\text{m})}$$

$$a = 30\,\text{MPa}$$

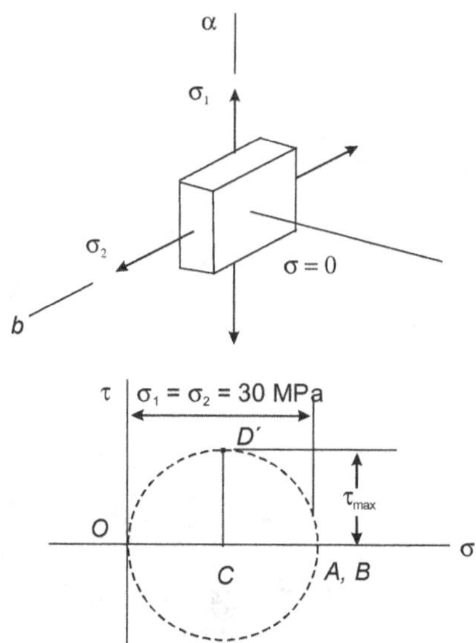

We note that for stresses in a plane tangent to the cap, Mohr's circle reduces to a point (A, B) on the horizontal axis and that all in-plane shearing stresses are zero.

On the surface of the cap the third principal stress is zero and corresponds to point O. On a Mohr's circle of diameter AO, point D′ represents the maximum shearing stress; it occurs on planes forming an angle of 45° with a plane tangent to the cap.

$$\tau_{max} = \frac{1}{2}(30\,\text{MPa})$$

$$\tau_{max} = 15\,\text{MPa}$$

b. *Cylindrical body of the tank* We first determine the hoop stress σ_1 and the longitudinal σ_2. Using equations. (17.1) and (17.2), we get

$$\sigma_1 = \frac{pr}{2t} = \frac{(1.2\,\text{MPa})(0.400\text{m})}{0.010\,\text{m}} = 48\,\text{MPa}$$

$$\sigma_2 = \frac{1}{2}\sigma_1 = 24\,\text{MPa}$$

Stresses at the weld   Noting that both the hoop stress and the longitudinal stress are principal stresses, we draw Mohr's circle as shown.

An element having a face parallel to the weld is obtained by rotating the face perpendicular to the axis OB counterclockwise through 25°. Therefore, on Mohr's circle we locate the point X′ corresponding to the stress components on the weld by rotating radius CB counterclockwise through $2\theta=50°$.

$$\sigma_w = \sigma_{ave} - R \sin 50° = 36 \text{ MPa} - (12 \text{ MPa}) \cos 50°$$
$$\tau_w = R \sin 50° = (12 \text{ MPa}) \sin 50°$$

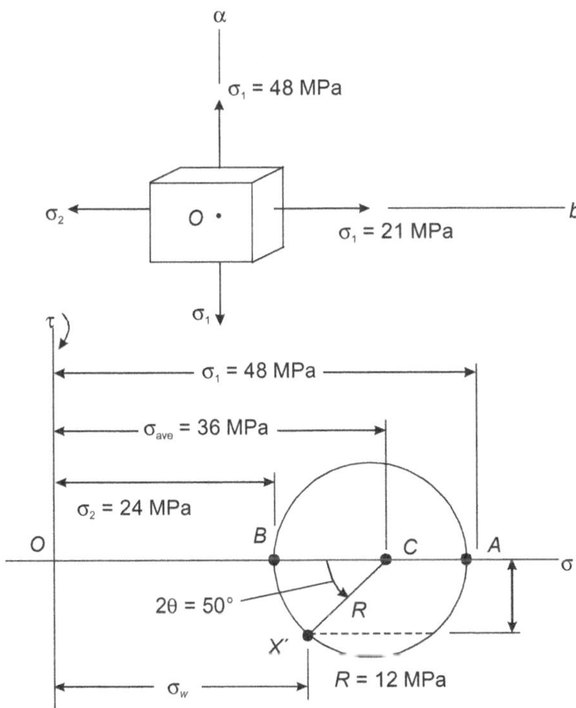

Since X′ is below the horizontal axis, τ_w tends to rotate the element counterclockwise.

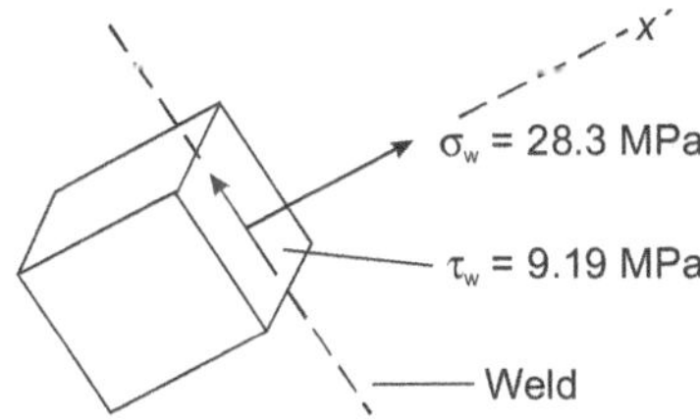

THICK CYLINDRICAL VESSELS

At any point in a cylindrical pressure vessel, there are three stresses in three perpendicular directions, namely,

a. longitudinal stress, σ_l

b. hoop stress σ_h, and

c. radial stress, σ_R

In deriving the equations for thin cylindrical shells, we made three assumptions.

i. σ_R is negligible compare to σ_l and σ_h.

ii. σ_h is uniform along the thickness, and

iii.   The longitudinal strain is uniform so that plane sections (transverse) remain plane before and after the application of internal pressure.

In the case of thick cylindrical vessels, assumptions (i) and (ii) are not valid, but assumption (iii) is assumed to be true.

A cylindrical shell can be classified as thick when its thickness is more than 1/20th its diameter. Such shells are analysed using Lamè's theory.

Lamè's Theory

Consider a thick cylindrical shell of internal radius r_i and external radius r_e. Let the pressures at these radial distances be P_i and P_e as shown in Figure. 17.7. We consider a ring at radius r and thickness δr. If we consider a unit length of the cylinder and the ring cut by a diametrical section, the free body diagram of the cylinder is as shown in Figure 17.7 (b). For this thin ring, the conditions are the same as that of a thin cylindrical shell. For the equilibrium of the thin ring shown in Figure 17.7 (b),

Total upward force = Total downward force

Upward force = $2r\,\sigma r$

Downward force = $2(r+\delta r)(\sigma r+\delta\sigma r)+2\delta r\sigma_h$

Equating the two and neglecting quantities of second order, we have

$$2\delta r\sigma_r + 2r\delta\sigma_r + 2\delta r\sigma_h = 0$$

Dividing δr by and as $\delta r \to 0$

$$r\frac{d\sigma_r}{dr}+\sigma_r+\sigma_h = 0$$

If σ_l is the longitudinal stress, then the longitudinal strain is given by

$$\varepsilon_1 = \frac{\sigma_1}{E} - \frac{v(\sigma_h - \sigma_r)}{E} \tag{17.12}$$

In this equation, if the plane sections are to remain plane before and after the deformation, then σ_l and $(\sigma_h - \sigma_r)$ must be constant. Setting $(\sigma_h - \sigma_r)$ = constant = 2A, we have

$$r\frac{d\sigma_r}{dr}+\sigma_r+(2A+\sigma_r)= 0$$

$$r\frac{d\sigma_r}{dr} = -(2A + \sigma_r)$$

$$\frac{2dr}{r} = \frac{d\sigma_r}{-(A + \sigma_r)}$$

Integrating both sides,

$$2\log_e r + \text{constant.} = -\log_e(A + \sigma_r)$$

Taking the constant as B, we get

$$\log_e\left[r^2(A + \sigma_r)\right] = B$$

$$\sigma_r = -A + \frac{B}{r^2}$$

From $\sigma_h = \sigma_r + 2A$

$$\sigma_h = -A + B/r^2 + 2A = A + B/r^2$$

We thus have

Radial $\sigma_r = -A + B/r^2$ and hoop stress $\sigma_h = A + \dfrac{B}{r^2}$

The constants A and B can be found from the boundary conditions in particular cases.

The equations for the radial stress σ_r and hoop stress σ_h are known as Lame´'s equations and the constants A and B are called Lame´'s constants.

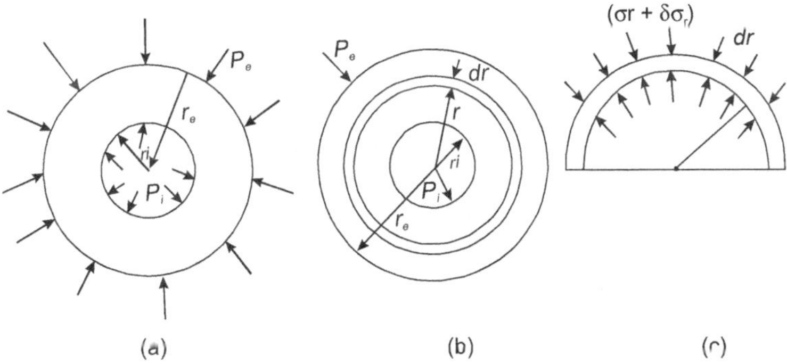

Figure 17.7

Stresses in a thick cylindrical shell The stresses in a thick cylindrical shell depend upon the nature of the loading which can result in (i) only an internal pressure P_i, (ii) only an external pressure P_e, and (ii) both P_i and P_e.

i. Only internal pressure, P_i

   a.   The hoop stress will be tensile.

   b.   The longitudinal stess will be tensile.

   c.   The radial stress will be compressive.

ii. Only external pressure, P_e

   a.   The hoop stress will be compressive.

   b.   The longitudinal stress will be compressive.

   c.   The radial stress will be compressive.

iii.   Both P_i and P_e applied

   a.   The hoop stress will be tensile if $P_i > P_e$

   b.   The longitudinal stress will be tensile for $P_i > P_e$

   c.   The radial stress will be compressive.

Considering a case where both internal and external pressures exist, we have $\sigma_r = -P_i$ at $r = r_i$ and $\sigma r = -P_e$ at $r = r_e$.

Substituting these values σ_r, for we get

$$P_i = -A + B/r^2 \text{ and } P_e = -A + B/r^2 \tag{17.13}$$

Solving for A and B,

$$B/r_i^2 - B/r_e^2 = P_i - p_e;$$

$$B(r_e^2 - r_i^2) = r_i^2 r_e^2 (P_i - P_e) \quad B = \frac{(P_i - P_e)r_i^2 r_e^2}{(r_e^2 - r_i^2)}$$

$$A = B/r_i^2 - P_i = \frac{P_i - P_e)r_e^2}{(r_e^2 - r_i^2)} - P_i = \frac{(P_i r_i^2 - P_e r_e^2)}{(r_e^2 - r_i^2)}$$

With these values A and B, the values of σ_r and σ_h at radius r are obtained as

$$\sigma_r = -\frac{(P_i r_i^2 - P_e r_e^2)}{(r_e^2 - r_i^2)} + \frac{(P_i - P_e)r_e^2 r_i^2}{(r_e^2 - r_i^2)r^2} = -\frac{(P_i r_i^2 - P_e r_e^2) + (P_i - P_e)(r_e^2 r_i^2/r^2)}{(r_e^2 - r_i^2)}$$

$$\sigma_h = A + B/r^2 = \frac{P_i r_i^2 - P_e r_e^2}{(r_e^2 - r_i^2)} + \frac{(P_i - P_e)r_e^2 r_i^2}{(r_e^2 - r_i^2)}$$

$$= \frac{(P_i r_i^2 - P_e r_e^2) + (P_i - P_e)r_e^2 r_i^2/r^2}{(r_e^2 - r_i^2)}$$

The above two equations give the values of hoop stress σ_h and radial stress σ_r at any radius r. The variations σ_h and σ_r of are parabolic and it will be seen that the maximum values of these stresses occur at the inner surface. These are given when $r = r_i$ by

$$\sigma_{r\,max} = P_i \text{ and } \sigma_{h\,max} = \frac{(P_i r_i^2 - P_e r_e^2) + (P_i - P_e)(r_e^2)}{(r_e^2 - r_i^2)}$$

$$= \frac{P_i(r_i^2 + r_e^2) - 2P_e r_e^2}{(r_e^2 - r_i^2)} \tag{17.14}$$

When $P_i = 0$ (only external pressure is acting on the pipe) or $P_e = 0$ (only internal pressure exists), the corresponding relationships can be obtained by putting the respective pressures to zero in the above relationships.

i. $P_e = 0$ (only internal pressure exists.)

At r,

$$\sigma_r = \frac{-P_i r_i^2 + P_i(r_e^2 r_i / r^2)}{(r_e^2 - r_i^2)} = \frac{P_i r_i^2}{(r_e^2 - r_i^2)}\left(\frac{r_e^2}{r^2} + 1\right)$$

$$\sigma_h = \frac{P_i r_i^2 + P_i(r_e^2 r_i^2 / r^2)}{(r_e^2 - r_i^2)} = \frac{P_i r_i^2}{(r_e^2 - r_i^2)}\left(\frac{r_e^2}{r^2} + 1\right) \tag{17.15}$$

$$At\ r = r_i, \sigma_r = P_i, \sigma_h = \frac{P_i(r_e^2 + r_i^2)}{(r_e^2 - r_i^2}$$

$$At\ r = r_e, \sigma_r = 0, \sigma_h = \frac{P_i r_i^2}{(r_e^2 - r_i^2)}$$

ii. $P_i = 0$ (only external pressure exists.)

$$\sigma_r = \frac{P_e r_e^2 + P_e(r_e^2 r_i^2 / r^2)}{(r_e^2 - r_i^2)} = \frac{P_e r_e^2\left[1 - \left(\frac{r_i^2}{r^2}\right)\right]}{(r_e^2 - r_i^2)}$$

$$\sigma_h = \frac{-P_e r_e^2 - P_e(r_e^2 r_i^2 / r^2)}{(r_e^2 - r_i^2)} = \frac{-P_e r_e^2\left[1 + \left(\frac{ri2}{r2}\right)\right]}{(r_e^2 - r_i^2)}$$

$$At\ r = r_i, \sigma_r = 0\ \sigma_h = -\frac{2 P_e r_e^2}{(r_e^2 - r_i^2)}$$

$$At\ r = r_e, \sigma_r = P_i, \sigma_h = \frac{P_e(r_e^2 + r_i^2}{(r_e^2 - r_i^2)}$$

The stress distributions in all cases are shown in Figure 17.8.

Figure 17.8 (a) shows the variation of σ_r and σ_h when both internal and external pressure are applied. The maximum stresses are in the interior fibres.

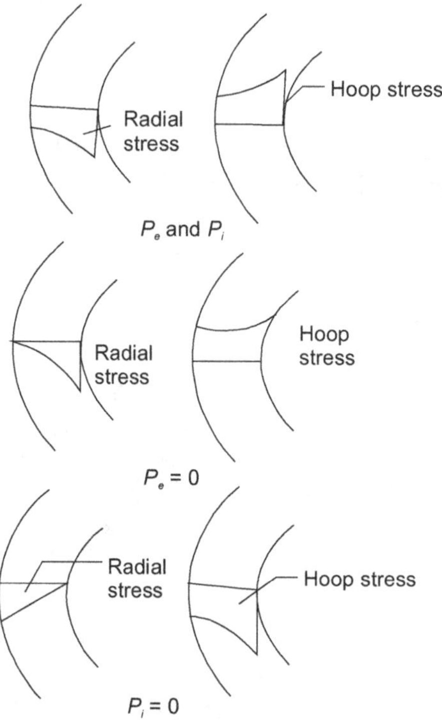

Figure 17.8

Figure 17.8 (b) shows the case when only internal pressure is applied, i.e., $P_e = 0$. The maximum stresses occur at the interior fibres at $r = r_i$. Detailed distribution is shown in shown in 17.8 (d). Figure 17.8 (c) shows the variation of these stresses when $P_i = 0$, i.e., when only external pressure P_e exists. In this case, the maximum radial stress occurs at $r = r_e$, while the maximum hoop stress occurs $r = r_i$. Both the stresses are compressive in nature here.

Longitudinal stresses are developed in thick-walled cylinders with closed ends. They are normally considered to be uniformly distributed. Using the same notations as before, the longitudinal stress may be derived as follows.

$$P_i \pi r_i^2 - P_e \pi r_e^2 = \sigma_l \pi (r_e^2 - r_i^2)$$

$$\sigma_l = \frac{P_i r_i^2 - P_e r_e^2}{(r_e^2 - r_i^2)}$$

It must be remembered that whenever thin cylindrical shells are subjected to external pressure, due to compressive hoop stresses shell can buckle. So, apart from stress compliance the stability against buckling or collapse must be checked. Ribs or longitudinal stiffeners and or ring stiffeners are used to improve stability against buckling.

THICK SPHERICAL SHELLS

Thick spherical shells are used in refineries and chemical plants to store LPG or LNG. The shells are constructed by welding several doubly curved panels to form the sphere. Thick spherical shells can be similarly analysed for stresses. In a thick spherical shell, the stress cannot be assumed to be uniform over the thickness. That is the main difference between a thin spherical shell and a thick one.

Consider the thick spherical shell shown in Figure 17.9.

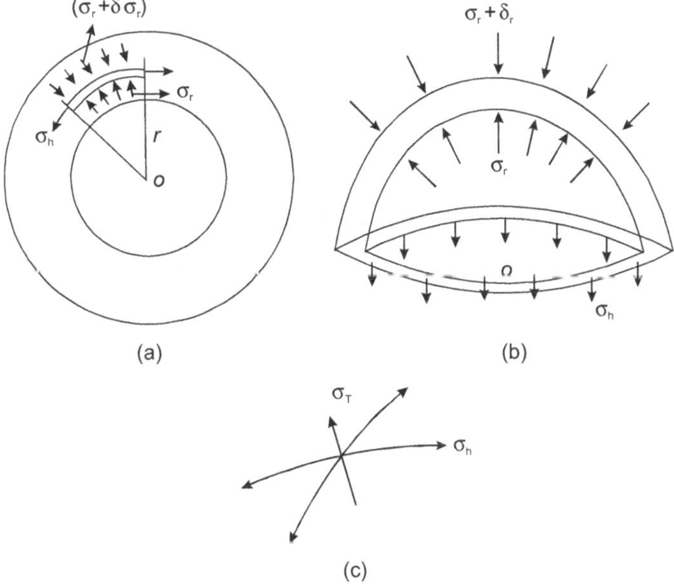

Figure 17.9 Thick spherical shell

The following symbols are used in the equation to follow:

r_i—Internal radius of the shell

r_e—External radius of the shell

P—Internal pressure in the shell

P_r—Pressure at radius

m—Poisson's ratio

ε_r—Radial strain

ε_h—Hoop or circumferential strain

σ_r—Radial stress

σ_h—Hoop stress

In Figure 17.9 (a), we have a diametrical section of the shell. Consider an element at radius r and of the thickkness δr, subtending an angle $\delta\theta$ at the centre. The pressure at r is P_r and the increase in pressure is δP_r as shown. Due to the pressure in the shell, let the radius increase to $(r + u)$. The thickness of the element will also increase from δr to $\delta r + \delta u$.

Radial strain, $\varepsilon_r = \dfrac{(\delta r + \delta u) - \delta r}{\delta r} = \dfrac{\delta u}{\delta r}$

Circumferential strain $\varepsilon_h =$ Change in circumference/Original circumference

$$\varepsilon_h = \frac{2\pi(r+u) - 2\pi r}{2\pi r} = \frac{u}{r}$$

$$u = r\varepsilon_h; \frac{\delta u}{\delta r} = \varepsilon_h + r\frac{\delta\varepsilon_h}{\delta r}$$

From Figure (b), the half spherical shell is in equilibrium due to the three forces acting on it.

i.   The force due to internal pressure $= \pi r^2 \sigma_r \uparrow$ (We use projected area to calculate the force)

ii. The force due to $(\sigma_r = \delta\sigma_r)$ acting from outside $= \pi(r + \delta r)^2 (p_r + \delta p_r) \uparrow$

iii. The force due to radial stress, $\sigma_r = 2\pi r \delta r \sigma r \downarrow$

The equilibrium condition requires that the total downward forces are equal to the total upward forces, i.e.,

$$\pi r^2 \sigma_r = \pi(r + \delta r)^2 (\sigma_r + \delta\sigma_r) + 2\pi r \delta r \sigma_h$$

Expanding the above expression and neglecting squares and products of δr and δP_r, we get

$$\sigma_h = -\sigma_r - r\delta\sigma_r / \delta r$$

At any point in the thick spherical shell, there are three principal stresses: (i) the radial stress σ_r, which is compressive, (ii) hoop stress σ_h, which is tensile, and (iii) hoop stress σ_h, which is tensile and is in a plane perpendicular to the plane on which the other hoop stress acts. These are shown in Figure 17.9(c). As δr approaches zero, the infinitesimal dimensions take the derivative form and lead to the differential equations.

The strains in the three principal directions can be calculated as

$$\varepsilon_r = \frac{\sigma_r}{E} + \frac{2\sigma_h}{mE}, \text{which is compressive}$$

$$\varepsilon_r = \frac{1}{mE}(m\sigma_r + 2\sigma_h) \tag{17.16}$$

Circumferential strain, $\varepsilon_h = \dfrac{\sigma_h}{E} + \dfrac{\sigma_r}{mE} - \dfrac{\sigma_h}{mE} = \dfrac{1}{E}\left[\dfrac{\sigma_h(m-1)\sigma_h}{mE} + \dfrac{\sigma_r}{mE}\right]$

We substitute this in the strain equation we derived earlier

$$-\varepsilon_r = \varepsilon_h + \frac{r\delta\varepsilon_h}{\delta r}$$

$$-\left[\frac{\sigma_r}{E} + \frac{2\sigma_h}{mE}\right] = \left[\frac{(m-1)\sigma_h}{mE} + \frac{\sigma_r}{mE}\right] + \frac{rd}{dr}\left[\frac{\sigma_h(m-1)}{mE} + \frac{\sigma_r}{mE}\right]$$

This equation reduces to

$$(m+1)(\sigma_h + \sigma_r) + (m-1)r\frac{d\sigma_h}{dr} + r\frac{d\sigma_r}{dr} = 0$$

We have from the equilibrium of half shell,

$$\sigma_h = -\sigma_r - (r/2)\delta\sigma_r/\delta r$$

$$\frac{d(\sigma_h)}{dr} = -\frac{d(\sigma_r)}{dr} - \frac{rd^2\sigma_r}{dr} \tag{17.17}$$

Substituting this value in the equation above, we have

$$(m+1)(\sigma_h + \sigma_r) + (m-1)r\left[\frac{-d\sigma_r}{dr} - \frac{rd2\sigma_r}{dr^2}\right] + \frac{rd\sigma_r}{dr} = 0$$

On opening brackets and after multiplication of terms, this reduces to

$$4\frac{d\sigma_r}{dr} + r\frac{d^2\sigma_r}{dr^2} = 0$$

To solve this equation, let $\dfrac{d\sigma_r}{dy} = Y$

$$4Y + rdY/dr = 0; \frac{dY}{Y} = -\frac{4dr}{r}$$

Integrating, $\log_e Y = -4\log_e r + \log_e C_1$

where $\log_e C_1$ is the constant of integration.

$$\log_e Y = \log_e r^{-4} + \log_e C_1 = \log_e\left[C_1/r^4\right]$$

or

$$Y = C_1/r^4 ; d\sigma_r/dr = C_1/r^4$$

Integrating again,

$$\sigma_r = -C_1/3r^3 + C_2 \quad \sigma_r = -C_1/3r^3 + C_2$$

Where C_2 is the constant of integration.

We have

$$\sigma_h = -\sigma_r - (r/2)\frac{d\sigma_r}{dr} = C_1/3r^3 - C_2 - (r/2)C_1/4r^4 (r/2)$$

$$\sigma_h = -\frac{C_1}{6r^3} - C_2 \tag{17.18}$$

This equation is generally written as $\sigma_h = A/r^3 + B$, putting $C_1 = -6\,A$, $C_2 = -B$

Hence, $\sigma_h = A/r^3 + B$ $\tag{17.19}$

The constants A and B are obtained from the boundary conditions.

In case of thick spherical shell under consideration, we have

i.   When $r = r_i; \sigma_r = p$ the internal pressure

ii. When $r = r_e; \sigma_r = 0$

Putting these values, we get the constants as

$$A = \frac{P r^3 i^r{}_e{}^3}{(r_e^3 - r_i^3)} \text{ and } B = \frac{P r_e^3}{r_e^3 - r_i^3} \tag{17.20}$$

PROBLEM 17.10

A thick spherical shell, of 250 mm internal diameter, is subjected to an internal pressure of 8 N/mm². If the maximum permissible tensile stress is 10 MPa, find the minimum thickness required. Find the stresses in the interior and exterior surfaces of the shell.

Solution

We have the radial compressive stress and hoop tensile stress given by

$$\sigma_r = (2A/r^3) - B \text{ and } \sigma_h = (A/r^3) - B$$

The constants A and B can be found from the following boundary conditions:

i.   When $r =$ internal radius $(250/2) = 125$ mm, $\sigma_r = 8\,N/mm^2$

ii. When $r = 125$ mm, $\sigma_h =$ permissible tensile stress, 10 N/mm²

The hoop stress will be maximum at the internal surface of the shell.

Now, we get

i. $8 = (2A \times 125^3) - B$

ii. $10 = (A \times 125^3) + B$

Adding the two equations,

$$18 = (3A/125^3) \text{ or } A = (6 \times 125^3)$$
$$B = (2A/125^3) - 8 = 12 - 8 = 4$$

Knowing A and B, we have the radial stress σ_r as zero at the external surface. If r_e is the outer radius, we have

$$0 = (2 \times 6 \times 125^3/r_e^3) - 4; 4r_e^3 = 2 \times 6 \times 125^3; r_e = 125 \times \sqrt{3} = 180.5 \, \text{mm}$$

Thickness of the shell = 180.5 – 125 = 55.5 mm

Hoop (tensile) stress at the outer surface = 2 × 125³/180.5³ + 4 = 4.7 N/mm²

COMPOUND CYLINDERS

In order to make pressure vessels economical in terms of thickness and ensure more uniform pressure distribution, cylinders are generally prestressed. We have discussed one method of doing so—using a wire-wound, thin cylindrical vessel. Another is to use a jacket over a cylinder. The jacket is shrunk on to the cylinder, causing prestressing. When the internal pressure is then applied, the stresses in the cylinder are much less than they would have been otherwise. The difference between the internal radius of the jacket and the outer radius of the cylinder are adjusted to achieve the required level of prestressing.

Figure 17.10 shows a compound cylinder. The shrinking of the jacket causes tensile stresses on the outer cylinder and compressive stresses on the inner cylinder.

In pressure vessel technology, we also have multilayered vessels where several layers of plates are bent into cylindrical form and tightly wrapped one over the other to make up a large thickness. The advantage is if the innermost layer cracks, the crack will not propagate to the other layers as there is a material discontinuity between two adjacent layers.

Now, we can find the radial displacement of a point due to the external jacket pressure and the internal pressure from the increase (or decrease) in the length of the circumference. If σ_h and σ_r are the hoop and radial stresses at any point A of the cylinder, the hoop strain is given by

$$\varepsilon_h = \frac{1}{E}(\sigma_h - v\sigma_r)$$

Note (v = m, Poisson's ratio)

$$= \frac{\text{Change in circumference}}{\text{Original length of circumference}}$$

$$= \frac{2\pi\delta r}{2\pi r} = \frac{\delta r}{r}$$

$$\frac{\delta r}{r} = \frac{1}{E}(\sigma_h - v\sigma_r)$$

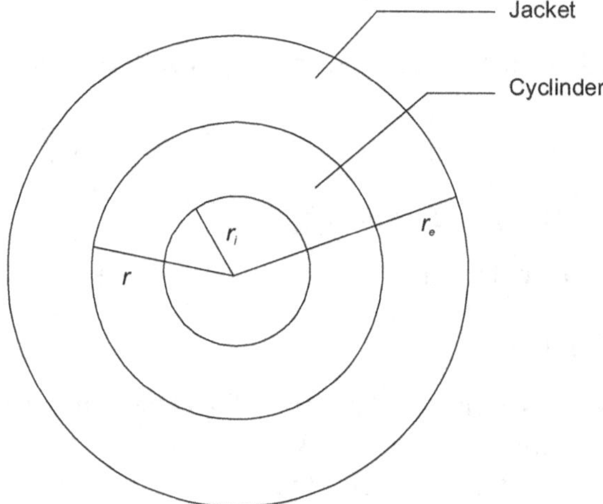

Figure 17.10

We have now two cases — one due to the external pressure alone because of the shrinking of the jacket and the second because of the internal pressure in the cylinder. Owing to external pressure P_e, the radius will decrease and due to the internal pressure the radius will increase.

On account of external pressure P_e alone, the decrease in length of the external radius is

$$\delta_e = -\frac{P_e r_e}{E}\left(\frac{r_e^2 + r_i^2}{r_e^2 - r_i^2} - v\right)$$

Due to internal pressure P_i, the increase in length of the external radius is given by

$$\delta_i = -\frac{P_i r_i}{E}\left(\frac{r_e^2 + r_i^2}{r_e^2 - r_i^2} + v\right)$$

The initial radial difference is given by

$$\delta = \delta_i + \delta_e$$

If r_3 = External radius, r_e = Radius at the junction, and r_i = Internal radius of the cylinder.

Shrinking allowance to be provided $= 2\delta = 2(\delta_i + \delta_e)$

$$\text{Shrinking allowance} = \frac{2P_s r_e}{E}\left(\frac{r_3^2 + r_e^2}{r_3^2 - r_e^2} + \frac{r_e^2 + r_i^2}{r_e^2 - r_i^2}\right)$$

where $= P_s$ is the shrinkage pressure.

The stress distribution in the cylinders and jacket due to the shrinking of the jacket on the cylinder, and the internal pressure are shown in Figure 17.11.

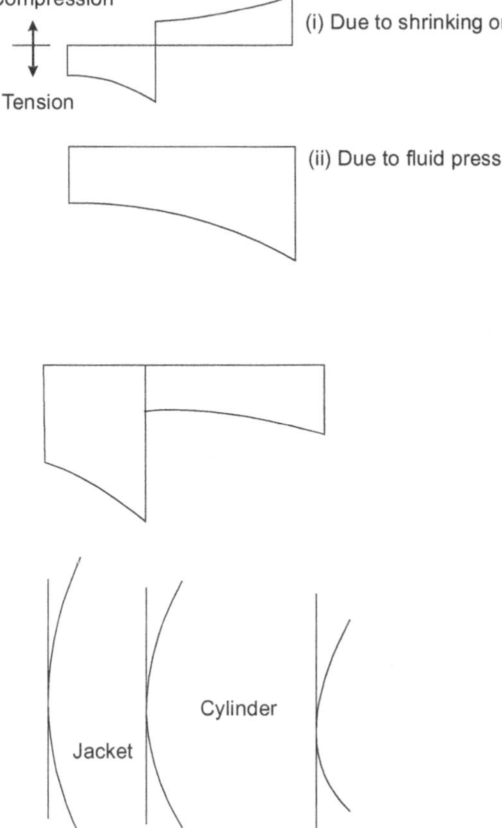

Figure 17.11

The following examples illustrates the use of these formulae.

PROBLEM 17.11

A compound cylinder is made by shrinking a jacket on to a cylinder. For the compound cylinder, the outer and inner radii are 100 mm and 60 mm, and the radius at the junction is 80 mm. Before the fluid pressure of 40 N/mm² is applied, the radial pressure at the junction is 10 N/mm² is applied. Determine the final stresses in the cylinder. Also calculate the difference in the diameters of tubes before the jacket is shrunk on to the cylinder and the temperature at which this can be done. Take $E = 200$ GPa and $\alpha = 12 \times 10^{-6}/°C$.

Solution

When the jacket is shrunk on to the cylinder, the cylinder is under compression and the jacket under tension. For the cylinder, the radial pressure is an external pressure, while for the jacket, it acts as an internal pressure. We have to determine the hoop stresses.

For the inner cylinder, with only the external pressure $P_e = 10$ N/mm² acting,

σ_{hmax} at $r = 60$ mm is given by

$$\sigma_{hmax} = \frac{P_e r_e}{r_e^2 - r_i^2}\left(1 + \frac{r_i^2}{r^2}\right) \qquad r = r_i$$

$$\frac{-2P_e r_e^2}{r_e^2 - r_i^2} = -\frac{2 \times 10 \times 80^2}{(80^2 - 60^2)}$$

$$= 45.71 \,\text{N}/\text{mm}^2 \,(\text{comp.})$$

$$= -\frac{P_e r_e}{r_e^2 - r_i^2}\left[1 + \left(\frac{r_i^2}{r_e^2}\right)\right] = -\frac{10 \times 80^2}{80^2 - 60^2}\left(1 + \frac{60^2}{80^2}\right)$$

$$= 34.72 \,\text{N}/\text{mm}^2 \,(\text{comp.})$$

For the jacket, with internal pressure

$$P_i = 10 \,\text{N}/\text{mm}^2,$$

$$\sigma_h = -\frac{P_i r_i^2}{r_e^2 - r_i^2}\left(1 + \frac{r_i^2}{r^2}\right)$$

$$r_i = 80$$

$$r_e = 100$$

σ_h is maximum at $r = r_i$

$$\sigma_{hmax} = \frac{10 \times 80^2}{100^2 - 80^2}\left(\frac{100^2}{80^2} + 1\right) = 45.55 \text{ N/mm}^2 \,(\text{tensile})$$

$$\sigma_h \text{ at } r = 100 = \frac{10 \times 80^2 \times 2}{100^2 - 80^2} = 35.55 \text{ N/mm}^2 \,(\text{tensile})$$

With the fluid pressure of 40 N/mm², the inner cylinder has an internal pressure of 40 N/mm², $\sigma_{h\,\mathrm{max}}$ will be at $r = 60$.

$$\sigma_{h\,\mathrm{max}} = \frac{P_i r_i^2}{r_e^2 - r_i^2}\left(\frac{r_e^2}{r_i^2} + 1\right)$$

Here $r_e = 100$ and $r_i = 60$.

$$\sigma_{h\,\mathrm{max}} = \frac{40 \times 60^2}{100^2 - 60^2}\left(\frac{100^2}{60^2} + 1\right) = 84.98\,\mathrm{N/mm^2}\ (\text{tensile})$$

At the junction, $r = 80$,

$$\sigma_h = \frac{40 \times 60^2}{100^2 - 60^2}\left(\frac{100^2}{80^2} + 1\right)$$

$$= 57.65\,\mathrm{N/mm^2}$$

At the outer end, $r = 100$.

$$\sigma_h = \frac{40 \times 60^2}{100^2 - 60^2}\left(\frac{100^2}{100^2} + 1\right)$$

$$= 45\,\mathrm{N/mm^2}$$

The stresses can be tabulated as follows.

Loading	Radius	Hoop stresses(N/mm²)			
		Inner tube		Jacket	
		r = 60	r = 80	r = 80	r = 100
Shrink on pressure 10 N/mm²		− 45.71	− 34.72	+ 45.55	+ 35.55
Fluid pressure 40 N/mm²		+84.98	+57.65	+57.65	+45
Total hoop stress		+39.27	+22.93	103.2	+80.55

The difference between the outer diameter of the inner tube and the inner diameter of the jacket is the shrinkage allowance.

$$\text{Shrinkage allowance} = \frac{2p_s r_2}{E}\left(\frac{r_3^2 + r_e^2}{r_3^2 - r_e^2} - \frac{r_e^2 + r_i^2}{r_e^2 - r_i^2}\right)$$

Here,

$$P_s = 10\,\text{N/mm}^2, r_3 = 100, r_2 = 80, r_i = 60\,\text{mm}$$

$$\text{Shrinkage allowance} = \frac{2 \times 10 \times 80}{2 \times 10^5}\frac{100^2 + 80^2}{100^2 - 80^2} - \frac{80^2 + 60^2}{80^2 - 60^2}$$

$$= 0.065 \text{ mm}$$

To find the temperature difference, the shrinkage is considered due to thermal expansion. The expansion or contraction due to temperature ΔT is given as $\alpha \Delta T d'$

$$\therefore 0.0605 = \alpha \Delta T 160$$

$$\text{Hence } \Delta T = \frac{0.0605}{12 \times 10^{-6} \times 160} = 33.8^\circ C$$

PROBLEM 17. 12

A high-pressure cylinder consists of a steel tube with inner and outer diameters of 20 and 40 mm respectively. It is jacketed by an outer steel tube having outer diameter of 60 mm. The tubes are assembled by a shrinking process in such a way that maximum principal stress induced in any tybe is limited to 100 N/mm².

(a) Calculate the shrinkage pressure and original dimensions of the tubes (E = 207 kN/mm²).
(b) Plot the distribution of stresses due to shrink fit in service, the cylinder is further subjected to an internal pressure of 300 MPa. Plot the resultant stress distribution.

Solution (a)

The maximum principal stress is tangential stress at the inner surface of jacket. From equation for σ_t.

$$\sigma_t = \frac{P(D_3^2 + D_2^2)}{D_3^2 - D_2^2} \text{ or } 100 = \frac{P(60^2 + 40^2)}{(60^2 - 40^2)}$$

$$\therefore P = 38.46\,\text{N/mm}^2$$

From Equation for δ

$$\delta = \frac{Pd_2}{E}\frac{2D_2^2(D_3^2 - D_1^2)}{(D_3^2 - D_2^2)(D_2^2 - D_1^2)}$$

$$or\ \delta = \frac{(38.46)(40)}{(207 \times 10^3)}\frac{2(40)^2(60^2 - 20^2)}{(60^2 - 40^2)(40^2 - 20^2)}$$

$$= 0.0317\,\text{mm}$$

The dimensions of the tubes are as follows:

Outer diameter of inner tube = 40 mm

Inner diameter of jacket = 40 − 0.0317

$$= 39.9683 \text{ mm}$$

Solution (b)

Stresses due to shrink fit

Jacket: The jacket is subjected to an internal pressure of 38.46 N/mm² due to shrink fit. From equation for σ_r and σ_t

$$\sigma_r = -\frac{PD_2^{\,2}}{(D_3^{\,2} - D_2^{\,2})}\left[\frac{D_3^{\,2}}{4r^2} - 1\right]$$

$$= -\frac{(38.46)(40)^2}{(60^2 - 40^2)}\left[\frac{60^2}{4r^2} - 1\right]$$

$$= -30.77\left[\left(\frac{30}{r}\right)^2 - 1\right]$$

$$\text{and } \sigma_t = +\frac{PD_2^{\,2}}{D_3^{\,2} - D_2^{\,2}}\left[\frac{D_3^{\,2}}{4r^2} - 1\right]$$

$$= +30.77\left[\left(\frac{30}{r}\right)^2 - 1\right]$$

The stresses in the jacket are tabulated as follows

r	20	22	24	26	28	30
σ_r	−38	−28	−17	−10	−5	0
σ_t	100	88	79	72	66	62

Inner tube The inner tube is subjected to an external pressure of 38.46 N/mm² due to shrink fit.

From Equations for σ_r and σ_t due to external pressure.

$$\sigma_r = -\frac{PD_2^{\,2}}{(D_2^{\,2} - D_1^{\,2})}\left[1 - \frac{D_1^{\,2}}{4r^2}\right]$$

$$= -\frac{(38.46)(40)^2}{(40^2 - 20^2)}\left[1 - \frac{20^2}{4r^2}\right]$$

$$= -51.28\left[1 - \left(\frac{10}{r}\right)^2\right]$$

$$\text{and } \sigma_t = -\frac{PD_2^{\,2}}{D_2^{\,2} - D_1^{\,2}}\left[1 + \frac{D_1^{\,2}}{4r^2}\right]$$

$$= -51.28\left[1 + \left(\frac{10}{r}\right)^2\right]$$

The stresses in the tube are tabulated as follows

r	10	12	14	16	18	20
σ_r	0	−16	−25	−31	−35	−38
σ_t	−103	−87	−77	−71	−67	−64

Stresses due to internal pressure

When the compound cylinder is subjected to an internal pressure of 300 MPa in service, the stresses are calculated from stresses due to internal pressure.

$$\sigma_r = -\frac{P_i D_1^{\,2}}{(D_3^{\,2} - D_1^{\,2})\,4r^2}\left[\frac{D_3^{\,2}}{4r^2} - 1\right]$$

$$= -\frac{(300)(20)^2}{(60^2 - 20^2)}\left[\frac{60^2}{4r^2} - 1\right] = -37.5\left[\left(\frac{30}{r}\right)^2 - 1\right]$$

$$\text{and } \sigma_t = +\frac{P_i D_1^{\,2}}{D_3^{\,2} - D_1^{\,2}}\left[\frac{D_3^{\,2}}{4r^2} + 1\right]$$

$$= +37.5\left[\left(\frac{30}{r}\right)^2 + 1\right]$$

The stresses in the compound cylinder are tabulated as follows

r	10	12	14	16	18	20	22	24	26	28	30
σ_r	−300	−197	−135	−94	−67	−47	−32	−21	−12	−6	0
σ_t	375	272	210	169	142	122	107	96	87	81	75

Resultant stresses   The resultant stresses in the stresses due to shrink fit and those due to internal compound cylinder are obtained by superimposition of pressure. They are tabulated as follows.

	Inner tube						Jacket					
r	10	12	14	16	18	20	20	22	24	26	28	30
σ_r	−300	−213	−160	−125	−102	−85	−85	−58	−38	−22	−11	0
σ_t	272	185	133	98	75	58	222	195	175	159	147	137

Figure shows the variation of these stresses.

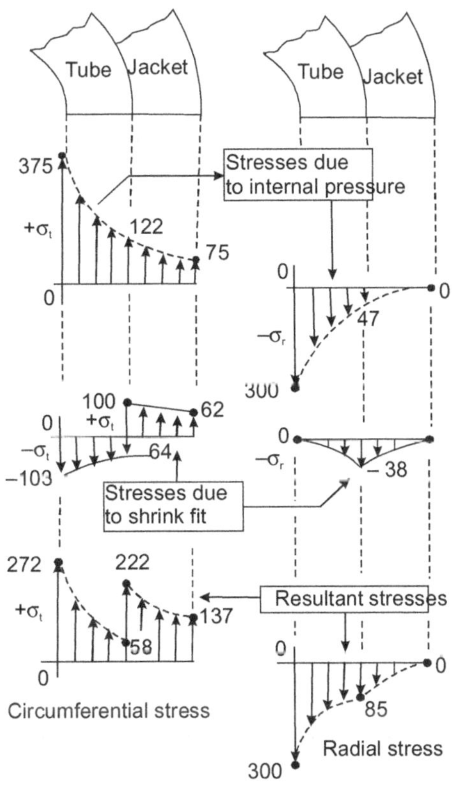

REVIEW QUESTIONS

SHORT QUESTIONS

1. Define " Thin Shell".

2. Mention the types of stresses produced in thin cylindrical shells.

3. What are the stresses developed in thin cylinders when they are subjected to internal fluid pressures?

4. List out the modes of failure in thin cylindrical shell due to an internal pressure.

5. Draw the state of stress at a point in the case of a thin cylindrical shell (ends closed) subjected to internal pressure and hence find the maximum principal stress.

6. A cylinder air receiver for a compressor is 3 m in internal diameter and made of plates 20 mm thick. If the hoop stress is not to exceed 90 N/mm² and the axial stress is not to exceed 60 N/mm², find the maximum safe air pressure.

LARGE QUESTIONS

1. A cylindrical shell 800 mm in diameter, 3 m long is having 10 mm metal thickness. If the shell is subjected to an internal pressure of 2.5 N/mm², calculate,

    i. the change in diameter

    ii. the change in length and

    iii. the change in volume.

    Assume the modulus of elasticity and Poisson's ratio of the material of the shell as 200 kN/mm² and 0.25 respectively.

2. A cylindrical shell 3 metre long which is closed at the ends has an internal diameter of 1 m and a wall thickness of 15 mm. Calculate the circumferential and longitudinal stress induced and also change in the dimensions of the shell if it is subjected to an internal pressure of 1.5 x 10⁶ N/m². Take E = 20 × 10¹⁰ N/m² and Poisson's ratio = 0.3.

3. i. A cylindrical vessel 2 m long and 500 mm in diameter with 10 mm thick plates is subjected to an internal pressure of 3 MPa. Calculate the change in volume of the vessel. Take E = 200 GPa and Poisson's ratio = 0.3 for the vessel material.

    ii. A spherical shell of 2 m diameter is made up of 10 mm thick plates. Calculate the change in diameter and volume of the shell, when it is subjected to an internal pressure of 1.6 MPa. Take E = 200 GPa and 1/m = 0.3.

4. i. A cylindrical thin shell 800 mm in diameter and 3 m long is having 10 mm metal thickness. If the shell is subjected to an internal pressure of 2.5 MPa, determine change in diameter and change in length. Take E = 200 GPa.

    ii. A boiler shell is to be made of 20 mm thick plate having a limiting tensile stress of 120 MPa. If the efficiencies of the longitudinal and circumferential joints are 70% and 30% respectively, determine the maximum permissible diameter of the shell for an internal pressure of 2 MPa. When the shell diameter is 1.5 m find the permissible intensity of internal pressure.

5. Derive the expressions for hoop stress and longitudinal stress in a thin cylinder with ends closed by rigid flanges and subjected to an internal fluid pressure. Take the internal diameter and shell thickness of the cylinder to be 'd' and 't' respectively.

INTERESTING POINTS TO PONDER

Tall cylindrical pressure vessels are employed in Refineries, Petrochemical and Fertilizer plants. The design for internal pressure is done using either thin or thick shell theory. However such tall vessels are subjected to wind and/or earthquake loads. These loads cause bending. Bending stresses are induced by such loads. These are determined by treating the tall vessel as a line element along the vessel centre line idealised as a cantilever beam. The loads can be lumped at different points along the height. Using beam formulation, the bending and shear stresses are determined at various sections. The direct net downward compressive stresses at each section due to the weight of the portion of vessel above the section is also to be considered and added to the longitudinal bending stress, longitudinal pressure stress, taking into account the net tensile, compressive stresses.

The stresses due to bending are more in the lower part of the vessel and designs may involve variable thickness shells with higher thickness provided at the bottom section. The welding of such vessels are done at site with cylindrical portions welded on top of one another.

Likewise combining the stresses in the three directions, the net stresses in x, y, z directions are determined. Then either maximum shear stresses or Vonmises stresses are determined. This is to be checked against codal values as per ASME sec VIII.

INTERESTING POINTS TO PONDER

Piping systems are generally treated as beams with multiple supports so long as bending due to self weight, Thermal expansion, windloads or earthquake loads are concerned. So far as internal pressure is concerned they are treated as thin walled pressure vessels or thickwalled vessels depending upon (d/t) ratio. The net stresses are produced due to both these effects. High pressure, High temperature piping systems used in chemical plants, thermal and nuclear power plants should satisfy the following two criteria:

i.   Stresses in the piping systems should be within allowable limits as per design code(ASME sec VIII/ ANSI sec B31.3/ ANSI B31.1).

ii. The reactions of the piping system exerted on the terminals (like Turbine or compressor or steam generators) between which the piping runs, should be within prescribed limits as per manufacturers' specification. The terminal reaction include six-forces F_x, F_y, F_z and Moments M_x, M_y, M_z.

The stresses in the piping systems to be considered are

a) *Pressure stresses.*

(Contd.,)

The hoop and longitudinal stresses are determined the same way as per thin/thick shell formulae for internal pressure.

b) *Stresses caused by thermal expansion*:

There are standard programs available like CAESAR-II, ANSYS etc. where the piping system is modeled like a beam represented by its centre line and stresses analysed for temperature and self weights. We need to use the value of E(Young's modulus) and α (coefficient of thermal expansion) at the appropriate temperature. The material properties can be obtained from ASME/ASTM handbooks.

c) *Stresses caused by Earthquake/Wind*

This analysis is also carried out mostly as static analysis treating the piping as a beam supported at various points. The supports could be of different types, some providing partial restraints and anchors provide full restraints.

For the wind load analysis, where a piping system is located on an open yard, the wind loads are lumped at several points on the piping system and analysed. For earthquake or seismic analysis, the earthquake forces are applied as lateral loads at base or support or ground points and stresses evaluated.

CHECKING STRESS COMPLIANCE

Each one of the above, after evaluation is checked for individual as well as combined limits as prescribed by the codes.

When combining stresses, the stresses should be added/grouped along the three directions (x, y, z) and later the principal stresses are determined. From this Vonmises Stresses are determined.

Depending upon the codal requirements either max shear stress or Vonmises stresses are computed and checked against limits.

What if stress compliance is not achieved?

Reroute the piping/relocate the supports and perform reanalysis till compliance is met. Rerouting the piping has practical constraints, as interference with equipment or other structures may take place. Relocating supports, providing bends/elbows or loops will tend to reduce thermal stresses and therefore is a better practical option.

18

FINITE ELEMENT ANALYSIS

HISTORICAL BACKGROUND AND DEVELOPMENT

This technique (FEA) was discovered about 60 years back the idea of assembling physical structures from their components gave impetus to the concept of extending to assembly of mathematical element model and their assembly to represent the reality. Earlier applications were confined to civil engineering as civil engineers were already familiar with matrix methods of structural analysis which bear a relation with finite element analysis.

With the advent of digital computers, which offers a vast scope for solving large systems of equations, the finite element solution techniques did not see computational issues as a problem at all.

The technique very soon expanded to cover areas of mechanical, aeronautical , electrical , civil, biomedical and other related fields.

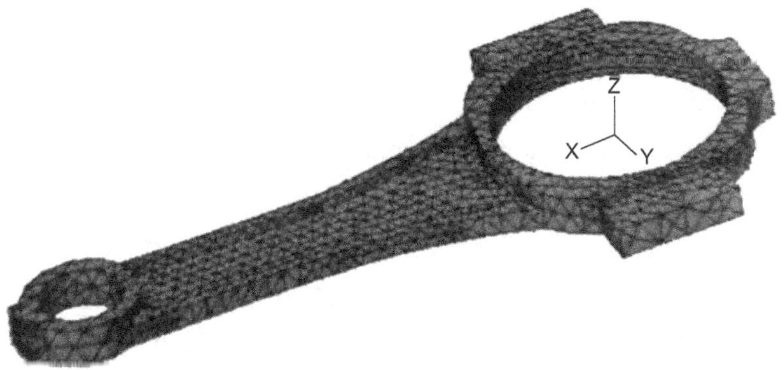

Figure 18.1

Interdisciplinary problems involving soil–structure interaction, fluid–structure interaction, inter–connection of complex structures and Heat-transfer problem are effectively and easily modelled and solved rapidly. Several iterations using altered design parameters are possible at once, thus making the design cycle, quick and accurate. This gave rise to CAD (computer aided design) which further links to CAM (computer aided manufacturing) to connect analysis, design and manufacturing processes.

The purpose of this chapter is only to expose the reader to some of the fundamental aspects of FEA, and therefore the coverage is extremely introductory in nature. For further details, the reader is urged to consult the many references cited in bibliography. Figure 18.1 shows a finite-element model of a crankshaft that was developed to study the effects of dynamic elastohydrodynamic lubrication on bearing and structural performance.

DISCRETIZATION

Simple two or three-dimensional geometric shapes are chosen to approximate.

    i. a given two-dimensional domain

    ii. the solution over the domain.

We not only seek an approximate solution to a given problem on a domain, but we also approximate the domain by a suitable finite element mesh. Consequently two approximation errors arise

    i. due to solution approximation

    ii. due to discretization leading to domain approximation .

We use elements such as triangles, rectangles, quadrilaterals and solid elements (brick elements) to construct FE models .

METHODOLOGY FOR FINITE ELEMENT (FE) MODELLING

This methodology consists of the following steps:

(a) Preparation of finite element model which represents the equipment adequately for the problem in hand .

(b) Definition of material properties and section properties .

(c) Boundary conditions and locations .

(d) Loading (distributed, concentrated, thermal, dynamic)

(e) Analysis (static, dynamic, stability).

(f) Review analysis/reanalyse results until codal or design requirements are satisfied .

Established computer program (e.g., ANSYS, NASTRAN, IDEAS, COSMOS etc.) can be used to generate the model with the help of geometric structure of the equipment with various loads and loading combinations.

Advantages

A very powerful analytical tool for solving design , structural , interdisciplinary problems and verify the design. It effectively deals with:

(a) complex geometries

(b) variation in materials properties

(c) numerical solution at all locations provided

(d) convergence accuracies

(e) model physical problem realistically

(f) interdisciplinary areas

(g) assembly of different geometries

(h) simulation of joints

(i) field problems in fluid mechanics/structural mechanics/thermal engineering/soil mechanics /electromagnetism

(j) extension to nonlinear problems

(k) fatigue, creep and fracture mechanics

FINITE ELEMENT ANALYSIS (FEA OR FEM)—DETAILED STEPS

❀ **Discretization of Continum**   (Problem – Physical Model – Idealization)

The physical model should be close to reality and should include all features which influence solution and adequate to provide reliable answers to the problem.

The method is described as (Piece – Wise Rayleigh – Ritz Method)

Discrete elements connected at corners (nodes) OR/AND midsides to adjacent elements to idealize the total structure (it could be piping, pressure vessel, support structure or heat exchanger or a building) (Complex Geometry).

Shape Function

❀ Displacement (or temperature or pressure) fields assumed (as polynomials) within the element such as line (1D), or triangular, rectangular, or quadrilaterial elements (2D) or thick shell or brick elements (3D). Element details are provided in Table 18.1 in elemental or local corrdinates with unknown coefficients for each term in the polynomial.

❀ Corner displacement (nodal)—Replace unknown coefficients in polynominals assumed for displacement, temperature, or flow field and express in terms of nodal variables called degrees of freedom.

❀ Strain–displacement relations (linear/non linear) to express in terms of strains for stress analysis problems.

❀ Stress–strain relations (constitutive relations) (Material behaviour) (linear/non linear) to express stress in terms of strain.

❀ Formulate strain energy over volume of element/or virtual work (Galerkin)

❀ Formulate stiffness and mass matrices derived from variational/Galerkin approach (Element level)

- ❀ Transform element matrices to global coordinates (Coordinate transformation from local axes of the element to global axes of the structure)

- ❀ Assemble at the global level to yield global or overall system stiffness and mass matrices (Apply constraints and loading on the global structure)

  upto this step, stress analysis procedure is kept in view. The following steps relate to stress, flow and kermel analysis.

- ❀ Solve for unknowns, usually displacements (temperatures or fluid pressures) using well-known matrix methods of solution.

- ❀ Back work and obtain strains and stresses, temperatures, relativities etc depending the problem.

- ❀ Post process to obtain plots (stress/mode shapes/temp fields/flow fields/pressure fields)

Basis

- ❀ Approximation of domain (discretization) use appropriate elements from the element library of the FEM program (FEMP). Refer Table 18.1 for the sample finite element library.

- ❀ Approximation of solution (variational principle/method of weighted residuals)

- ❀ Approximation of function (interpolation function).

- ❀ Generation of simultaneous algebraic equation in nodal coordinates (primary variable).

- ❀ Transformation from local to global coordinates.

- ❀ Imposition of boundary conditions (secondary variable).

- ❀ Solution of equations (Guass elimination/LDL decomposition).

Element Geometries

Many geometric shapes of elements are used in finite-element analysis for specific applications. The various elements used in a general-purpose commercial FEM software code constitute what is referred to as the element library of the code. Elements can be placed in the following categories: line elements, surface elements, solid elements and special-purpose elements. Table 18.1 provides some of the elements used for structural applications.

Some Guidelines in Solving FE Problems

(a) Give the inputs correctly with appropriate and compatible units as required by FEMP (load, geometry, material, constrains)

(b) Use a coarse mesh to start with using element aspect ratio not larger than 2. Aspect ratio refers to the ratio of largest dimension to the smallest dimension in an element. Ascertain symmetric planes in the structure and use either half or quarter or axymmetric section for modelling the structure. (Elements should not be too long). Coarse mesh means larger element size and number of elements not too large.

(c) Use finer mesh at stress concentration zones (to make sure, run the solution with still more finer mesh and achieve convergence) results of two consecutive iterations not varying by more than 5% indicate convergence. At joining points of two or more different elements the common node should bear the same nodal number .

Table 18.1   Sample finite-element library

Element type	None	Shape	Number of nodes	Applications
Line	Truss		2	Pin-ended bar in tension or compression
	Beam		2	Bending
	Frame		2	Axial, torsional, and bending with or without load stiffening
Surface	4-node quadrilaterial		4	Plane stress or strain, axisymmetry, shear panel, thin flat plate in bending
	8-node quadrilaterial		8	Plane stress or strain, thin plate or shell in bending
	3-node triangular		3	Plane stress or stain, axisymmetry, shear panel, thin flat plate in bending prefer quad where possible. Used for transitions of quads.
	6-node triangular		6	Plane stress or strain, axisymmetry, thin plate or shell in bending. Prefer quad where possible. Used for transitions of quads.

Contd.

Table 18.1   Sample finite-element library

Element type	None	Shape	Number of nodes	Applications
Solid	8-node hexagonal (brick)		8	Solid, thick plate
	6-node pentagonal (wedge)		6	Solid, thick plate. Used for transitions.
	4-node tetrahedron (tet)		4	Solid, thick plate. Used for transitions
Special purpose	Gap		2	Free displacement for prescribed compressive gap
	Hook		2	Free displacement for prescribed tension. Gap
	Rigid		Variable	Rigid constraints between nodes

(d)   For large size and complex problems use coarse mesh initially

  ❋   Make a section with fine mesh which requires refined analysis for sharp corners/holes/ nozzles, etc.

  ❋   Match the displacements at the boundaries between coarse (large model) and section (detailed) model to achieve a smooth merger between coarse and fine mesh model .

(e)   Make equilibrium checks to ascertain correctness of solutions

  (sum of reactions (x,y,z) at constraint points) = sum of applied loads (x,y,z)).

(f)   In dealing with fine meshes (say for example stress distribution around a bolt hole), in case abnormal values of stress or displacement obtained at one or two stray nodes/elements, ignore them so long as the values at other locations are reasonable (may be due to computational errors accumulated due to small size of mesh). Remember experimental verification add strength to analytical method.

(g) In verification problems, try to create a model as close to reality as possible (field/test conditions) so that gaps between real and simulated situations are minimized. One should run a known problem first, where text book solutions are available to ascertain correctness and gain confidence over the model chosen for the complex problem.

(h) Make logical assumptions and their implications on the final results should be kept in mind.

(i) Study results display and compare and incorporate design improvements in model if needed to bring down or alter the results as desired. Re-examine and conclude once acceptance is obtained from codal compliance.

(j) Iterate till results are acceptable as per code prescriptions vendor requirements and conforming to design specifications.

Design and Analysis Cycle

We refer to this figure again to illustrate the connection between design and analysis.

The role of FEA is shown in terms of modeling, analysis, checking and network if nessitated by acceptability criteria. This give a picture of how FEA is intergrated with design and analysis cycle.

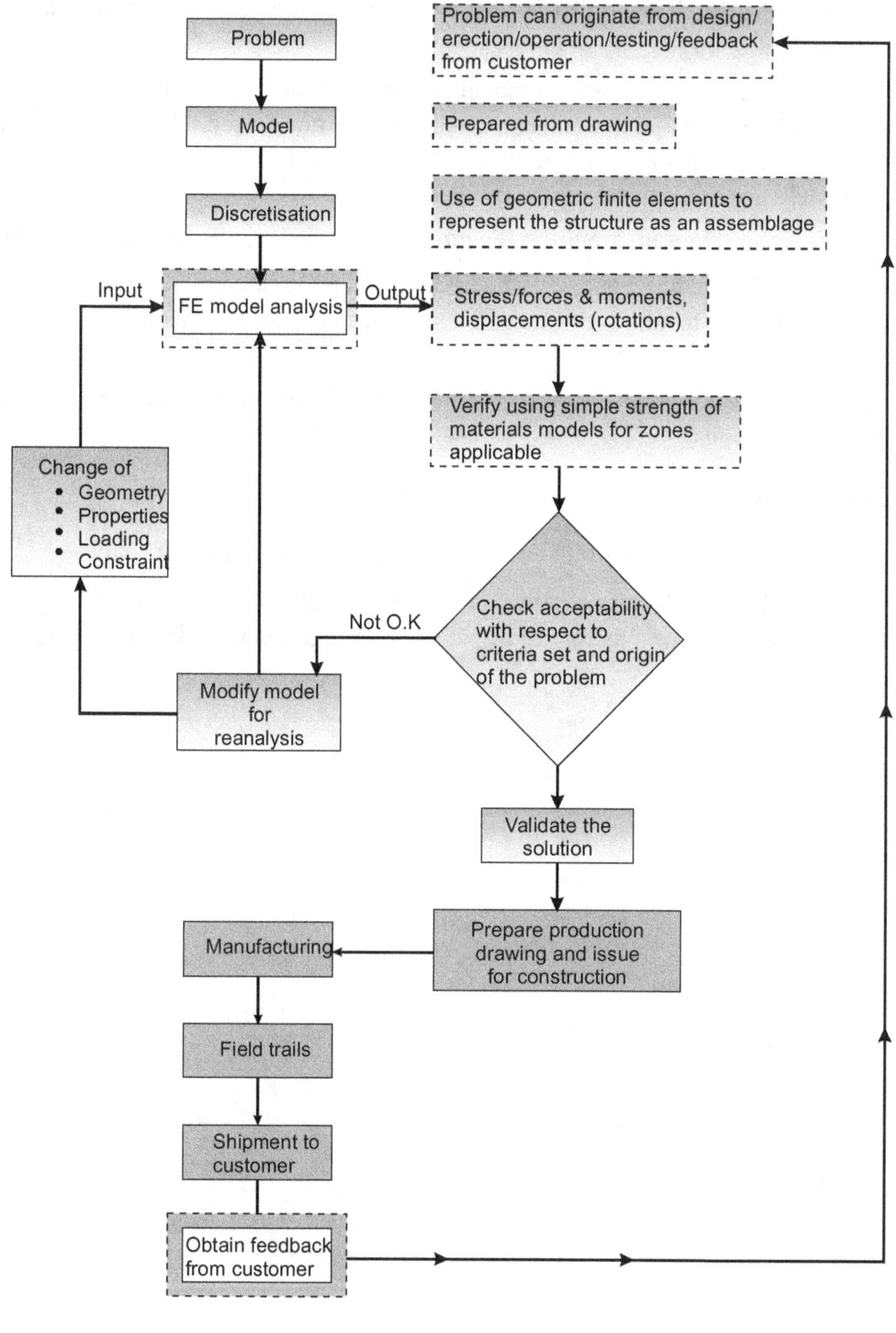
Problem
Model
Discretisation
Input
FE model analysis
Output
Problem can originate from design/erection/operation/testing/feedback from customer
Prepared from drawing
Use of geometric finite elements to represent the structure as an assemblage
Stress/forces & moments, displacements (rotations)
Verify using simple strength of materials models for zones applicable
Change of
• Geometry
• Properties
• Loading
• Constraint
Not O.K
Check acceptability with respect to criteria set and origin of the problem
Modify model for reanalysis
Validate the solution
Manufacturing
Prepare production drawing and issue for construction
Field trails
Shipment to customer
Obtain feedback from customer

A paper published by the author is enclosed as a case study. This involves strength of materials solution for

a) Pressure stresses in the piping (treated as a pressure vessel).

b) Self weight and Temperature stresses in the piping (treated as a beam).

The combined stresses are determined giving due weightage to stress concentration at bends and 'T' joints in the piping system. The paper discusses from a piping stress point of view, the permissible rate of warming of the steam pipeline or so as to decide at what rate should steam be admitted in the piping system so as to keep the total piping stresses within limits for a safe start up of the turbo generator.

The study was conducted for a typical 500 MW Thermal power station in view of safety requirements and the same got actually implemented.

The enclosed paper describes the details of study and the results obtained.

The paper highlights certain practical stress analysis solutions to plant problems or any other applied mechanics problem. It also deals with the formulation, methodology, codal requirements, design and operating data that would need to be considered simultaneously so as to arrive at an implementable solution at the site.

The paper is published in the proceedings of the 40th meeting of Mechanical Failure Prevention Technology at Gaithersberg, Maryland, USA.

STRESS ANALYSIS OF PIPING SYSTEM TO AID TG START-UPS IN LARGE THERMAL POWER STATIONS

Abstract: In large thermal power stations, it is quite likely that the start-up and shut-down operations of the turbo-generator could be governed by the maximum stress produced in the main steam piping system owing to its large wall thickness. While the equipment designers could provide automated supervisory systems to check the safety of their equipment against overstressing, such a feature was not attempted from the piping designer's end. This paper descusses how, with the help of a detailed stress analysis of piping systems, necessary information could be generated for the plant operator to control the rate of warming up/cooling down to keep the maximum stress in the piping system within permissible limits. While this is so from the design point of view, actual operations at the plant pose situations where such limits may have to be marginally exceeded. The paper also discusses the implications of such operations on the remaming life of components which undergo overstressing for short durations.

Keywords   Creep-fatigue; critical stress locations; finite element method; life expectancy; main steam piping; stress analysis; thermal power plant.

Introduction

In a thermal power station, especially in a larger unit generating 500 MW and above, it is important for the operator to know how quickly the turbine can be started up and what changes in load he may make without fear of overstressing the components and thereby causing excessive fatigue. Normally, all leading turbine manufacturers provide a Turbine Stress Evaluator (TSE), a device which, by controlling the temperature difference across the wall thickness of turbine components like the Emergency Stop Valve (ESV), turbine casing and shaft at different pressures, maintains the stresses within allowable limits and thus aids in controlling the rate of warming up or cooling down of the turbine.

While a TSE takes care of the turbine from the stresses due to startups, it was necessary to know if stresses produced in the main steam piping could also govern start-up/shut-down operations. Such an investigation is especially warranted in large units like a 500 MW sub-critical unit and more so in super-critical units as a result of employing pipes of large wall thicknesses and high operating pressures. The objective of the investigation is to establish a relation between permissible temperature differential (ΔT) perm. across the wall thickness for the most critically stressed location in the main steam piping and the corresponding operating line pressures. This information together with similar information for the TSE and the strainer housing are inputted along with measured values of temperature differentials and line pressure to the warm-up controller. This device selects the minimum out of the above (ΔT) perm. and compares it with actual temperature differentials. Based on this, it directs a control signal to the operator on the correct rate of steam admission to be adopted.

The succeeding section on stress analysis describes how the stresses in piping due to different loadings are computed, how the most critically stressed location is identified from such computations and, finally, the manner in which the required information is generated. This general philosophy is illustrated through the example of a detailed stress analysis performed on the typical main steam piping of A 500 MW thermal power plant.

Stress analysis   The relevant portion of the main steam piping between the superheater outlet header and the turbine nozzle is shown in Figure 1. Stress analysis of this system has been performed by considering the combination of stresses generated as a result of:

(a) constrained thermal expansion (commonly termed as flexibility analysis)

(b) internal pressure

(c) thermal differential across pipewall thickness

Each one of these aspects is discussed below.

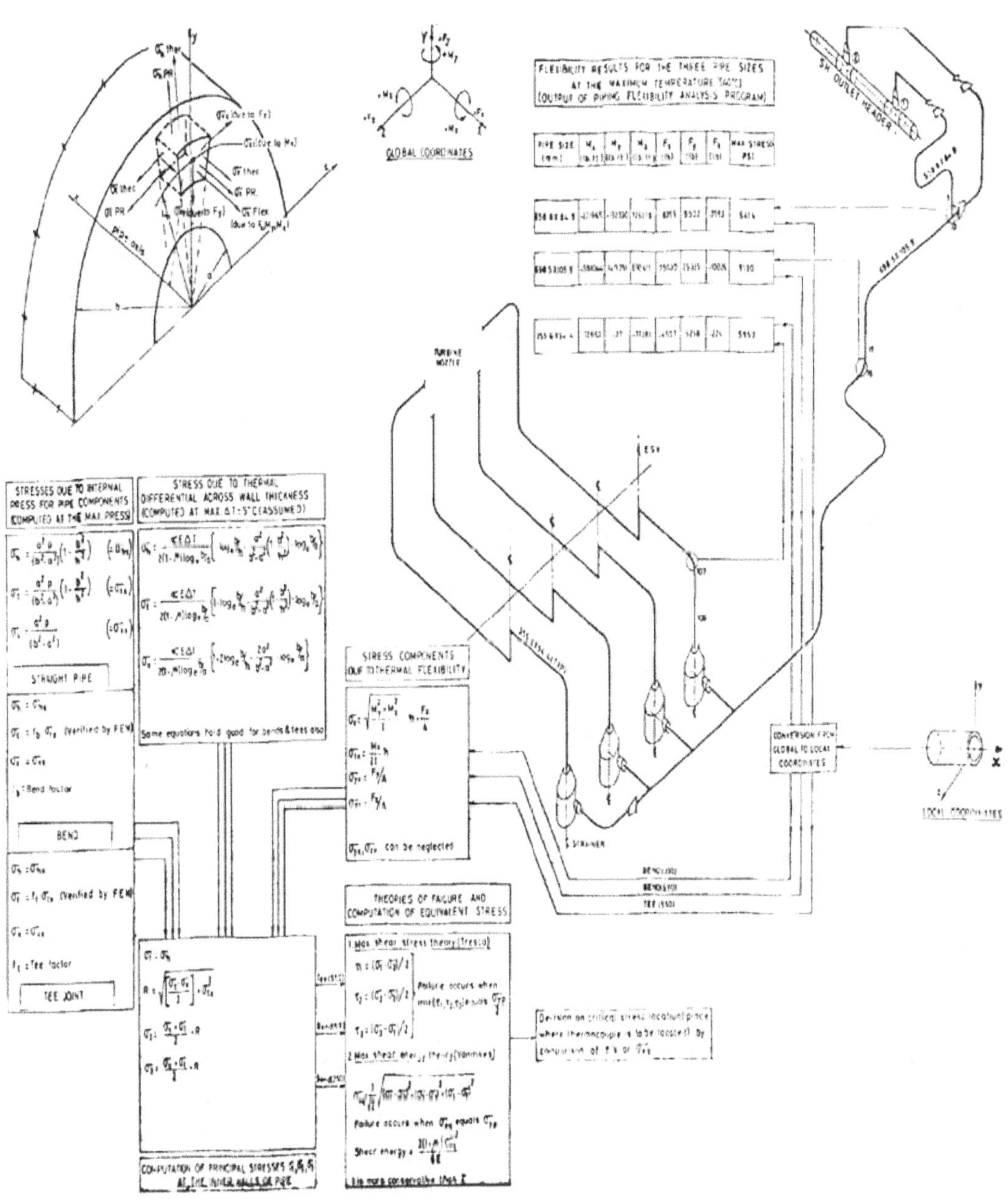

Figure 1   Flow diagram for decoding the critical stress location in the main steam piping

Flexibility analysis   Piping operating at elevated temperatures in chemical and power plants is subjected to stresses due to constrained thermal expansion. A computer program based on flexibility analysis is used to estimate the stresses, displacements, etc., in the piping. At any given location, the program outputs in the three global directions X, Y, Z.

(a) Forces Fx, Fy, Fz

(b) Moments Mx, My, Mz

(c) Displacements Dx, Dy, Dz

(d) Rotations Rx, Ry, Rz

and the maximum stress for the mainstream piping isometric inputs. These forces and moments act on the pipe and produce direct and bending stresses and torsional and transverse shear stresses. The flexibility analysis performed on the main steam piping system yields locations of maximum stress in the three different pipe sizes and are given below:

No.	Pipe size OD in mm and Th. in mm	Location of maximum stress from results of flexibility analysis (Refer Fig.1)	Nature of location
1.	698.5 × 105.9	Bend at 18 (branch 18–17)	Bend
2.	558.8 × 84.9	Tee at 10 (branch 10–9)	Tee
3.	355.6 × 54.4	Bend at 107 (branch 106–107)	Bend

Stresses due to internal pressure and thermal differential for straight pipes   While in flexibility analysis the pipe is treated as a beam. In this analysis the straight pipe is idealised as a long thick cylinder. Expressions for radial, tangential and axial stresses for internal pressure loading and steady state temperature difference across the thickness are based on the theory of elasticity for axisymmetric bodies. Results for straight pipes form the basis of stress calculations in bends and tees.

Hoop stresses in bends due to internal pressure   Bends in the main steam piping needing to be stress analysed are thick and are of long radius.

Hoop stresses in such bends are intensified by a factor called 'Torus hoop stress factor' over the nominal values. For bends at junctions 18 and 107 in Figure 1, this factor has been used to evaluate the maximum hoop stress.

For the bend at 18 (Refer to Figure 1) a finite element analysis is performed to determine the stresses due to internal pressure. This is done to verify the applicability of the Hoop Stress expression given in reference-1 to bends in the present piping system. The agreement between results obtained by the above is excellent.

Hoop stresses in tees due to internal pressure   The empirical hoop stress concentration factor based on experimental and analytical investigations as reported in the literature for unreinforced tees has been suitably modified for reinforced tees and used for the tee at junction 10 to evaluate the stresses due to internal pressure. Again, a two dimensional axisymmetric finite element analysis performed in the longitudinal section of the tee for this loading yields stress values that compare quite well with the empirical ones given in reference-2. The finite element model and the distribution of hoop stress across different layers for the tee are furnished in Figures 2 and 3.

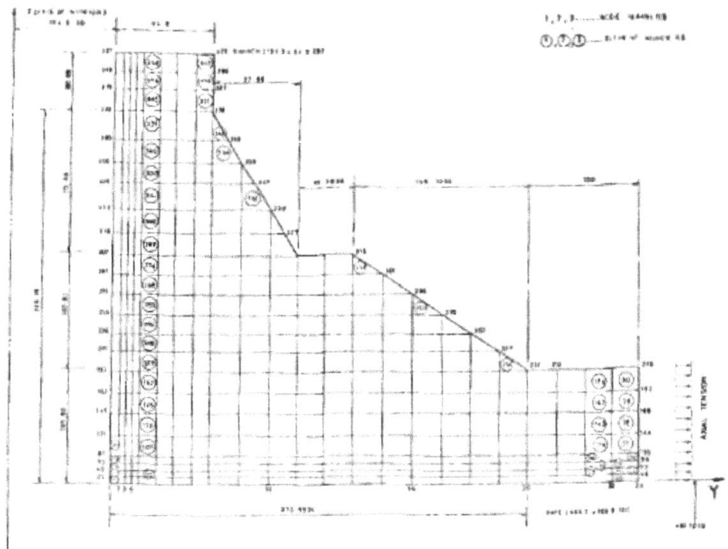

Figure 2   Finite element model of the longitudinal section of the tee for axis symmetric analysis for stresses due to internal pressure

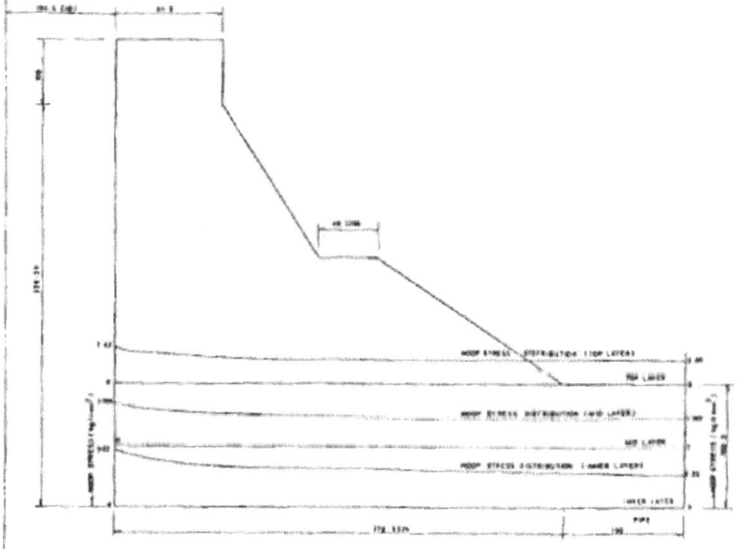

Figure 3   Hoop stress distribution along the pipe across three sections in the TEE

Thermal stress in bends and tees   Studies have indicated that the manner of distribution of thermal stresses due to steady state heat conduction across the thickness in straight pipes holds good for tees and bends also by using the appropriate thicknesses for these.

Stress calculations   All stress components in three cylindrical coordinate directions r, t and x for the three locations mentioned earlier are calculated at the inner walls for the following cases:

(a) Operating temperature of 540°C

(b) Operating internal pressure of 178 kgf/cm^2

(c) An estimated value of 5°C for the temperature difference between the inner and outer walls at the highest operating temperature based on transient heat conduction equations.

Principal stresses are evaluated using well known expressions from the above.

Equivalent stresses based on the maximum shear strain energy theory of failure are computed from principal stresses for the three locations and compared. The location corresponding to the maximum equivalent stress is the critical stress location.

For this location, the variation of permissible thermal differential, ΔT, with pressure, P, is computed by treating both these parameters as variables in the expression for equivalent stress and equating it to the allowable stress of the pipe material as recommended by Sec. VIII, DIV. *2* of the ASME Pressure Vessel Code.

It may be noted that the final relations needed are only between T and P. Therefore all stress components which depend on temperature need to be expressed in terms of pressure. The relation between temperature and pressure is derived from boiler start-up curves for startup after 72 hours shutdown given by the boiler manufacturer in drawings by eliminating the time parameter. The resulting pressure temperature relations are used to express temperature in terms of pressure.

Criteria and values for allowable stresses of the pipe material are taken from Figure 4-130 in Appendix-4 of ASME Sec. VIII, Div. 2. The function connecting the two parameters viz. ΔT and P and allowable stress is set to zero and this equation yields the desired variation of (ΔT) permissible versus line pressure. Curves furnishing such a variation are presented in Figures 4 & 5 for the critical as well as at an alternate location proximate to this. These curves also furnish (ΔT) permissible for cooling conditions, in which case, the differential temperature should be interpreted as the one between the mid and inner walls.

The locations referred to above are for placing the thermocouples which would measure the actual thermal gradients, and the comparison of measured values vis-a-vis permissible values computed serve as the basic input for controlling warming up and cooling down operations.

In the actual unit, the critical stress location has been identified as the 'Tee', and thermocouples were inserted at the appropriate locations. The actual differential temperature vs the permissible was also available for taking proper control actions.

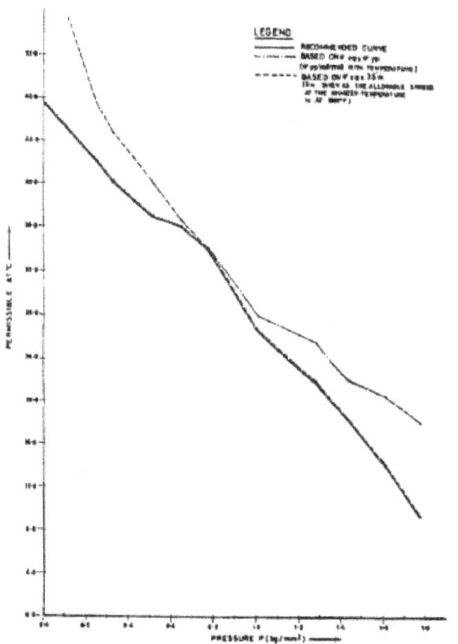

Figure 4 Permissible thermal differential versus line pressure at reinforcement of TEE

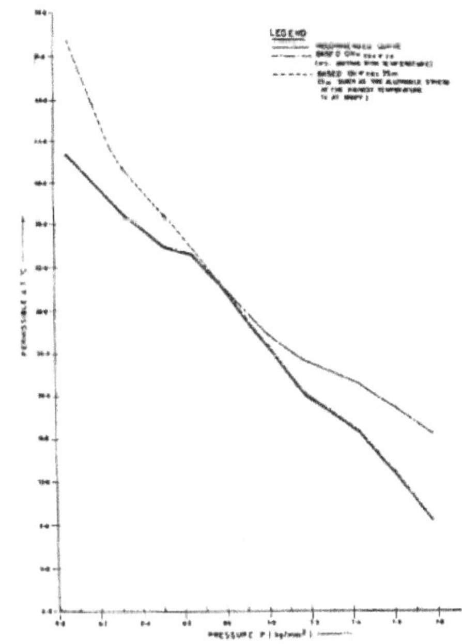

Figure 5   Permissible thermal differential versus line pressure at Pipe portion (unreinforced of TEE))

Actual plant operation versus design criteria   On actual operation of the plant, it was observed that the above curves are very useful to guide an operator for a cold start but, for a hot start, which occurs after a trip, the operator has to wait until the 'Tee' cools down to such a temperature as to keep an acceptable level of temperature differential to admit steam. This is because the boiler gets ready much earlier to generate steam. Hence, it is necessary to know what the consequences are if the prescribed limits of ΔT are exceeded marginally and also occasionally for a short duration as demanded by the system to save waiting time for hot start-ups.

Interaction of creep and low cycle fatigue   The repeated operations of start-up and shutdown of a power plant subject its components to thermal fatigue of a low-cycle nature. In addition, between two cycles of operation, there is a considerable time of 'hold' at a high operating temperature where a creep phenomenon plays a significant role. Both low cycle fatigue and creep contribute to a slow deterioration of the material strength of the components. Although this phenomenon is well known, the calculated life of components based on an anticipated schedule of operations of the plant always indicated conservative values and could not precisely reveal at the end of a certain number of years of operations, what useful life is left. In the current context of energy crisis and increasing cost of spare parts, it is extremely important to go into deeper investigation on this subject.

There is no doubt that quite a large number of publications on fatigue, creep and fracture mechanics have appeared recently. However, on problems of real engineering applications, such as the one discussed in this paper, concrete recommendations or guidelines are rarely published. One of the primary reasons could be that the material property data available are based primarily on laboratory investigations under controlled environments and hold good only for that specific material.

Therefore in order to develop a proper workable scheme to compute the life of a component expended and the balance of life left, the following steps -are involved, basically:

(a) Acquire continuous data on the thermal and pressure cycles that are introduced by the cyclic operation

(b) Record time and temperature at which a component is held during operation

(c) Perform a proper stress analysis and decide on the steady and alternating components of stresses

(d) Obtain data on the stress (or strain) vs number of cycles to failure at the appropriate temperatures.

(e) Obtain data on the stress vs rupture time curve at the appropriate temperatures.

(f) Sum up the "damage" components introduced by the above accounting for other possible agencies like corrosion using the most appropriate rule for summing up 'damage ratios'.

(g) The cumulative damage index reveals the desired information for the plant engineers to plan their next course of action.

These are only broad steps in which the work could progress to achieve the desired results and attempts are being made to further the study and our future publications would deal with ths subject more elaborately.

Such an approach, though adopting simpler analytical techniques, would be more advantageous to operating power plants in taking proper steps for preventing mechanical failures, as real plant data acquired from the power plant on operation; as well as on material conditions, have been well taken into account in the computations.

Conclusion

The paper highlights the application of stress analysis of piping systems as a powerful tool in the power plant engineering field for preventing mechanical failure. By suitable input from associated fields of fatigue, creep and fracture, the capability of such a tool can be enhanced to predict the balance of life left for critical components at any given time of plant operation.

References

Rodabaugh, E. C. et al, '*Effect of Internal Pressure of Flexibility and Stress Intensity Factors of Curved Pipe welded Elbows*", ASME journal of Applied Mechanics, Vol. 79, 1957 pp 939-948.

Redekop, D. and Schroeder, J., "*Approximate prediction of Hoop Stresses at major Section of Tee Intersections of Cylindrical Shells Subjected to Internal Pressure*", ASME Journal of Pressure Vessel Technology, Vol. 101, Aug. 1979, pp 186-193.

APPENDIX

LIST OF NOTATIONS/SYMBOLS USED

1.	a,b,c,d	Dimensions
2.	e	Volumetric strain
3.	g	Acceleration due to gravity
4.	h, t	Thickness
5.	l	Length
6.	l, m, n	Direction cosines
7.	r	Radius
8.	u, v, w	Displacements, deflections
9.	ω	Distributed load per unit length (beams)
10.	A	Area of cross section
11.	C	Constant
12.	E	Young's modulus
13.	F	Force
14.	G	Shear or rigidity modulus
15.	I, J	Moments of inertia of sections, Polar moment of inertia
16.	K	Stress concentration factor
17.	M	Moment/bending moment
18.	P	Load
19.	R	Radius of curvature
20.	T	Temperature/also torque
21.	W	Concentrated load (Beams)
22.	U	Strain energy
23.	X, Y, Z	Body forces (force/unit volume)

24. α — Coefficient of linear expansion(*alpha*), angle

25. β — Angle(beta)

26. γ — Shear strain (gamma)

27. δ — Differential or incremental prefix (e.g., $\partial_x, \partial_y, \partial_z$) deflection (Delta)

28. ε — Normal strain (epsilon)

29. λ — Constant (lambda)

30. μ — Micron(mu)

31. ν — Poisson's ratio(nu)

32. π — Circumference/diameter of circle (pi)

33. ρ — Mass density (rho)/specific resistance

34. σ — Normal stress (sigma)

35. τ — Shear stress (tao)

36. θ, ϕ — Angles (Theta, phi)

37. Δ — Operator (delta)

38. Σ — Summation (sigma)

39. Δ^2 — $$= \frac{\partial^2}{\partial x^2} + \frac{\partial^2}{\partial y^2} + \frac{\partial^2}{\partial z^2}$$ (Del square, Laplacian operator)

GLOSSARY

Allowable stress is prescribed by design codes and is arrived at by providing a safety margin on the ultimate or yield stress depending upon the ductile or brittle behaviour of the material.

Anisotropic material is a material whose mechanical properties like Young's Modulus, Shear Modulus, Poisson's Ratio etc vary with directions. The Material is called Isotropic if the properties are same in all the directions.

Area is a vector having magnitude equal to the extent occupied by a closed contour in a plane and its direction is normal to the plane.

Autofrettage This relates to built-up shrink-fit thick walled cylindrical pressure vessels. A high internal pressure is applied to produce plastic flow in the inner part of the cylinder. After removal of this internal pressure, residual stresses are created due to plastic flow or deformation, with the inner part in compression and the outer part in tension. This is called Autofrettage.

Bar One dimensional structure or body acted upon by forces acting along the longitudinal axis.

Bauschinger Dependence of yield stress on the loading path and direction. i.e Yield stress in compression differs from yield stress in tension. This phenomenon is usually ignored in plasticity theory

Beam One dimensional structure or body acted upon by forces transverse to the longitudinal axis in addition to the forces, acting along the axis.

Bending moment at any section is the moment of all forces that act on the beam to the left of that section, taken about the longitudinal axis of that section.

Biaxial stress Stresses acting along the coordinate axes in a plane namely σ_x, σ_y and τ_{xy}.

Boundary condition Two types of Boundary conditions are specified. One is natural or force type and the other is geometric or kinematic, relating to displacements and slopes. Force include axial force, transverse force or shear force or moment causing Bending or twisting. A free end is not acted upon by any external force or moment.

Brittle fracture Abrupt fracture of material when load reaches ultimate value without exhibiting any yielding or ductility. Even normally Ductile materials fail by brittle fracture due to notch effects contributed by high local stresses, tri-axial tensile state of stress, high load strain hardening and cracking and locally increased strain rates.

Buckling/Buckling load Phenomenon of elastic instability caused in axially loaded member or loaded in plane in compression mode. When columns are axially loaded in compression, when load reaches critical or Euler load, columns deflect laterally under no lateral load.

Centroid A point in the plane of area about any axis through which the moment of area is zero. Also commonly called as centre of gravity.

Corrosion Environmental reaction of salt like chemicals on metallic materials leading to destruction or deterioration of material.

Corrosion fatigue aggravated by corrosion as in the case of parts or components repeatedly stressed while exposed to salt water or moist air containing salt.

Crack Represents a discontinuity in a material. Behaviour of materials under tensile stress having a crack or discontinuity or defect show crack growth at a rapid rate causing fracture under certain conditions.

Creep   A special case of inelasticity which relates to the stress induced time-dependent deformation under load. Small time dependent deformation may occur even after removal of all applied loads. Phenomenon is more pronounced at high temperatures coupled with high stresses.

Damping capacity   Relates to the amount of work dissipated as heat energy in cyclic loading and unloading of material(Eg. Vibrations). The area contained within the hysteresis loop obtained from cyclic stress-strain curve is a measure of damping capacity of the material. Gray Cast iron has very good damping property.

Deformation   Changes in form or dimension of the body under the action of the load or temperature change.

Eccentricity/Eccenteric laoding   Deviation between actual point of application of load and the centroidal axis of a beam or column causes eccentric loading.

Elastic curvre   assumed by an elastic beam or beam column under transverse loads which cause the beam stressed within elastic limit.

Elastic limit   The least stress that will cause permanent set.

Elasticity   Property of material showing capability of recovering to the original shape and dimensions after removal of all loads. Elasticity is a 'reversible' process.

Endurance limit/endurance strength   The largest stress a material can withstand with repeated application or load reversal without rupture for a number of cycles is the endurance strength. Fatigue strength or endurance limit is the maximum stress which can be reversed for an indefinitely large number of cycles without producing fracture.

Equivalent bending moment   Action of this will produce the same normal stress as would be produced by the simultaneous action of bending and twisting moments on a circular shaft.

Equivalent twisting moment   Action of this will produce the same shear stress on a circular shaft as would be produced by the simultaneous action of twisting moment and bending moment.

Factor of safety (F.S)   F.S is the ratio of load causing failure to the actual service load and is a measure of safety margin provided in design, which is needed to account for several uncertainities like loads, environmental condition, material inhomogeneity etc.

Factor of stress concentration (S.C.F)   Abrupt change or discontinuity in material or geometry(like presence of holes, keyways, grooves, splines, notches, threads, sharp corners) produce high localised stresses and this phenomenon is called stress concentration. S.C.F is the ratio of actual stress at such discontinuities to the stress that would have existed in the absence of such discontinuities.

Factor of stress concentration in fatigue   Ratio of fatigue strength without stress concentration to the fatigue strength with stress concentration.

Failure   denotes inability of a part or component or system to meet the intended design requirements. Machine or structural members can fail to perform the intended functions on account of the following

   i.   Excessive elastic deformation

   ii.   Yielding or excessive plastic deformation

   iii.   Fracture

Fatigue   Tendency of materials to fracture under repeated loading, stress being considerably lower than ultimate static strength is classified as Low cycle fatigue(Characterised by high strains) and high cycle Fatigue (low stresses as seen in vibrations of rotating machines)

Fiber stress   Term used to designate longitudinal tensile or compressive stress in a beam. Fiber stress is zero along the longitudinal direction of the neutral axis or neutral surface.

Flexure   Alternate term used for bending(in the context of beam).

FMEA   Is the abbreviation for Failure Modes and Effects Analysis. Normally carried out for process as well as product. In product design development activity, this analysis constitutes a significant part and helps in quantitatively assessing product improvement or new product development.

Fracture toughness   Toughness is the property of a material to absorb energy during plastic deformation. Fracture toughness denotes the limiting value of Stress Intensity Factor(S.I.F) corresponding to unstable propagation of crack.

Fretting fatigue   Fatigue aggravate by surface rubbing as in shafts with press fitted collars.

Inelasticity   Inelasticity is a general characteristics of a material which does not allow return to original shape after removal of load. Plasticity and Creep are special cases of inelasticity.

Isotropic   Isotropic material has same properties in all directions(like young's modulus, shear modulus, poisson's ratio etc)

Modulus of elasticity   Referred as Young's modulus for isotropic materials and is the slope of the linear portion of stress-strain curve within the elastic limit of the material.

Modulus of rigidity   This is also called shear modulus. It is the ratio of shearing stress to shearing strain in the case of pure shear within elastic limit.

Mohr's circle   A geometrical construction used to determine principal stresses, principal strains, principal moments of inertia from the corresponding quantities along coordinate axes. viz σ_x, σ_y and σ_{xy} (OR) ε_x, ε_y and γ (OR) I_{xx}, I_{yy} and I_{xy}.

Moment of an area   Refers to the first moment of an area. For eg $\int dAy$ and $\int dAx$

Moment of inertia (M.I) of area or section   Refers to the second moment of an area for eg $\int dAx^2$ and $\int dAy^2$ (product of inertia is $\int dAxy$)

Neutral axis or neutral surface   Line or surface of zero fiber stress in any given section of member subject to bending.

Plasticity   Plasticity is the special case of inelasticity in which material undergoes time independent non recoverable deformation. Thermodynamically, plasticity is classified as a irreversible process.

Poisson's ratio   Ratio of lateral unit strain to longitudinal unit strain under condition of uniform and uniaxial longitudinal stress within proportional limit.

Principal stresses/planes   Principal planes contain only principal stresses. Shear stresses are zero on principal planes. Principal stresses are maximum and minimum and designated as σ_1 and σ_2 (2D case) and σ_1, σ_2 and σ_3 (3D)

Radius of gyration   Radius of gyration of an area A is given by,

$\sqrt{\dfrac{I}{A}}$, where I is the section of moment of inertia.

Slenderness ratio   refers to the length of a uniform column divided by the least radius of gyration of the cross section.

Standards   Normally refer to approved international or national codes for designs covering plant piping, pressure vessels, storage tanks etc against various loading situations like pressure, thermal, wind loads, seismic loads and hydrostatic test conditions eg ASME, ANSI, ASTM, ISO, BIS, BS, DIN

Strain corrosion cracking (SCC)   SCC is the failure of an alloy from the combined effects of corrosive environment and a static tensile stress. The stress may result from applied loads or "Locked-in" residual stresses(usually produced in a manufacturing or fabrication).

Strain   Dimensionless quantity referring to elongation per original length. This is also a Tensor. Strain at a point is defined by three Normal strains and three shear strains.

Strain hardening Is caused by dislocation interacting with each other and with barriers which impede their motion through the crystal lattice.

Strain rosette Three intersecting strain gauges arranged to facilitate calculation of principal stresses on the surface of a stressed body over which the rosette is cemented.

Stress cycle Stress cycle is a plot of variation of stress over time showing a maximum and minimum during a period or time interval and the process repeated as seen in a vibrating beam. A single service or operational cycle like start-up and shutdown may involve one or more stress cycles.

Stress intensity Is twice the maximum shear stress which is the difference between algebraically largest and smallest principal stresses at a given point.

Stress Unit of stress is N/mm2. It is a Tensor. Stress at a point is specified by three normal stresses and three shear stresses.

Stress-strain diagram A plot made from the data obtained from the tension test(Monotonic stress-strain curve) and cyclic loading (cyclic stress-strain diagram) for different materials.

Thermal stress is a self balancing stress produced by a non-uniform temperature distribution or by differing values of coefficients of thermal expansion. Thermal stress is developed in a solid body whenever a volume of material is prevented from assuming the size and shape that it normally should under a change in temperature. Thermal stresses are of two types one is general like what is seen in a hot piping system and the other is local as in the case of welding of a small zone in a component.

True stress For an axially loaded bar, the load divided by the corresponding actual cross sectional area.

Twisting moment At any given section of a member, the moment of all forces that act on the member to the left of that section, taken about a polar axis through the flexural centre of that section.

Ultimate strength of a material in tension, compression or shear is the corresponding maximum stress a can sustain, computed on the basis of ultimate load and the original unconstrained dimensions

Yield point/yield strength Denotes the lowest stress at which strain increases without increase in stress. For Ductile materials, the stress-strain plot indentifies upper and lower yield point beyond which irrecoverable permanent deformation occurs.

BIBLIOGRAPHY

Strength of Materials/Mechanics of Materials/Mechanics of Solids

Bansal, Sr.R.K., '*A Text book of Strength of Materials*', Laxmi Publications(P) Ltd.,New Delhi, 2001.

Basavarajaiah, B.S, '*Strength of Materials*, 2nd Edn, CBS Publications & Distributors, New Delhi, 1984.

Bhavikatti, S.S., '*Strength of Materials*', 3rd Edn, Vikas Publishing Houes Pvt. Ltd., New Delhi, 2003.

Case, John & Chilver AH, '*Strength of Materials & Structures*', 'ELBS', 1985.

Cook, R.D., and W.C.Young: "*Advanced Mechanics of Materials*," Macmillan Publishing Co., 1985.

Egor P. Popove & Toader.A.Balan.,'*Engineering Mechanics of Solids*', Pearson Education (P) Ltd.,

Ferdinand, P. Beer, Johnston Russel, E, '*Mechanics of Materials*', Tata McGrawHill Co, 2005.

Hibbeler, R.C., '*Mechanics of Materials*', Prentice Hall International, 2002.

Junnarkar, S.B., '*Mechanics of Structures*', Volume 1, Charotar Book Stall.

Kazimi, S.M.A, '*Solid Mechanics*', Tata McGraw-Hill Publishing Company Ltd., New Delhi, 1976.

Kurmi, R.S., , '*Strength of Materials*, 1st Edn, S.Chand & Company Ltd., New Delhi 1999.

Norman, E. Dowling, '*Mechanical Behavior of Materials*', Prentice Hall International Inc New Jersey, 1999.

Rajput, R.K, '*Strength of Materials*', S.Chand & Company Ltd., New Delhi, 1996.

Ramamrutham, S and Narayanan, R ; '*Strength of Materials*', Dhanpat Rai & Sons, New Delhi, 1996.

Riley, William.F, '*Mechanics of Materials*', John Wiley and Sons, 5th Edn, 1999.

Sadhu Singh, '*Strength of Materials*,'Khanna Publishers, New Delhi, 1984.

Stephen Timoshenko, '*Strength of Materials*', CBS Publishers.

Subramanian, R., '*Strength of Materials*', Oxford University Press, New Delhi, 2011.

V.Feodesyev '*Strength of Materials*', Mir Publishers, 1968

Vaidhyanathan, R., Perumal, P., Lingeshwar, S., '*Mechanics of Solids*', Scitech Publications (India) Pvt. Ltd., 2011.

Victor Dias da Silva, "*Mechanics and Strength of Materials*", Springer, Berlin, India Edition, New Delhi, 2010.

William Nash, '*Strength of Materials*, 4th Edn, Tata Tata McGrawHill Publishing Co. Ltd., 2007.

Theory of Elasticity

Cook, R.D., and W.C.Young: "*Advanced Mechanics of Materials*," Macmillan Publishing Company, 1985.

Love, A.E.H.: "*Mathematical Theory of Elasticity*," 2d ed., Cambridge University Press. 1906.

Timoshenko,S., and J.N.Goodier: "*Theory of Elasticity*," 2d ed., Engineering Society Monograph, McGraw-Hill Book Company, 1951.

Stress Concentration

Pattabiraman. J., Ramamurthi, V. And Reddy, D.V., "*Statics and Dynamics of Shells with cut-outs* – A Review Journal of Ship Research, June 1974.

Peterson, R.E.: "*Stress Concentration Factors*," John Wiley & Sons, Inc., 1974.

Savin, G.N., "*Stress Concentration Around Holes*", Pergamon Press, 1961.

Warren C Young., "*Roarks Formulas for Stress and Strain*" 6th Edn., McGraw-Hill Book Company, International Edition, 1989.

Mechanical Design / Engineering Design / Machine Design

Anid R.Eide etal, "*Introduction to Engineering Design*", McGraw-Hill's Best, 1998.

Bhandari, V.B., "*Design of Machine Elements*", 2nd Edn., Tata McGraw-Hill Publishing Company Limited, New Delhi, 2007.

Eliahu Lahari, "*Fatigue Design: Life Expectancy of Machine Parts*", CRC Press, 1996.

George. E, Dieter., "*Engineering Design–A Material and Processing Approach*", McGraw-Hill Book Company, 1991.

George. E.Dieter, "*Mechanical Metallurgy*", McGraw-Hill Book Company, 1988.

Joseph E Shigley etal, "*Mechanical Engineering Design*", Eighth Edn in SI units, Tata McGraw-Hill Publishing Company Ltd., New Delhi, 2008.

PSG College of Technology, "*Design Data*", Coimbatore, India, 2002.

Thomas A.Cruse, "*Reliability Based Mechanical Design*", Marcel Dekker, Inc, 1997.

Pressure Vessels, Plates and Shells

ASME, "*Rules for Construction of Pressure Vessels*," Section VIII, ASME Boiler and Pressure Vessel Code

Brownell, L.E and E.H.Young, "*Process Equipment Design*" : Vessel Design", John Wiley & Sons, New York, 1959.

Flugge, W. "*Stress in Shells*, "Springer-Verlag, 1960.

Henry H. Bednar, P.E, *Pressure vessel design Handbook* 2R, Krieger Publishing company florida, 1986.

John F. Harvey, "*Theory and Design of Pressure Vessels*", CBS Publishers & Distributors Pvt. Ltd., Chennai, India, 2001.

Timoshenko, S and S.Woinowsky Kreiger, "*Theory of Plates and Shells*", Second Edn., McGraw-Hill Book Company, 1959.

Timoshenko, S., and J.M.Gere: "*Theory of Elastic Stability*," 2d ed., McGraw-Hill Book Company, 1961.

Experimental Stress Analysis-Strain Measurement

Dally, J.W., and W.F.Riley: "*Experimental Stress Analysis*," 2d ed., McGraw-Hill Book Company, 1978.

Dove, R.C., and P.H.Adams: "*Experimental Stress Analysis and Motion Measurement*," Charles E.Merill Books, Inc., 1955

Durelli, A.J., and V.J.Parks" "*Moire Analysis of Strain*," Prentice-Hall, Inc., 1970.

Durelli, A.J.,E.A.Phillips, and C.H.Tsao: "*Introduction to the Theoretical and Experimental Analysis of Stress and Strain*," Prentice-Hall, Inc., 1958.

Durelli, A.J.: "*Applied Stress Analysis*," Prentice-Hall Inc., 1967.

Durelli, A.J.,W.F.Riley: "*Introduction to Photomechanics.*" Prentice-Hall, Inc., 1965.

Frocht, M.M.: "*Photoelasticity*," vols. 1and 2, John Wiley & Sons, Inc.,1941,1948.

Hetenyi, M.(ed.): "*Handbook of Experimental Stress Analysis*, " John Wiley & Sons, Inc., 1950.

Holister, G.S.: "*Experimental Stress Analysis: Principles and Methods*," Cambridge University Press, 1967.

Kobayashi, A.S.(ed): "*Handbook on Experimental Mechanics*," Prentice-Hall Inc. 1987.

Proc. Soc.Exp. Stress Anal., 1943-1960; Exp. Mech., J.Soc.Exp.Stress Anal., from Jan .1, 1961 to June 1984 ; and Exp. Mech., J.Soc.Exp. Mechanics after June 1984.

Warren C.Young , "*Roarks Formulas for Stress and Strain*", 6th Ed., McGraw-Hill Book Company 1989.

Finite Element Method / Finite Element Analysis

Cook, R.D., Malkus, D.S.,Plesha,M.E., and Witt,R.J. "*Concepts and Applications of Finite Element Analysis*", Wiley Student Edition, 4th Edition, Wiley India(P) Ltd., New Delhi.

George R.Buchanan and Rudramoorthy, R., "*Finite Element Analaysis*", Tata McGraw-Hill Publishing Company Ltd., New Delhi, 2006

Heubner, K.H., Dewhirst, D.L., Smith, D.E & Byron, T.G., *"The Finite Element Method for Engineers"*, Wiley Student Edition, Fourth Edition 2004, John Wiley & Sons (Asia) Pvt. Ltd.,

Nitin S.Gokhale etal,*"Practical Finite Element analysis"*, Finite to infinite, Pune, India, 2008.

Pattabiraman, J., "Certain Typical Applications of Finite Element Method to Practical Structural Analysis Problems", Paper presented at the International Conference on Engineering Application of the Finite Element Method, Paper A2,PP.A3.1-A 3.11 (first 10 pages of the paper), Vol.2 of the Proceedings, A.S.Computas, Hovik, Oslo, Norway, May, 1979.

Ramamurti V., *"Finite Element Method in Machine Design"*, Narosa Publishing House, New Delhi 2009.

Rao, S.S., *"The Finite Element Method in Engineering"*, Butterworth-Heinsmann—Elsevier India Pvt. Ltd., New Delhi.

Reddy, J.N., *"Introduction to Non-Linear Finite Element Analysis"*, Oxford University Press, 2008

Reddy,J.N., *"Introduction to the Finite Element Method"*, 3rd Edn, Tata McGraw-Hill Publishing Company Limited, New Delhi, 2008

Reddy, J.N., Gartling, D.K., *"The Finite Element Method in Heat Transfer and Fluid Mechanics"*, CRC Press, 1994.

Seshu, P.,*"Text Book of Finite Element Analysis"*, Prentice hall of India Learning Pvt.Ltd, New Dehli, 2010.

Tirupathi R.Chandrupatla and Ashok D.Belegundu, *"Introduction to Finite Elements in Engineering"*, 3rd Edn, Prentice Hall of India Private Limited, New Delhi, 2005.

Zienkiewics.O.C, Taylor.R.L, & Zhu, J.Z. *"The Finite Element Method: Its Basis & Fundamentals"*, Butterworth-Heinemann—Elsevier India Pvt.., New Delhi.

Zienkiewicz.O.C,Taylor.R.L *"The Finite Element Method"* McGraw-Hill International Editions, Fourth Edition, 1991, Volume 2.

INDEX

www.ingramcontent.com/pod-product-compliance
Lightning Source LLC
Chambersburg PA
CBHW081246130726
47998CB00010B/2689